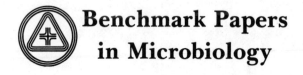

Benchmark Papers in Microbiology

Series Editor: Wayne W. Umbreit
Rutgers University

Published Volumes and Volumes in Preparation

MICROBIAL PERMEABILITY
   John P. Reeves
CHEMICAL STERILIZATION
   Paul M. Borick
INFLUENCE OF TEMPERATURE ON MICROORGANISMS
   J. L. Stokes
MICROBIAL GENETICS
   Morad Abou-Sabé
MOLECULAR BIOLOGY AND PROTEIN SYNTHESIS
   Robert Niederman
MICROBIAL GROWTH
   P. S. S. Dawson
MICROBIAL METABOLISM
   H. W. Doelle
ANTIBIOTICS AND CHEMOTHERAPY
   M. Solotorovsky
MICROBIAL PHOTOSYNTHESIS
   June Lascelles
INDUSTRIAL MICROBIOLOGY
   Richard W. Thoma
MICROBIAL VIRUSES
   S. P. Champe
TISSUE CULTURE AND BASIC VIROLOGY
   R. J. Kuchler
MARINE MICROBIOLOGY
   C. D. Litchfield

**Benchmark Papers
in Microbiology**

—————A *BENCHMARK* TM Books Series—————

MICROBIAL
PERMEABILITY

Edited by
JOHN P. REEVES
University of Texas Southwestern Medical School

**Dowden, Hutchinson
& Ross, Inc.**
Stroudsburg, Pennsylvania

Copyright © 1973 by **Dowden, Hutchinson & Ross, Inc.**
Library of Congress Catalog Card Number: 73–4299
ISBN: 0–87933–032–5

Library of Congress Cataloging in Publication Data

Reeves, John P   comp.
    Microbial permeability.

    (Benchmark papers in microbiology)
    Bibliography: p.
    1. Cells--Permeability--Addresses, essays, lectures.
2. Biological transport--Addresses, essays, lectures.
3. Membranes (Biology)--Addresses, essays, lectures.
I. Title. [DNLM: 1. Biological transport--Collected
works. 2. Cell membrane permeability--Collected works.
3. Microbiology--Collected works. QW 4 R332m 1973]
QH601.R43   574.8'75'08   73-4299
ISBN 0-87933-032-5

Manufactured in the United States of America.

Exclusive distributor outside the United States and
Canada: John Wiley & Sons, Inc.

Acknowledgments
and Permissions

The following papers have been reprinted with the permission of the author and the copyright owners.

American Chemical Society—*Biochemistry*
"Glycine Transport by Hemolyzed and Restored Pigeon Red Cells"

The American Physiological Society—*American Journal of Physiology*
"Effect of Digitalis on Active Intestinal Sugar Transport"

The American Society of Biological Chemists, Inc.—*The Journal of Biological Chemistry*
"The Role of Energy Coupling in the Transport of β-Galactosides by *Escherichia coli*"
"Mechanisms of Active Transport in Isolated Membrane Vesicles. I. The Site of Energy Coupling Between D-Lactate Dehydrogenase and β-Galactoside Transport in *Escherichia coli* Membrane Vesicles"
"Mechanisms of Active Transport in Isolated Membrane Vesicles. II. The Mechanism of Energy Coupling Between D-Lactate Dehydrogenase and β-Galactoside Transport in Membrane Preparations from *Escherichia coli*"
"Restoration of Active Transport of Glycosides in *Escherichia coli* by a Component of a Phosphotransferase System"
"The Role of the Phosphoenolpyruvate–Phosphotransferase System in the Transport of Sugars by Isolated Membrane Preparations of *Escherichia coli*"
"A Binding Site for Sulfate and Its Relation to Sulfate Transport into *Salmonella typhimurium*"
"The Release of Enzymes from *Escherichia coli* by Osmotic Shock and During the Formation of Spheroplasts"
"Amino Acid-Binding Protein Released from *Escherichia coli* by Osmotic Shock"
"The Reduction and Restoration of Galactose Transport in Osmotically Shocked Cells of *Escherichia coli*"
"The Histidine-Binding Protein *J* Is a Component of Histidine Transport"

American Society for Microbiology—*Bacteriological Reviews*
"Bacterial Permeases"

Elsevier Publishing Company—*Biochimica & Biophysica Acta*
"Thermodynamic Analysis of the Permeability of Biological Membranes to Non-electrolytes"
"The Role of Permease in Transport"
"Studies on the Mechanism of Intestinal Absorption of Sugars. V. The Influence of Several Cations and Anions on the Active Transport of Sugars *in vitro*, by Various Preparations of Hamster Small Intestine"
"Studies on the Mechanism of Intestinal Absorption of Sugars. VI. The Specificity and Other Properties of Na⁺-Dependent Entrance of Sugars into Intestinal Tissue Under Anaerobic Conditions, *in vitro*"

The Faraday Society and contributors—*Transactions of the Faraday Society*
"The Permeability of Plant Protoplasts to Non-electrolytes"

Federation of American Societies for Experimental Biology—*Federation Proceedings*
"Na⁺-Dependent Transport in the Intestine and Other Animal Tissues"

The Macmillan Company—*The Permeability of Natural Membranes*
  "The Theory of Penetration of a Thin Membrane"

Macmillan (Journals) Ltd.—*Nature New Biology*
  "Role of the Galactose Binding Protein in Chemotaxis of *Escherichia coli* Toward Galactose"

National Academy of Sciences—*Proceedings of the National Academy of Sciences*
  "Specific Labeling and Partial Purification of the M Protein, a Component of the β-Galactoside Transport System of *Escherichia coli*"
  "β-Galactoside Transport in Bacterial Membrane Preparations: Energy Coupling via Membrane-Bound D-Lactate Dehydrogenase"
  "Phosphate Bound to Histidine in a Protein as an Intermediate in a Novel Phospho-transferase System"
  "Two Classes of Pleiotropic Mutants of *Aerobacter aerogenes* Lacking Components of a Phosphoenol-pyruvate-Dependent Phosphotransferase System"
  "Genetic Evidence for the Role of a Bacterial Phosphotransferase System in Sugar Transport"
  "Components of Histidine Transport: Histidine-Binding Proteins and *hisP* Protein"

Physiological Society—*The Journal of Physiology*
  "Inability of Diffusion to Account for Placental Glucose Transfer in the Sheep and Consideration of the Kinetics of a Possible Carrier Transfer"
  "Effect of Ionic Environment on Intestinal Sugar Transport"

The Rockefeller University Press—*Journal of General Physiology*
  "Uphill Transport Induced by Counterflow"
  "Kinetic Relations of the Na–Amino Acid Interaction at the Mucosal Border of the Intestine"

Society for Experimental Biology Symposia—*Symposia of the Society for Experimental Biology*
  "The Evidence for Active Transport of Monosaccharides Across the Red Cell Membrane"

Springer-Verlag—*European Journal of Biochemistry*
  "Close Linkage Between a Galactose Binding Protein and the β-Methylgalactoside Permease in *Escherichia coli*"

Series Editor's Preface

While it is true that the modern student of any science should read the classic papers in his field in the original sources, it is increasingly difficult to do so. This is partly because there are so many papers; they are scattered; and the time involved in hunting down the many journal volumes becomes excessive. For these reasons, among others, we have ventured to collect within single volumes selections of classic papers in the field of microbiology—each volume to cover a specific subject or "slice" of microbiology.

One should remember that even a classic concept may first appear in more than one paper, and it may be more neatly phrased or more dramatically bolstered by experiment in some than in others. Once an illuminating concept is grasped there may be hundreds of papers exploiting the breakthrough—each with some claim to originality; each with some justification for being the classic contribution in which the concept was first clearly demonstrated. Judgment must be exerted or we would be overwhelmed by sheer volume, and the selection of a particular benchmark paper does not mean that there are not others equally classic, possibly even surpassing the one chosen. The objective of this volume of "Benchmark Papers in Microbiology" is to present, by reproduction of selected original papers, the development and present status of our knowledge of microbial permeability; this, we believe, has been accomplished by the papers chosen.

This volume is devoted to microbial permeability, a subject that has proved to be more complex, more confusing, and more exasperating (if this is the word) the more we learn about it. It is, therefore, a pleasure to have Dr. Reeves serve as an experienced guide to the subject—to select the critical papers and provide a guide to their meaning.

Wayne W. Umbreit

Contents

I. DIFFUSION

II. CARRIERS

III. THE GALACTOSIDE PERMEASE OF *ESCHERICHIA COLI*

IV. THE PHOSPHOTRANSFERASE SYSTEM

V. BINDING PROTEINS

VI. THE CO-TRANSPORT THEORY

Contents by Author

Introduction

Membrane transport has entered its scientific adolescence, a stormy period of rapid development when even the mildest assertions are likely to provoke an argument. A poignant index of the controversies that still surround even the fundamental tenets of the field is the frequent use of quotation marks to offset such terms as "active transport" and "mobile carrier." To publish a collection of "definitive" papers while basic definitions are still subject to debate might therefore seem incongruous at best. Despite such considerations, I hope that this book will be useful, perhaps even stimulating, not only to scientists wishing an introduction to membrane transport but also to established investigators in the field who might appreciate having many of the seminal papers in this area collected into a single volume.

The book is divided into six separate sections, each emphasizing a particular transport system or mechanism. The sections on diffusion and carrier-mediated transport do not deal specifically with microorganisms, but the ideas discussed are assumed to apply to nearly all natural membrane systems. The restriction to only a few areas of research may involve some sacrifice in scope but it leads, in my opinion, to a fuller appreciation of possible transport mechanisms and to a more unified and readable volume. The individual papers were chosen on the basis of their influence on subsequent research and their contributions toward understanding the still poorly defined molecular events associated with transport. The particular research areas that constitute the different sections of the book were selected with an eye toward tracing the development of concepts fundamental to the field as a whole.

As implied above, no attempt has been made to represent all areas of membrane transport research. In particular, I have deliberately omitted papers concerned solely with ionic permeability because an adequate treatment of the subject would require another volume the size of the present one. Even within the areas selected, many important and influential papers could not be included because of space limitations. Moreover, several of the views set forth in the selected papers have been disputed vigorously by others, but space limitations prevented the inclusion of the opposing viewpoints in their original form. For these reasons, each section has been prefaced by a short introduction that attempts to provide a semihistorical account of the developments within that area of research. These introductions are *not* to be considered reviews of the literature; for that purpose, the reader should consult the recent reviews of Kaback (1970), Lin (1971), and Oxender (1972).

I would like to express my appreciation to several colleagues, Drs. R. K. Crane, R. M. Dowben, H. R. Kaback, F. J. Lombardi, R. A. Niederman, S. A. Short, and W. W. Umbreit (Series Editor), all of whom reviewed the manuscript and offered valuable advice on the selection of papers. I would also like to thank Drs. H. V. Rickenberg and A. Kepes, who reviewed the English translations of their papers and suggested many improvements. The translations were carried out by The Language Center, Inc., South Orange, New Jersey.

References

Kaback, H. R. (1970). Transport. *Ann. Rev. Biochem.*, **39**, 561–598.
Lin, E. C. C. (1971). The molecular basis of membrane transport systems. In L. I. Rothfield (ed.), *Structure and Function of Biological Membranes*. Academic Press New York, pp. 285–341.
Oxender, D. L. (1972). Membrane transport. *Ann. Rev. Biochem.*, **41**, 777–814.

Diffusion

I

Editor's Comments on Papers 1, 2, and 3

1   **Collander:** *The Permeability of Plant Protoplasts to Non-electrolytes*

2   **Danielli:** *The Theory of Penetration of a Thin Membrane*

3   **Kedem and Katchalsky:** *Thermodynamic Analysis of the Permeability of Biological Membranes to Non-electrolytes*

The assumption underlying most early permeability studies was that the properties of the membrane as a diffusion barrier could be related, albeit indirectly, to membrane structure. The starting point for this approach was Overton's (1899) demonstration that the permeability of nonelectrolytes in plant cells increased with the substance's lipid solubility. He proposed that cells were surrounded by a thin layer of lipoidal material, the plasma membrane, in which molecules must dissolve in order to enter the cell. Collander (Paper 1) and his co-workers (Collander and Barlund, 1933; Wartiovaara and Collander, 1960) demonstrated a direct relation between permeabilities and oil–water partition coefficients for a large number of nonelectrolytes. An effect of molecular size was also observed, leading these workers to suggest that permeabilities were to some extent determined by a sieve-like membrane structure. Höber and Ørskov (1933) argued for a porous structure in the erythrocyte membrane on the grounds that molecular size was more important than lipid solubility in determining relative permeabilities within groups of related nonelectrolytes.

Danielli (Paper 2) pointed out that the reported effects of molecular size did not necessarily imply the existence of aqueous pores in the membrane; they might simply reflect the fact that larger molecules form more hydrogen bonds to water than smaller ones, bonds that must be broken for the molecules to dissolve in the membrane. Danielli's analysis of permeability data from a wide variety of cells provided indirect support for the Davson–Danielli model of membrane structure, in which the core of the membrane was thought to be a nearly continuous bimolecular layer of lipid.

More recently, Solomon and co-workers (Paganelli and Solomon, 1957; Sidel and Solomon, 1957) have provided strong support for the concept of a porous membrane with their elegant demonstration that osmotic volume changes in erythrocytes occur several times more rapidly than permitted by simple diffusion of water. Satisfactory agreement with the results was obtained by a model invoking laminar flow of

4

water through pores, 3.5 Å in radius, occupying less than 1 percent of the surface area of the cell. By a completely independent method, which involved monitoring osmotic volume changes induced by permeable nonelectrolytes, Goldstein and Solomon (1960) obtained an effective pore radius of 4.1 Å for the erythrocyte membrane.

A basic assumption of most early permeability studies was that there existed a simple proportionality between the rate at which a substance entered a cell and its concentration difference across the cell membrane. A more general approach to permeability phenomena is provided by nonequilibrium thermodynamics, a method of relating irreversible processes to local gradients of intensive thermodynamic parameters. In this approach, the flux of a single component is a function of the free-energy differences across the membrane of every component in the system. Thus, interactions between different components within the membrane are described explicitly. Kedem and Katchalsky (Paper 3) have cast the equations of nonequilibrium thermodynamics into a form suitable for physical measurements. Their paper laid the groundwork for most modern studies of passive diffusion in biological membranes. Further information on nonequilibrium thermodynamics may be found in papers by Kedem and Katchalsky (1961), Katchalsky and Curran (1965), and de Groot and Mazur (1962).

References

Collander, R., and H. Barlund (1933). Permeabilitäts-studien an *Chara ceratophylla. Acta Botan. Fennica*, **11;** 1–14.

Goldstein, D. A., and A. K. Solomon (1960). Determination of equivalent pore radius for human red cells by osmotic pressure measurement. *J. Gen. Physiol.*, **44;** 11–17.

de Groot, S. R., and P. Mazur (1962). *Non-equilibrium Thermodynamics*. North-Holland, Amsterdam; Wiley, New York.

Höber, R., and S. L. Ørskov (1933). Untersuchengen über die Permiergeschwindigkeit um Anelektrolyten bei den roten Blutkörperchen vershiedener Tierarten. *Arch. Ges. Physiol.*, **231;** 599–615.

Katchalsky, A., and P. F. Curran (1965). *Nonequilibrium Thermodynamics in Biophysics*. Harvard University Press, Cambridge, Mass.

Kedem, O., and A. Katchalsky (1961). A physical interpretation of the phenomenological coefficients of membrane permeability. *J. Gen. Physiol.*, **45;** 143–179.

Overton, E. (1899). Über die allgemeinen osmotischen Eigenshaften der Zelle, ihre vermutlichen Ursachen und ihre Bedeutung für die Physiologie. *Vjschr. Naturforsch. Ges. Zürich*, **44;** 88–135.

Paganelli, C. V., and A. K. Solomon (1957). The rate of exchange of tritiated water across the human red cell membrane. *J. Gen. Physiol.*, **41;** 259–277.

Sidel, V. W., and A. K. Solomon (1957). Entrance of water into human red cells under an osmotic pressure gradient. *J. Gen. Physiol.*, **41;** 243–257.

Wartiovaara, V., and R. Collander (1960). Permeabilitätstheorien. In *Protoplasmotologia: Handbuch der Protoplasmaforschung*, Vol. II. C. Springer, Vienna.

Copyright © 1937 by the Faraday Society and contributors

Reprinted from *Trans. Far. Soc.*, **33**, 985–990 (1937)

THE PERMEABILITY OF PLANT PROTOPLASTS TO NON-ELECTROLYTES.

By Runar Collander.

Received 23rd February, 1937.

Our knowledge of the permeability of plant protoplasts to non-electrolytes has increased with remarkable rapidity during the last ten years. Although in the nineties Ernst Overton, the founder of the modern study of protoplasmic permeability, carefully investigated the permeability of a very large number of different plant and animal cells to a variety of non-electrolytes as well as of electrolytes, his projected great publication on protoplasmic permeability was unfortunately never completed. We therefore know his results, especially as far as plant cells are concerned, only from some short papers [1] which, although the main results are excellently summarised, contain only scanty concrete facts. The first plant objects of which the permeability was the subject of detailed papers were the bacterium *Beggiatoa mirabilis* studied in 1925 by Ruhland and Hoffmann [2] and later by Schönfelder,[3] and the epidermal cells of *Rhoeo discolor* studied in 1929 by Bärlund.[4] Since then the progress in this field has been very rapid, so that at present more than thirty very different kinds of plant cells have been fairly closely studied with regard to their permeability to various non-electrolytes.[5]

In this paper we shall first consider more closely the permeability properties of one particular object and then compare with each other the different objects hitherto studied.

As the standard object to be treated more in detail we choose the large, multinucleate cells of an alga, viz. *Chara ceratophylla*.[6] This object is so far the only one the permeability of which to non-electrolytes has been closely studied, using a direct method of work, *i.e.*, by analysing quantitatively the cell sap squeezed out from single cells which have been placed for known time intervals in solutions of the substances to be studied. Obviously such a method of permeability measurement involves a minimum of theoretical assumptions, whereas the method most widely used, *viz.* the plasmolytic method, is theoretically open to several objections, though it also seems to give quite useful results in the hands of a skilful worker.

All experimental facts seem to indicate that the penetration of the *Chara* cells by dissolved non-electrolytes is a simple diffusion process in which the only significant concentration gradient in most cases is that

---

[1] Overton, E., *Vierteljahrsschr. Naturforsch. Ges. Zürich*, 1895, **40**, 159; 1896, **41**, 383; 1899, **44**, 88.

[2] Ruhland, W., and C. Hoffmann, *Planta*, 1925, **1**, 1.

[3] Schönfelder, S., *ibid.*, 1930, **12**, 414.

[4] Bärlund, H., *Acta Bot. Fennica*, 1929, **5**, 1.

[5] The two most recent comprehensive publications in this field are: Hofmeister, L., *Bibliotheca Bot.*, 1935, **113**, 1, and Marklund, G., *Acta Bot. Fennica*, 1936, **18**, 1.

[6] Collander, R., *ibid.*, 1930, **6**, 1; Collander, R., and H. Bärlund, *ibid.*, 1933, **11**, 1.

across the protoplasm itself. Thus Fick's law is directly applicable in the following form :

$$P = \frac{v}{qt} \ln \frac{C}{C - c},$$

where v is the volume (cm.³) and q the surface of the cell (cm.²), c the internal concentration of the penetrating substance at the time t, C its equilibrium internal concentration (almost identical with its external

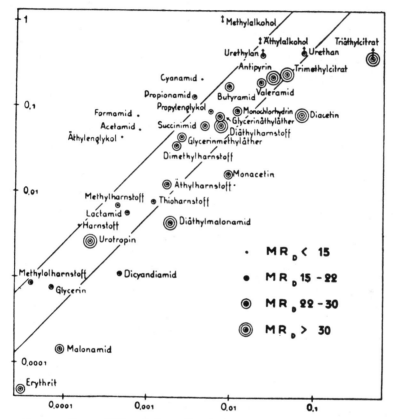

FIG. 1.—Permeability of the cells of *Chara ceratophylla* to various non-electrolytes. Ordinates indicate the permeability constants (cm./hours). The abscissa gives the oil/water distribution coefficients of the substances studied. After Collander and Bärlund (somewhat changed).

concentration), and P a constant which measures permeability and which is therefore termed the permeability constant.

To test the validity of the current hypotheses concerning proto-plasmic permeability, *viz.* (1) the lipoid-solubility hypothesis, (2) the ultrafiltration hypothesis, and (3) a combination of these, *i.e.*, the lipoid-sieve hypothesis, measurements were made of the permeability of the *Chara* cells to 45 organic non-electrolytes of very different lipoid solu-bility and different molecular size. In addition, the distribution of all these substances between ethyl ether and water, as well as between olive oil and water was determined so that it is possible to use the distribution coefficients thus obtained as a provisional measure of their relative lipoid-solubility. As a measure of their molecular volume the molar refraction (MR_D) can be used. The result of this study is represented in Fig. 1 from which, however, those substances are omitted which either penetrate the *Chara* cells so slowly that their permeability constants could not be determined or the oil-solubility of which is so slight that it could not be determined experimentally. To some substances the protoplasm is so permeable that they could probably permeate faster, were it not for the comparatively slow diffusion through the cell sap; such cases are indicated by upwardly directed arrows.

There is clearly a somewhat close correlation between the oil-solubility of the substances on the one hand, and their permeability constants, on the other; this is not merely a general concordance but, at least approximately, a direct proportionality. This is shown by the fact that most of the points in Fig. 1 fall between the two parallel lines cutting the axes at an angle of 45°. On the other hand, the smallest molecules obviously permeate faster than would be expected on account of their oil-solubility alone.

We thus see that both the lipoid-solubility and the molecular volume (or some other properties varying in close accordance with these) are factors involved in the permeation process. It seems, therefore, natural to conclude that the plasma membranes of the *Chara* cells contain lipoids, the solvent power of which is on the whole similar to that of the olive oil. But, while the medium-sized and large molecules penetrate the plasma membrane only when dissolved in the lipoids, the smallest molecules can also penetrate it in some other way. Thus, the plasma membrane seems to act both as a selective solvent and as a molecular sieve.

Against this view, for which the name " lipoid-sieve hypothesis " has been proposed,[7] at least three objections have been raised:

(1) Some authors[8] are of the opinion that it is not necessary to assume the occurrence of lipoids in the plasma membrane as, for example, hydrophobic proteins also would favour the passage of lipoid-soluble, *i.e.*, hydrophobic solutes through the membrane. However, so far as the present writer is aware, there has never been observed a preferential permeability of pure protein membranes to lipoid-soluble substances which would be even remotely similar to that of the living protoplasts. So far it seems, therefore, most reasonable to ascribe the favoured permeation of lipoid-soluble substances through the protoplasm to the occurrence of some sort of lipoids in the plasma membranes.

(2) Provided that the plasma membrane contains lipoids as an essential component, it does not necessarily follow that the lipoid-soluble substances must penetrate the membrane truly dissolved in the lipoids. It is well known that lipoid-soluble substances are in most cases also surface-active and their great penetration power might therefore perhaps be ascribed, not so much to their solubility in lipoids, as to their adsorption at the water/lipoid interfaces and, it may be difficult, and perhaps even unwise, to distinguish very sharply between these two effects. If the plasma membrane is only a few molecules thick and consists of strongly oriented molecules, then evidently such expressions as " solubility in the plasma membrane substances " or " adsorbability

[7] Poijärvi, L. A. P., *Acta Bot. Fennica*, 1928, **4**, 1.
[8] Brooks, S. C., *Arch. exper. Zellforsch.*, 1934, **15**, 236; Frey-Wyssling, A., *Die Stoffausscheidung der höheren Pflanzen*, Berlin, 1935.

by the lipoid particles of the plasma membrane " must clearly be only very cautiously used. Nevertheless, the observed proportionality between oil-solubility and penetration power seems to make it preferable to interpret the relations between the physico-chemical character and the penetrating power of different substances in terms of solubility rather than in those of adsorption.

(3) Finally, it has been suggested [9] that the faster penetration of the smallest molecules may be due simply to the well-known fact that the diffusibility of dissolved substances always (*i.e.*, when there is free diffusion) decreases with increasing molecular size. This suggestion does not necessarily involve the assumption of a special sieve-effect in the case of protoplasmic permeability; it, however, disregards the fact that the rate of free diffusion decreases rather slowly with increasing molecular size, while the penetrating power often decreases fairly abruptly when the MR_D increases, for example from 10 to 20.

We thus reach the conclusion that among the permeability hypotheses so far put forth the lipoid-sieve hypothesis seems best to explain the permeability of the *Chara* cells to non-electrolytes. Let us now turn to the permeability of other plant cells.

Fig. 2 gives an idea of the permeability of the sixteen plant objects so far studied in this laboratory.[10] Each object is represented by a vertical line on which the permeability constants of seven substances are denoted by points using a logarithmic scale. The objects are arranged according to increasing permeability to erythritol.

It should be noted that the sixteen objects in Fig. 2 represent extremely different types of plant cells. Thus, such very different taxonomical groups as flowering plants, mosses, green algae, diatoms, brown algae, red algae, blue-green algae and bacteria are represented. Also the physiological character of the cells examined varies greatly. Nevertheless, all these cells agree in the main features of their permeability. If the lipoid-sieve hypothesis is accepted as an explanation for the permeability of the *Chara* cells, then it seems logical to accept it also in the case of the other objects.

This impression is even strengthened by a closer examination of the permeability differences, admittedly very great, between some of the objects studied, for it is easy to show that these differences are, at least for the most part, such as could be predicted from the standpoint of the lipoid-sieve hypothesis. Some examples will make this clear. (1) *Spirogyra* and *Chara* are seen from Fig. 2 to agree very closely as to their permeability, except that the permeability of the *Chara* cells is about three to ten times greater than that of *Spirogyra* cells. Obviously such a difference can, at least theoretically, be explained on the assumption that the plasma membrane of *Spirogyra* is correspondingly thicker than that of *Chara*. (2) The epidermal cells of *Rhoeo* have an exceptionally low permeability to all amides. (Thus *Rhoeo* is unique among the sixteen objects in being less permeable to malonamide than to erythritol and also in being less permeable to methyl urea than to glycerol.) This peculiarity can easily be explained along lines first put forth by Höber and his school.[11] It can be shown experimentally that the addition of an oil-soluble acid to a neutral oil largely increases its

[9] Traube, I., and F. Dannenberg, *Biochem. Z.*, 1928, **198**, 209.

[10] Except objects studied by Bärlund,[4] Collander and Bärlund,[6] and Marklund,[5] Fig. 2 also comprises some hitherto unpublished results of Mr. J. E. Elo.

[11] Höber, R., *Biol. Bull.*, 1930, **58**, 1 ; Wilbrandt, W., *Pflüger's Arch.*, 1931, **229**, 86.

solvent capacity for amides.   This being known, we have only to assume
that the plasma membrane lipoids of most plant cells are more or less
acidic in character while those of *Rhoeo* are approximately neutral.
(3) The root cells of *Lemna* differ from most other plant cells in being
more permeable to urea than to the more lipoid-soluble methyl urea.
Perhaps this can be explained on the assumption that the plasma membrane of the cells in question contains a considerable number of pores of
such a diameter as to be just penetrable by the urea molecules but not
by the somewhat greater molecules of methyl urea.   (4) The two diatoms so far studied, viz. *Melosira* and *Licmophora*, are both characterised

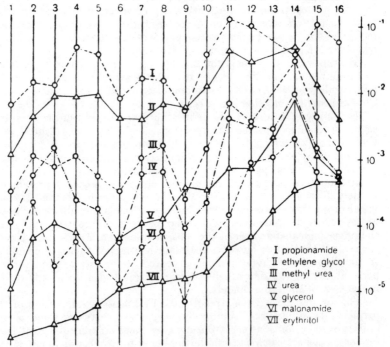

FIG. 2.—Permeability of sixteen different kinds of plant cells to some non-
electrolytes : 1. Leaf cells of *Plagiothecium denticulatum*, 2. *Oedogonium* sp.,
3. root cells of *Lemna minor*, 4. *Pylaiella litoralis*, 5. *Zygnema cyanosporum*,
6. subepidermal cells of *Curcuma rubricaulis*, 7. *Spirogyra* sp., 8. leaf cells of
Elodea densa, 9. epidermal cells of *Rhoeo discolor*, 10. epidermal cells of *Taraxa-
cum pectinatiforme*, 11. " leaf " cells of *Chara ceratophylla*, 12. internodal cells of
Ceramium diaphanum, 13. *Bacterium paracoli*, 14. *Oscillatoria princeps*, 15. *Melosira* sp., 16. *Licmophora* sp.

by their remarkably high permeability to erythritol and sucrose, *i.e.*,
to substances which have an extremely low lipoid-solubility and a considerable molecular volume.   This fact points to the occurrence of plasma
membrane pores of an extreme width in these cells.   (5) A great abund-
ance of somewhat smaller plasma membrane pores may be assumed in
the case of *Oscillatoria* which differs from most other cells in that the
sieve principle is more dominant and the effect of the lipoid-solubility
less so than in other cases.   (6) When *Chara* cells are placed in a glycerol solution it takes about two days before the glycerol concentration
in the interior of the cell reaches one-half of its concentration in the

external solution. With the cells of *Bacterium paracoli*, however, the same state is reached in less than a minute. The first impression is, of course, that the permeability of the bacterial cells is of quite a different order of magnitude than that of the *Chara* cells. This impression is however erroneous, for if the permeability is expressed, as in Fig. 2, per unit cell area it is found that there is in reality only a rather slight permeability difference between *Chara* and the bacterium.

It is not possible to present here in detail the results of the numerous permeability measurements carried out in other laboratories. Suffice it to say that the results are, for the most part, easily explicable in terms of the lipoid-sieve hypothesis. This applies even to that object, viz., *Beggiatoa mirabilis*, which, among all plant cells so far studied, deviates most strikingly from the main type. While in most plant cells the lipoid-solubility is the factor primarily responsible for the different permeation capacity of different solutes, their molecular volume being a factor of minor importance, just the reverse is true in the case of *Beggiatoa*. It is however, interesting to note that *Beggiatoa* has not a completely isolated position in regard to its permeability, for *Oscillatoria* forms an obvious connecting link in this respect between *Beggiatoa* on the one side and most plant cells on the other.

It still remains to consider the question what kind of structure would best explain the permeability qualities of the plasma membrane.

The rather strongly accelerated penetration of the smallest molecules makes the existence of a homogeneous lipoid layer consisting of random oriented molecules unlikely. It seems, on the contrary, that some sort of a sieve-like structure must be attributed to the plasma membrane. Perhaps the simplest assumption of this kind is that the plasma membrane consists of a few layers of regularly oriented lipoid molecules. The plasma membrane would thus have a structure like that of the phosphatide double films recently described by Bungenberg de Jong and Bonner [12] or like the oil films invented by Danielli.[13] Unfortunately, almost nothing seems to be known at present about the permeability properties of such films, but from a purely theoretical point of view it seems probable that they will show a preferential permeability (*a*) to lipoid-soluble substances and (*b*) to extremely small molecules, thus corresponding at least in a general way to the demands of the lipoid-sieve hypothesis. It is very much to be hoped that the permeability properties of these artificial films should be cleared up as soon as possible. Perhaps it will then turn out that some of these films correspond in their permeability so closely to the plasma membrane that an essential conformity, also in structure, must be assumed to exist between them.

Finally, it should perhaps be pointed out that the present writer is quite aware of the fact that not only electrolytes but probably also non-electrolytes may often be actively absorbed or excreted by living cells, the simple laws of diffusion being not directly applicable to such cases. For this active transport of matter Overton, already in the nineties, proposed the term "adenoid activity." It seems in fact important to distinguish carefully between this activity on the one side and the simple permeability processes on the other, for they probably have very little in common.

University of Helsingfors.

[12] Bungenberg de Jong, H. G., and J. Bonner, *Protoplasma*, 1935, **24**, 198.
[13] Danielli, J. F., *J. Cell. Comp. Physiol.*, 1936, **7**, 393.

Reprinted from *The Permeability of Natural Membranes*, 341–352 (1943)

2

APPENDIX A

THE THEORY OF PENETRATION OF A THIN MEMBRANE

By J. F. DANIELLI

DEDUCTION OF A FUNDAMENTAL EQUATION

THE plasma membrane of most cells is, to a first approximation, a thin layer of lipoid not more than 10^{-6} cm. in thickness. In the following matter discussion will be restricted to such thin layers of lipoid, though many of the conclusions reached are valid for membranes of other materials.

To a penetrating molecule a membrane presents three sites of resistance: (1) the membrane interface, for diffusion in the direction water → lipoid; (2) the membrane interface, for diffusion in the direction lipoid → water; (3) the interior of the membrane. All previous studies have assumed that (1) and (2) are so much less important than (3) that (1) and (2) may be neglected except in so far as they enter into partition coefficients. This is not a justifiable assumption, for such thin membranes.

The gross rate of passage of molecules from a solution of concentration C into the interior of the membrane is

$$\frac{dN}{dt} = aC, \qquad \ldots\ldots(42)$$

and in the reverse direction, from the membrane into water,

$$\frac{dN'}{dt} = aC', \qquad \ldots\ldots(43)$$

where C' is the concentration in the membrane. a and b are constants varying for different substances, and with temperature. These two equations define the rate of diffusion across the membrane interface. We now need to deal with the interior of the membrane.

One method is to use Fick's equation:

$$\frac{\partial S}{\partial t} = -D\frac{\partial C}{\partial x}, \qquad \ldots\ldots(19)$$

where ∂S is the amount of substance diffusing across an area of 1 cm.^{2} in time ∂t under a concentration gradient $\partial C/\partial x$. D is a constant for a given medium, substance and temperature. With the three constants a, b and D, we can obtain an expression for the permeability P of the membrane in gram moles crossing unit area per sec. per gram mol. per litre concentration difference. Mr F.J. Turton has made such an analysis, finding that, if

$$\frac{bl_2}{D} \ll 1 \quad \text{and} \quad \frac{dS}{dt} = P(C - C_i),$$

then
$$\frac{C - C_i}{C} = \frac{m_2}{m_1 - m_2} e^{m_1 t} - \frac{m_1}{m_1 - m_2} e^{m_2 t}, \qquad \ldots\ldots(44)$$

where
$$m_1 = \frac{b}{l_2}\left[-\frac{al_2}{2bl_3} + \frac{1}{2}\left(\frac{al_2}{2bl_3}\right)^2 \ldots \right],$$

$$m_2 = \frac{b}{l_2}\left[-2 - \frac{al_2}{2bl_3} - \frac{1}{2}\left(\frac{al_2}{2bl_3}\right)^2 \ldots \right],$$

$C_i =$ concentration of the molecular species under consideration inside a cell of constant volume,

$C =$ concentration outside the cell,

$l_2 =$ thickness of the cell membrane,

$l_3 =$ "equivalent thickness" of the cell, i.e. volume per sq. cm. of membrane.

This result (equation (44)) is valid within the range of values of a, b and D possible in cases where the rate of permeation can be observed experimentally, provided diffusion through the membrane is so much slower than through the aqueous phase that the membrane may be regarded as constituting the only significant barrier to diffusion. If, in addition,

$$\frac{2bt}{l_2} \gg 1 \quad \text{and} \quad \frac{al_2}{bl_3} \ll 1,$$

$$P = \frac{a}{2}. \qquad \ldots\ldots(28)$$

In this case diffusion through the interior of the membrane, and from membrane into water, is so much faster than diffusion from water into membrane, that only the latter factor is of importance.

Unfortunately, Fick's equation is not necessarily applicable to such thin layers. Fick (1855) assumed that the resistance to diffusion of a solute molecule was due to a medium of particle size negligible compared with that of a diffusing molecule. But in fact in the cell membrane the resistance to diffusion is due to large molecules such as cholesterol, much larger than most penetrating molecules. In a medium of molecules of the same or

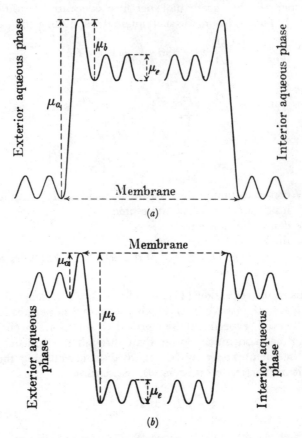

FIG. 73. Potential energy diagrams of the cell membrane: the potential energy barriers met (a) by a molecule such as glycerol, (b) by a molecule such as benzene or propane.

greater size than the solute molecule, the diffusing solute molecule does not encounter a smooth continuous resistance, but an intermittant resistance, which may be represented to a first approximation by the potential energy diagrams of Fig. 73. In this medium not every molecule will be diffusing simultaneously.

Many molecules will be vibrating about a mean position between adjacent potential energy barriers, and only those having more than the minimum kinetic energy μ_e necessary to pass over a potential energy barrier will in fact be capable of doing more than vibrate. The rate of diffusion through such a medium must now be considered.

Let the distance between adjacent potential energy minima be λ. Assume that the height of the potential energy maxima and minima does not vary with time, and that the diffusing molecules diffuse only when possessed of the minimum kinetic energy μ_e, necessary to break the bonds restricting diffusion.* Consider diffusion through a section of the medium 1 cm.² in area, containing n maxima. Let the concentration at one side of the section be kept at C_0, at the other side at C_n, and the concentrations in successive minima be $C_1, C_2, C_3, \ldots$ etc. When the condition of steady flow is reached the concentration across each maximum will be the same, $\partial S / \partial t$. The rate of flow from C_0 to C_1 is

$$\frac{dN_0}{dt} = eC_0, \qquad \ldots\ldots(45)$$

where e is a constant and is a function of μ_e. Similarly, the flow from C_1 to C_0 is

$$\frac{dN_1}{dt} = eC_1. \qquad \ldots\ldots(46)$$

Hence the net flow from C_0 to C_1 is

$$\frac{dN_0}{dt} - \frac{dN_1}{dt} = \frac{dS}{dt} = e(C_0 - C_1)$$
$$= e(C_1 - C_2) = \ldots = e(C_{n-1} - C_n)$$
$$= \frac{e}{n}(C_0 - C_n). \qquad \ldots\ldots(47)$$

But Fick's equation for the state of steady flow is

$$\frac{dS}{dt} = -D\,\frac{C_n - C_0}{(n+1)\,\lambda}. \qquad \ldots\ldots(19\cdot1)$$

Combining (19·1) and (47):

$$D = \frac{n+1}{n}\,\lambda e. \qquad \ldots\ldots(48)$$

* Due to Van der Waals' forces, hydrogen bonds, etc.

The expression $D = \{(n+1)/n\}\,\lambda e$ reduces to λe if n is large. But if the membrane is thin, n is small, and D is a function of n, i.e. of the membrane thickness. With this established we may now proceed to deal with the membrane as a whole.

To different molecules the cell membrane will present different potential energy barriers, e.g. Fig. 73 (a) (for a polar molecule such as glycerol $CH_2OH\,.\,CHOH\,.\,CH_2OH$); Fig. 73 ($b$) (for a non-polar molecule such as propane $CH_3\,.\,CH_2\,.\,CH_3$). But in all cases there are three types of barrier, one μ_a met on entering the membrane, one μ_b met on leaving the membrane, and n of height μ_e met in the interior of the membrane.

Consider the case where the concentration on the two sides is kept steady at C and C_i respectively. When the condition of steady flow is reached, the rate of flow across every barrier is the same, dS/dt. From the minor barriers we obtain equation (47). For the major barriers

$$\frac{dS}{dt} = aC - bC_0 = bC_n - aC_i$$

$$= \frac{a}{2}\,(C - C_i) - \frac{b}{2}\,(C_0 - C_n). \qquad \ldots\ldots (49)$$

Combining (47) and (49):

$$C_0 - C_n = \frac{a}{b + 2e/n}\,(C - C_i). \qquad \ldots\ldots (50)$$

Substituting for $(C_0 - C_n)$ in (47) from (50):

$$\frac{dS}{dt} = \frac{e}{n}\cdot\frac{a}{b + 2e/n}\,(C - C_i). \qquad \ldots\ldots (51)$$

But

$$\frac{dS}{dt} = P\,(C - C_i).$$

$$\therefore\ P = \frac{ae}{nb + 2e}. \qquad \ldots\ldots (26)*$$

This is the fundamental equation defining the permeability of a homogeneous lipoid membrane.

The equation $P = ae/(nb + 2e)$ has been obtained for a homogeneous membrane. For an inhomogeneous membrane we must

* I am greatly indebted to Mr F. J. Turton for this simple analysis.

treat each area separately, i.e. obtain a term $a_r e_r/(n_r b_r + 2e_r)$ for the rth homogeneous element and then sum, obtaining

$$P = \Sigma \frac{ae}{nb + 2e}.$$

The deduction of the equation $P = ae/(nb + 2e)$ is based on a few simple assumptions about the structure of matter. In its general form it is undoubtedly correct, and further advance in this matter must consist in filling in the details of our knowledge of the way in which a, b, e and n vary with molecular species, time, temperature, etc.

Special Cases

Consider a homogeneous lipoid membrane to which the equation $P = ae/(nb + 2e)$ applies. The value of μ_b is greater than μ_e for the great majority of substances which penetrate cell membranes very slowly; hence for such substances b is much less than e, and with thin membranes, i.e. when n is small, nb may be neglected compared with $2e$. Hence

$$P = \frac{a}{2}, \qquad \qquad \text{......(28)}$$

i.e. for molecules which penetrate a thin membrane very slowly, the rate of passage over the oil-water interface in the direction water $\rightarrow$ oil is often the term dominating the rate of penetration. This is the same result as was obtained by Turton, assuming that Fick's equation could be applied to thin membranes. The reason why the same result is obtained from both sets of assumptions is that the effective resistance to free diffusion is almost entirely located at the oil-water interface, so that the nature of diffusion in the interior of the membrane is, in this case, unimportant.

On the other hand, for many very rapidly penetrating molecules nb is equal to, or often greater than, $2e$. In this latter case we obtain

$$P = \frac{a}{b} \cdot \frac{e}{n}. \qquad \qquad \text{......(27)}$$

Thus we obtain simplified versions of equation (26) for very slowly and for very rapidly penetrating molecules.

An Approximate Method for Evaluating a, b, e and n

(1) *Theoretical Aspects.*

There is no accurate method yet available for evaluating a, b and e. The following procedure yields an approximate result.

Consider equation (42), $dN/dt = eC_0$, which gives the rate of flow of molecules from one side to the other of a single potential energy barrier of height μ_e calories. If the media on either side of the barrier were perfect gases, e would be given by

$$e = \sqrt{\frac{RT}{2\Pi M}} e^{-\frac{\mu_e}{RT}}, \qquad \ldots\ldots(52)$$

where R is the gas constant per gram mol., T is absolute temperature, M is molecular weight. If the medium is not a perfect gas, the numerical coefficient of the terms on the right will be different, and we shall multiply the right-hand side by a constant r to compensate for this. This correction is purely empirical. Then, in addition, not every molecule having the minimum kinetic energy μ_e will in fact be able to diffuse; let ϕ_e be the probability that a molecule having the activation energy μ_e will actually diffuse across a potential energy barrier. Then

$$e = r\phi_e \sqrt{\frac{RT}{2\Pi M}} e^{-\frac{\mu_e}{RT}} \qquad \ldots\ldots(53)$$

and $$(Q_{10})_e = \frac{e_{T+10}}{e_T} = \sqrt{\frac{T+10}{T}} e^{\frac{\mu_e}{RT}\left(\frac{10}{T+10}\right)}. \qquad \ldots\ldots(54)$$

This semi-empirical equation needs testing to ascertain its limits of accuracy. This can be done in the following ways: (1) keeping M constant and varying the temperature: then any failure in e to follow the predicted course must be due to variation in $r\phi_e$, or some inadequacy in the terms $\sqrt{T}e^{-\mu_e/T}$. A suitable method for this study is the rate of diffusion of molecules across the liquid-vapour interface at various temperatures; it is found that, if allowance is made for the variation in μ_e with temperature, the terms $\sqrt{T}e^{-\mu_e/RT}$ are accurate, to a first approximation, and the value of e at any temperature may be calculated from the value at any other temperature within a range of $50°$ C. with an accuracy of $\pm 10\%$. (2) Keep the temperature constant, and vary M. This tests the terms $\sqrt{(1/M)}\, e^{-\mu_e}$. This may be done using the Fick diffusion constant for diffusion in liquids, for, as we saw above, $e = D/\lambda$, so that

$$D = \lambda r \phi \sqrt{\frac{RT}{2\Pi M}} e^{-\frac{\mu_e}{RT}}, \qquad \ldots\ldots(55)$$

and at constant T: $\qquad \dfrac{DM^{\frac{1}{2}} e^{\mu_e}}{\lambda} = \text{constant}. \qquad \ldots\ldots(56)$

Or, since λ is approximately constant also, $DM^{\frac{1}{2}} e^{\mu_e}$ should be a constant, say β, if the terms $M^{\frac{1}{2}} e^{\mu_e}$ in equation (53) are correct. Unfortunately there are very few accurate values for μ_e under these conditions, but with the values available we can show that from the average value of β, the value of D for any other molecule up to the size of glucose can be calculated, with an error not greater than $\pm 100\%$ in solvents such as water and alcohol.

The validity of Thovert's equation, $DM^{\frac{1}{2}} = \text{constant}$ (see Chapter v), is probably due to the fact that e^{μ_e} does not vary greatly in solvents of low viscosity, so that from (56) we have $DM^{\frac{1}{2}} = \text{constant}$, approximately.

Summing up, equation (53) must be regarded as approximate only, but it can be used for approximate work. The error involved in the calculation of e by this equation is probably much less than a factor of fivefold, so that if we can assume the calculated value of e may be five times less or five times greater than the true value, we shall probably be within safe limits.

This leaves the terms a, b and n. Of these a and b will be of the same form as e. n cannot be estimated accurately, but cannot be less than 10^{-8} cm. Hence if the thickness of a membrane is divided by 10^{-8} we obtain a *maximum* value of $n = 50$, in the case of a membrane 50 Å. thick. The relative values of a and b can also be obtained from the partition coefficient B, for

$$B = \frac{a}{b}. \qquad \ldots\ldots(57)$$

Now consider the special case of equation (28). Substituting from (53) we have

$$P = \frac{a}{2} = \tfrac{1}{2} r \phi_a \sqrt{\frac{RT}{2\Pi M}} e^{-\frac{\mu_a}{RT}}. \qquad \ldots\ldots(58)$$

Hence $\qquad (Q_{10})_P = \sqrt{\frac{T+10}{T}} e^{\frac{\mu_a}{RT}\left(\frac{10}{T+10}\right)}. \qquad \ldots\ldots(59)$

And substituting for $e^{\mu_a/RT}$ in (57):

$$PM^{\frac{1}{2}}Q_{10}^{(T+10)/10} = r\phi_a\sqrt{\frac{RT}{8\Pi}}\left(\sqrt{\frac{T+10}{T}}\right)^{\frac{T+10}{10}} \qquad \ldots\ldots(60)$$

= constant, at constant temperature.

Similarly, in the special case of equation (27):

$$P = \frac{a}{b}\cdot\frac{e}{n} = \frac{r\phi_a\phi_e}{n\phi_b}\sqrt{\frac{RT}{2\Pi M}}e^{\frac{\mu_b-\mu_a-\mu_e}{RT}}. \qquad \ldots\ldots(61)$$

Hence

$$(Q_{10})_P = \sqrt{\frac{T+10}{T}}e^{-\frac{\mu_b-\mu_a-\mu_e}{RT}\left(\frac{10}{T+10}\right)}. \qquad \ldots\ldots(62)$$

And

$$PM^{\frac{1}{2}}Q_{10}^{(T+10)/10} = \frac{r\phi_a\phi_e}{n\phi_b}\sqrt{\frac{RT}{2\Pi}}\left(\sqrt{\frac{T+10}{T}}\right)^{\frac{T+10}{10}} \qquad \ldots\ldots(63)$$

= constant, at constant temperature.

Probably $\phi_a \approx \phi_b \approx \phi_e$, so that ϕ_e and ϕ_b in equations (61) and (63) cancel out.

So we find that, if equation (53) is approximately correct, then for the two special cases of equations (27) and (28) $PM^{\frac{1}{2}}Q_{10}^{(T+10)/10}$ is constant at constant temperature, and the temperature variation of P may be calculated. $PM^{\frac{1}{2}}$ and Q_{10} are negatively correlated. For a membrane made up of areas of varying properties, Q_{10} and P will differ from place to place. If each area were treated separately, we should find $\Sigma PM^{\frac{1}{2}}Q_{10}^{(T+10)/10}$ = constant. But experimentally we measure an *average* Q_{10} and an *average* P, and hence for such a membrane $PM^{\frac{1}{2}}Q_{10}^{(T+10)/10}$ will not be constant.

For homogeneous membranes, owing to the deficiencies in equation (53), permissible variations in $PM^{\frac{1}{2}}Q_{10}^{(T+10)/10}$ will lie between values five times less and five times greater than the average value.

Since $PM^{\frac{1}{2}}$ and (Q_{10}) are negatively correlated for a homogeneous membrane, if this negative correlation is not observed in a particular series of experiments, the membrane involved cannot be homogeneous. This applies to all types of molecules, not only those obeying (27) and (28).

(2) Practical Aspects.

In practice one can make only two direct observations, giving (1) the permeability constant P; (2) the Q_{10} of P. Hence to obtain values of a, b, e and n we must resort to a number of devices. In the first place we shall not be able to make the fullest quantitative use of equation $P = ae/(nb + 2e)$, but shall be restricted for quantitative work to the special case of slowly penetrating molecules obeying the equation $P = a/2$. To prove that this equation is obeyed we must prove that $nb \ll 2e$, i.e. we must know the minimum value of e, and the maximum value of n and b.

TABLE LXXI

All figures here refer to flow per sq. cm. per sec. per gram mol. per c.c. concentration.

Substance	D_{water}	Min. value of $2e$	Q_{10}	Min. value of μ_a	Max. value of a	B	Max. value of nb ($= 50a/B$)
Propyl alcohol	$9 \cdot 3 \times 10^{-3}$	3·72	1·37	5,100	$1 \cdot 22 \times 10^{5}$	0·005*	$1 \cdot 22 \times 10^{8}$
Urea	$9 \cdot 4 \times 10^{-3}$	3·76	1·86	10,200	$1 \cdot 81 \times 10^{1}$	0·05	$6 \cdot 0 \times 10^{5}$
Thiourea	$8 \cdot 25 \times 10^{-3}$	3·3	2·14	12,750	$2 \cdot 0 \times 10^{-1}$	0·015*	$6 \cdot 7 \times 10^{2}$
Glycol	$9 \cdot 3 \times 10^{-3}$	3·72	2·92	18,400	$1 \cdot 41 \times 10^{-5}$	0·00049	1·44
$\alpha\beta$ Dioxypropane	$8 \cdot 25 \times 10^{-3}$	3·3	3·75	22,500	$1 \cdot 03 \times 10^{-8}$	0·0059	$8 \cdot 7 \times 10^{-5}$
Glycerol	$7 \cdot 2 \times 10^{-3}$	2·88	3·65	22,200	$1 \cdot 62 \times 10^{-8}$	0·00007	$1 \cdot 15 \times 10^{-2}$
$\alpha\gamma$ Dioxypropane	$8 \cdot 25 \times 10^{-3}$	3·3	3·31	20,300	$4 \cdot 28 \times 10^{-7}$	0·001	$2 \cdot 14 \times 10^{-2}$
Diethylene glycol	$6 \cdot 9 \times 10^{-3}$	2·76	3·42	20,800	$1 \cdot 26 \times 10^{-7}$	0·005*	$1 \cdot 26 \times 10^{-3}$
Triethylene glycol	$5 \cdot 75 \times 10^{-3}$	2·3	3·34	20,500	$2 \cdot 22 \times 10^{-7}$	0·02*	$5 \cdot 55 \times 10^{-4}$

* Approximate values.

The interior of the cell membrane is a hydrocarbon liquid and its viscosity cannot be more than 10^4 to 10^5 times greater than that of water—if the membrane were of olive oil, for example, the viscosity would be only 10^2 times greater than water. Hence the minimum value of the Fick diffusion constant D in the membrane interior cannot be less than about 10^{-4} of the value in water (D_w). Hence

$$e = \frac{D}{\lambda} = \frac{D_{water}}{\lambda} \times 10^{-4}.$$

The maximum possible value of λ is 5×10^{-7} cm. (it is probably nearer to 10^{-8} cm.). Hence the *minimum* possible value of e is given by

$$e = \frac{10^{-4} D_{water}}{5 \times 10^{-7}} = 2 \times 10^{2} D_{w}. \qquad \ldots\ldots(64)$$

The *maximum* possible value of n, for a membrane about 50 Å. thick, has been given above as 50.

Lastly, b is given by $\qquad B = \dfrac{a}{b}, \qquad\qquad \ldots\ldots(57)$

where B is the oil-water partition coefficient. Hence to find b we need a, which must be obtained from equation (53). Now in the equation $P = ae/(nb + 2e)$ the term having the largest Q_{10} is a. Hence the Q_{10} of P is either equal to the Q_{10} of a, if the molecular species obeys the simplified form $P = a/2$, or else is less than the Q_{10} of a, i.e. the Q_{10} of $P \leq Q_{10}$ of a. The higher the Q_{10} of a, the smaller is a. Hence, if we calculate $e^{\mu_a/RT}$ from (59), we shall obtain the maximum possible value of a when this value of $e^{\mu_a/RT}$ is substituted in equation (58). The term $\sqrt{RT/2\Pi M}$ of (58) is also known, so that now we only need the value of $r\phi_a$ to get the maximum possible value of a. In solvents such as water, benzene and alcohol, $r\phi_a$ varies very little, and is about 10^5 to 10^6. In solvents consisting of larger molecules, $r\phi$ is probably smaller, so that if we use the value 10^5 we have the maximum possible value of $r\phi$. Hence the maximum possible value of a is

$$a = 10^5 \sqrt{\frac{RT}{2\Pi M}} e^{-\frac{\mu_a}{RT}} \text{ per cm.}^2 \text{ per sec. per mol. per litre,}$$

$$\ldots\ldots(65)$$

where $e^{\mu_a/RT}$ is calculated from the Q_{10} of P. Then the maximum value of b can be obtained from equation (57).

TABLE LXXII

Substance	P	Q_{10}	$PM^{\frac{1}{2}}Q_{10}^{(T+10)\cdot 10}$
Propyl alcohol	$10\cdot6 \times 10^{-16}$	$1\cdot37$	$1\cdot1 \times 10^{-10}$
Urea	$7\cdot8 \times 10^{-16}$	$1\cdot86$	$8\cdot5 \times 10^{-7}$
Thiourea	$0\cdot019 \times 10^{-16}$	$2\cdot14$	$1\cdot7 \times 10^{-7}$
Glycol	$0\cdot209 \times 10^{-16}$	$2\cdot92$	$2\cdot06 \times 10^{-2}$
$\alpha\beta$ Dioxypropane	$0\cdot405 \times 10^{-16}$	$3\cdot75$	$82\cdot0$
Glycerol	$0\cdot0017 \times 10^{-16}$	$3\cdot65$	$0\cdot17$
$\alpha\gamma$ Dioxypropane	$0\cdot105 \times 10^{-16}$	$3\cdot31$	$0\cdot48$
Diethylene glycol	$0\cdot075 \times 10^{-16}$	$3\cdot42$	$1\cdot16$
Triethylene glycol	$0\cdot0333 \times 10^{-16}$	$3\cdot34$	$0\cdot28$

We now have the minimum value of e, maximum values of n and b. If $nb \ll 2e$, then if the membrane is homogeneous $PM^{\frac{1}{2}} Q_{10}^{(T+10)/10} = \text{constant}$. If this quantity is not constant for molecules for which $nb \ll 2e$, the membrane is not homogeneous. Tables LXXI and LXXII show an example of this treatment applied to the data of Jacobs *et al.* (1935) for ox red cells, assuming the partition coefficient membrane/water = partition coefficient olive oil/water. It will be seen that only $\alpha\beta$ dioxypropane, $\alpha\gamma$ dioxypropane, glycerol, diethylene glycol and triethylene glycol have maximum values of $nb \ll 2e$. Of these molecules $\alpha\beta$

dioxypropane is unsymmetrical and will not have the same ϕ value as the other molecules. This leaves the last four molecules, which should obey the equation $P = a/2$, so that if the membrane is homogeneous $PM^{\frac{1}{2}} Q_{10}^{(T+10)/10}$ should be approximately constant. Table LXXII shows that this quantity is within the permissible limits of variation. Hence the membrane must be homogeneous, to a first approximation.

The Quantitative Relationships between Permeability and Partition Coefficients. In the case of very slowly penetrating molecules for which $P = a/2$, we can use (57) to relate P to the partition coefficient B: then

$$P = \frac{a}{2} = \tfrac{1}{2} bB$$

$$= \tfrac{1}{2} B r \phi_b \sqrt{\frac{RT}{2\Pi M}}\, e^{-\frac{\mu_b}{RT}}$$

or $\qquad PM^{\frac{1}{2}} e^{\frac{\mu_b}{RT}} = B \times \text{constant, at constant temperature.}$

$$\dots\dots (66)$$

Hence $PM^{\frac{1}{2}} e^{\mu_b/RT}$ is a linear function of the partition coefficient B. μ_b is given approximately by $\mu_b = 750x$ calories, where $x =$ number of unscreened CH_2 groups per molecule. CH_2 groups screened from the solvent by polar groups such as OH do not contribute to x.

In the other extreme case, very rapidly penetrating molecules,

$$P = \frac{a}{b} \cdot \frac{e}{n} = B \times \frac{e}{n}$$

$$= \frac{B}{n} r \phi_e \sqrt{\frac{RT}{2\Pi M}}\, e^{-\frac{\mu_e}{RT}}.$$

Or, since μ_e probably varies only slightly from substance to substance in liquids,

$PM^{\frac{1}{2}} = B \times \text{constant, at constant temperature.} \qquad \dots\dots (67)$

Hence in this case $PM^{\frac{1}{2}}$ is, to a first approximation, a linear function of the partition coefficient. Both these relationships (66) and (67) are subject to the accuracy limits of a factor of fivefold mentioned earlier in connection with equation (53).

REFERENCES

FICK. 1855. *Pogg. Ann.* **94**, 59.

JACOBS, GLASSMAN & PARPART. 1935. *J. Cell. Comp. Physiol.* **7**, 197.

23

Copyright © 1958 by Elsevier Publishing Company
Reprinted from *Biochim. Biophys. Acta*, **27**, 229–246 (1958)

3

THERMODYNAMIC ANALYSIS OF THE PERMEABILITY OF BIOLOGICAL MEMBRANES TO NON-ELECTROLYTES

O. KEDEM AND A. KATCHALSKY

Weizmann Institute of Science, Rehovot (Israel)

I. INTRODUCTION

In spite of the large amount of information which has accumulated on permeability phenomena, the conventional equations of volume and solute flow (equations (1) to (6)) do not completely describe the physical behaviour of membranes, and the permeability data obtained by different methods are not quantitatively comparable. The insufficiency of the permeability equations was felt by previous authors and several attempts have been made to supplement them. Thus FREY-WYSSLING[1] and LAIDLER AND SHULER[2] took into account the contribution of the solute to volume flow; USSING[3] claimed that a force exerted by solvent flow "enters into the escaping tendency of all substances present in the solutions in contact with the membrane"; PAPPENHEIMER[4] treated the flow of solute through membranes as two flows—a flow by filtration and a flow by diffusion. These attempts did not however develop into a self-consistent and general set of equations able to cover the whole range of phenomena. A solution to this problem can be obtained through the methods of irreversible thermodynamics. STAVERMAN[5, 6] has recently given a complete treatment of osmotic pressure measurements applying these methods and KIRKWOOD[7] has similarly treated the transport of ions through membranes. The expressions obtained by these authors are however not directly applicable to the physiological measurements described in the literature. The present work is devoted to a suitable modification and extension of the equations derived by the methods of the thermodynamics of irreversible processes in order to apply them to biological permeability data. It follows the approach of STAVERMAN.

The equations obtained are applied to the analysis of several commonly used experimental methods and it is shown how the coefficients defined by the thermodynamic equations can be evaluated from the data. Only these coefficients are of significance in the analysis of membrane structure and the mechanism of permeation.

Moreover, an examination, with the aid of the thermodynamic equations, of the results reported in the literature, reveals that in spite of numerous measurements carried out, additional data are needed in many cases to obtain the pertinent coefficients.

II. THE INSUFFICIENCY OF THE CONVENTIONAL PERMEABILITY EQUATIONS

The conventional description of transport through membranes makes use of two equations, one for solute flow and one for volume flow. Consider an isothermal system

References p. 246.

24

consisting of two compartments, an inner one designated by the superscript i and an outer one designated by the superscript o, the two compartments being separated by a membrane. Only non-electrolyte solutes are considered.

The equation for solute flow is written analogously to Fick's equation

$$\frac{dN_s^i}{dt} = k_s A (c_s^o - c_s^i) \tag{1}$$

where N_s^i denotes the number of moles of permeable solute in the inner compartment, k_s is the permeability coefficient of the solute which includes the thickness of the membrane, A is the membrane area, and c_s is the concentration of solute s in moles per unit volume. If the volume of the inner compartment is denoted by V^i, $c_s^i = N_s^i/V^i$ and eqn. (1) may be written in the form[8]

$$\frac{dN_s^i}{dt} = k_s A \left(c_s^o - \frac{N_s^i}{V^i} \right) \tag{2}$$

The equation for volume flow when no hydrostatic pressure difference exists between the inner and outer compartments is based on the proportionality between the flow dV^i/dt of volume (usually identified with the flow of water), and the difference $\pi^i - \pi^o$ of osmotic pressure between the inner and outer solutions, *i.e.*

$$\frac{dV^i}{dt} = k_w' A (\pi^i - \pi^o) \tag{3}$$

where k_w' denotes the permeability coefficient of water. Putting $\pi = RTc$ (c is the osmotic concentration) and absorbing RT into the proportionality coefficient, equation (3) can also be written

$$\frac{dV^i}{dt} = k_w A (c^i - c^o) \tag{4}$$

In (4), c denotes the sum of the concentrations of all the solutes whether the membrane is permeable for them or not.

If the system contains only one permeable solute and if the total amount of the non-permeable solutes in the inner compartment is denoted by N_m^i, c^i becomes

$$c^i = \frac{N_m^i + N_s^i}{V^i}$$

Denoting the concentration of non-permeable solutes in the outer compartment by c_m^o and that of permeable solute by c_s^o, c^o can be written

$$c^o = c_s^o + c_m^o$$

Introducing into (4) we obtain the equation used by Jacobs[9], namely

$$\frac{dV^i}{dt} = k_w A \left\{ \frac{N_m^i + N_s^i}{V^i} - (c_s^o + c_m^o) \right\} \tag{5} *$$

When a hydrostatic pressure difference $\Delta p = p^o - p^i$ exists between the compartments in addition to the osmotic pressure difference $\Delta \pi = \pi^o - \pi^i$ considered above,

---

* Jacobs writes $c_0 V_0$ for N_m^i where c_0 and V_0 are the initial values of non-permeable solute concentration and cell volume.

References p. 246.

equation (3) according to Starling's hypothesis (c/. PAPPENHEIMER[4]) assumes the form

$$\frac{dV^i}{dt} = k'_w A (\Delta p - \Delta \pi) \tag{6}$$

Permeability to water

The inadequacy of the equations presented above for the consistent description of the behaviour of biological systems will be first demonstrated on the basis of the measurements of ZEUTHEN AND PRESCOTT[10] on the permeability of fish and frog eggs.

ZEUTHEN AND PRESCOTT suspended the eggs in a frog Ringer solution. After osmotic equilibrium was reached, the volume flow dV^i/dt became zero. Introducing into equation (5)*, this leads to the condition

$$\frac{N^i_m}{V^i} = c^o_m \tag{7}$$

as in this part of the experiment no permeable solute was present.

The equilibrated cells were then transferred into a Ringer solution of corresponding solute composition, in which however 10–15% of the water was substituted by heavy water (D_2O). It was found that the heavy water penetrates the cells following equation (2) exactly, thus proving that it behaves as any solute. At the same time it was observed that the cell volume remains *constant*, so that $dV^i/dt = 0$.

Further experiments were carried out on water flow in hyper- or hypotonic solutions of non-penetrating solutes and it was found that equation (5) accurately describes the volume changes of the eggs. The values of k_w obtained should thus enable us to calculate the volume flow in any solution. In particular, we should find that the flow in the experiments carried out in isotonic Ringer solutions containing heavy water is given by

$$\frac{dV^i}{dt} = k_w A \left(\frac{N^i_s}{V^i} - c^o_s\right) \qquad (8) \qquad \text{as} \qquad \frac{N^i_m}{V^i} = c^o_m \tag{9}$$

Now, as demonstrated above, heavy water behaves as an ordinary solute which in this case has in the outer solution a very high molar concentration compared with the other solutes of the Ringer solution. It would be expected therefore that dV^i/dt would assume large negative values at the beginning of the experiment and would approach zero when N^i_s/V^i approaches c^o_s. The observation that $dV^i/dt = 0$ throughout the experiment proves therefore that equation (5) is incomplete. As shown in the following, this is due to the fact that no distinction is made between permeable and non-permeable solutes. ZEUTHEN AND PRESCOTT found moreover, that the penetration of hevay water in non-equilibrated solutions is not represented adequately by equation (2). It is more rapid in hypotonic solution and slower in hypertonic, thus proving that cross relations exist between diffusion and filtration flows which are not expressed in the conventional equations (1) to (6).

Plasmolytic measurements

The insufficiency of the equations is revealed also when less permeable solutes than heavy water are used. Especially instructive are the observations made on the

---

* It is generally assumed that between such cells and the surroundings, no pressure difference can be maintained.

References p. 246.

threshold concentration for plasmolysis of plant cells. In order to understand what underlies the concept of threshold concentration, let us consider the behaviour of a plant cell in solutions of different concentrations. Fig. 1 represents schematically the dependence of the cell volume V^i on time t in solutions of a single permeable solute of various concentrations c_s^o. Consider first a cell immersed in pure water. The cell swells to its maximal volume V_{max}^i and a hydrostatic pressure difference Δp is set up between the cell sap and the surrounding medium. Δp is maintained by the rigidity of the cell wall and is equal to the osmotic pressure difference $\Delta \pi$. V_{max}^i is independent of time and corresponds to the upper horizontal straight line.

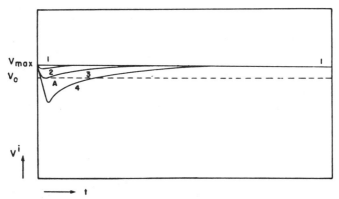

Fig. 1. Schematic representation of volume changes of a plant cell immersed in solutions of one permeable solute at different concentrations. Curve 1 in pure water; curves 2, 3, and 4 in increasing concentrations of permeable solute. The difference between V_{max} (the volume in pure water) and V_0 (the equilibrium volume of the cellulose walls) is exaggerated. A is the plasmolytic point.

The maximally swollen cell is now introduced into a solution of low concentration (Curve 2). The stress in the cell wall will be slightly relieved as a small amount of quickly permeating solvent immediately leaves the cell. Later, as both solute and solvent penetrate into the cell, the stress recovers. As long as c_s^o is sufficiently small, the volume changes are insignificant and not observed under the microscope. However, above a definite concentration, the plasmolytic concentration c_s^{o*}, the rapid initial escape of solvent will become sufficiently large so as to cause the plasma to shrink away from the cell wall and thereby relieve the stress completely, making $\Delta p = 0$ (curve 3, Fig. 1). After the cell volume goes through a minimum, at which $dV^i/dt = 0$, the penetration of the solute will cause deplasmolysis and bring the volume slowly to its initial state at which dV^i/dt again equals zero. The phenomenon becomes even more pronounced as the concentration of the external solution is still further increased.

At the plasmolytic point both Δp and dV^i/dt equal zero (see Fig. 1). Thus one should expect eqn. (5) to be valid and, as $c_m^o = 0$ in these experiments, we find:

$$0 = k_w A \left(\frac{N_m^i + N_s^i}{V^i} - c_s^{o*} \right) \quad \text{or} \quad c_s^{o*} = \frac{N_m^i + N_s^i}{V^i}$$

As in this case the permeation of water is much faster than that of solute, the amount of solute already penetrated at the point of minimum volume is rather small. We find therefore that N_s^i/V^i is much smaller than c_s^{o*} and may be neglected.

References p. 246.

Moreover, as pointed out, the change in volume at the plasmolytic point is very small so that N_m^i/V^i equals the initial cell concentration c_{int}^i, hence

$$c_s^{o*} = c_{int}^i \qquad (10)$$

Equation (10) shows that the plasmolytic threshold concentration might be expected to be independent of the nature of the external solute. Experience, however, has shown that the ratio c_s^o/c_{int}^i differs from unity within a range of more than a thousand and the quantity $\mu = c_s^{o*}/c_{int}^i - 1$ has even been defined as a coefficient characterizing permeability—the Lepeschkin constant[11].

III. PERMEABILITY EQUATIONS DERIVED ON THE BASIS OF IRREVERSIBLE THERMODYNAMICS

The thermodynamic approach to the problem of permeability of membranes leads to the conclusion that the incompleteness of the equations (1)–(6) is due to the fact that they involve only two of the three coefficients, required to characterize permeability for a solute–solvent system. The necessity for three coefficients in the case of membranes permeable to solvent and only one of the solutes, may be understood in a qualitative way as follows:

In the case of free diffusion, solvent and solute migrate only relatively to each other. Hence the hydrodynamic resistance of diffusion flow is due to the friction between solute and solvent alone, so that diffusion in a solution of a single solute is determined by a single diffusion coefficient. The passage through a membrane, however, involves two additional factors, namely, the friction between solute and membrane and the friction between solvent and membrane. A full description thus has to take account of three coefficients whose values will depend on the nature of the three processes involved. Not in all cases, however, will all three processes be equally important. For example, in coarse membranes with large pore dimensions, the solvent–solute friction will contribute more than the other factors, and permeability will approach the behaviour in free diffusion (as is the case in the fritted disk method of NORTHROP AND ANSON[12]). In dense membranes on the other hand, in which part of the solute penetration takes place say through dissolution in the membrane, the contribution of the friction between solute and the membrane becomes predominant.

Derivation of the equations of flow in two-component systems

The starting point of the thermodynamic description of non-equilibrium systems is a calculation of the entropy production during the process. In the case of a two component system in which two solutions of the same solvent and solute are separated by a membrane, the entropy production per unit time d_iS/dt is given by the equation

$$\frac{d_iS}{dt} = \frac{1}{T}\,(\mu_w^o - \mu_w^i)\,\frac{dN_w^i}{dt} + \frac{1}{T}\,(\mu_s^o - \mu_s^i)\,\frac{dN_s^i}{dt} \qquad (11)$$

where μ denotes the chemical potential (of the solvent, subscript w, and solute, subscript s), dN_w^i/dt and dN_s^i/dt represents the number of moles of solvent and solute respectively entering the inner compartment per unit time[13].

It is often convenient to use the "dissipation function" which is given by $T\,d_iS/dt$. We shall use the dissipation function per unit area *i.e.*

References p. 246.

$$\Phi = T \cdot \frac{1}{A} \frac{d_i S}{dt} = (\mu_w^o - \mu_w^i) \dot{n}_w + (\mu_s^o - \mu_s^i) \dot{n}_s \qquad (12)$$

where

$$\dot{n}_w = \frac{1}{A} \frac{dN_w^t}{dt} \text{ and } \dot{n}_s = \frac{1}{A} \frac{dN_s^i}{dt}$$

It will be observed that the dissipation function (12) is composed of the sum of products of flows per unit area ($\dot{n}_w$ and $\dot{n}_s$) and corresponding "forces"—the differences in chemical potential.

Equation (12) constitutes a special case of the general expression $\Phi = \Sigma_i J_i X_i$ where J_i denotes a flow and X_i the generalized conjugated force. The choice of flows and forces is arbitrary to a certain extent, so long as their products sum up to the same dissipation function.

In the following we shall make the approximation that the chemical potentials for ideal solutions may be used, so that

$$\mu^o - \mu^i = \bar{v} \Delta p + RT \Delta \ln \gamma \qquad (13)$$

where $\bar{v}$ is the partial molar volume, Δp is the difference in pressure between the outer and inner compartment, and γ the molar fraction of the constituent. In the case of dilute solutions, where the volume fraction φ of the solute is small, $\varphi = c_s \bar{v}_s \ll 1$ and equation (13), written for the solute, becomes

$$\mu_s^o - \mu_s^i = \bar{v}_s \Delta p + RT \Delta \ln c_s = \bar{v}_s \Delta p + RT \frac{\Delta c_s}{c_s} \qquad (14)$$

where $\Delta c_s = c_s^o - c_s^i$ and c_s is a mean of the concentrations of the solute in the two compartments given by

$$\frac{\Delta c_s}{c_s} = \ln \frac{c_s^o}{c_s^i}$$

If

$$\frac{\Delta c_s}{c_s^i} \ll 1, \ c_s = \frac{c_s^i + c_s^o}{2}$$

The corresponding equation for the solvent is

$$\mu_w^o - \mu_w^i = \bar{v}_w \Delta p - RT \frac{\Delta c_s}{c_w} \qquad (15)$$

where $c_w = (1 - \varphi)/\bar{v}_w$ or, to a good approximation, $c_w = 1/\bar{v}_w$. Introducing equations (14) and (15) into equation (12) and rearranging we get

$$\Phi = (\dot{n}_w \bar{v}_w + \dot{n}_s \bar{v}_s) \Delta p + \left(\frac{\dot{n}_s}{c_s} - \frac{\dot{n}_w}{c_w} \right) RT \Delta c_s \qquad (16)$$

It will be observed that in (16) the dissipation function is represented by a new set of forces and flows. The new forces $X_v = \Delta p$ and $X_D = RT \Delta c_s$ are the forces usually employed in permeability studies, Δp is the hydrostatic pressure while $RT \Delta c_s$ is the driving force in Fick's equation. The conjugate flows are the total volume flow per unit area:

$$J_v = \dot{n}_w \bar{v}_w + \dot{n}_s \bar{v}_s \qquad (17)$$

and the relative velocity of solute versus solvent which is a measure for exchange flow:

$$J_D = \frac{\dot{n}_s}{c_s} - \frac{\dot{n}_w}{c_w} \qquad (18)$$

References p. 246.

The general theory of irreversible thermodynamics is based on the assumption that the flows J are functions of all the forces operative in the system and that, if the forces are sufficiently small, the dependence is linear. Thus in the case of two flows J_1 and J_2 dependent on two forces X_1 and X_2, the relation between the J's and X's is given by

$$J_1 = L_{11}X_1 + L_{12}X_2$$
$$J_2 = L_{21}X_1 + L_{22}X_2 \tag{19}$$

where the L's are called the phenomenological coefficients.

The phenomenological coefficients are correlated by the law of Onsager which requires the equality of the cross-coefficients $L_{ik} = L_{ki}$ or in our case

$$L_{12} = L_{21} \tag{20}$$

Writing (19) in the notation applying to our system, we obtain:

$$J_v = L_p \Delta p + L_{pD} RT \Delta c_s \tag{21}$$
$$J_D = L_{Dp} \Delta p + L_D RT \Delta c_s$$

with $$L_{Dp} = L_{pD}$$

The second law of thermodynamics requires that entropy production be always positive, from which we may conclude that the straight coefficients L_p and L_D must be positive, while L_{pD} may be either positive or negative. The magnitude of L_{pD} is restricted by the condition, $L_p \cdot L_D - L_{pD}^2 > 0$.

The physical meaning of (21) may be seen in the following way: in very coarse membranes, volume flow and exchange flow are independent. Each of the flows is determined only by its conjugate force: J_v by the pressure gradient Δp and J_D by the concentration gradient Δc_s. However in many less permeable membranes, the flows are interdependent and the gradient in solute concentration produces a volume flow, even when $\Delta p = 0$; this is known as osmotic flow. Similarly, a pressure difference causes not only a total volume flow but also a relative velocity in the solute–solvent flow—this is known as ultrafiltration. These interdependences are incorporated in the coefficient L_{pD}.

Consideration of some special cases

The volume flow at *zero concentration difference*,

$$J_v = L_p \Delta p \qquad (\Delta c_s = 0)$$

measures the mechanical permeability of the membrane for a given solution, and L_p is the filtration coefficient.

In an ideal semipermeable membrane $n_s = 0$ so that $J_D = -\dot{n}_w/c_w$ and $J_v = n_w \bar{v}_w$. However, as pointed out, for dilute solutions $c_w = 1/\bar{v}_w$ and hence

$$J_D = -J_v \text{ (ideal semipermeable membrane)} \tag{22}$$

This is obvious since in this case both volume and exchange flows are due to solvent only. Introducing (21) into (22) and rearranging terms, we obtain

$$(L_p + L_{pD}) \Delta p + (L_D + L_{pD}) RT \Delta c_s = 0 \qquad (\dot{n}_s = 0) \tag{23}$$

As (23) must hold for all pressure and concentration differences, it can only be satisfied

References p. 246.

30

if both bracketed expressions equal zero. Hence

$$L_p = -L_{pD} = L_D \text{ (ideal semipermeable membrane)} \qquad (24)$$

The system is then fully characterized by the filtration coefficient L_p alone. It will be shown below (V) that in the case of permeable membranes, L_{pD} should be smaller than L_p. In the case of completely non-selective membranes, L_{pD} is zero, as is readily seen: with these latter membranes a pressure difference alone ($\Delta c_s = 0$) does not produce an exchange flow, so that

$$J_D = 0 = L_{pD} \Delta p \text{ or } L_{pD} = 0 \text{ (non-selective membrane)} \qquad (25)$$

The same conclusion may also be derived from the fact that in non-selective membrane no volume flow is produced by a concentration difference alone *i.e.*

$$J_v = 0 = L_{pD} RT \Delta c \text{ or again } L_{pD} = 0 \text{ (non-selective membrane)}$$

In the intermediate cases L_{pD} is negative and lies between 0 and $-L_p$. When however the membrane is more permeable to solute than to solvent L_{pD} is positive.

Besides volume flow, the flow generally measured is not J_D but the solute flow $\dot{n}_s$. In terms of J_v and J_D, $\dot{n}_s$ is given by

$$\dot{n}_s = (J_v + J_D) c_s \qquad (\varphi = c_s \bar{v}_s \ll 1) \qquad (26)$$

$\dot{n}_s$ is often measured at constant volume, *i.e.* at $J_v = 0$ and Δp in this case is given by

$$\Delta p = -\frac{L_{pD}}{L_p} RT \Delta c_s \qquad (J_v = 0) \qquad (27)$$

(see (21)).

Introduction of (21) and (27) into (26) gives

$$\dot{n}_s = \frac{L_p L_D - L_{pD}^2}{L_p} c_s RT \Delta c_s \qquad (J_v = 0) \qquad (28)$$

Thus at constant volume the solute flow is proportional to the concentration difference.

The equations for practical calculation

For comparison with experimental data it is convenient to pass from the system L_p, L_D and L_{pD} to another set of coefficients. STAVERMAN has introduced the reflection coefficient, σ, defined in our terms by

$$L_{pD} = -\sigma L_p \qquad (29)$$

It is clear that $\sigma = 0$ applies to a non-selective membrane and $\sigma = 1$ to an ideally selective one, permeable to solvent only.

We further define the mobility of the solute, ω, as

$$\omega = \frac{L_p L_D - L_{pD}^2}{L_p} c_s = (L_D - L_p \sigma^2) c_s \qquad (30)$$

It is seen from equation (28) that ωRT is the proportionality coefficient between the solute flow at constant volume and Δc_s. Therefore ωRT is the equivalent of the solute permeability constant k_s measured at *constant volume*. For an ideally semipermeable membrane $\omega = 0$ (see (24)). We shall now transcribe our equations in terms of L_p, σ and ω. Introducing (29) and (30) into (21) and (26) we obtain

References p. 246.

$$J_v = L_p \left(\Delta p - \sigma RT \Delta c_s \right) \tag{31}$$

$$\dot{n}_s = c_s L_p \left(1 - \sigma \right) \Delta p + \left[\omega - c_s L_p \left(1 - \sigma \right) \sigma \right] RT \Delta c_s \tag{32}$$

Another convenient formulation which permits us to correct the equation for solute flow directly for volume flow readily derives from (31) and (32):

$$\dot{n}_s = \omega RT \Delta c_s + \left(1 - \sigma \right) c_s J_v \tag{33}$$

Equations (31) and (32), derived on a thermodynamic basis, will now be compared with the conventional expressions (1)–(6).

The equation for $J_v = \mathrm{d}V^i / \mathrm{d}t \cdot 1 / A$ corresponds to equation (6), and L_p is the equivalent of k'_w. It will be observed that, as $RT\Delta c_s = \Delta \pi$, equation (6) becomes identical with (31) only if $\sigma = 1$, *i.e.* if the membrane is impermeable to the given solute. However, as will be shown later, for many solutes σ is only a small fraction of unity.

The simultaneous use of (1) and (6) or (4) is therefore self-contradictory.

Comparison of (1) and (33) shows that the expressions become identical, and $k'_s = \omega RT$, only if $J_v = \delta$. (The case $1 - \sigma = 0$ is not an exceptional one, as then also $\omega = 0$).

It is seen from equations (32) and (33) that the solute flow can be described as a function of Δc_s alone if either Δp or J_v are zero. It should however be emphasized that for the same Δc_s, the flow in these two cases may be quite different. In designing physiological measurements, it is thus of the utmost importance that the conditions under which the experiments are performed are fully known. For example for hydrodynamic calculations the significant coefficient is ω so that the measurements of $\dot{n}_s$ have to be carried out at constant volume.

Polycomponent systems with a single permeable solute

In biological studies, we are not dealing with simple two-component systems containing only one solute and solvent. On one side of the membrane at least, a large number of components is to be found even if only two—the solvent and one solute—are permeable. In the latter case, the number of streams and coefficients remains the same as described previously, but the presence of the non-permeable solutes modifies the chemical potentials and hence the driving forces.

We continue the derivation of the pertinent equations on the assumption that the solutions are ideal. The case of real solutions can be treated by introducing activity factors. Let the system contain solvent (w) a permeable solute (s) and several non-penetrating solutes (i) with molar fractions $\gamma_w, \gamma_s, \gamma_i$ which are different on the two sides of the membrane.

In correspondence with the previous notations

$$\Delta \mu_w = RT \ln \frac{\gamma_w^o}{\gamma_w^i} + \bar{v}_w \Delta p \tag{34}$$

so that, assuming the volume fractions of all the solutes to be small,

$$\Delta \mu_w = -\frac{RT \Delta c_s}{c_w} - \frac{RT \Sigma \Delta c_i}{c_w} + \bar{v}_w \Delta p \tag{35}$$

and

$$\frac{+}{-} RT \Sigma \Delta c_i = \Delta \pi_i$$

References p. 246.

where $\Delta \pi_i$ is the contribution of the non-permeable solutes to the difference in osmotic pressure.

As in the case of the two-component system

$$\Delta \mu_s = \frac{RT \, \Delta c_s}{c_s} + \bar{v}_s \Delta p$$

If the same flows are considered as before, the conjugated forces X_v and X_D may be derived from the dissipation function as follows:

$$\Phi = \dot{n}_s \Delta \mu_s + \dot{n}_w \Delta \mu_w = (\dot{n}_s \bar{v}_s + \dot{n}_w \bar{v}_w) \, X_v + \left(\frac{\dot{n}_s}{c_s} - \frac{\dot{n}_w}{c_w} \right) X_D \qquad (36)$$

From the coefficients of $\dot{n}_s$ and $\dot{n}_w$ on both sides of equation (36) one obtains:

$$X_v = c_s \Delta \mu_s + c_w \Delta \mu_w$$
$$X_D = (1 - \varphi) \, c_s \Delta \mu_s \mp \varphi c_w \Delta \mu_w$$

Introducing the values of $\Delta \mu_s$ and $\Delta \mu_w$ we get:

$$X_v = \Delta p - \Delta \pi_i$$
$$X_D = RT \, \Delta c_s + \varphi \Delta \pi_i \qquad (37)$$

We see that at equilibrium, both forces are zero, because the equilibrium pressure head is $\Delta \pi_i$ and the equilibrium distribution of the solute is such that

$$\Delta \mu_s = - \Delta \pi_i \bar{v}_s \quad i.e. \, RT \, \Delta c_s = - \varphi \Delta \pi_i \qquad (38)$$

With the driving forces (37), the equations of flow corresponding to (31) and (32) in the presence of non-permeable solutes become

$$J_v = L_p \, (\Delta p - \Delta \pi_i) - \sigma L_p \, (RT \, \Delta c_s + \varphi \Delta \pi_i) \qquad (39)$$

$$\dot{n}_s = L_p \, (1 - \sigma) \, c_s \, (\Delta p - \Delta \pi_i) + [\omega - \sigma L_p \, (1 - \sigma) \, c_s] \, (RT \, \Delta c_s + \varphi \Delta \pi_i) \qquad (40)$$

and

$$\dot{n}_s = J_v \, (1 - \sigma) \, c_s + \omega \, (RT \, \Delta c_s + \varphi \Delta \pi_i) \qquad (41)$$

The term $\varphi \Delta \pi_i$ in (39), (40) and (41) in most cases represents only a small correction to $RT \Delta c_s$, but the contribution of $\Delta \pi_i$ to the first driving force in (39) and (40) is very important. Actually in many biological systems $\Delta p = 0$ whereas $\Delta \pi_i \neq 0$.

Comparing (39) with (6) it is seen that equation (6) describes the volume flow correctly only if none of the solutes present can pass the membrane (*i.e.* if $\sigma = 1$). The contribution of a solute characterized by reflection coefficient σ to the volume flow is σ times smaller than that of a non-penetrating solute at the same concentration.

We may now summarize the relation between the three coefficients L_p, σ, ω and the conventional constants k'_w and k_s.

(1) k'_w as defined by equation (6) can be identified with the filtration coefficient L_p if no permeable solute is present or if the latter is at equilibrium (equation (38)).

(2) The coefficient of solute flow k_s can be identified with ωRT if there is no volume flow.

(3) σ is a third independent coefficient ignored by the conventional equations. It will be shown that it is closely related to Lepeschkin's constant.

References p. 246.

33

IV. APPLICATION TO SPECIAL SYSTEMS

Filtration and diffusion of water

The inconsistencies in the application of the conventional equations to the permeability data for heavy water are removed by the introduction of the reflection coefficient.

It is evident that membranes will have little ability to differentiate between ordinary and heavy water. σ for heavy water will therefore be either zero or have a very small positive value. If we introduce $\sigma = 0$ into equations (39) (41) we get

$$J_v = L_p \, (\Delta p - \Delta \pi_i) \tag{42}$$

$$\dot{n}_s = J_v c_s + \omega \, RT \, \Delta c_s \tag{43}^*$$

as the flow equations applying to the present case. We see therefore that for this case two coefficients, the filtration constant L_p and ω, corresponding to the diffusion constant at zero volume flow, suffice to describe the permeability behaviour.

Equation (43) demonstrates the pronounced dependence of solute on volume flow flow when $\sigma = 0$.

In the measurements with frog and fish eggs, for which, as we have seen, $\Delta p = 0$, J_v changes sign with the sign of $\Delta \pi_i$ (see (42)). The flow of heavy water $\dot{n}_s$ will thus be different depending on whether a simultaneous bulk flow of water enters or leaves the cell, an effect observed by ZEUTHEN AND PRESCOTT[10].

If σ for heavy water is to be determined, the experiments have to be made under special conditions. Thus if we arrange that $\Delta p - \Delta \pi_i = 0$ then $J_v = - \sigma L_p \, RT \Delta c_s$ and it becomes possible to evaluate σ from the magnitude of the small volume flow.

Permeability of plant cells

In part II, we pointed out that while the plasmolytic threshold concentration of the penetrating solute, c_s^{o*}, should equal, according to the conventional equations, the concentration c_{int}^i of the non-permeable cell constituents, c_s^{o*}/c_{int}^i differs enormously from unity. The following consideration will show that the coefficient actually measured by plasmolytic experiments is the reflection coefficient, σ, of the cell membrane.

At the plasmolytic point $\Delta p = 0$ and $J_v = 1/A \cdot dV^i/dt = 0$. Introducing into equation (37) we thus get:

$$J_v = 0 = L_p \, (\Delta \pi_i + \sigma \, RT \, \Delta c_s)$$

or

$$- \Delta \pi_i = \sigma \, RT \, \Delta c_s$$

As pointed out in II moreover the amount of solute penetrated at the plasmolytic point is negligible. This implies that $\Delta c_s = c_s^{o*}$ and hence

$$- \Delta \pi_i = \sigma \, RT \, c_o^{s*}$$

As

$$- \Delta \pi_i = RT \, c_{int}^i, \quad \sigma \, RT \, c_s^{o*} = RT \, c_{int}^i$$

or

$$\frac{c_s^{o*}}{c_{int}^i} = \frac{1}{\sigma} \tag{44}$$

---

* We have neglected $\varphi \Delta \pi_i$ against $RT \, \Delta c_s$ despite the fact that φ is rather large in experiments with heavy water and may approach 0.1. However Δc_s is so large, throughout the greater part of the experiment, that the neglect is justified. For example in the case of the 10 % D_2O in water solutions Δc_s was 5.5 moles/liter at the beginning of the experiment. The initial value of $\Delta \pi_i$ for a cell immersed in frog Ringer × 0.5 expressed in moles/liter, is approximately 0.1, so that $\varphi \Delta \pi_i = 0.01$ as compared with 5.5.

References p. 246.

The plasmolytic threshold concentration therefore gives a direct measure of σ. Similarly the Lepeschkin constant $c_s^{o*}/c_{int}^i - 1 = \mu$ is another function of the reflection coefficient

$$\sigma = \frac{1}{1 + \mu}$$

From the measurements of RUHLAND AND HOFFMANN on the permeability of *Beggiatoa mirabilis*, quoted by HÖBER[11], it is seen that the threshold concentration of urea is $1.75 \cdot 10^3$ times larger than that of sucrose, *i.e.* $\sigma_{urea}/\sigma_{sucrose} = 5.7 \cdot 10^{-4}$.

There is no way of obtaining the solute mobility ω from the plasmolytic concentration. It should be stressed however that the classical measurements of OVERTON, and of COLLANDER AND BÄRLUND[14] give this constant in a clear and definite manner. Permeability measurements on plant cells under non-plasmolytic conditions are here carried out at constant cell volume so that $J_v = 0$. Hence from equation (41), the solute flow is determined by the single coefficient ω:

$$\dot{n}_s = -\frac{V^i}{A}\frac{d\Delta c_s}{dt} = \omega\,RT\,\Delta c_s \tag{45}$$

Stationary state measurements

The evaluation of the coefficients becomes relatively simple if one of the flows vanishes. This condition is known in the thermodynamics of irreversible processes as a "stationary" state and is characterized by several important features. In the case of permeability studies, the flow which can be made to vanish by experimental conditions is generally the volume flow *i.e.* $J_v = 0$. In the case of this stationary state we get from equation (39) and (38)

$$(\Delta p - \Delta\pi_i) = \sigma\,RT\,\Delta c_s \text{ (for negligible } \varphi\Delta\pi_i) \tag{46}$$
and
$$\dot{n}_s = \omega\,RT\,\Delta c_s \tag{47}$$

respectively, so that σ and ω can be obtained directly from pressure measurements and from solute flow.

The first to carry out systematic stationary state studies on biological material were PAPPENHEIMER, RENKIN AND BORRERO[15]. These workers circulated plasma through a dog's hind leg maintaining a hydrostatic pressure in the blood vessels which kept the weight of the limb constant. It can be shown that the condition of constant weight is practically equivalent to zero volume flow.

With plasma fluid alone, the "isogravimetric" hydrostatic pressure was found to be nearly equal to the osmotic pressure of the plasma proteins (actually 97% of the osmotic pressure as measured by artificial membranes). On addition of permeable solute to the plasma fluid, the pressure had to be raised to a value Δp_i in order to avoid filtration. The increase was sudden and large and declined only slowly with time.

In these experiments it was impossible to determine the volume of the tissue fluid into which diffusion takes place during the experiment. It was also difficult to determine the effective membrane surface. Therefore an exact relation between $\dot{n}_s$ and Δc_s could not be obtained. PAPPENHEIMER *et al.* gave the value of $\dot{N}_s$, *i.e.* the number of moles of solute penetrating per unit time through the membrane area A^g per 100 g tissue

$$\frac{\dot{N}_s}{A^g} = \dot{n}_s$$

References p. 246.

From equation (47) therefore

$$\dot{N}_s = A^g \omega RT \, \Delta c_s \tag{48}$$

It was observed that $\dot{N}_s/(\Delta p_i - \Delta \pi_i)$ is constant during a given experiment. From (48) and (46) we see that

$$\frac{\dot{N}_s}{\Delta p_i - \Delta \pi_i} = \frac{A^g \omega}{\sigma} \tag{49}$$

which is indeed a constant. PAPPENHEIMER *et al.* assumed however that $\Delta p - \Delta \pi_i = RT \Delta c_s$ instead of equation (46) and deduced that

$$\frac{\dot{N}_s}{\Delta p_i - \Delta \pi_i} = A^g \frac{k_s}{RT}$$

$$(\text{At } J_v = 0, \quad k_s = \omega RT)$$

The value of the permeability constant obtained in this way may be many times larger than the true value as already demonstrated by GRIM[16]. Nevertheless, the data of PAPPENHEIMER *et al.* permit the evaluation of σ and $A^g \omega$, as σ can be determined from equation (46) at zero time, when $\Delta c_{s,0}$ is known, (provided the time required for the distribution in the plasma fluid is short compared to the time for penetration). Δp_i cannot be measured exactly at $t = 0$; however, it can be deduced by extrapolating the plot of $\ln \Delta p_i$ against t (which is a straight line) to $t = 0$.

Introducing $\Delta p_{i,0}$ and $\Delta c_{s,0}$ into equation (46) we derive the σ values given in the following Table. With the aid of equation (49) the corresponding values of $A^g \omega$ are obtained.

TABLE I

REFLECTION COEFFICIENTS AND SOLUTE MOBILITIES
(calculated from PAPPENHEIMER, RENKIN AND BORRERO[15])

	σ	$A^g \omega \cdot 10^{12}$ $cm^2 \cdot \frac{1}{sec} / \frac{dyn}{mol}$
Glucose	0.04	1.05
Sucrose	0.058	0.99
Inulin	0.375	0.55

As expected, the selectivity increases with molecular weight while ω decreases.

In experiments with artificial membranes in which $\Delta \pi_i$ was equal to zero, GRIM was able to demonstrate that there exists a very large difference between stationary state pressure Δp_i and $RT \Delta c_s$. He interpreted the ratio between Δp_i and $RT \Delta c_s$ on the basis of the kinetic theory of LAIDLER AND SHULER[2], which will be discussed in the next paragraph in the light of irreversible thermodynamics.

Relaxation measurements and the kinetic theory of LAIDLER AND SHULER

In relaxation experiments, one starts with a system in a non-equilibrium state and subsequently allows it to go to equilibrium. The observer does not interfere with the run of the process and only records the mode of approach to the final state. Normally, the evaluation of coefficients from the results of relaxation experiments is not as simple as that from stationary state studies, however the data give more information and all the coefficients can be obtained from a single curve.

References p. 246.

A typical permeability relaxation experiment was carried out by SHULER, DAMES AND LAIDLER[17] on the permeation of water and non-electrolyte through collodion membranes. SHULER et al. immersed a collodion bag filled with the solution under consideration into a large thermostatted water bath and followed the changes in the volume and pressure by means of an attached narrow capillary. Immediately after the immersion, the solution rose rapidly in the capillary, passed a maximum and descended slowly to its equilibrium level. The pressure difference between the bag and surroundings is proportional to the height h of the column in the capillary, while the volume flow is proportional to the change of this height with time. Calculations of flows as function of time are simplified by the fact that the bag volume remains practically constant during the experiment. Furthermore the external bath is so much larger than the collodion bag that

$$-\frac{d\Delta c_s}{dt} = \dot{n}_s \frac{A}{V^i} \tag{50}$$

where A is the membrane area and V^i the volume of the bag.

The pressure difference $\Delta p = -hg$, if the density of the solution is assumed to be close to unity, and

$$\frac{d\Delta p}{dt} = -\frac{dh}{dt} g \tag{51}$$

where g is the gravitational acceleration.

If the cross section of the capillary is a, the volume flow per unit membrane area is

$$J_v = \frac{a}{A} \cdot \frac{dh}{dt} \tag{52}$$

or introducing (51)

$$J_v = -\frac{a}{Ag} \frac{d\Delta p}{dt} \tag{53}$$

LAIDLER AND SHULER recognized the difference between volume flow and water flow and used the correct expression

$$J_v = \dot{n}_w \bar{v}_w + \dot{n}_s \bar{v}_s \tag{17}$$

However, they calculated the flows on a two-coefficient system according to the equations:

$$\dot{n}_w = \frac{Q_1 c_t \bar{v}_w}{d'RT} (\Delta p - RT\Delta c_s) \tag{54}$$

$$\dot{n}_s = \frac{Q_2 c_t \bar{v}_w}{d'} \Delta c_s \tag{55}$$

where Q_1 and Q_2 are permeability coefficients for solvent and solute respectively, d' is the thickness of the membrane, c_t is "the total concentration of all species present" assumed to be constant, $c_t \bar{v}_w$ being close to unity.

Introducing (54) and (55) into (17) and (50) and making use of (53) one obtains

$$-\frac{d\Delta p}{dt} = \lambda_1 (\Delta p - RT \Delta c_s) + \lambda_2 RT \Delta c_s \tag{56}$$

$$= \lambda_1 \Delta p - (\lambda_1 - \lambda_2) RT \Delta c_s$$

and
$$\frac{d\Delta c_s}{dt} = \lambda_3 \Delta c_s \qquad (57)$$

respectively, where

$$\lambda_1 = \frac{A \cdot g}{a} \frac{Q_1 c_t \bar{v}_w^2}{d'RT} \qquad (58)$$

$$\lambda_2 = \frac{Ag}{a} \frac{Q_2 c_t \bar{v}_w \bar{v}_s}{d'RT}$$

and
$$\lambda_3 = \frac{A}{Vi} \frac{Q_2 c_t \bar{v}_w}{d'}$$

It will be observed that despite the use of three parameters λ_1, λ_2 and λ_3, there are in LAIDLER AND SHULER's system only two independent coefficients. The coefficient λ_2 is derived from λ_3 by multiplication with a constant independent of the membrane.

Integration of (56) and (57) carried out by LAIDLER AND SHULER leads to the expressions

$$\Delta p = RT \Delta c_{s,o} \frac{\lambda_1 - \lambda_2}{\lambda_1 - \lambda_3} (e^{-\lambda_3 t} - e^{-\lambda_1 t}) \qquad (59)$$

$$\Delta c_s = \Delta c_{s,o} \cdot e^{-\lambda_3 t} \qquad (60)$$

in which the initial condition $\Delta p = 0$ and $\Delta c_s = \Delta c_{s,o}$ at time $t = 0$ was taken into account. The authors found that equation (59) describes the experimental results satisfactorily. It was found that a short while after maximal pressure was attained, $\ln \Delta p$ *versus* t gives a straight line. This fact shows that equation (59) can apply if λ_3 differs widely from λ_1 and after a certain time one of the terms $e^{-\lambda_3 t}$ or $e^{-\lambda_1 t}$ will vanish in comparison with the other.

SHULER, DAMES AND LAIDLER assumed $\lambda_3 > \lambda_1$. It is seen immediately that in the limiting case of an ideal semipermeable membrane, λ_1 must be larger than λ_3, as in this case $Q_2 = 0$ and λ_3 is thus zero. That the assumption $\lambda_1 > \lambda_3$ is reasonable in the case of a penetrating solute, can be shown as follows. From (59) and (60)

$$\frac{\Delta p}{RT \Delta c_s} = \frac{\lambda_1 - \lambda_2}{\lambda_1 - \lambda_3} (1 - e^{-(\lambda_1 - \lambda_3)t})$$

If $(\lambda_1 - \lambda_3)$ is a sufficiently large positive number, $e^{-(\lambda_1 - \lambda_3)t}$ becomes negligible as compared to unity after a short time, and thus the ratio of Δp and Δc becomes constant. If, on the other hand, $\lambda_1 - \lambda_3$ is an equally large negative number, $e^{-(\lambda_1 - \lambda_3)t}$ becomes much larger than 1 and the ratio of Δp to Δc would increase logarithmically.

The values given for λ_1 and λ_3 have therefore to be interchanged. In Table II the experimental parameters from the measurements of SHULER *et al.* for the penetration of some sugars through the same membrane are given, assuming $\lambda_1 > \lambda_3$.

The values of λ_2/λ_3 calc. given in the last column of Table II were calculated from equation (58) introducing for V^i and a the values given by SHULER, DAMES AND LAIDLER. The partial molar volumes were derived from the densities of the solutions, cited in Landoldt-Börnstein's Tables.

The large difference between the calculated and experimental ratio λ_2/λ_3 indicates that the permeability of the synthetic membrane cannot be expressed by two straight coefficients alone and that the neglect of the cross coefficient leads to intrinsic contradictions.

References p. 246.

TABLE II

VALUES OF λ, AS DEFINED BY EQUATION (58), FOR A NUMBER OF SUGARS[17]

From SHULER, DAMES AND LAIDLER's paper[17] (their Table II), λ_1 and λ_3 interchanged.

	$\lambda_1 \cdot 10^2$ min^{-1}	$(\lambda_1 - \lambda_2) \cdot 10^4$ min^{-1}	$\lambda_3 \cdot 10^2$ min^{-1}	$\lambda_2 \cdot 10^2$ min^{-1}	λ_2/λ_3 exp.	λ_2/λ_3 calc.
Sucrose	7.8	1.4	1.00	7.49	7.79	0.0936
Lactose	9.1	2.9	1.00	9.07	9.07	0.094
Raffinose	8.0	1.9	0.89	7.98	8.95	0.19

The following values of partial molar volume were used in the derivation of λ_2/λ_3 calc. Sucrose: 202 c.c./mol. Lactose: 203 c.c./mol. Raffinose 405 c.c./mol.

If we introduce (50) and (53) into the thermodynamically derived equations (31) and (33), we find

$$-\frac{d\Delta p}{dt} = \frac{Ag}{a} L_p \ (\Delta p - \sigma RT \Delta c_s) \tag{61}$$

$$-\frac{d\Delta c_s}{dt} = \frac{A}{Vi} \ [\omega RT \Delta c_s - (1 - \sigma) J_v c_s] \tag{62}$$

As is easily verified $(1 - \sigma) J_v c_s$ can be neglected as compared with $\omega RT \Delta c_s$ in these experiments. Consequently comparison of (61) and (62) with (56) and (57) leads to the following identifications:

$$\lambda_1 = \frac{Ag}{a} L_p$$

$$\lambda_3 = \frac{Ag}{Vi} \omega RT \tag{63}$$

$$\lambda_1 - \lambda_2 = \frac{Ag}{a} \sigma L_p$$

It is clear from this that λ_1, λ_2 and λ_3 should be independent. In particular $\lambda_2/\lambda_3 = gV^i/aRT \cdot (1 - \sigma)L_p/\omega$ is seen to depend on the membrane system. From (63) and Table II, the following values are obtained for σ and $A\omega$:

TABLE III

REFLECTIVITY CONSTANTS AND SOLUTE MOBILITIES FOR SOME SUGARS IN COLLODION MEMBRANES

From SHULER, DAMES AND LAIDLER's measurements.

	σ	$A\omega \cdot 10^{13}$ cm$^2 \cdot \frac{1}{sec} / \frac{dyn}{mol}$
Sucrose	0.0018	3.96
Lactose	0.0029	3.96
Raffinose	0.0024	3.53

The values show that the collodion membranes are less selective than the natural membranes investigated by PAPPENHEIMER et al. The reflection coefficient of collodion membranes can also be determined from the measurements of GRIM[16]. GRIM measured the ratio of Δp and $RT\Delta c_s$ at zero volume flow, which gives directly the reflection coefficient (equation (31)). From these experiments the reflection coefficient of sucrose penetrating through a collodion membrane was 0.004, which is of the same order of magnitude as the corresponding value in Table III.

References p. 246.

The permeability of collodion membranes differs of course according to the mode of preparation.

V. DISCUSSION OF THE REFLECTION COEFFICIENT

As pointed out previously, the values of the coefficients σ, ω and L_p are independent of each other, subject to a restriction corresponding to the condition $L_p L_D - L_{pD}^2 > 0$ Any explicit correlations between them have to be derived kinetically on the basis of models for the transport mechanism. However, without considering a detailed model, it is possible to delimit the range of σ, for a given pair of ω and L_p, more closely on the basis of the following general assumption.

We assume that solvent and solute interact with each other and this interaction endows each of them with a velocity component *in the direction of the force acting on the other*. The extent to which this interaction takes place in the passage through the membrane depends on the nature of the system. Cases of lowest interaction are systems where solute and solvent follow different paths through the membrane, as encountered in aqueous solutions of liquid-soluble substances passing through a mosaic membrane. Highest interaction of solute and solvent occurs in free diffusion and is approached in coarse capillary membranes.

Let us now derive σ for given L_p and ω in a system where the solute passes the membrane by dissolution and the solvent goes separately through the membrane capillaries. The driving force on the solute is

$$\Delta \mu_s = \bar{v}_s \Delta p + \frac{RT \Delta c_s}{c_s}$$

and the velocity of solute penetration in this type of system will evidently be determined only by $\Delta \mu_s$ and will be independent of $\Delta \mu_w$. Let us now consider the solute flow under two different experimental conditions. One: $\Delta p = 0$, $RT \Delta c_s / c_s = a$ and another: $\bar{v}_s \Delta p = a$, $\Delta c_s = 0$. As $\Delta \mu_s$ is the same in both cases, $\dot{n}_s^I$ for the first case and $\dot{n}_s^{II}$ for the second will be equal. Introducing the values for cases I and II into equation (40) we get

$$\dot{n}_s^I = a c_s \left[\omega - \sigma L_p (1 - \sigma) c_s \right] = \dot{n}_s^{II} = \frac{L_p (1 - \sigma)}{\bar{v}_s} a c_s$$

which, rearranging terms and neglecting $\sigma \varphi$ as compared to 1, becomes

$$\sigma = 1 - \frac{\omega \bar{v}_s}{L_p} \tag{64}$$

Passing to systems where hydrodynamic interaction occurs in the membrane, we can easily show that $\dot{n}_s^I$ will no longer equal $\dot{n}_s^{II}$. In case I, the driving force on the solute is opposite in direction to that on the solvent and the latter will thus tend to diminish solute flow. In case II, on the other hand, the pressure difference Δp operates on both solute and solvent in the same direction. Hence

$$\dot{n}_s^I < \dot{n}_s^{II}$$

or, translated in terms of σ

$$\sigma < 1 - \frac{\omega \bar{v}_s}{L_p} \tag{65}$$

The inequality (65) together with equation (64) thus delimits the range of values of

References p. 246.

40

the reflection coefficient in dependance on the extent of hydrodynamic interaction between solute and solvent in the membrane.

The condition $\sigma = 1$ may thus be regarded as sufficient experimental evidence for semi-permeability, as in view of (65) σ can approach unity only when $\omega \to 0$.

Finally it is clear that for very coarse membranes, σ goes to zero. For given ω and L_p, we have in general

$$0 \leq \sigma \leq 1 - \frac{\omega \bar{v}_s}{L_p} \tag{66}$$

The above consideration may thus help to decide from measurements of σ, ω and L_p what mechanism of solute transfer is involved. In fairly permeable membranes values of σ close to $(1 - \omega \bar{v}_s/L_p)$ indicate independent passage of solute and solvent, while $\sigma \ll 1 - \omega \bar{v}_s/L_p$ indicates capillary mechanism.

ACKNOWLEDGEMENT

The authors are indebted to Prof. A. J. STAVERMAN for helpful discussions.

SUMMARY

The application of the conventional permeability equations to the study of biological membranes leads often to contradictions. It is shown that the equations generally used, based on *two* permeability coefficients—the solute permeability coefficient and the water permeability coefficient—are incompatible with the requirements of thermodynamics of irreversible processes.

The inconsistencies are removed by a thermodynamic treatment, following the approach of STAVERMAN, which leads to a *three* coefficient system taking into account the interactions: solute–solvent, solute–membrane and solvent–membrane.

The equations derived here have been applied to various permeability measurements found in the literature, such as: the penetration of heavy water into animal cells, permeability of blood vessels, threshold concentration of plasmolysis and relaxation experiments with artificial membranes.

It is shown how the pertinent coefficients may be derived from the experimental data and how to choose suitable conditions in order to obtain all the required information on the permeability of the membranes.

The significance of these coefficients for the elucidation of membrane structure is pointed out.

REFERENCES

[1] A. FREY-WYSSLING, *Experientia*, 2 (1946) 132.
[2] K. J. LAIDLER AND K. E. SHULER, *J. Chem. Phys.*, 17 (1949) 851, 857.
[3] H. H. USSING, *Advances in Enzymol.*, 13 (1952) 21.
[4] J. R. PAPPENHEIMER, *Physiol. Revs.*, 33 (1953) 387.
[5] A. J. STAVERMAN, *Rec. trav. chim.*, 70 (1951) 344.
[6] A. J. STAVERMAN, *Trans. Faraday Soc.*, 48 (1948) 176.
[7] J. G. KIRKWOOD, in T. CLARKE, *Ion Transport Across Membranes*, Academic Press, New York, 1954, p. 119.
[8] H. DAVSON AND J. F. DANIELLI, *The Permeability of Natural Membranes*, University Press, Cambridge, 1952.
[9] M. H. JACOBS, in E. S. G. BARRON, *Trends in Physiology and Biochemistry*, Academic Press, New York, 1952, p. 149.
[10] E. ZEUTHEN AND D. M. PRESCOTT, *Acta physiol. Scand.*, 28 (1953) 77.
[11] R. HÖBER, *Physical Chemistry of Cells and Tissues*, The Blakiston Comp., Philadelphia, 1945, pp. 233, 229.
[12] I. H. NORTHROP AND M. L. ANSON, *J. Gen. Physiol.*, 12 (1929) 543.
[13] S. R. DE GROOT, *Thermodynamics of Irreversible Processes*, North-Holland Publishing Comp., Amsterdam, 1952.
[14] *cf.* [11], p. 229.
[15] J. R. PAPPENHEIMER, E. M. RENKIN AND L. M. BORRERO, *Am. J. Physiol.*, 167 (1951) 13.
[16] E. GRIM, *Proc. Soc. Exptl. Biol. Med.*, 83 (1953) 195.
[17] K. E. SHULER, C. A. DAMES AND K. J. LAIDLER, *J. Chem. Phys.*, 17 (1949) 860.

Received June 14th, 1957

Carriers

II

Editor's Comments on Papers 4, 5, and 6

It has long been recognized that the movement of certain ions and metabolites across natural membranes cannot be explained in terms of diffusion alone (see Höber, 1911). Human erythrocytes, for example, are known to be highly permeable to glucose, and yet this sugar, in high concentrations, does not equilibrate across the membrane (Klinghoffer, 1935). Moreover, many cells perpetually maintain high concentration gradients of sodium and potassium ions, and specialized tissues such as kidney and intestine bring about the net movement of molecules *against* a concentration gradient. The basis of most modern attempts to understand these phenomena is the concept of mobile carriers, an idea developed gradually throughout the late 1940s and early 1950s. According to this view, metabolites bind specifically to certain components of the membrane, which, by virtue of their rotational or translational mobility, convey the transported substance across the membrane.

Lefevre (Paper 4) summarized much of the early evidence supporting the idea that carriers are involved in glucose transport across the red cell membrane. (At the time, the term "active transport" in the title of Lefevre's paper did not necessarily imply transport against a concentration gradient.) The hypothetical carriers provided a straightforward explanation for the special characteristics of sugar entry into cells: (1) structural specificity for certain sugars, (2) competition between sugars of similar structure, (3) saturation kinetics, and (4) sensitivity to enzyme inhibitors. The entry of many other metabolites into a variety of cellular systems showed similar characteristics. These kinetic properties were taken to be the necessary, but not sufficient, experimental criteria for carrier-mediated transport (Rosenberg and Wilbrandt, 1952; Wilbrandt, 1954; Wilbrandt and Rosenberg, 1961).

The kinetic treatment of Widdas (Paper 5) provided a simple but elegant description of carrier-mediated transport and, in addition, predicted the phenomenon of counterflow. Counterflow refers to the net transfer of one substance against a concentration gradient by virtue of an oppositely directed gradient of a second substance that competes with the first for a common carrier. Counterflow was first demonstrated experimentally by Park et al. (1956) and Rosenberg and Wilbrandt (Paper 6), the latter authors emphasizing its importance in confirming the mobile carrier hypothesis.

Although the concept of mobile carriers is now accepted, in principle if not in detail, by most workers in the field, recent kinetic data have pointed to a major difficulty. The crux of the problem is that the kinetic parameters of transport vary by more than a factor of 10 in some instances, depending upon the method used for their determination (Miller, 1968, 1971). The rate of exchange of glucose across the red cell membrane, for example, is half of the maximal rate at a concentration of 38 mM. (This value, called the K_m for transport by analogy with the Michaelis constant for enzyme activity, provides an approximate measure of the affinity of the carriers for the transport substrate.) In contrast, the K_i for the inhibition of net glucose efflux by external glucose is only 1.8 mM. According to Lieb and Stein (1972), the difference in these two values cannot be accounted for by *any* mechanism invoking a mobile carrier. Several alternative schemes have been proposed in an effort to account for these kinetic discrepancies (Lieb and Stein, 1971, 1972; Naftalin, 1970). As the history of model building makes clear, however, models supported solely by kinetic data are not likely to gain widespread acceptance until they are also backed by biochemical or genetic evidence.

References

Höber, R. (1911). *Physikalische Chemie der Zelle und Gewebe,* 3rd ed. Engelmann, Leipzig.

Klinghoffer, K. A. (1935). Permeability of the red cell membrane to glucose. *Amer. J. Physiol.,* **111,** 231–242.

Lieb, W. R., and W. D. Stein (1971). New theory for glucose transport across membranes. *Nature New Biol.,* **230,** 108–109.

Lieb, W. R., and W, D. Stein (1972). Carrier and non-carrier models for sugar transport in the human red blood cell. *Biochim. Biophys. Acta,* **265,** 187–207.

Miller, D. M. (1968). The kinetics of selective biological transport. IV.

Miller, D. M. (1971). The kinetics of selective biological transport. V. Further data on the erythrocyte–monosaccharide transport system. *Biophys. J.,* **11,** 915–923.

Naftalin, R. J. (1970). A model for sugar transport across red cell membranes without carriers. *Biochim. Biophys. Acta,* **211,** 65–78.

Park, C. R., R. L. Post, C. F. Kalman, J. H. Wright, Jr., L. H. Johnson, and H. E. Morgan (1956). The transport of glucose and other sugars across cell membranes and the effect of insulin. *Ciba Found. Colloq. Endocrinol.,* **9,** 240–260.

Rosenberg, T., and W. Wilbrandt (1952). Enzymatic processes in cell membrane penetration. *Intl. Rev. Cytol.,* **1,** 65–92.

Wilbrandt, W. (1954). Secretion and transport of nonelectrolytes. *Symp. Soc. Exptl.-Biol.,* **8,** 136–162.

Wilbrandt, W., and T. Rosenberg (1961). The concept of carrier transport and its corollaries in pharmacology. *Pharmacol. Rev.,* **13,** 109–183.

Copyright © 1954 by the Society for Experimental Biology
Reprinted from *Symp. Soc. Exp. Biol.*, **8**, 118–135 (1954)

THE EVIDENCE FOR ACTIVE TRANSPORT OF MONOSACCHARIDES ACROSS THE RED CELL MEMBRANE

4

By PAUL G. LeFEVRE

U.S. Atomic Energy Commission, Washington, D.C.

In the passage of the blood sugar between the human red cell and the plasma, certain complicating peculiarities have been apparent from the earliest investigations. Among mammalian erythrocytes, those of the primates appear to be unique in showing an appreciable degree of permeability to the hexoses. Moreover, even these cells fail to haemolyse appreciably when suspended in pure isosmotic glucose solutions, so that the entrance of the glucose appears to be limited in some manner. This is not evident, however, in the normal distribution of the human blood sugar, which appears to be uniform throughout the water of the cells and plasma (Kozawa, 1914; Ege & Hansen, 1927). Also, Klinghoffer showed in 1935 that there was rapid equilibration of glucose added in small amounts to that already present in the blood. Ege & Hansen concluded that the totality of information on glucose distribution in the blood was 'impossible to explain' in keeping with the natural assumption of the sugar's free solution in the two water phases.

Klinghoffer's investigations of the apparent paradox revealed that ready penetration of the sugar occurred only if the glucose concentration did not exceed about 2 %. At higher concentrations, an extracellular excess was maintained almost indefinitely; and this unbalance was of sufficient degree to account for the failure of haemolysis to appear in isosmotic solutions. Bang & Ørskov (1937) measured this divergence from simple diffusion behaviour by showing, in a few experiments with varying glucose concentration in the neighbourhood of M/20, that the conventional red cell 'permeability constant' was approximately inversely proportional to the glucose concentration. Guensberg (1947) greatly extended this observation, finding that the variation of the 'constant', inversely with the glucose concentration, is over a range of at least a thousandfold.

Such behaviour implies some limitation on the absolute rate at which the glucose can move into the red cell; one suggestion is that the process requires participation of some ingredient of the barrier through which the sugar must pass to enter the cell interior. In recent years, much additional evidence has appeared in support of this view. Over the period 1946–52, while at the University of Vermont, I have frequently returned to this problem, and would like now to summarize the lines of evidence which indicate that the monosaccharides, in passing through the human red cell surface in either direction, temporarily combine with a 'carrier' molecule which is confined to that membrane or cortex layer. This evidence is in general along three lines:

(1) the kinetics of the sugar movements;
(2) the mutual interference with the movements in mixtures of sugars;
(3) the action of inhibitory substances.

In my own work, each of these lines was studied almost entirely by means of a single basic method, that of Ørskov (1935). This involves photometric recording of light transmittance through a very dilute suspension of red cells, as a means of following osmotic volume changes reflecting the movements of water across the cell surfaces. The general procedures and the operation of the recording system have been described elsewhere (LeFevre, 1948; LeFevre & Davies, 1951). The records show, as a function of time, the changes in direct transmittance which occur in response to various osmotically significant alterations in the composition of the medium (which consists of a buffered balanced salt solution to which the test substances are added). Since the suspension volume is at least 200 times as large as the total cell volume in these dilute suspensions, the concentrations in the medium are not appreciably altered by the cellular events, and may be treated as constant. Also, since the passage of water across the cell membranes, under a diffusion gradient, is much more rapid than the movements of the sugars with which we are concerned, it is legitimate to consider the osmotic pressure within the cells as identical with that of the medium at all times; the relatively slow volume changes recorded are then taken as a measure of the passage of glucose across the cell surface. Excellent linearity is found between the recorded quantity and the haematocrit or the calculated cell volumes in saline media of varying tonicity. When the sugars are present, small empirical corrections (LeFevre & LeFevre, 1952) must be taken into account if precise estimation of the cell-volume changes is to be attempted; but for any but the most critically quantitative work, direct inspection of the records is satisfactory for general analysis of the train of events. By arrangement of a suitable sequence of sudden alterations of the total osmotic pressure and the concentration of the penetrant in the medium, one can follow not only the entry of the substance into the cell, but also its subsequent exit.

The anomalous behaviour of glucose is apparent in the simplest series of this sort which was first attempted, in which glucose was simply added at various concentrations to a suspension of washed cells previously glucose-free. The pattern of the volume changes recorded in such an experiment is shown in Fig. 1. Certain clear deviations from the predictions of simple diffusion, as expressed in Fick's law, are immediately apparent. Approach to the equilibrium state is decidedly the more delayed, the more glucose is added; in fact, the initial rate of swelling actually decreases as the concentration of sugar is increased. At still higher concentrations, the rate of swelling

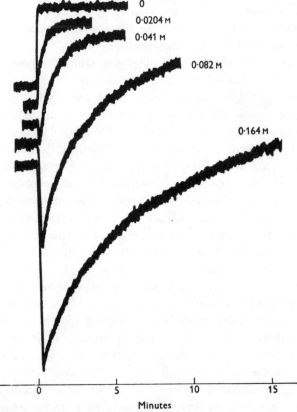

Fig. 1. Kinetics of swelling in glucose-saline mixtures. At zero time, 1 ml. of saline medium, containing glucose at 11 times final concentration shown, was added to 10 ml. of cell suspension (⅓ vol. %) in saline medium. (Medium here was only 0·6 × isotonic, so as to render rate differences more distinct.) 38° C. Immediate deflexion at zero time is resultant of dilution of suspension (upward deflexion) and cell-volume change (shrinkage downward); subsequent upward deflexion records cell swelling with uptake of sugar and water.

diminishes markedly after the first few minutes, and drops to a nearly imperceptible rate, while the cells are still much too small for an even distribution of glucose to have been effected (for records see LeFevre, 1948). This latter special complication will be taken up later; the point of special interest in the pattern shown is that the apparent uptake of the sugar is not proportional to the gradient, but is limited to a maximum rate dictated by some other factor necessary for the translocation of the sugar.

Such series of tests were run with all the hexoses and pentoses readily available: D-dextrose, D-laevulose, D-mannose, L-sorbose, D-galactose, L-arabinose and D-xylose. Among these, a clear dichotomy was apparent: all the aldoses (dextrose, mannose, galactose, and the two pentoses) behaved as just described; Fig. 2b shows, for instance, the behaviour of galactose. The two ketoses, laevulose and sorbose, as in Fig. 2a, on the other hand, seemed to obey reasonably well the predictions of Fick's law, and there was no reason to suppose any limiting factor other than the passive permeability of the cell membrane and the existing gradient for the sugar.

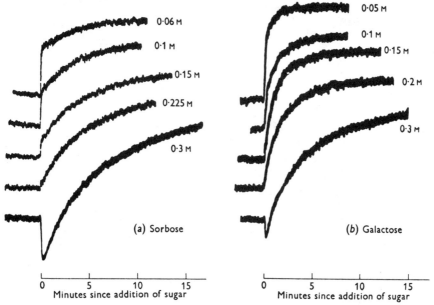

Fig. 2. Kinetics of swelling in sugar-saline mixtures. At zero time, 2 ml. saline medium, containing sugar at 6 times final concentration shown, was added to 10 ml. cell suspension (⅓ vol. %) in saline medium. 37° C. Deflexions interpreted as in Fig. 1.

This dissimilarity in behaviour of the aldoses and ketoses explains the discrepancy between the data of Kozawa (1914) and those of Wilbrandt (1938), with respect to the comparative rates of penetration of these sugars into the red cell. Kozawa, who worked with approximately $\frac{2}{3}$-isosmotic solutions at room temperature, found (by haematocrit and direct chemical analytic methods) the following sequence, from fastest to slowest:

arabinose, xylose > galactose, mannose, sorbose > dextrose > laevulose;

while Wilbrandt found, with much lower concentrations of the sugars (and at body temperature), with an optical method:

xylose, arabinose > mannose > galactose > dextrose > sorbose ≫ laevulose.

The contrast in the pattern of dextrose and sorbose penetration as a function of concentration immediately accounts for the major disagreement between these two series. The other minor discrepancies are also attributable to the lesser differences between the sugars in this respect, in view of the differing concentrations at which the two investigators were working.

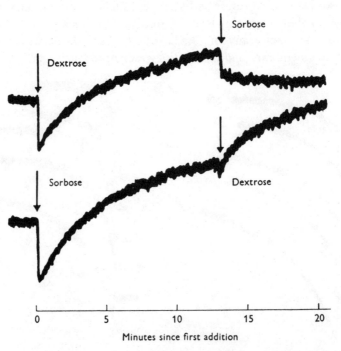

Fig. 3. Unilateral inhibition of uptake between sugars. At zero time, 1 ml. of saline medium, with sugar indicated at 1·8 M, was added to 10 ml. of cell suspension ($\frac{1}{3}$ vol. %) in medium; about 13 min. later, at time marked, a second 1 ml. was added, with sugar indicated at same concentration. Final concentration of each sugar was thus 0·15 M, or half-isosmotic. 37° C. Deflexions interpreted as in Fig. 1.

The obvious hypothesis to be derived from these observations was that the aldoses penetrate by a process involving participation of a cell component, while the ketoses penetrate simply by passive diffusion. This interpretation did not survive further experimentation concerned with the influence of the presence of one sugar on the rate of penetration of another. If, as seemed likely, all the aldoses shared a common transport system, there should be mutual interference with their entry when two or more aldoses are mixed; whereas the rate of entry of a ketose into the cells should be unaffected by the presence of other sugars of either type. This did not prove to be the true situation; instead, *all* the sugars appeared to be involved in a common reaction, so that in mixtures of any two the rate of swelling was always less than would be predicted on the basis of addition of the separate entries of each sugar. This was as true when ketoses were involved as with the aldoses. (For records, see LeFevre & Davies, 1951.)

Further light was shed on the situation by the procedure of adding the sugars serially rather than simultaneously, awaiting equilibration of the cells with the first before adding the second. In such experiments, the effect of the presence of the first sugar on the rate of entry of the second could be readily estimated in a quasi-quantitative manner. An example is provided in Fig. 3, which shows the characteristic situation between any aldose and either ketose. In the mixture of sorbose and glucose, each at 0·15 M, the entry of the sorbose is essentially completely prevented; whereas, prior addition of the sorbose had no effect on the later uptake of glucose, other than that attributable simply to its osmotic pressure. Similar relations were demonstrable between the two ketoses, and between any pair of the aldoses, except that the inhibitions were not always so overwhelmingly unilateral nor so absolute. Depending on the particular pair of sugars involved, the effect varied all the way from no detectable influence to apparently complete inhibition. The results of this entire series of experiments are summarized in Table 1, which indicates the relative effectiveness of each of the seven monosaccharides tested against each of the others. From this information, slightly modified by secondary factors discussed elsewhere (LeFevre & Davies, 1951), the avidity of the several sugars in attachment to the carrier molecule was considered to decrease in the order in which they are listed in Table 1, with the two pentoses being indistinguishable. The largest gaps appear to be between the aldoses and the ketoses, and between the two ketoses. Note that these relative affinities for the carrier do not define the relative rates of penetration, although such correlation improves as the sugar concentration is lowered.

Table 1. *Mutual inhibition in uptake of sugars*

In presence of	Inhibition of uptake of						
	Dext.	Mann.	Gal.	Xyl.	Arab.	Sorb.	Laev.
Dextrose	—	+ + +	+ + +	+ + +	+ + +	+ + + +	+ + + +
Mannose	+ + +	—	+ + +	+ +	+ +	+ + + +	+ + +
Galactose	+ +	+ +	—	+	+ +	+ + +	+ + + +
Xylose	+ +	+ +	+ +	—	+	+ +	+ + +
Arabinose	+ +	+ +	+ +	+	—	+ +	+ + +
Sorbose	o	o	o	o	o	—	+ + +
Laevulose	o	o	o	o	o	o	—

o No, or doubtful, effect.
+ Just noticeable inhibition.
+ + Moderate inhibition.
+ + + Very marked inhibition.
+ + + + Essentially complete block of uptake.

Another procedural variation, verifying the interpretation placed on these experiments, involved the simultaneous setting up of equal and opposite gradients for two sugars in a mixture. This was effected by equilibration with one sugar in a somewhat hypotonic saline medium, and subsequent addition of the second sugar together with a quantity of concentrated saline calculated to reduce the cell volume to the point that the outward gradient for the first sugar momentarily exactly equalled the inward

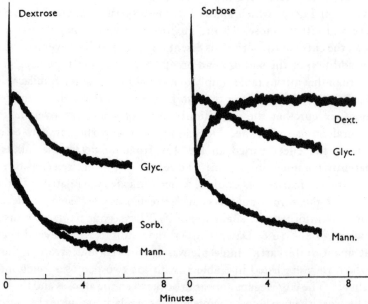

Fig. 4. Unilateral inhibition between sugars of movement in opposing gradient. Just prior to zero time, 10 ml. of cell suspension ($\frac{1}{2}$ vol. %) in 0·7 × isotonic saline medium had been equilibrated with either sorbose or dextrose (as labelled) at 0·262 × isosmotic. At zero time, 2 ml. was added, containing either glycerol, mannitol, sorbose, or dextrose, as labelled, at 1·5 × isosmotic, in saline medium at 4·0 × isotonic. This was calculated to reduce the cell-water volume from 1·43 to 0·80 × the 'normal', and thus set up an outward gradient for the original sugar of 0·25 isosmotic units, just equalling the inward gradient for the second non-electrolyte. 38° C. Deflexions interpreted as in Fig. 1. For significance of records see text.

gradient for the second. Thus the immediate cell volume assumed would correspond to the final equilibrium volume, and any intervening changes in volume reflect the net gain or loss as the two sugars move in opposite directions. Fig. 4 shows the behaviour of the cells in one such experiment; in this instance, opposing gradients for sorbose and dextrose are set up in each of the two possible arrangements. In both cases, the equilibration of glucose proceeded without impediment, while the sorbose movement was reduced to a small fraction of its uninhibited rate. The accessory records in Fig. 4 show the uncomplicated exit of the original sugar in the presence of corresponding concentrations of mannitol (non-penetrating) and glycerol (penetrating much more rapidly than the sugars). Comparison with these control records makes it clear that in the one case the intracellular dextrose moves outward with almost complete exclusion of the sorbose; while, in the other instance, the intracellular sorbose cannot escape, so that after entry of the dextrose the cell volume remains near the maximum level attained at the beginning of the glycerol record (the record closely resembles the mirror image of that for glucose exit in the presence of mannitol). With either situation, the sorbose movement is reflected by only a very slow drift of the record back toward the final equilibrium level.

These indications of competition among all of the sugars tested led to early abandonment of the notion that only the aldoses shared the carrier system. Additional indication that the ketoses were similarly involved was found in the common sensitivity to inhibitory agents (discussed in a later section of this report); also, all showed a similar Q_{10} of the order of 3·0 (LeFevre & Davies, 1951). The fact that the ketoses failed to show the limitation on rate of uptake into the cells, which was observed with the aldoses, does not in itself militate against the hypothesis that the same carrier system is shared by both sorts of sugars. The appearance of a pattern resembling that of passive diffusion does not imply that necessarily no limiting reaction with the cell surface is involved in the movement of the ketoses into the cells, but might reflect simply a difference in relative velocity constants as compared with the case of the aldoses.

This is immediately apparent in considering the properties of the simplest model of the 'carrier system' that might be proposed. In the absence of any preliminary demonstration of the degree of complexity that might appear in (1) the formation of the sugar-carrier complex, (2) the movement or reorientation of the complex, or (3) the uncoupling of the sugar from the carrier, the least involved situation was first assumed. Steps (1) and (3) may be treated grossly in terms of only the net ingredients, so that any enzymic participation, or rate-limiting factors arising from step (2), are reflected only in the overall velocity constants; the following diagrammatic presentation emerges as representing the minimal essentials:

Outside	Cell surface	Inside

$$P + A \underset{k_2}{\overset{k_1}{\rightleftharpoons}} A-P \underset{k_4}{\overset{k_3}{\rightleftharpoons}} A + P$$

C_s = conc. of P A_s = amount of complex $A-P$ S = amount of P

A = total amount of carrier S/V = conc. of P

$A - A_s$ = amount of uncombined carrier

in which V is the cell-water volume, and k_1, k_2, k_3 and k_4 are the velocity constants for the several steps as labelled; the equilibrium constant for the reaction at the outer surface, K_1, is then equal to k_2/k_1, and similarly $K_2 = k_3 k_4$ for the interior reaction. Mass action law would then give the relations

$$\frac{dS}{dt} = k_3 A_s - k_4 (A - A_s) \frac{S}{V},$$ (1)

and

$$\frac{dA_s}{dt} = k_1 C_s (A - A_s) - k_2 A_s - \frac{dS}{dt}.$$ (2)

No explicit solution of these equations to express S in terms of t appears to be possible; however, with glucose and the other aldoses, the observations noted above allow special restrictions on the system which simplify these relations. In the early stages of the process, while S is still a negligible factor, the rate of entry is essentially $k_3 A_s$, i.e. it is proportional to the amount of sugar-carrier complex. The finding that the process is limited so that no increase in initial rate occurs with increased concentration indicates therefore that the amount of this complex, A_s, remains nearly constant in the face of variation in C_s over the experimental range. (In Fig. 1, the initial slopes are approximately inversely proportional to total osmotic pressure.) This constancy of A_s indicates that the velocity constant k_3 is the factor limiting the overall transfer rate; that the reactions at the outer interface must be significantly faster than at the inner interface, so that A_s is nearly in equilibrium with the sugar in the external medium. Furthermore, k_1 must be considerably larger than k_2, since K_1 is evidently small compared to the lowest C_s at which the rate clearly ceases to increase with C_s.

The sequence of the sugars with respect to their competitive prowess in utilizing the carrier system, discussed above, is interpretable in the same terms. It presumably reflects the order of increasing K_1, the dissociation constant of the sugar-carrier complexes. The range of C_s in which the experiments of the type represented by Fig. 2 were carried out (about 0·05–0·3 M) defines a range of magnitude of K_1 apparently exceeding that of the aldoses, but not that of the ketoses. More precise calculation of this constant for the various sugars will be considered later from an entirely different experimental approach.

Assumption that K_1 is negligible compared to C_s seems then to be justified for the aldoses in the experimental range of C_s, at least in the case of glucose, the natural blood sugar in which we have the most interest. This, together with the conclusion that the outer reactions are near equilibrium by reason of the lower order of velocity constants at the interior, permits explicit solution of the equations. Also, since we are dealing experimentally with osmotic volume changes, we may treat C_s and S/V more properly as thermodynamic activities than as concentrations, and in these terms it is impossible for K_1 and K_2 to be unequal. With these simplifications, the earlier equations may be reduced to

$$\frac{dS}{dt} = Ak_3 \left[\frac{C_i V_i - SC_m/C_s}{C_i V_i + S} \right], \tag{3}$$

in which V_i is the volume of the cell water at isotonicity (C_i), and C_m is the concentration of the non-penetrating components (salts) in the medium (all concentrations being expressed in osmotic terms). This may be integrated directly to give the relation of S and t; but since we are dealing with volume records, and since $S = V(C_m + C_s) - C_i V_i$, it is convenient to express the relation in terms of V:

$$\frac{dV}{dt} = Ak_3 \left[\frac{C_i V_i - C_m V}{C_s V(C_m + C_s)} \right], \tag{4}$$

which may be integrated to give

$$t = \frac{C_s(C_m + C_s)}{Ak_3 C_m} \left[V_0 - V + \frac{C_i V_i}{C_m} \ln \frac{C_i V_i - C_m V_0}{C_i V_i - C_m V} \right], \tag{5}$$

in which V_0 is the cell-water volume when $t = 0$.

This equation predicts that, *in any given mixture*, the course of volume changes will follow the pattern dictated by the laws of passive diffusion; but that, in the comparison of rates in *different* situations, the pattern will be entirely different from that derived from Fick's law.

The general applicability of this system to all situations with respect to glucose movements across the red cell membrane, in either direction, was tested in a wide variety of experiments involving as many contrasting situations as could be arranged. Usually, several factors were held constant while another was varied several times; for example, the initial cell volume, the initial glucose gradient, the initial cell glucose level, the total glucose transferred, the glucose level of the medium, or its total osmotic pressure. A number of examples of the results of such experiments, involving both outward and inward movements, have been illustrated elsewhere (LeFevre & LeFevre, 1952); space limitations here allow only one example, in Fig. 5. The match with the predictions from equation (5), which are shown for comparison, is evident, and was equally good for all circumstances tested, provided the extracellular glucose concentrations (C_s) were not allowed to exceed about 70% of isosmotic.

The significance of the rate equation (3) above, which gives this fit with experiment, is much more apparent after conversion to the following forms:

$$\frac{dS}{dt} = Ak_3 \frac{C_s - S/V}{C_s}, \tag{6}$$

or

$$\frac{dS}{dt} = Ak_3 \left(1 - \frac{S/V}{C_s}\right). \tag{7}$$

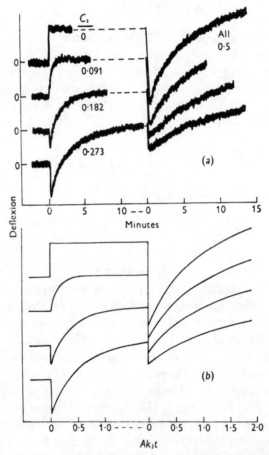

Fig. 5. Glucose entry in two stages, with fixed final C_s. (a) At zero time, to 10 ml. of cell suspension ($\frac{1}{3}$ vol. %), 1 ml. was added containing glucose at 11 × the concentration labelled; after equilibration, at second 'zero' time, an additional 1 ml. was added, containing glucose sufficient to bring the final concentration in each case to 0·5 × isosmotic, as labelled. All solutions contained salt mixture at 0·6 × isotonic. 37·5° C. (b) Pattern for same experiment, on basis of the hypothetical carrier system; scale for deflexions is matched approximately to fit lowest record in (a).

Equation (6) calls attention to the fact that the rate is at any instant *directly proportional to the existing gradient*, at a given extracellular concentration, but that with a given gradient it is *inversely proportional to the extracellular concentration.* The rearrangement in equation (7) combines these variables into a single term, i.e. the ratio of the intracellular glucose level to the extracellular; *the rate of transfer is proportional to the difference from unity in this ratio.* Thus the *inward* rate (positive dS/dt) can never exceed Ak_3, since the ratio cannot fall below zero; but the *outward* rate is not so restricted (the negative value of dS/dt will exceed Ak_3 whenever S/V is more than twice C_s).

Wilbrandt & Rosenberg (1950) found that the rate increased with increasing concentration on the upper end of the gradient only up to a certain maximum; and that the rate decreased with increasing concentration on the lower end of the gradient, but to a much greater degree than predicted by Fick's law. It is interesting to note that this is exactly what would occur according to the system developed above, if these experiments were performed by varying the external-sugar concentration, holding fixed the cell-sugar level. Table 2 illustrates how the carrier system would produce the results reported by Wilbrandt & Rosenberg, if this procedure were used.

Table 2. *Contrast of carrier and diffusion systems:*
effect of varying sugar concentration

	C_s	Relative dS/dt by	
		Fick's law	Equation (7)
Entry: $S/V=0.1$,	0.2	100—reference level	
upper end of	0.3	200	133
gradient varied	0.4	300	150
	0.5	400	160
	0.6	500	167
Exit: $S/V=0.6$,	0.1	−500	−1000
lower end of	0.2	−400	−400
gradient varied	0.3	−300	−200
	0.4	−200	−100
	0.5	−100	−40

On the other hand, if the reverse procedure were followed, so that the cell-sugar concentration became the experimental variable, with a fixed level in the medium, then Fick's law and equation (7) would be indistinguishable, and the peculiarities seen by Wilbrandt & Rosenberg should not appear. Since these experiments have not been fully described, it is uncertain whether they offer additional support for the scheme developed here, or invalidate it.

The simple system definitely breaks down at sugar concentrations approaching isosmotic, which are of course far above the physiological norm. The transfer of glucose slows down markedly after the first few minutes, and may come essentially to a standstill while there is still a considerable gradient across the cell surface. Several possible explanations of this have been previously discussed (LeFevre & LeFevre, 1952). It was possible to reject on experimental grounds the suggestions of loss of major cell constituents, or of 'fixation' of the cells so as to preclude osmotic volume changes. The most likely interpretation consistent with the facts seems to be that the high glucose concentrations block the carrier reactions themselves. Wilbrandt & Rosenberg (1950) have in fact taken this view in a wider sense, claiming that the entire pattern of glucose movements suggests a case of enzyme inhibition by an excess of substrate. Although it does not appear from the experiments described above that this factor is involved appreciably in the operation of the system at reasonable sugar concentrations, it may well be the basis of its failure to function when C_s becomes excessive. In interpreting this type of inhibition as observed with DNAase activity, Cavalieri & Hatch (1953) point out that a molecule of water is involved in the cleavage of the sugar-phosphate bond, and suggest that the substrate may compete with water for a site on the enzyme. This hypothesis could equally well be applied in the present instance.

That the complex formed in the membrane is in fact a sugar phosphate has in no way been indicated directly by the work reviewed here; but involvement of some enzymic factor is implied. The operation of inhibitors has been suggestive; inhibition of the uptake of glucose into red cells was effected by very small concentrations of Hg^{++}, Hg_2^{++}, or p-chloromercuribenzoate (LeFevre, 1947, 1948), and by chloropicrin, bromacetophenone, allyl mustard oil, or gold (Wilbrandt, 1950). (Iodine is also an effective inhibitor, but only at concentrations which also lead to an obvious discoloration of the haemoglobin.) The efficacy of this group of substances suggests that some part of the transport process involves sulphydryl groups; if so, these groups are evidently of the not easily available type characterized by Barron & Singer (1945), since there appears to be no inhibition at all by Cu^{++}, alloxan, mapharsen, iodoacetate or arsenite (LeFevre, 1948).

Use of another class of inhibitors has more recently been particularly fruitful in the analysis of the carrier mechanism; I refer to the glucoside phlorizin (generally considered to be rather specifically active against phosphorylation transfer systems), and its aglucon, phloretin (β-(p-hydroxyphenyl) 2, 4, 6-trihydroxypropiophenone). Either of these agents acts as a block to the transfer of the monosaccharides across the human red cell surface; but as Wilbrandt (1950) has shown, the simpler molecule, phloretin, is many times more effective than its glucoside phlorizin. Wilbrandt expressed the conviction that these agents act on the process by which the sugar *emerges* from the membrane (whether this be on the inside or the outside) rather than on the step of entry into the membrane. He could show inhibition of glucose exit from the cell without any disturbance of its entry, when (by reason of slow penetration) the agent was more concentrated in the external medium than within the cell. Wilbrandt suggests that phosphorylation by hexokinase is concerned in the initial step, and dephosphorylation by a phosphatase in the second step, and that it is this latter step which is sensitive to phlorizin and phloretin. In Wilbrandt's scheme, the system does not simply consist of a reversible set of reactions, but involves different operating units according to whether the sugar is entering or leaving the cell. With such a system it is difficult to account for the apparent failure of glucose ever to accumulate against a concentration gradient in these cells; whether a reasonable fit with the observed kinetics could be achieved with this system has not been considered.

Wilbrandt's published statements with regard to the peculiar action of phloretin in selective inhibition of the exit process have been so far only qualitative descriptions; a complete statement of procedure would be helpful, as without this it is impossible to determine whether the observations actually refute the simpler interpretation of the inhibition under the scheme offered here. My own experiments are entirely in accord with the hypothesis that the phloretin acts on the first reaction involved, by *direct competition* with the sugars for combination with the carrier molecule (or the limiting molecule involved in the chain leading to formation of the carrier complex). The following analysis, derived from this hypothesis, has in fact permitted rough calculation of the dissociation constants of some of the carrier complexes.

If the inhibitor acts at low concentrations by combining with the carrier in the same manner as do the sugars, it must have considerably higher affinity for the carrier (a much smaller K). Thus, when an extra ingredient of this type is added to the former system,

$$\frac{dS}{dt} = \frac{A(k_3 C_s/K_s - k_4 S/V)}{1 + C_I/K_I + C_s/K_s},$$ (8)

in which C_I and K_I are respectively the concentration and equilibrium constant for the inhibitor and K_s is the equilibrium constant for the sugar (equal to K_1 or K_2 of the original system). The ratio of the uninhibited rate (R_0) to the inhibited rate (R_I) is then given by the relation

$$\frac{R_0}{R_I} = 1 + \frac{K_s C_I}{K_I(K_s + C_s)}.$$ (9)

Thus, in a plot of this ratio against C_I, in a series of tests in which only C_I is varied, a straight line should be obtained, the slope of which is

$$K_I^{-1}(1 + C_s/K_s)^{-1}.$$

This rectilinearity is observed experimentally, as shown in Fig. 6. The records of cell shrinkage during glucose exit, in a series of concentrations of phloretin, under an otherwise constant set of conditions, are given in Fig. 6a. From such records the relative initial rates of glucose loss may be estimated and compared as a function of the inhibitor concentration. Fig. 6b shows the data of this same experiment, plotted in the manner prescribed above; a similar set of data for inhibition by Hg^{++} is included for comparison, showing that with this agent the inhibition is clearly not of the competitive type.

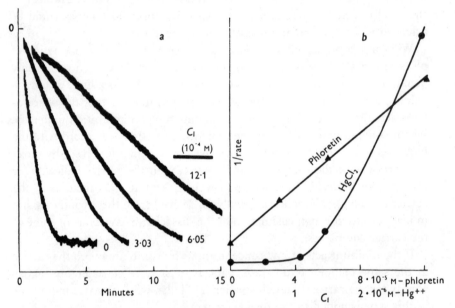

Fig. 6. Inhibition of glucose exit as a function of inhibitor concentration. (a) A 3 % cell suspension was equilibrated at 38° C. for 1 hr. with 0·5 × isosmotic dextrose in 0·7 × isotonic saline medium. Then, at zero time, 2 ml. of this was added to 10 ml. of the saline medium, containing phloretin so as to make the final concentration of the inhibitor as labelled in the figure. (b) The data of (a), and a similar experiment with HgCl$_2$ in place of the phloretin, plotted as suggested in the text.

More convincing evidence of the competitive nature of the phloretin inhibition is obtained from consideration of the effect of varying the sugar concentration, with a fixed inhibitor concentration. Equation (9) may be rearranged

$$\frac{R_I}{R_0 - R_I} = \frac{K_I}{C_I}\left(\frac{C_s}{K_s} + 1\right);\qquad (10)$$

so that if $R_I(R_0 - R_I)^{-1}$ is plotted against C_s at a fixed C_I, it should yield a straight line with $-K_s$ as the x-intercept and K_I/C_I as the y-intercept. By this means, then, both K_I and K_s can be estimated. Such a graph, for inhibition of glucose exit by phloretin, is presented in Fig. 7; this experiment gives a glucose K_s of 0·009 M, and for phloretin a K_I of $4·9 \times 10^{-6}$ M. Thus the inhibitor's 'affinity' for the carrier appears to be about 1800 times that of the sugar.

The useful measurements obtained by this approach are summarized in Table 3. The work was necessarily cut short soon after the initiation of this phase in August 1952, and it was impossible to gather a full complement of

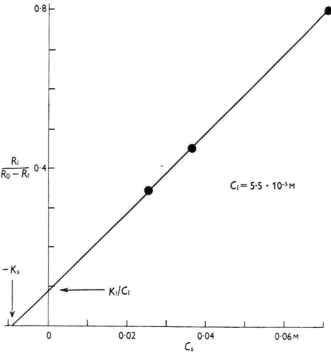

Fig. 7. Inhibition of glucose exit by phloretin as a function of glucose concentration. Procedure as with Fig. 6a, except that glucose was added in varying amounts to the final mixture; at each C_s, two runs were taken, one with and one without phloretin at $5·5 \times 10^{-5}$ M. For rationale of system of plotting data, see text.

data for estimation of K_s of each of the sugars and to check the K_I for phloretin and phlorizin using each of the sugars as test penetrant. However, the legitimacy of the interpretation of the observed rectilinearity in the plotted relations is attested by the finding of reasonably similar values in the constants with different experimental procedures. It is especially to be noted that the dissociation constants for sugar and inhibitor are of similar magnitude in experiments with outward movement as with inward movement. Most reassuring perhaps is the finding of the same range of value for K_I with different sugars having quite different K_s.

These results lend new support to the previously postulated interpretation of the differences in behaviour between the aldoses and ketoses, and of the pattern of competitive inhibition between the various sugars. Thus, the value found for glucose K was appreciably smaller than the C_s range useful in the experimental procedures; that for galactose was at the lower edge of this range, whereas the K for the two ketoses was in excess of the upper experimental limit for C_s. The practical limitations on C_s, in fact, made it impossible to ascertain the ketoses' equilibrium constants with any satisfactory degree of precision. In contrast to the situation with the aldoses, the concentration of the ketoses (C_s) did not affect appreciably the degree to which a given concentration of inhibitor would act. The plot of $R_I(R_0 - R_I)$ against C_s therefore gave for the ketoses a line of such low slope that the location of its x-intercept was a matter of enormous uncertainty. Nevertheless, the experiments with the ketoses gave a similar K_I for phloretin, and showed equally good rectilinearity in the plot of R^{-1} against C_I. All this is in complete accordance with the theoretical relation derived above, in the contrast of the situation $K_s \gg C_s$, with the situation $K_s \ll C_s$.

Table 3. *Estimation of carrier-complex dissociation constants*

Inhibition of	Sugar K_s	Phloretin K_I*	Phlorizin K_I†
Dextrose entry	$7 \cdot 5 \times 10^{-3}$ M	$4 \cdot 5 \times 10^{-6}$ M	—
Dextrose entry	10×10^{-3} M		—
Dextrose exit	9×10^{-3} M	$4 \cdot 9 \times 10^{-6}$ M	—
Dextrose exit	8×10^{-3} M		—
Dextrose exit	8×10^{-3} M		—
Dextrose exit	$7 \cdot 5 \times 10^{-3}$ M		$1 \cdot 45 \times 10^{-4}$ M
Galactose entry	$5 \cdot 0 \times 10^{-2}$ M	—	—
Galactose exit	$4 \cdot 4 \times 10^{-2}$ M	$4 \cdot 8 \times 10^{-6}$ M	—
Sorbose entry	$1 \cdot 3 - 2 \cdot 0$ M‡	—	$1 \cdot 27 \times 10^{-4}$ M
Sorbose exit	*ca.* 2 M‡	$4 \cdot 4 \times 10^{-6}$ M	—
Laevulose entry	*ca.* 2 M‡	$4 \cdot 4 \times 10^{-6}$ M	—

* Five values given represent complete series permitting plot as in Fig. 7; where no value is listed, phloretin K_I was taken as $4 \cdot 7 \times 10^{-6}$ M in calculation of K_s from plot as in Fig. 6b.

† Two values given are on basis of parallel tests with phloretin, taking K_I of latter as $4 \cdot 7 \times 10^{-6}$ M.

‡ Sorbose and laevulose K cannot be satisfactorily estimated in these experiments; see text for discussion.

In summary, then, the several lines of attack have all fitted into the schematic system illustrated above. This does not mean that the actual mechanism may not be considerably more complicated, with extra steps involving additional components, perhaps enzymic, which are not specifically included in the postulated system. It does indicate, however, that any such additional steps do not represent separate rate-limiting factors, so that for kinetic analysis they can be lumped together into the two overall reactions dealt with here.

Beyond the suggestiveness of the nature of the inhibitors found to be effective, there has not been any indication in this work of the nature of the carrier-complex or of the probable enzymic factors involved. Phosphorylation is of course the obvious suggestion; glucose-6-phosphate does not measurably penetrate the red cell, however, and it seems unlikely that the complex which is supposed to be confined to the surface layer would show this inability to enter that layer from the medium. Wilbrandt (1950) has described a possible form of hexose-metaphosphate which should be unionized and fairly fat-soluble, and has suggested that this could well be the complex involved; there does not seem to be any direct evidence of this at the moment.

Finally, it should perhaps be emphasized that, whatever the details of the mechanism may prove to be, there is no evidence that the red cell is equipped with a hexose 'pump' that can provide the energy for transporting sugar against a concentration gradient. The data merely indicate that there is a temporary complex formed between the sugars and a cell-surface component during the transfer (in a somewhat circumscribed manner); and this is apparently the basis for the peculiar ability of the primate erythrocyte to take up these substances.

REFERENCES

BANG, O. & ØRSKOV, S. L. (1937). *J. Clin. Invest.* **16**, 279.
BARRON, E. S. G. & SINGER, T. P. (1945). *J. Biol. Chem.* **157**, 221, 241.
CAVALIERI, L. F. & HATCH, B. (1953). *J. Amer. Chem. Soc.* **75**, 1110.
EGE, R. & HANSEN, K. M. (1927). *Acta med. scand.* **65**, 279.
GUENSBERG, E. (1947). Die Glukoseaufnahme in menschliche rote Blutkörperchen. Inauguraldissertation, Bern, Gerber-Buchdruck, Schwarzenburg.
KLINGHOFFER, K. A. (1935). *Amer. J. Physiol.* **111**, 231.
KOZAWA, SHUZO (1914). *Biochem. Z.* **60**, 231.
LeFEVRE, P. G. (1947). *Biol. Bull., Woods Hole*, **93**, 224.
LeFEVRE, P. G. (1948). *J. Gen. Physiol.* **31**, 505.
LeFEVRE, P. G. & DAVIES, R. I. (1951). *J. Gen. Physiol.* **34**, 515.
LeFEVRE, P. G. & LeFEVRE, M. E. (1952). *J. Gen. Physiol.* **35**, 891.
ØRSKOV, S. L. (1935). *Biochem. Z.* **279**, 241.
WILBRANDT, W. (1938). *Arch. ges. Physiol.* **241**, 289.
WILBRANDT, W. (1950). *Arch. exp. Path. Pharmak.* **212**, 9.
WILBRANDT, W. & ROSENBERG, T. (1950). *Helv. physiol. acta*, **8**, C82.

Copyright © 1952 by The Physiological Society
Reprinted from *J. Physiol.*, **118**, 23–39 (1952)

5

INABILITY OF DIFFUSION TO ACCOUNT FOR PLACENTAL GLUCOSE TRANSFER IN THE SHEEP AND CONSIDERATION OF THE KINETICS OF A POSSIBLE CARRIER TRANSFER

By W. F. WIDDAS

*From the Department of Physiology,
St Mary's Hospital Medical School, London*

(*Received* 6 *December* 1951)

The passage of glucose across biological membranes has long been recognized to be incapable of adequate explanation on a basis of diffusion. The position in regard to intestinal absorption is emphasized by Verzár & McDougall (1936). The extensive literature in this connexion, and in regard to absorption in the renal tubules, has been reviewed by Höber (1946), and more recently by Davson (1951).

The entry of glucose into human erythrocytes has been more extensively studied in a quantitative sense. Ege is quoted by Bang & Ørskov (1937) as having shown in 1919 that the permeability constant was diminished at high glucose concentrations. Wilbrandt, Guensberg & Lauener (1947) showed quantitatively the change in the constant over a hundred-fold concentration range. The present position in regard to this subject is well reviewed by LeFevre (1948), whose own results confirm the inability of kinetics based on simple diffusion to fit experimental data. LeFevre showed that the critical factor in diminishing the permeability was the raised internal concentration of glucose, and postulated a carrier mechanism in which the rate of transfer depended upon the difference between the internal concentration and some 'limiting' concentration.

In regard to placental glucose transfer from mother to foetus less is known, but it was suggested by Huggett, Warren & Warren (1951) that simple diffusion was unable to account satisfactorily for qualitative circumstances associated with placental glucose transfer, and Widdas (1951) reported that their quantitative results were in closer agreement with kinetics predicted from a postulated carrier system.

The purpose of this paper is to review the qualitative and quantitative circumstances associated with placental glucose transfer in the sheep, which are incapable of explanation on a basis of diffusion, and to describe a simple hypothesis of carrier transport which has been used in deriving kinetic relationships which more closely agree with the quantitative experiments.

The experimental data in such sheep experiments are not so precise as is possible with erythrocyte studies, but the same kinetic relationships have been applied to the erythrocyte problem and give results not inconsistent with those obtained by LeFevre and by Wilbrandt *et al.*

The basic hypothesis is used to develop a more generalized kinetics for carrier transport systems of this type.

EXPERIMENTAL DATA

To investigate the appearance of fructose in the foetal circulation Huggett *et al.* (1951) produced hyperglycaemia by the injection of glucose solution into the maternal circulation or the foetal circulation and followed the subsequent course of maternal and foetal blood sugar levels for 3 hr or more. Their results for three different types of experiments are relevant and these are summarized below for ease of reference.

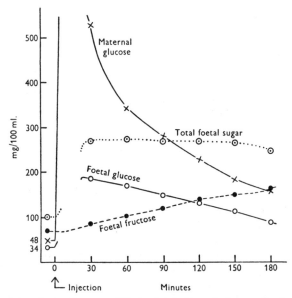

Fig. 1. Hyperglycaemia in the ewe. Effect of injection of glucose, 2 g/kg body wt. into pregnant ewes. Points represent means of four such experiments (Huggett *et al.*). For further details see text.

Type 1. *Hyperglycaemia in the ewe.* Injection of glucose into pregnant ewes in doses of 2 g/kg body wt. causes characteristic changes in the maternal and foetal sugar levels which are illustrated in Fig. 1. The points in this graph represent the mean of four such experiments by Huggett *et al.* (Data supplied by private communication.)

During the intravenous injection the maternal blood glucose level probably rises to between 700 and 800 mg/100 ml. but falls off over the subsequent 3 hr. The foetal glucose rises steeply during the first half-hour to about 200 mg/100 ml. and then starts to decline, while the foetal fructose shows a steady and more prolonged rise.

Type 2. Hyperglycaemia in the foetus, placental circulation intact. Injection of glucose into the umbilical vein in doses of 2 g/kg foetal body wt. gives rise to a picture which varies markedly according to whether the foetus is left attached to the placenta or (if viable) whether it is detached. This is illustrated in Fig. 2, from Huggett *et al.*

When the placental circulation is intact the hyperglycaemia in the foetal circulation is short-lived, the normal excess of maternal over foetal sugar levels being re-established in 1 hr. There is a transient rise in the maternal glucose level, suggesting a back transfer from the foetal circulation across the placenta.

Type 3. Hyperglycaemia in the foetus, foetus detached. Injection of glucose into the umbilical vein, followed by tying off the umbilical cord and detaching the foetus, gives rise to a hypergly-caemia of the type illustrated in Figs. 2 and 3. The hyperglycaemia reaches a higher level than in *Type 2* and is followed by a fall in glucose concentration which is approximately linear with time.

Placental glucose transfer

To see whether diffusion could adequately account for placental glucose transfer in these experi-ments a quantitative approach was attempted. The quantitative feature of experiments of *Type 1* which seemed of the greatest significance, and which it is desired to emphasize here, is the way in which the foetal glucose concentration starts to fall after the first half-hour in spite of the existence at that time of a concentration difference between maternal and foetal circulations of over 300 mg/100 ml.

Such a concentration difference is more than 10 times the magnitude of the physiological con-centration gradient usually found, and if the transfer of glucose across the placenta is by simple diffusion (that is, proportional in rate to the concentration difference), amounts of sugar far in excess of normal physiological requirements must be presumed to be entering the foetal circulation. That the concentration in the foetal circulation is actually falling after 30 min is thus surprising. The fall in glucose concentration indicates that the rate of removal of glucose from the foetal circulation has come to exceed the rate of accumulation. This may be due to an increased rate of removal from the foetal blood or a decreased rate of glucose transfer across the placenta or a com-bination of these two factors.

Removal of glucose from the foetal blood can be due to increased tissue utilization or storage, loss to the urine, amniotic or allantoic fluids, or conversion to fructose in the placenta, as shown by Huggett *et al.* (1951).

Using ^{14}C labelled glucose, Alexander, Huggett & Widdas (1951) confirmed the rapid incorpora-tion of labelled carbon in the foetal fructose and, in view of these findings, the increase in foetal fructose must be taken into account when attempting to explain the fall in glucose. For this reason the level of total sugar (glucose plus fructose) is included in Fig. 1, and this keeps a more or less constant value from 30 to 150 min. Assuming that changes in the fluid spaces in which these sugars are distributed to be small during the same period, the constant total sugar level would indicate no net increase in the reserve of these sugars in the foetal blood and tissue spaces during this time. There remains, however, the question of foetal utilization or possible excretion.

In experiments of *Type 3*, where a foetus has a hyperglycaemia induced by the injection of glucose into the umbilical vein prior to tying off the cord, there is no evidence of a rapid rate of glucose fixation or excretion bringing about a precipitous fall in sugar levels when these have high values. On the contrary, the evidence is of a relatively slow and approximately linear fall with time of both glucose and fructose levels.

The rate of glucose assimilation and excretion by the foetus would be expected to increase at high concentrations by analogy with the work of Wierzuchowski (1936), but the latter showed in dogs that there was an upper limit beyond which the rate could not proceed. The relatively linear fall in glucose and fructose concentrations of separated foetuses may thus represent the upper limits of combined assimilation and excretion for foetal sheep.

At any rate, the processes which bring about a decline of 0·6 mg/100 ml./min in the 135-day foetus, and 1·5 mg/100 ml./min in the near full-term foetus of 144 days, when separated from the placenta, could only effect a smaller decline in the larger system represented by placenta plus foetus.

Although a rapid fall occurs in attached foetuses when the maternal glucose concentration is low, as in experiments of *Type* 2, this can be accounted for by back transfer across the placenta to

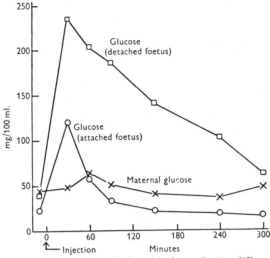

Fig. 2. Hyperglycaemia in twin foetuses, 135 days foetal age, showing difference between course of events in an attached and a detached twin (from Huggett *et al.*). See text.

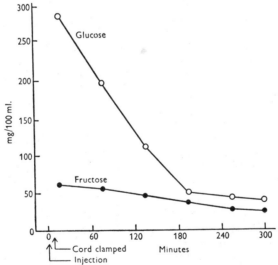

Fig. 3. Hyperglycaemia in foetus, 144 days foetal age, showing decline of glucose and fructose levels following separation from placenta (from Huggett *et al.*).

the much larger maternal circulation. An estimate of the relative fluid volumes in which sugars are distributed can be made by following the rise produced by the injection of a known weight of fructose. This suggests the foetal and foetal placental fluid space to be about 1·5 l. for foetuses of 135 days, as in the experiment of Fig. 2. The volume for the ewe is from 24–30% body weight

(37 kg) and is thus six to seven times the foetal volume. The maternal glucose rose from 43 mg/100 ml. to 62 mg/100 ml. at 1 hr and would account for a fall in the foetal circulation of 114–135 mg/100 ml. in the foetal circulation. The experiments of Alexander *et al.* (1951) with ^{14}C labelled glucose confirm that back transfer occurs very rapidly in terms of isotopic activity, but isotopic exchange and net transfer will clearly have different values.

Where the maternal glucose concentration is higher than that in the foetal circulation, as in experiments of *Type* 1, and net transfer back across the placenta is not possible, there is no reason to believe that foetal sugar fixation or excretion will differ greatly from that of detached foetuses of equivalent age and size. Quantitatively we may summarize the events in experiments of *Type* 1 as follows.

0–30 *min.* Rise in foetal glucose concentration from 34 to 185 mg/100 ml. Rise in foetal fructose from 70 to 86 mg/100 ml. Combined rise 5·5 mg/100 ml./min with average concentration gradient of 600 mg/100 ml.

30–150 *min.* Rise in foetal fructose from 86 to 151 mg/100 ml. Foetal excess assimilation and excretion estimated from fall in sugar concentrations in experiments of *Type* 3 to be 0·2–0·5 mg/100 ml./min for glucose and 0·04 mg/100 ml./min for fructose. (Fluid volume of foetus aged 115 days represents only one-third fluid space of foetus and placenta.)

Combined increment to foetal circulation 0·78–1·08 mg/100 ml./min.

Decline in foetal glucose from 185 to 113 mg/100 ml./min represents a decrement of 0·6 mg/100 ml./min, leaving a net gain of from 0·18 to 0·48 mg/100 ml./min.

The gradient of glucose at 30 min was 343 mg/100 ml. and at 150 min was 70 mg/100 ml. Taking the 0–30 min rate of transfer at 0·9 mg/100 ml./min/100 mg gradient, the expected transfer would vary from 3·0 to 0·63 mg/100 ml./min.

There is to date no direct evidence that the foetus utilizes fructose, and the only increase in fructose loss which may occur with increasing concentration is a possible increase in the loss by excretion. The order of magnitude of fructose loss is small, and if it should increase in rate proportional to concentration, the effect would still be negligible from a quantitative point of view and could not help to narrow the gap between observed and expected rates of placental transfer.

Thus the fall of foetal glucose concentrations after the first half-hour in experiments of *Type* 1 cannot be attributed to excessive foetal utilization or loss. One is drawn to conclude that placental glucose transfer during the period from 30 to 150 min has fallen to a rate approximately equivalent to that of foetal utilization (a relatively small and constant quantity). Simultaneously the decrease in glucose concentration is balanced by an increase in fructose production and accumulation.

The constancy of the glucose transfer which is inferred from 30 to 150 min is in marked contrast to the change in magnitude of the concentration difference between maternal and foetal circulations during the same time. At the beginning of the period the concentration difference is 343 mg/100 ml. and at the end about 70 mg/100 ml. If placental transfer was by simple diffusion one would expect evidence of a marked change in transfer rate during the same time.

Examined in this way, it is clear that simple diffusion will not adequately account for the quantitative results of placental glucose transfer. Its failure may be briefly summarized as follows:

(i) In experiments of *Type* 1, after 30 min during which the foetal glucose concentration may have risen by 150 mg/100 ml., it starts to fall again in spite of the existence of a concentration difference between mother and foetus of 300 mg/100 ml. This concentration difference is more than 10 times greater than that normally found, and there is no evidence of excessive foetal utilization or loss. Thus the rate of glucose transfer cannot be in simple proportion to the concentration difference.

(ii) From 30 to 150 min the experimental evidence suggests a relatively constant rate of glucose transfer, although the concentration difference across the placenta shows a fivefold change during the same period.

In addition to the quantitative circumstances described, there is considerable qualitative evidence against simple diffusion in the experiments of Huggett *et al.* and Alexander *et al.*, namely, that although glucose can pass readily from maternal to foetal circulation and vice versa when a foetal hyperglycaemia is artificially produced, the same is not true of fructose. It is difficult to imagine a physical sieve across which simple diffusion occurs which yet would so completely separate glucose and fructose molecules.

CARRIER TRANSFER HYPOTHESIS

The inability of simple diffusion to explain the quantitative evidence described, and the qualitative circumstances associated with the different placental behaviour towards glucose and fructose, suggested that a carrier system was involved. A carrier system was therefore postulated to be as simple as would fit the qualitative requirements and was based on the three following assumptions:

(i) The carriers can adsorb glucose but not fructose.

(ii) The carriers are in adsorption equilibrium at each interface with glucose in the respective circulations and are distributed equally between maternal and foetal interfaces.

(iii) The carriers pass backwards and forwards across the membrane due to thermal agitation irrespective of whether they are saturated with glucose or not.

It follows from these assumptions that isothermally the net rate of transfer will be proportional to the difference in the fraction of saturated carriers at the two sides.

The adsorption equilibrium is defined by application of the Adsorption Equation evolved by Langmuir (1918). Thus, the adsorption equilibrium at any interface can be expressed as

$$\theta = \frac{kC}{kC + \phi},\tag{1}$$

where θ is the fraction of carriers saturated with glucose molecules, C is the concentration of glucose in the solution at the interface, k being a constant. ϕ (also a constant) is proportional to the reciprocal of the average life of the glucose-carrier adsorption complex.

It is worth noting here that in the Michaelis-Menten equation an expression for the fraction of enzyme combined with substrate takes on the same form as eqn. (1). Thus, whether the carrier-substrate adsorption is thought of as physical or as a reversible chemical combination, the nature of the expression for the equilibrium will be the same. The new placental transfer rate will be given by

$$\text{Transfer rate} = K\{\theta_m - \theta_f\}\tag{2}$$

$$= K\left\{\frac{kC_m}{kC_m + \phi} - \frac{kC_f}{kC_f + \phi}\right\},\tag{3}$$

where the subscripts m and f refer to maternal and foetal sides respectively, and K is a constant.

The assumption that the carriers are in equilibrium with the respective glucose concentrations at either interface is an approximation which will hold so long as the rate of attaining adsorption equilibrium is rapid relative to the movement of carriers. In practice, transfer of carriers will effect a small perturbation of maternal and foetal equilibria, and in opposite senses, but this may be taken to be trivial. In any case, eqn. (3) is of limited use without knowledge of the values of the respective constants, but it can be simplified in the case of the two extremes. If the concentrations are small such that $kC \ll \phi$,

$$\text{Transfer rate} \simeq K\frac{k}{\phi}\{C_m - C_f\}. \qquad (4)$$

If the concentrations are large such that $kC \gg \phi$,

$$\text{Transfer rate} \simeq K\frac{\phi}{k}\left\{\frac{1}{C_f} - \frac{1}{C_m}\right\}. \qquad (5)$$

TABLE 1. Concentration data from Fig. 1. If transfer across placenta is by diffusion, transfer rate should be proportional to concentration difference. If by carrier transport and approximation of eqn. (5) holds, transfer rate should be proportional to $\left\{\frac{1}{C_f} - \frac{1}{C_m}\right\}$ (col. 5)

Time (min) (1)	Maternal glucose concn. (C_m) (mg/100 ml.) (2)	Foetal glucose concn. (C_f) (mg/100 ml.) (3)	$\{C_m - C_f\}$ (4)	$\left\{\frac{1}{C_f} - \frac{1}{C_m}\right\}$ (5)
30	528	185	343	0·0035
60	344	169	175	0·0030
90	280	150	130	0·0031
120	228	132	96	0·0032
150	183	113	70	0·0035
180	157	91	66	0·0046

Under normal physiological conditions a saturation of carriers to the extent of about 50% would probably be required to give the necessary practical efficiency and, under such conditions, the magnitude of ϕ would be of the same order as that of kC_r, where C_r is the resting concentration.

In the hyperglycaemia experiments of *Type* 1, both maternal and foetal glucose concentrations were high during the period from 30 to 150 min, being of the order 10 times normal resting values at 30 min and 3 times resting values at 150 min. Since $C \gg C_r$ and $kC > kC_r \simeq \phi$, it seemed justified to try the approximation of eqn. (5). This has been done in Table 1.

This approximation is in close agreement with the experimental evidence of a relatively constant net glucose transfer during the period from 30 to 150 min. This also suggests that, under such high glucose levels, the carriers are approaching full saturation on both sides and that the net transfer would be smaller than expected from the concentration difference.

The dependence of the fraction of saturated carriers upon concentration defined in eqn. (1) is indicated by a curve such as that shown in Fig. 4 which demonstrates how, at lower concentrations, a small difference in concentrations such as normally exists may be equally effective in maintaining the difference in the fraction of saturated carriers and thus an adequate transfer rate.

So far as experimental evidence to date goes there is no net 'against the gradient' placental glucose transfer, and, in the kinetics derived, the direction of transfer is determined by which concentration is the greater. The hypothesis, although speculative and permissive only, has therefore led to a type of kinetics which could explain the quantitative circumstances associated with placental glucose transfer in the sheep.

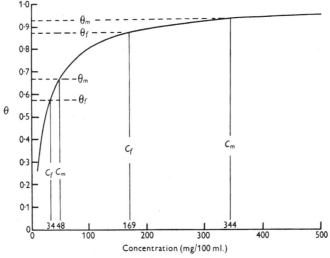

Fig. 4. Type of curve obtained from eqn. (1) relating saturation fraction (θ) to concentration. In this illustration the value of the constants of eqn. (1) have been arbitrarily fixed such that $\phi/k = 25$ mg/100 ml. to show how it would be possible for the difference in saturation of maternal and foetal carriers to be less when $C_m = 344$, $C_f = 169$ ($C_m - C_f = 175$ mg/100 ml.) than when $C_m = 48$, $C_f = 34$ ($C_m - C_f = 14$ mg/100 ml.).

Entry of glucose into erythrocytes

The sheep experiments described lack the precision possible in experiments with erythrocytes, but there are features in them resembling those demonstrated in the erythrocyte by LeFevre. For example, the chief factor in the diminution in placental transfer appears to be the raised glucose level on the foetal side of the placental barrier; this is analogous to the raised internal concentration in the erythrocyte. It would thus be of considerable theoretical interest to see if the approximation of eqn. (5) could be applied to erythrocyte studies.

LeFevre investigated erythrocytes in isotonic saline, to which 20% by volume of isotonic saline plus six times the required concentration of glucose were added. The cells rapidly shrank in the hypertonic mixture and then more slowly returned towards their original volume (V_0) as glucose entered. This change in volume was followed by a photoelectric method.

Using isotonicity units, diffusion across the red cell membrane would lead to the following expression relating time, volume, extracellular glucose concentration and the permeability constant K:

$$Kt = 1 + (1 + C_s)\left\{\left(\ln \frac{C_s}{(1-V)(1+C_s)}\right) - V\right\}. \tag{6}$$

In this expression C_s = extracellular concentration of glucose in multiples of the isotonic concentration, V = volume of cells in fractions of the isotonic volume, and K is the permeability constant.

Applying the approximation of eqn. (5) in the form

$$\frac{dQ}{dt} = K'\left\{\frac{V}{Q} - \frac{1}{C_s}\right\}, \tag{7}$$

where Q is the amount of sugar transferred, the expression obtained is

$$K't = C_s{}^2\left[(1-V)\left(\frac{1}{C_s}+1\right) + (1+C_s)\left\{\left(\ln \frac{C_s}{(1-V)(1+C_s)}\right) - V\right\}\right], \tag{8}$$

where the same convention is used, but K' has different dimensions by virtue of the use of reciprocals of the concentration.

Eqns. (6) and (8) can be used to derive swelling time curves in different glucose concentrations. Those obtained from eqn. (6) were shown by LeFevre to be inconsistent with the experimental results.

Curves derived from eqn. (8) can be made to give a more reasonable fit, although they give too rapid initial swelling rates. This would be expected since the approximation does not hold at low internal concentrations.

Eqns. (6) and (8) make it possible to suggest a theoretical basis for the way in which the apparent permeability constant K, of eqn. (6), should change with concentration. For, if it be presumed that eqn. (8) more closely represents the kinetics of penetration and K' is a true constant, then K must change so that

$$\frac{K}{K'} = \frac{\text{R.H.S. eqn. (6)}}{\text{R.H.S. eqn. (8)}},$$

and values of this fraction can be calculated for different concentrations and volumes to which the cells return in the measured time on which the estimate of K is based. These values are given in Table 2. Although the fraction K/K' depends to a small extent upon the degree of return to normal volume taken, the chief difference between the two equations (6) and (8) is in the term $C_s{}^2$ which is independent of the volume.

Taking 0·95 as a representative value for V/V_0, it is possible to predict a range of values of K for different concentrations which can be compared with those obtained by Wilbrandt *et al.* Thus, arbitrarily fixing a point for isotonicity at $K=1$, $K'=0.76$, other values of K can be predicted and should be as shown in Fig. 5, where the predicted values have been superimposed on the published data of Wilbrandt *et al.*

TABLE 2. The ratio K/K' which would be obtained if eqns. (6) and (8) were used on the same data

	C_s isotonicity units						
V/V_0	1·0	0·777	0·555	0·500	0·333	0·250	0·1667
0·5	—	—	—	—	—	—	—
0·55	11	—	—	—	—	—	—
0·60	5·3	23	—	—	—	—	—
0·65	3·6	9·6	—	—	—	—	—
0·70	2·8	6·3	26	54	—	—	—
0·75	2·3	4·75	14·3	22	—	—	—
0·80	1·9	3·76	10	14	66	—	—
0·85	1·7	3·14	7·5	10	34	99	—
0·90	1·5	2·67	6	7·7	22	49	180
0·95	1·32	2·3	4·9	6·2	15·7	31	86
0·99	1·16	2·0	4·0	5·0	11·9	22	53

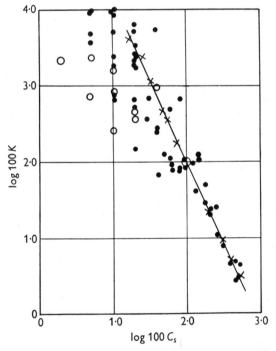

Fig. 5. Variation of apparent permeability constant with glucose concentrations. Experimental results of Wilbrandt *et al.* showing variation of permeability constant with glucose concentration. ○ and ● refer to results obtained by a direct and indirect photoelectric method respectively. Taking ⊕ as reference point where $K=1$, points × represent calculated values of K on the assumption that K' of eqn. (8) is constant. For fuller details see text.

Plotted on a double logarithmic scale the predicted values fall close to a straight line with a slope of -2, demonstrating the preponderance of the C_s^2 term. This slope of predicted values is in reasonable agreement with experiment, particularly at high glucose concentrations.

These investigations therefore suggest that the carrier kinetics derived are not inconsistent with erythrocyte measurements, and it is felt that theoretical considerations of a more generalized kinetics, based on the same hypothesis, may merit record.

A more generalized kinetics

In the derivation of eqn. (3) it was assumed that the carrier molecules were equally distributed between maternal and foetal interfaces. Implicit in this assumption is the further assumption that the energy required for saturated carriers to cross from maternal to foetal interfaces is the same as from foetal to maternal. Similarly, the energy required for unsaturated carriers to cross in either direction is presumed equal.

While this may be true where the carriers are in a neutral membrane and are themselves unaffected by the ionic environment at either interface, it would not, in general, be true for membranes across which an electrical potential exists or in which the free energy of the carriers is not the same at the two interfaces. Further, the fact of saturation may alter the free energy of the carrier. Allowance may have to be made for four different energy requirements: the energies required by saturated and unsaturated carriers to cross from one interface (S_1) to a second (S_2), and the energy requirements of saturated and unsaturated carriers crossing in the reverse direction (S_2 to S_1).

If the four energy requirements are different from one another, the number of carriers at each interface would no longer tend to remain the same, and this fact, as well as the energy requirements, would influence the rate of transfer.

To derive the kinetics appropriate to this more general case, the assumption is made that the carriers are insignificant in numbers relative to the main structural elements of the membrane, and that there are no structural factors impeding the redistribution of carriers between the two interfaces.

On this basis the redistribution of carriers would rapidly occur until the numbers leaving each interface were equal over short intervals of time.

Let the number of carriers in S_1 be n_1 per cm^2, and let the number of carriers in S_2 be n_2 per cm^2 and let $n_1 + n_2$ equal N a constant.

Let the fraction of saturated carriers having energy to cross from S_1 to S_2 in time dt be α and the fraction of unsaturated carriers having energy to cross from S_1 to S_2 in time dt be $p\alpha$.

Similarly, let the fraction of saturated carriers having energy to cross from S_2 to S_1 in time dt be $q\alpha$, and the fraction of unsaturated carriers having energy to cross from S_2 to S_1 be $s\alpha$.

Let the equilibrium at S_1 and S_2 be θ_1 and θ_2 defined as in eqn. (1).

When the redistribution of carriers has been effected, the total loss of carriers from S_1 in time dt will equal the loss of carriers from S_2 in time dt. Therefore

$$n_1\theta_1\alpha + n_1(1-\theta_1)p\alpha = n_2\theta_2 q\alpha + n_2(1-\theta_2)s\alpha, \tag{9}$$

$$\frac{n_1}{n_2} = \frac{q\theta_2 + s(1-\theta_2)}{\theta_1 + p(1-\theta_1)}. \tag{10}$$

But

$$n_1 + n_2 = N, \tag{11}$$

therefore

$$n_1 = N\frac{q\theta_2 + s(1-\theta_2)}{\theta_1 + p(1-\theta_1) + q\theta_2 + s(1-\theta_2)}, \tag{12}$$

$$n_2 = N\frac{\theta_1 + p(1-\theta_1)}{\theta_1 + p(1-\theta_1) + q\theta_2 + s(1-\theta_2)}. \tag{13}$$

The net transfer, however, will be proportional to the difference between the loss of saturated carriers from S_1 and the loss of saturated carriers from S_2:

$$\text{Transfer rate} = K'\{n_1\theta_1\alpha - n_2\theta_2 q\alpha\}, \tag{14}$$

where K' is a constant.

Substituting the values for n_1 and n_2 from eqns. (12) and (13) and inserting values for θ_1 and θ_2, the following relationship is obtained:

$$\text{Transfer rate} = \frac{K'\alpha N\frac{\phi}{k}(sC_1 - pqC_2)}{\overline{1+q}\,C_1C_2 + \frac{\phi}{k}\{\overline{1+s}\,C_1 + \overline{p+q}\,C_2\} + \overline{p+s}\,\frac{\phi^2}{k^2}}. \tag{15}$$

This expression shows that at equilibrium, when the transfer rate is zero,

$$sC_1 = pqC_2, \tag{16}$$

or

$$\frac{C_1}{C_2} = \frac{pq}{s}. \tag{17}$$

Thus where s, p and q are different from 1, equilibrium will not necessarily occur when $C_1 = C_2$ and 'against the gradient' transfer may occur where $C_2 > C_1$ so long as $sC_1 > pqC_2$.

Working in a restricted range of concentrations, such that changes in the denominator may not be large, eqn. (15) reduces approximately to the form used in isotope studies (for example by Harris & Burn, 1949) where the inward and outward fluxes are presumed proportional to the respective concentrations.

Making $s = p = q = 1$, eqn. (15) reduces to eqn. (3) representing the simplified kinetics applied to placental glucose transfer and to glucose entry into erythrocytes.

Competitive adsorption. The possibility that a carrier may adsorb more than one species of substrate would lead to additional complications, but it is worth while examining the effects upon the kinetics of the last section.

The effect of adsorption of two species (A and B) can be considered on the following lines. Let the state of adsorption equilibrium of each species separately be defined by

$$\theta_A = \frac{kC_A}{kC_A + \phi_A} \quad \text{and} \quad \theta_B = \frac{kC_B}{kC_B + \phi_B}.$$

To simplify equations write

$$x = \frac{kC_A}{\phi_A} \quad \text{and} \quad y = \frac{kC_B}{\phi_B},$$

so that

$$\theta_A = \frac{x}{x+1} \quad \text{and} \quad \theta_B = \frac{y}{y+1}.$$

Then it can be shown that in a competitive system

$$\theta_A = \frac{x}{1+x+y}, \quad \theta_B = \frac{y}{1+x+y}. \tag{19}$$

Let α, $p\alpha$, $q\alpha$, $s\alpha$, have the same significance as in the previous section—the saturated carriers being saturated with species A. Let $u\alpha$ and $v\alpha$ be the fraction of carriers saturated with species B having the energy necessary to cross from S_1 to S_2 and S_2 to S_1 respectively.

Then, as before, the carriers will be redistributed until the numbers leaving S_1 and S_2 are equal over a given time. This will occur when

$$n_1 = N \frac{(1+x_1+y_1)(qx_2+vy_2+s)}{(1+x_1+y_1)(qx_2+vy_2+s)+(1+x_2+y_2)(x_1+uy_1+p)}, \tag{20}$$

$$n_2 = N \frac{(1+x_2+y_2)(x_1+uy_1+p)}{(1+x_1+y_1)(qx_2+vy_2+s)+(1+x_2+y_2)(x_1+uy_1+p)}. \tag{21}$$

The transfer rate of species A from S_1 to S_2 will be given by

$$\text{Transfer rate} = K'\alpha N \frac{x_1(vy_2+s)-qx_2(uy_1+p)}{(1+x_1+y_1)(qx_2+vy_2+s)+(1+x_2+y_2)(x_1+uy_1+p)}. \tag{22}$$

At equilibrium

$$\frac{x_1}{x_2} = \frac{quy_1+pq}{vy_2+s}. \tag{23}$$

When $p=q=s=u=v=1$

$$\frac{x_1}{x_2} = \frac{y_1+1}{y_2+1}, \tag{24}$$

or, on reinserting concentrations,

$$\frac{C_{A1}}{C_{A2}} = \frac{C_{B1}+(\phi_B/k)}{C_{B2}+(\phi_B/k)} < \frac{C_{B1}}{C_{B2}}. \tag{25}$$

In this latter case the equilibrium ratio of concentrations of species A approaches but never reaches the ratios of the concentrations of species B at the two sides. If the concentration of species B was maintained high at S_1 and low at S_2 by metabolic activity, then a competitive species A would tend to be transferred to S_1 to maintain its concentration high at S_1 and low at S_2.

Thus a competitive system could effect 'against the gradient' transfer across the membrane in which there was no difference in the free energies of the saturated and unsaturated carriers at the two sides.

Competitive inhibition. If species B is unable to be transported across the membrane but competes at interface S_1, the effect on the transfer rate of species A from S_1 to S_2 can be obtained from eqn. (22) by making $y_2=u=v=0$, giving

$$\text{Transfer rate} = \frac{K'\alpha N [sx_1 - pqx_2]}{1+qx_1x_2+\overline{1+s}\,x_1+\overline{p+q}\,x_2+\overline{p+s}+y_1(qx_2+s)}. \tag{26}$$

On inserting concentrations this expression only differs from eqn. (15) by virtue of the terms in the denominator involving y_1. The effect then is to immobilize a portion of the carriers and so effect a slowing down of the transfer rate. The magnitude of $y_1 = kC_B/\phi_B$ may be large by virtue of the adsorption complex being long-lived so that ϕ_B is small. This term could thus predominate in the denominator and cause an inhibition of greater magnitude than might be expected from considerations of concentration alone.

3—2

DISCUSSION

The transfer of substances across biological membranes presents a difficult problem, since, under physiological conditions, any actual absorption or transfer through membranes of epithelial or other cells must comprise transfer across relatively large distances by physical diffusion as well as across the more limited regions of cell membranes by processes which may or may not follow diffusion laws.

The rate of transfer will therefore always contain a diffusion element even where different kinetics apply to the passage across cell membranes. Thus, if carrier transfer does occur across a membrane, deviation from diffusion laws will only be apparent where the carrier transfer determines the rate.

Where the diffusion gradient is large and the net transfer small, as in the experiments on hyperglycaemic sheep, it may be presumed that the carrier system is the chief rate-determining step and the deviation from diffusion kinetics is marked. Moreover, there is reasonable agreement with kinetics derived from the hypothesis of a carrier mechanism.

Where concentrations are small, then the diffusion gradient through cells and intercellular spaces may be so small as to make diffusion of comparable importance to the carrier system in determining the overall rate. Further, when the concentrations are small the carrier system itself may approach eqn. (4) which is of the diffusion type. Thus, it is clear that at small concentrations, the fitting of experimental results to kinetics based on diffusion may be satisfactory and the presence of a carrier mechanism escape detection. Experiments with increasingly large gradients, or of constant gradients at increasingly high concentration levels, may be required to demonstrate whether there is a deviation from diffusion kinetics.

Where diffusion kinetics do not hold and a transport by a selective carrier system is being investigated, it may not be justifiable to assume that the inward and outward fluxes will be proportional to the respective concentrations. Eqn. (15), derived in a general way from the carrier hypothesis, contains concentration terms in the denominator which must be taken into account when attempting to integrate the transfer over a range of concentrations.

No exact description of the mechanism of transfer is involved in deriving the kinetic relationships, but a model which adequately fits in with the postulates used is that advanced by Lundegardh (1940) and reviewed by Krogh (1946). On the Lundegardh model the membrane is to be regarded as an orientated bimolecular lipid and protein layer, the individual molecules of which may acquire sufficient thermal energy to leave one orientated interface and by rotation of polar and non-polar portions of the molecule enter the second orientated interface. The carriers on this model are molecules of the same general type which may specifically combine with molecules of smaller species at one interface and release them at the second, or vice versa.

The carrier hypothesis used to derive the kinetics discussed in this paper is essentially passive in nature. Where the membrane has a potential difference across it, however, more work may be done on carriers crossing in one direction than in another. Thus 'against the gradient' transfer may result, although the carriers themselves may still be regarded as acting passively.

Similarly, in the competitive system, the change in chemical potential of the one species transferred from an area of greater to an area of less concentration provides the energy which enables the second species to be transferred against the gradient, and there is, as may be expected, no circumvention of the Second Law of Thermodynamics. However, metabolic activity could supply the necessary energy for a continuing process based on either mechanism to work in a biological system.

Other possible mechanisms for 'against the gradient' transfer have been advanced. For example, Franck & Mayer (1947) suggested a mechanism for an osmotic diffusion pump. What methods are actually employed by living cells and biological systems must await further research and elucidation of the problem.

Energy requirements in any 'against the gradient' transfer would cause a close correlation between such transfer and glucose metabolism. Where the substance transferred is itself glucose, this correlation makes elucidation of the underlying mechanism more difficult.

The association of phosphorylation with glucose absorption in the intestine has already been mentioned. There are various possible explanations of this: the extreme views may be expressed as (1) phosphorylation of every glucose molecule is a prerequisite for its passage across a membrane; (2) since intestinal absorption can occur 'against the gradient' the phosphorylation is evidence of the metabolism supplying the necessary energy.

If the first extreme is taken there is some difficulty in explaining certain quantitative anomalies which have been observed. Thus Fisher & Parsons (1949) found that the rate of glucose utilization was not in the same proportion to the rate of glucose absorption for different levels of the small intestine of the rat. They suggest that glucose translocation may be a property of different fractions of the populations of mucosal cells at different levels. Other anomalies are discussed by Davson (1951).

In studying the uptake of glucose by muscle in a medium containing glucose phosphate labelled with ^{32}P, Sacks (1944, 1945) came to the conclusion that glucose from the glucose phosphate entered the muscle cells but that the phosphate remained extracellular. He postulated a hydrolysis in the cell membrane which allowed the glucose to enter the cell. However, he found this process much slower if the animal was in a post-absorptive state. Thus the rate of entry of glucose appeared to depend upon the cell requirements. This evidence would tend to the view that phosphorylation of each glucose molecule is a prerequisite of its transfer across the membrane.

The finding of diminished entry of glucose when the internal concentration is presumed to be raised is in accordance with the general pattern emphasized by LeFevre and, if the foetal circulation be accepted as being analogous in this respect to the internal concentration, it is a feature common to placental transfer as well. It is a necessary corollary to the kinetics described in this paper.

In placental glucose transfer it is not possible to exclude phosphorylation as an essential factor. The borders of the foetal trophoblastic cells and the adjacent maternal tissue give phosphatase reactions histochemically. The production of fructose by the placenta has been investigated by Parr & Warren (1951), who have studied the phosphatase activity with a view to explaining the production of fructose along the lines of glucose phosphate intermediaries, but evidence that phosphorylation is a prerequisite for placental transfer is not yet decisive.

It may be that the postulated carriers form complexes with the more reactive glycosidic radicals, supplied by glucose 1-phosphate, rather than with glucose itself. Such a possibility would not necessarily affect the kinetics, since, in a chain reaction, the observed velocity is that of the slowest component. As has been pointed out, the expression for the equilibrium at either interface is the same whether physical adsorption or reversible chemical combination is involved.

If the transfer of glucose across biological membranes is to be regarded as a process involving enzyme systems in conjunction with a carrier system on the Lundegardh model, then it is clearly a complicated system, and which factor determines the rate may well vary with experimental conditions.

In consequence, kinetics which go some way to explain a particular type of experiment in placental glucose transfer in the sheep may not apply to glucose transfer under other circumstances. That they appear to apply reasonably well to the erythrocyte suggests that the same factors determine the rate in this site too. The satisfactory fitting of results to kinetics based on a carrier hypothesis, such as the one described, does not necessarily confirm the underlying assumptions made in their derivation. It is conceivable that enzymic or other processes may so combine as to give essentially similar rate equations. The kinetics do, however, emphasize the differences from diffusion which may be looked for and the similarities to diffusion which need to be guarded against, and may prove of value in suggesting new experimental approaches.

SUMMARY

1. The inability of simple diffusion to account for quantitative and qualitative circumstances associated with placental glucose transfer in the sheep is reviewed.

2. The possible kinetics expected under an alternative hypothesis (i.e. of a carrier mechanism) is derived, and is in closer agreement with experiment. The same kinetics are not inconsistent with experiments by LeFevre and others on glucose entry into erythrocytes.

3. When concentrations are large the theory predicts that the transfer rate will be proportional to the difference of the reciprocals of the concentrations:

$$\frac{\text{Transfer rate}}{S_1 \rightarrow S_2} = K\left\{\frac{1}{C_2} - \frac{1}{C_1}\right\}.$$

4. A more general kinetics derived from the same hypothesis is described showing conditions under which 'against the gradient' transfer could occur.

5. The possible admixture of diffusion, enzymic action and carrier kinetics in transfers across biological membranes is discussed in order to emphasize the possibility of confusion in looking for deviations from diffusion equations.

The author is indebted to Prof. A. St G. Huggett, Dr F. L. Warren and Mrs N. V. Warren for the use of quantitative data obtained in their experiments.

REFERENCES

Alexander, D. P., Huggett, A. St G. & Widdas, W. F. (1951). *J. Physiol.* **115**, 36 P.
Bang, O. & Ørskov, S. L. (1937). *J. clin. Invest.* **16**, 279.
Davson, H. (1951). *A Textbook of General Physiology.* London: Churchill.
Ege, T. (1919). Thesis: Copenhagen. Cited by Bang, O. & Ørskov, S. L. in *J. clin. Invest.* 1937, **16**, 279.
Fisher, R. B. & Parsons, D. S. (1949). *J. Physiol.* **110**, 281.
Franck, J. & Mayer, J. E. (1947). *Arch. Biochem.* **14**, 297.
Harris, E. J. & Burn, G. F. (1949). *Trans. Faraday Soc.* **45**, 508.
Höber, R. (1946). *Physical Chemistry of Cells and Tissues.* London: Churchill.
Huggett, A. St G., Warren, F. L. & Warren, N. V. (1951). *J. Physiol.* **113**, 258.
Krogh, A. (1946). *Proc. Roy. Soc.* B, **133**, 140.
Langmuir, I. (1918). *J. Amer. chem. Soc.* **40**, 1361.
LeFevre, (1948). *J. gen. Physiol.* **31**, 505.
Lundegardh, H. (1940). *Ann. agric. Coll. Sweden*, **8**, 234.]
Parr, C. W. & Warren, F. L. (1951). *Biochem. J.* **48**, xv.
Sacks, J. (1944). *Amer. J. Physiol.* **142**, 145.
Sacks, J. (1945). *Amer. J. Physiol.* **143**, 157.
Verzár, F. & McDougall, E. J. (1936). *Absorption from the Intestine.* London: Longmans, Green and Co.
Widdas, W. F. (1951). *J. Physiol.* **115**, 36 P.
Wierzuchowski, M. (1936). *J. Physiol.* **87**, 311.
Wilbrandt, W., Guensberg, E. & Lauener, H. (1947). *Helv. physiol. acta*, **5**, C 20.

Reprinted from *J. Gen. Physiol.*, **41**, 289–296 (1957)

6

UPHILL TRANSPORT INDUCED BY COUNTERFLOW

By THOMAS ROSENBERG and W. WILBRANDT*

(*From The Danish Atomic Energy Commission, Copenhagen, Denmark, and The Marine Biological Laboratory, Woods Hole, Massachusetts, and The Department of Pharmacology, University of Berne, Berne, Switzerland*)

(Received for publication, November 19, 1956)

ABSTRACT

1. In a membrane transport system containing a mobile carrier with affinities for two substrates a concentration gradient with respect to one of the substrates under certain conditions is able to induce an "uphill" transport (against the concentration gradient) of the other.

2. In a kinetic treatment quantitative conditions for such a "flow-induced uphill transport" and some of its characteristics are derived.

3. Experimentally the uphill transport of labelled glucose induced by a concentration gradient for mannose or unlabelled glucose is demonstrated in the human red cell.

4. It is shown that the flow-induced uphill transport is a feature characteristic for mobile carrier systems only and is not to be expected in systems in which the substrate is bound to a fixed membrane component ("adsorption membrane"), although such a system may yield identical transport kinetics. Also with respect to Ussing's flux ratio the two systems are different, the adsorption membrane meeting Ussing's criterion, the carrier membrane not.

5. It is concluded that the transport system in the human red cells must contain a mobile carrier, identical for glucose and mannose.

Through the epithelial cells of kidney (1) and intestine (2) sugars may be transported "actively" in the thermodynamic sense; *i.e.*, against the concentration gradients. Rosenberg (3) has discussed transports of this kind in terms of thermodynamic quantities and potentials. He showed that the mere fact of a transport occurring against the concentration gradient (hereafter termed "uphill") allows the conclusion that a "carrier" in a broader sense is involved in the transport mechanism. In addition, in the case of sugar absorption, other indirect criteria (4) likewise point to carrier mechanisms (limited capacity of the transport, competition, kinetics, effect of enzyme inhibitors).

In the erythrocyte, the transport of sugars under conventional experimental

* This work was carried out with partial support from the Lalor Fellowship Foundation which is gratefully acknowledged.

J. GEN. PHYSIOL., 1957, Vol. 41, No. 2

conditions does not occur uphill. From a number of similar indirect criteria however, in this case also a carrier mechanism has been decided upon (3, 5–7). Although this indirect evidence as a whole is fairly suggestive, none of the observations as such is strictly conclusive.

However, the assumption of a carrier transport has a corollary which can be tested experimentally. If the carrier mechanism can transport two molecular species R and S, it should be possible under certain conditions to induce an uphill transport of R by establishing a concentration gradient for S. In this paper a treatment of some aspects of this induced active transport will be given and the phenomenon will be demonstrated experimentally. Finally its conclusiveness for a carrier mechanism as compared to that of other criteria will be discussed.

Kinetics of the Induced Uphill Transport

Kinetic treatments of the supposed carrier transport as given by various authors (8–10) differ mainly in the assumptions concerning the reactions between substrate and carrier (equilibrium or steady state, enzymatic or nonenzymatic reactions). For the purpose of this discussion it appears advantageous to choose simple conditions. Thus equilibrium between substrate and carrier on both sides of the membrane will be assumed.

The following notations will be used.

R, S, substrates reacting with the carrier C.

C, free carrier.

CR, CS, carrier substrate complexes.

$(C_T) = (C)_I + (CR)_I + (CS)_I = (C)_{II} + (CR)_{II} + (CS)_{II} =$ total concentration of the carrier (free + combined).

v_R, v_S, v_C, transport velocities for R, S, and C (positive sign in the direction I → II).

K_{rR}, K_{rS}, dissociation constants of the carrier substrate complexes.

$$S' = \frac{(S)}{K_{rS}}$$

$$R' = \frac{(R)}{K_{rR}}$$

D', diffusion coefficient of C, CR, and CS in the membrane, divided by the thickness and multiplied by the surface area of the membrane.

Concentrations are denoted by symbols in parentheses.

Indices I and II refer to the two sides of the membrane.

Under the conditions of a steady state in the membrane ($v_R + v_S + v_C = 0$) the following equations can be derived.

$$v_R = (C_T)D' \left\{ \frac{R'_I}{R'_I + S'_I + 1} - \frac{R'_{II}}{R'_{II} + S'_{II} + 1} \right\} \tag{1}$$

$$v_S = (C_T)D' \left\{ \frac{S'_I}{R'_I + S'_I + 1} - \frac{S'_{II}}{R'_{II} + S'_{II} + 1} \right\} \tag{2}$$

Introducing

$$p = \frac{R'_I}{R'_{II}} = \frac{(R)_I}{(R)_{II}} \tag{3}$$

and

$$q = \frac{S'_I + 1}{S'_{II} + 1} = \frac{(S)_I + K_{rS}}{(S)_{II} + K_{rS}} \tag{4}$$

we may transform equation (1) to

$$v_R = (C_T)D' \frac{R'_{II}(S'_{II} + 1)(p - q)}{(R'_I + S'_I + 1)(R'_{II} + S'_{II} + 1)} \tag{5}$$

If for both R and S the concentrations at the side I are higher than at the side II, ($p > 1$, $q > 1$), equation (5) shows that R will move along its gradient from I to II ($v_R > 0$) only if $p > q$, whereas for $p < q$, v_R will be negative and consequently the transport of R will be uphill. The general condition for this event is, that p lies between 1 and q.

If $R'_I = R'_{II} = R'$ and $S'_I > S'_{II}$ the flow of S will induce a transport of R. For this case, we may also from (1) and (2) derive the relation

$$v_R = - v_S \frac{R'}{R' + 1} \tag{6}$$

yielding the limiting values

$$v_R = -v_S \tag{7}$$

and

$$v_R = -v_S R' \tag{8}$$

for $R' \gg 1$ and $R' \ll 1$ respectively.

The concentration ratio for which

$$v_R = 0 \tag{9}$$

is according to (3), (4), (5), and (9)

$$p = q = \frac{(R)_I}{(R)_{II}} = \frac{S'_I + 1}{S'_{II} + 1} \tag{10}$$

The limiting values of this ratio are $\frac{(S)_I}{(S)_{II}}$ (for S'_I, $S'_{II} \gg 1$) and 1 (for S'_I, $S'_{II} \ll 1$).

Under conditions of slowly changing values for $(R)_I$ and $(R)_{II}$ (due to accumulation or depletion of R, if the volumes at I and/or II are limited as for instance in experiments with cell suspensions) this ratio is the maximal accumulation ratio for R.

Experimental Methods and Results

The experiments were carried out with human red cells. They were equilibrated with one penetrating sugar R in a low concentration. A second penetrating sugar S

was then added in high concentration to the external solution. The movement of R was followed by determinations of its concentration in the external solution. In control experiments a corresponding volume of saline was added instead of S to produce the same degree of dilution.

The substrates used were (1) unspecifically labelled C_{14}-glucose (as R) and (2) mannose or unlabelled glucose (as S). Mannose and glucose were chosen because, according to LeFèvre and Davies (11), they have a relatively high affinity for the transport system.

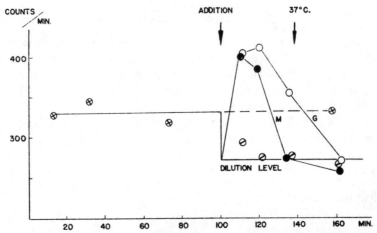

Fig. 1. Experiment showing uphill transport of labelled glucose across the human red cell membrane induced by counterflow of mannose (filled circles) or unlabelled glucose (open circles). Ordinate, activity in c.p.m., in 10 μl. of the external medium. Abscissa, time in minutes. Circle with x in center, activity before addition of 0.16 volume of unlabelled sugar (0.72 M in saline). Circle with bar, activity after addition of 0.16 volume saline. Temperature 0°C. until second arrow, then 37°C. (The temperature was raised in order to accelerate the penetration which, however, proved to be unnecessary.) The calculated maximal concentration ratio for labelled glucose is approximately 4.

It was found that the penetration of labelled glucose across the red cell membrane is very rapid. The experiments (with the exception of the equilibration phase) were therefore carried out at 0°C.

7.5 ml. defibrinated blood were equilibrated with 0.2 ml. labelled glucose solution, 5.6 mM with an activity of 1 μc./ml. The mixture was stored in the cold for 6½ hours. During this time several samples were centrifuged and the activity of the supernatant was counted. The hematocrit value was found to be 0.48 (extrapolated to infinite time of spinning at 3500 R.P.M.). The blood (6.7 ml.) was then centrifuged 9 minutes at 3500 R.P.M., 2.5 ml. supernatant were siphoned off, and the remaining 4.2 ml. were thoroughly mixed. The (calculated) hematocrit reading was now 0.77. Three additional samples were taken within 2 hours for determination of the activity in the external medium.

Three samples of 0.9 ml. blood each were then pipetted into ice cold centrifuge tubes and 144 μl. of one of the following (ice cold) solutions added: (1) mannose 2.4 isotonic (0.72 M) in isotonic saline; (2) glucose 2.4 isotonic (0.72 M) in isotonic saline; (3) isotonic saline. The samples were thoroughly mixed, kept at 0°C., and aliquots taken at definite time intervals for the determination of the activity in the external medium.

The activities were counted with a gas flow counter. 10 μl. of external medium were used for each determination. The self-absorption was determined separately and corrected for.

Fig. 1 shows the results obtained. The activity of the external solution is plotted against time. Before the addition of S it is approximately constant. After addition of NaCl it attains a new level which corresponds closely to that calculated for the degree of dilution brought about by the added volume ("dilution level").

In the experiment with mannose or glucose the activity of the external solution first rises to a higher level, where it remains for some time. This means that R moves from the cells into the external solution against a concentration gradient, the maximum value of which can be calculated to be about 4:1.

Later the external activity falls and finally reaches the dilution level, as is to be expected, if the sugars penetrate into the cells.

The penetration was followed in a separate experiment at 0°C. under identical conditions using conventional osmotic methods. It was found that the penetration of mannose is somewhat faster (equilibrium after about 30 to 35 minutes) than that of glucose (equilibrium after about 40 minutes).

The observations thus show the occurrence of an induced uphill transport of R and are in qualitative agreement with predictions derived from the penetration characteristics of mannose and glucose.

DISCUSSION

The experiments reported here clearly demonstrate the occurrence of induced uphill transports of glucose across the red cell membrane. Similar observations on phosphate and arsenate transports in bacterial cells have been reported by Mitchell (12).

In the treatment given for the induced active transport the type chosen for the reaction-diffusion system (non-enzymatic reactions and equilibrium between carrier and substrate on both sides) is a special (and the simplest) case.

The question arises therefore whether the results of the treatment have more general validity and how far they are influenced by the simplifying assumptions used. In general, those results which also would follow from thermodynamic considerations, especially the maximal possible values for the concentration ratio of R and for the ratio v_R/v_S, will be independent of special assumptions concerning the reactions and will be generally valid for carrier systems of the type treated here. On the other hand conclusions with respect to velocities, *e.g.* equation (5) will depend on the characteristics of the reactions involved.

The induced uphill transport as observed in the red cell has been derived from the transport kinetics of a simple carrier system. Rosenberg (3) has shown by thermodynamic arguments that a mobile carrier is indeed an indispensable condition for such a transport. This may be illustrated by a comparison of the system treated here and a model membrane consisting of two adsorption layers. These two models which show a close resemblance in their transport kinetics differ in the essential point, that the substrate is bound to a mobile membrane component in one case, to a fixed component in the other.

It will first be shown that the steady state kinetics of the two systems are formally identical. However, as will be derived later only the carrier model is able to induce an uphill transport. Another characteristic difference in the two systems which will finally be discussed is the flux ratio introduced by Ussing (13).

In the system discussed here, if $(S)_I = (S)_{II} = 0$, equation (1) may be written

$$v_R = D'(C_T) \left\{ \frac{R'_I}{R'_I + 1} - \frac{R'_{II}}{R'_{II} + 1} \right\} \tag{11 a}$$

or

$$v_R = D'(C_T) \frac{R'_I - R'_{II}}{(R'_I + 1)(R'_{II} + 1)} \tag{11 b}$$

It can be shown that an equation closely resembling equation (11b) is obtained for the "adsorption membrane." The transfer from layer I to layer II is assumed to be the rate-limiting step (velocity factor = a). Using Langmuir's adsorption equation $\left(\text{equilibrium constant} = K_L \text{ and } R' = \frac{(R)}{K_L} \right)$, we then obtain

$$v_R = an^2 \frac{R'_I - R'_{II}}{(R'_I + 1)(R'_{II} + 1)} \tag{12}$$

in which n is the total number of adsorption places in each layer. Equations (11 b) and (12) are formally identical. Near saturation both equations yield the special kinetic type which has been termed E, characterized by

$$v_R = K_E \left(\frac{1}{(R)_{II}} - \frac{1}{(R)_I} \right) \tag{13}$$

and which actually was observed for transports of sugars in erythrocytes (6, 7, 14). Thus a differentiation of the two mechanisms by a kinetic analysis of the transport of a single substance appears impossible as was pointed out previously (15).

With respect to the flow-induced uphill transport however, the two mechanisms differ. For the carrier system containing both R and S the flow-induced active transport emerged from equation (5). The corresponding equation for the adsorption membrane is

$$v_R = an^2 R'_{II} \frac{p - 1}{(R'_I + S'_I + 1)(R'_{II} + S'_{II} + 1)} \tag{14}$$

Since here v_R always has the same sign as $(p - 1)$, the transport always occurs along the gradient.

The essential point in this discussion is whether or not the membrane component combining with the substrate is mobile. For example a transfer by combination with a rotating protein or enzyme molecule, as occasionally suggested for active transport (e.g. reference 16), would not only follow kinetics similar to those of the aforementioned cases but would also be able to perform an induced uphill transport in the same manner as the carrier system. On the other hand, the "carrier" assumed by LeFévre for sugar transports (5), is actually a fixed membrane component and consequently would not be able to perform flow-induced uphill transport.

Another characteristic difference between the two types of membrane components emerges from the application of Ussing's flux ratio criterion (13) for passive diffusion:

$$\frac{m_I}{m_{II}} = -\frac{a_I}{a_{II}} \tag{15}$$

(m = flux; a = activity).

It was pointed out (15) that as a general condition for the validity of equation (15) the two diffusion streams which are called flux, must be independent of each other. This is certainly not true for the carrier system, since here both fluxes are interdependent by mediation of the movement of the carrier, and the effect of this movement on the two fluxes is opposite. In the adsorption system, however, although there also is an interdependence by mediation of the number of free places, the role of this number is that of a resistance which reduces the two fluxes by the same fraction so that their ratio remains unchanged.

In accordance with this argument the flux ratio in a transport through an adsorption membrane following equation (12) is (activity coefficients being neglected):

$$\frac{m_I}{m_{II}} = -\frac{(R_I)}{(R_{II})} \tag{16}$$

thus meeting the above-mentioned criterion for passive diffusion. For the carrier membrane equation (11 a) yields for the flux ratio:

$$\frac{m_I}{m_{II}} = -\frac{(R)_I}{(R)_{II}} \frac{(R'_{II} + 1)}{(R'_I + 1)} \tag{17}$$

not obeying Ussing's criterion. Actually the experiments reported with glucose as S show that Ussing's criterion cannot hold for the transport of glucose.

The conclusion from these considerations and from the experimental demonstration of the flow-induced uphill transport of glucose in red cells is that here a mobile carrier is involved in the transport of mannose and glucose. It might be argued that more than one system may be involved in sugar transfer across the red cell membrane. The induced uphill transport demonstrated here might be due to a system which can only perform exchange. For the net transfer system involved in osmotic experiments, then, a mobile carrier would not necessarily be required, although the kinetics and other criteria exclude free diffusion.

With respect to quantitative details the experiments have preliminary value. It is hoped that a study of the quantitative relationships between the transfer of labelled sugar and the net transport will shed more light on this question.

BIBLIOGRAPHY

1. Smith, H. W., The Kidney, New York, Oxford University Press, 1951.
2. Barany, E., and Sperber, E., *Skand. Arch. Physiol.*, 1939, **81**, 290.
3. Rosenberg, T., *Acta Chem. Scand.*, 1948, **2**, 14.
4. Rosenberg, T., and Wilbrandt, W., *Internat. Rev. Cytol.*, 1952, **1**, 65.
5. LeFèvre, P. G., *J. Gen. Physiol.*, 1948, **31**, 505.
6. Wilbrandt, W., and Rosenberg, T., *Helv. Physiol. Acta* 1950, **8**, C82.
7. Widdas, W. F., *J. Physiol.*, 1952, **118**, 23.
8. Wilbrandt, W., and Rosenberg T., *Helv. Physiol. Acta*, 1951, **9**, C86.
9. LeFèvre, P. G., and LeFèvre, M. E., *J. Gen. Physiol.*, 1952, **35**, 891.
10. Rosenberg, T., and Wilbrandt, W., *Exp. Cell Research*, 1955, **9**, 49.
11. LeFèvre, P. G., and Davies, R. I., *J. Gen. Physiol.*, 1951, **34**, 515.
12. Mitchell, P., *Symp. Soc. Exp. Biol.*, 1954, **8**, 254.
13. Ussing, H. H., *Acta Physiol. Scand.*, 1949, **19**, 43.
14. Wilbrandt, W., Frei, S., and Rosenberg, T., *Exp. Cell. Research*, 1956, **11**, 59.
15. Rosenberg, T., *Symp. Soc. Exp. Biol.*, 1954, **8**, 27.
16. Lundegårdh, H., *Lantbruks-Högskol. Ann.*, 1940, **8**, 233.

The Galactoside Permease
of <u>Escherichia coli</u>

III

Editor's Comments on Papers 7 through 15

Initially, the mobile carrier hypothesis rested almost exclusively on kinetic evidence gathered by mammalian physiologists; its most substantial support, however, was to come from very different quarters. The existence of specific mechanisms for the entry of metabolites into microbial cells was strongly implied by the phenomenon of *specific crypticity*, in which an organism loses, by mutation, its ability to metabolize a particular substance even though activity of the requisite enzymes can be demonstrated in cell extracts (Doudoroff et al., 1949). Rickenberg, Cohen, Buttin, and Monod (Paper 7) showed that cryptic *Escherichia coli* mutants that were unable to utilize lactose or hydrolyze the lactose analogue, *o*-nitrophenyl-β-D-galactoside (ONPG), had specifically lost the ability to transport galactosides into the cell. The mutation had not affected the activity of β-galactosidase, nor had it affected the cell's ability to metabolize other sugars.

The system responsible for the specific uptake of galactosides in the wild-type organism was termed the "galactoside permease" by Rickenberg et al. [see also Cohen and Monod (Paper 8)]. The enzyme-like name, which engendered considerable criticism (Kleinzeller and Kotyk, 1961; Christensen, 1960), emphasized the catalytic nature of the uptake process and its kinetic similarities to enzyme activity. As pointed out by the authors, however, it did not necessarily imply that transport involved the formation or rupture of covalent bonds in the substrate. The inducibility of both the permease and β-galactosidase was shown to be controlled by a distinct regulatory gene. These observations, familiar to today's student of elementary biology, provided the foundation for the operon theory of Jacob and Monod (1961). In its general form, the permease hypothesis also provided a single explanation for many hitherto perplexing and seemingly unrelated observations concerning amino acid utilization in *E. coli* (Cohen and Monod, Paper 8).

The kinetic studies of Rickenberg et al. (Paper 7), Kepes (Paper 9), Koch (Paper 10), and Winkler and Wilson (Paper 11) established the main features of the galactoside transport system in *E. coli*: (1) Galactosides are accumulated inside the cell in unmodified form at concentrations 50–100 times that in the external medium. Accumulated galactosides exchange readily with external galactosides, indicating that the concentration gradient is maintained by a steady-state process. (2) Uptake is nearly abolished by uncoupling agents (e.g., azide, 2,4-dinitrophenol) or sulfhydryl reagents. (3) Efflux is mediated by the same carriers as influx and is stimulated by uncouplers but inhibited by sulfhydryl reagents. (4) The carriers appear to be the products of the *y* gene, as shown by the absence of counterflow in *y*⁻ mutants (Winkler and Wilson, Paper 11).

Koch (Paper 10) observed that uncoupling agents do not markedly inhibit ONPG hydrolysis by whole cells, although they abolish the concentrative uptake of galactosides. The rate of ONPG hydrolysis by whole cells is limited by the movement of the galactoside across the cell membrane; in this case transport is *down* a concentration gradient, since ONPG is hydrolyzed as soon as it enters the cell. The implication of this result, that energy is not required to facilitate the diffusion of galactosides down a concentration gradient, has recently become a matter of controversy. Winkler and Wilson (Paper 11), in support of Koch (Paper 10), observed rapid equilibration of galactosides in the presence of uncouplers, and demonstrated by counterflow that the carriers remain active under these conditions. Furthermore, they found that uncouplers induce a dramatic reduction in the K_m for efflux but have little effect on the kinetics of influx. Thus armed, they argued that the effect of energy coupling is to decrease the affinity of the carriers for the transport substrate on the inner surface of the membrane, thereby allowing a high concentration of galactosides to be built up within the cell. More recently, "coupling" mutants have been isolated that are defective in their ability to accumulate galactosides but retain normal carrier properties, as shown by their high rates of ONPG hydrolysis and by counterflow (Wong, Kashket, and Wilson, 1970; Wilson and Kusch, 1972).

The contrary hypothesis, that energy is required for movement of carriers across the membrane, is supported by the results of several different laboratories. Koch (1971) retracted his earlier observations and showed that extensive energy depletion markedly lowers the rates of ONPG hydrolysis by whole cells. Manno and Schachter (1970) observed an 800-fold decrease in the rate of entry of galactosides into cells that had been treated with dinitrophenol and fluoride. Barnes and Kaback (Paper 13) reported that the initial rate of lactose uptake in isolated membrane vesicles was stimulated nearly 20-fold by the addition of the specific energy source, D-lactate. The reasons for the discrepancies between the results just cited and the earlier observations of Koch (Paper 10) and Winkler and Wilson (Paper 11) are not clear. Schachter and Mindlin (1969), however, have stressed that in certain experiments the latter authors were measuring exchange, rather than net transport, a difference that may be of critical importance when considering the effect of energy coupling.

In 1965, Fox and Kennedy (Paper 12) described a method for the specific labeling and partial purification of the galactoside carrier protein. After a preliminary non-specific reaction with nonlabeled N-ethylmaleimide (NEM), in which the carriers were protected by thiodigalactoside, a transport substrate, cells were exposed to labeled NEM. By comparing the labeling patterns of induced and uninduced cells, the authors were able to demonstrate the specific combination of radioactive NEM with a component of the membrane (the M protein) that was later shown to be the product of the y gene (Fox, Carter, and Kennedy, 1967). Unfortunately, the labeling procedure results in irreversible inactivation of the M protein, so the functional properties of the purified carrier cannot be studied.

There has been no shortage of hypothetical models for the galactoside permease system. The early models of Kepes (Paper 9) and Koch (Paper 10), in which the permease catalyzes the combination of the transport substrate with a nonspecific carrier, the *"transporteur,"* have lost favor in the light of results suggesting that the permease is itself the carrier (Winkler and Wilson, Paper 11; Fox, Carter, and Kennedy, 1967). More recent approaches envision the carriers as facilitating the passage of galactosides across the membrane, with energy coupling reducing the affinity of the carriers for internal galactosides (Winkler and Wilson, Paper 11; Fox and Kennedy, Paper 12). Schachter and Mindlin (1969) have altered this model somewhat to include an effect of energy coupling on the rate of passage of the loaded carriers across the membrane. Kepes (1971) has suggested that the movement of the *unloaded* carriers across the membrane may be the energy-dependent step during net uptake.

A common feature of the above models is their almost embarrassing vagueness in characterizing the energy-coupling step. Indeed, only very recently has any information at all come to light concerning the energy-coupling process. Scarborough, Rumley, and Kennedy (1968) reported that ATP stimulated ONPG hydrolysis in osmotically shocked cells but cautioned that ATP might be acting through a general effect on cell metabolism rather than serving as the specific energy source for the permease. Pavlasova and Harold (1969) showed that uncouplers inhibit transport of galactosides under anaerobic conditions without depressing ATP levels, suggesting a direct interference by uncouplers in the energy-dependent step. The possibility that phosphoenolpyruvate provides energy for galactoside accumulation through the phosphotransferase system in *E. coli* (see Part IV) appears to be ruled out by the finding that certain strains with mutations in Enzyme I transport galactosides in a normal fashion (see Kaback, 1970; Kennedy, 1970) and by the experiments to be described below with membrane vesicles.

In 1966, Kaback and Stadtman described the preparation of membrane-bound vesicles by the lysis of *E. coli* spheroplasts; the vesicles retained the cell's ability to take up proline against a gradient, although they were devoid of RNA, DNA, and the soluble components of the cell cytoplasm. It was later shown that the vesicles also transport α-methylglucoside and glucose by the phosphotransferase system when supplied with phosphoenolpyruvate (Kaback, Paper 20). Recently, Barnes and Kaback (Paper 13) reported that the uptake of lactose and other galactosides in membrane vesicles is coupled primarily to the oxidation of D-lactate. Other oxidizable substrates, such as L-lactate, α-hydroxybutyrate, and succinate also support lactose uptake, but not as well as D-lactate. None of the 36 other possible energy sources tested, including

ATP and phosphoenolpyruvate, were effective in stimulating transport of galactosides. Transport appears to be coupled directly to the electron-transfer reactions and does not involve ATP or a high-energy phosphorylated intermediate. Studies of the effects on transport of different oxidizable substrates and electron-transfer inhibitors has indicated that the site of energy coupling lies between the flavin-linked D-lactate dehydrogenase and cytochrome b_1 in the electron-transfer chain (Barnes and Kaback, Paper 14; Kaback and Barnes, Paper 15).

Oxidation of D-lactate also provides the primary driving force for the concentrative uptake of 15 different amino acids (Kaback and Milner, 1970), galactose (Kerwar, Gordon, and Kaback, 1972), arabinose, glucuronate, gluconate, and glucose-6-phosphate (Kaback, 1972), and, in the presence of valinomycin, of potassium or rubidium (Bhattacharyya, Epstein, and Silver, 1971; Lombardi, Kaback, and Reeves, 1972). Each of these transport systems, as well as others in vesicles from different organisms for which the physiological electron donor has yet to be identified, can also be driven by the artificial electron donor ascorbate-phenazine methosulfate (Konings, Barnes, and Kaback, 1971).

The model suggested by Kaback and Barnes (Paper 15) to account for these results pictures the carriers as electron-transfer intermediates. A change from the oxidized to the reduced state results in translocation of the carrier–substrate complex to the inner surface of the membrane and a concomitant decrease in the carrier's affinity for the substrate. Harold (1972) has criticized this model on the grounds that it fails to explain the action of uncouplers, which block transport without affecting D-lactate oxidation, and it does not account for transport under anaerobic conditions. He has argued vigorously in support of Mitchell's (1963) suggestion that the immediate driving force for transport consists of a transmembrane hydrogen ion gradient. In this view, the hydrogen ion gradient is generated during respiration as a result of the geometrical arrangement of hydrogen carriers vis-à-vis electron carriers in the respiratory chain; i. e., the arrangement is such that protons are consumed on one side of the membrane and released on the other (Mitchell, 1966, 1967). There is a considerable body of evidence indicating that uncouplers increase the hydrogen ion conductivity of membranes, which, if Mitchell's hypothesis were accepted, would explain their inhibitory effect on transport (Harold, 1972).

Kaback (1972) has recently summarized a series of observations which appear to cast doubt on the participation of hydrogen ion gradients in transport, at least in membrane vesicles. A detailed exposition of the arguments and counterarguments in this controversial area will not be attempted here. Perhaps just one point can be mentioned, however, as it appears to transcend the immediate issue of hydrogen ion gradients. Lombardi and Kaback (1972) have shown that different oxidizable substrates support transport with different relative efficiencies, depending on the substance being transported. For example, NADH will not support uptake of lactose by membrane vesicles but will stimulate the uptake of proline at a rate nearly 30 percent of that for D-lactate. Again, succinate is 70 percent as effective as D-lactate in driving lysine uptake but has only a marginal effect on proline transport. Many other examples could be cited. As these transport systems are alike in nearly every other respect (Lombardi and Kaback, 1972), the conclusion seems inescapable that energy coupling involves a local interaction of the carriers with individual electron-transfer chains rather than the production of a central "high-energy" intermediate or state, be it a hydrogen ion gradient, ATP, or an "energized" state of the membrane, for, if such were the case, the relative efficiencies of the various energy sources should be the same for each transport system.

References

Bhattacharyya, P., W. Epstein, and S. Silver (1971). Valinomycin-induced uptake of potassium in membrane vesicles from *Escherichia coli*. *Proc. Natl. Acad. Sci. USA*, **68**, 1488–1492.

Christensen, H. N. (1960). Reactive sites and biological transport. *Advan. Protein Chem.*, **15**, 234–314.

Doudoroff, M., W. Z. Hassid, E. W. Putnam, A. L. Potter, and J. Lederberg (1949). Direct utilization of maltose by *Escherichia coli*. *J. Biol. Chem.*, **179**, 921–934.

Fox, C. F., J. R. Carter, and E. P. Kennedy (1967). Genetic control of the membrane protein component of the lactose transport system of *Escherichia coli*. *Proc. Natl. Acad. Sci. USA*, **57**, 698–705.

Jacob, F., and J. Monod (1961). Genetic regulatory mechanisms in the synthesis of proteins. *J. Mol. Biol.*, **3**, 318–356.

Kaback, H. R. (1970). Transport. *Ann. Rev. Biochem.*, **39**, 561–598.

Kaback, H. R. (1972). Transport across isolated bacterial cytoplasmic membranes. *Biochim. Biophys. Acta*, **265**, 367–416.

Kaback, H. R., and L. S. Milner (1970). Relationship of amembrane-bound D-(−)-lactic dehydrogenase to amino acid transport in isolated bacterial membrane preparations. *Proc. Natl. Acad. Sci. USA*, **66**, 1008–1015.

Kaback, H. R., and E. R. Stadtman (1966). Proline uptake by an isolated cytoplasmic membrane preparation of *Escherichia coli*. *Proc. Natl. Acad. Sci. USA*, **55**, 920–927.

Kennedy, E. P. (1970). The lactose permease system of *Escherichia coli*. In J. R. Beckwith and D. Zipser (eds.), *The Lactose Operon*. Cold Spring Harbor Laboratory, New York, pp. 49–92.

Kepes, A. (1971). The β-galactoside permease of *Escherichia coli*. *J. Membrane Biol.*, **4**, 87–112.

Kerwar, G. K., A. S. Gordon, and H. R. Kaback (1972). Mechanisms of active transport by isolated membrane vesicles. IV. Galactose transport by isolated membrane vesicles from *Escherichia coli*. *J. Biol. Chem.*, **247**, 291–297.

Kleinzeller, A., and A. Kotyk (eds.) (1961). *Membrane Transport and Metabolism*. Academic Press, New York.

Koch, A. L. (1971). Energy expenditure is obligatory for the downhill transport of galactosides. *J. Mol. Biol.*, **59**, 447–459.

Konings, W. N., E. M. Barnes, Jr., and H. R. Kaback (1971). Mechanisms of active transport in isolated membrane vesicles. III. The coupling of reduced phenazine methosulfate to the concentrative uptake of β-galactosides and amino acids. *J. Biol. Chem.*, **246**, 5857–5861.

Lombardi, F. J., and H. R. Kaback (1972). Mechanisms of active transport in isolated bacterial membrane vesicles. VIII. The transport of amino acids by membranes prepared from *Escherichia coli*. *J. Biol. Chem.*, **247**, 7844–7857.

Lombardi, F. J., H. R. Kaback, and J. P. Reeves (1972). Valinomycin-induced Rb⁺ transport by bacterial membrane vesicles. *Fed. Proc.*, **31**, 457 abs.

Manno, J. A., and D. Schachter (1970). Energy-coupled influx of thiomethylgalactoside into *Escherichia coli*. *J. Biol. Chem.* **245**, 1217–1223.

Mitchell, P. (1963). Molecule, group and electron translocation through natural membranes. *Biochem. Soc. Symp.*, **22**, 142–168.

Mitchell, P. (1966). Chemiosmotic coupling in oxidative and photosynthetic phosphorylation. *Biol. Rev.*, **41**, 445–502.

Mitchell, P. (1967). Proton translocation, phosphorylation in mitochondria, chloroplasts and bacteria: natural fuel cells and solar cells. *Fed. Proc.,* **26,** 1370–1379.

Pavlasova, E., and F. M. Harold (1969). Energy coupling in the transport of β-galactosides by *Escherichia coli*: effect of proton conductors. *J. Bacteriol.,* **98,** 198–204.

Scarborough, G. A., M. K. Rumley, and E. P. Kennedy (1968). The function of adenosine-5′-triphosphate in the lactose transport system of *Escherichia coli*. *Proc. Natl. Acad. Sci. USA,* **60,** 951–958.

Schachter, D., and A. J. Mindlin (1969). Dual influx model of thiogalactoside accumulation in *Escherichia coli*. *J. Biol. Chem.,* **244,** 1808–1816.

Wong, P. T. S., E. R. Kashket, and T. H. Wilson (1970). Energy coupling in the lactose transport system of *Escherichia coli*. *Proc. Natl. Acad. Sci. USA,* **65,** 63–69.

Wilson, T. H., and M. Kusch (1972). A mutant of *Escherichia coli* K12 energy-uncoupled for lactose transport. *Biochim. Biophys. Acta,* **255,** 786–797.

Translated from *La galactoside-perméase d'Escherichia coli*
Ann. Inst. Pasteur, Paris, **91,** 829–857 (1956)

7

Galactoside Permease
in *Escherichia coli**

H. V. RICKENBERG, G. N. COHEN,
G. BUTTIN, and J. MONOD

Introduction

In this report we describe a system characterized by the property of accumulating exogenous galactosides in the cells of *Escherichia coli.* The discovery of this inducible system, distinct from the β-galactosidase, but which controls the in vivo activity of this enzyme, as well as its induction, provides a solution to numerous problems posed by the metabolism of galactosides and by the induction of β-galactosidase in *E. coli,* and furnishes experimental confirmation of the hypothesis (frequently advanced) that stereospecific and functionally specialized catalytic systems, distinct from metabolic enzymes themselves, govern the penetration of certain substrates in microbial cells.

The very general validity of this hypothesis has also been confirmed by the discovery of systems which ensure the penetration and accumulation of various amino acids in *E. coli,* systems comparable to the one discussed here in their kinetic properties and their degree of specificity [5, 6]. Some of the results in this report have been summarized in a preliminary note [4]. The role of the system which concentrates galactosides in the induction of galactosidase has been considered in other publications [34, 8].

Chemicals, Strains, and Techniques

Chemicals

The radioactive methyl-β-D-thiogalactoside was synthesized in our laboratory (from tetraacetylbromogalactose and ^{35}S-labeled methylmercaptan), using the tech-

---
*This work has been subsidized by the Rockefeller Foundation of New York, by the "Jane Coffin Child's Memorial Fund," and by the Atomic Energy Commission.

nique of D. Turk [47], developed in Helferich's laboratory in Bonn. The initial specific activity was 5 mC/mmole. Other thiogalactosides were also synthesized by D. Turk [47]. Other products were of commercial origin. The commercial maltose was recrystallized twice from alcohol at 80°C.

Strains

We used various normal and mutant strains of *E. coli* ML and K12. The ML mutants were isolated in our laboratory [35, 37, 7], while most of the K12 mutants were sent to us by J. Lederberg of Madison, Wisconsin. In the experiments below, whenever the strain is not specified, *E. coli* ML30 (wild-type) is used.

Media and Conditions of Growth

We used synthetic medium 56 [36] without CaCl₂ with succinate (4 mg/ml) or with maltose (2 mg/ml) as the source of carbon. The cultures were agitated in conical flasks at 34°C. The cultures used were in the exponential stage of growth and were centrifuged and resuspended in the desired medium at the desired density. Culture density was determined by reading the optical density at 600 nm; it is expressed in micrograms of bacterial dry weight per milliliter.

Measurement of the β-Galactosidase Activity of the Suspensions

Toluene was added to the suspensions, and they were agitated for 15 min at 34°C. The activity was measured in the presence of $M/350$ ONPG and $M/10$ Na$^+$ at pH 7 and 28°C. The unit of β-galactosidase is the amount of enzyme which, under these conditions frees 1 nM of *o*-nitrophenol in 1 min [36]. The liberation of *o*-nitrophenol was determined by measuring the increase in optical density at 4200 Å in a Beckman spectrophotometer.

Hydrolysis of ONPG in Vivo

The rate of in vivo hydrolysis was determined in medium 56, in the presence of succinate (0.2 percent) with $M/400$ ONPG, in agitated flasks at 34°C. The reaction was stopped by the addition of $M/2$ Na$_2$CO$_3$ [26].

Measurement of TMG Accumulation

The technique generally adopted (except where specified otherwise)was as follows.
The suspensions, adjusted to a concentration of about 150–200 μg/ml were agitated at 34°C in medium 56, in the presence of maltose (0.2 percent) or succinate (0.2 percent) chloromycetin (50 μg/ml) and TMG ($5\times10^{-4}\,M$). (The radioactive product was mixed, in the desired proportion, with nonradioactive TMG so as to obtain about 20,000 counts/min per μM with the counter used.) After 10 min incubation, a

sample (5 ml) was removed, cooled to 0°C, and then centrifuged at 15,000g for 5 min. The supernatant was aspirated off with a pipet connected to a vacuum pump, and the walls of the tubes were carefully dried with filter paper, right down to the sample. The pellet was then resuspended in 1.5 ml of water, heated to 100°C for 5 min, and recentrifuged. The supernatant liquid was decanted and its radioactivity determined.

Radioactivity Measure

A Geiger counter was used, connected to a scale of 100. Samples were counted at infinite thinness. For each determination we counted 1000 counts in two samples. Taking into account the volumetric errors, the error is about ±5 percent.

Experiments

Accumulation of TMG by Induced Bacteria

When induced bacteria (i.e., bacteria grown in the presence of an inducing galactoside) are incubated for 5 min in the presence of methyl-β-D-thiogalactoside (radioactive TMG) and then centrifuged, we find that they contain a much greater amount of radioactivity than noninduced bacteria treated in the same way. In the

Table 1. Accumulation of thiomethyl-β-D-galactoside by induced and noninduced bacteria

Bacteria[a]	Counts/min for 100 μg of bacteria		TMG accumulated in	
	obs.	corrected[b]	% dry wt. bacteria	relative values
Noninduced	17	2	0.0	0
$d°$ + NaN$_3$ ($2 \times 10^{-2} M$)	15	0	0.0	0
Induced	422	407	2.7	100
$d°$ + NaN$_3$ ($2 \times 10^{-2} M$)	35	20	0.13	5
$d°$ + 2,4-dinitrophenol ($10^{-3} M$)	39	24	0.16	6
$d°$ + phenyl-β-D-thiogalactoside ($10^{-3} M$)	56	41	0.27	10
$d°$ + phenyl-β-D-thioglucoside ($10^{-3} M$)	415	400	2.6	96

[a]The experiment was performed under conditions given on page 97 in the presence of $5 + 10^{-4} M$ TMG and of chloramphenicol at 50 μg/ml; strain ML30.
[b]The "observed" values were corrected by at least 15 counts/min, corresponding to the radioactivity carried by the noninduced bacteria (see p. 99).

98

presence of sodium azide or of 2,4-DNP, the accumulation of TMG by the induced bacteria is inhibited by over 95 percent (Table 1).

The presence of a metabolizable carbon substrate increases the TMG accumulation by variable amounts (10 to 100 percent), depending on whether the bacteria have been previously deprived of a carbon source for a shorter or longer time.

Assuming a density of 1 for the bacteria and a hydration of 75 percent, we see that the intracellular concentration of TMG in induced bacteria is about 70 times its concentration in the external medium. In other experiments, this concentration factor was over 100. Thus there is in these bacteria a mechanism that can remove TMG from the external medium and accumulate it in the cells. The inhibition of this mechanism by azide, by DNP, and by the absence of a carbon source indicates that the accumulation of TMG is coupled to reactions donating metabolic energy.

This mechanism does not function in noninduced bacteria, since the small amounts of radioactivity taken up are not diminished in the presence of the azide. Thus we are dealing here with a passive transport, doubtless primarily from the TMG contained in the residual liquid film adhering to the glass and to the bacteria themselves. In order to calculate the accumulation of TMG, the passive transport must be subtracted. On the basis of a series of tests performed with noninduced bacteria in the presence of NaN₃, we have adopted a uniform correction of 1 μliter assumed to be "passively transported" per 100 μg dry weight.

Figure 1 shows that after taking this correction into account, the amount of TMG accumulated by the bacteria per unit volume is proportional to the bacterial concentration, provided, however, that the concentration of TMG in the external medium remains essentially constant.

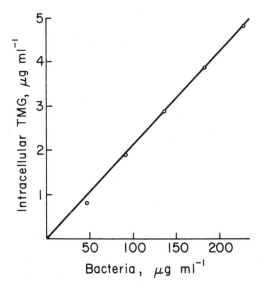

Figure 1. Intracellular accumulation of TMG per unit volume of culture versus bacterial mass in suspension. The accumulation experiment was performed under conditions given on page 97 with different dilutions of the same bacterial suspension. The bacterial mass per unit volume is expressed here in micrograms of dry weight per milliliter.

The radioactivity accumulated by the bacteria is completely extracted by boiling water. TMG is not hydrolyzed by β-galactosidase and is not metabolized by the bacteria, in the sense that it is not used as the carbon source or as a sulfur source. The radioactivity extracted from bacteria incubated at 34°C, from between 10 min to 1 h, in the absence of an exogenous energy source and in the presence of radioactive TMG, is always associated with a single substance having the chromatographic properties of TMG. However, when the incubation occurs in the presence of an exogenous energy source, a radioactive substance with a higher Rf than TMG in the solvent used (butanol: acetic acid: water, 4: 1: 1), accumulates slowly in the bacteria and in the medium. The nature of this substance and of the reaction forming it will be the subject of a separate paper [20]. It is enough to emphasize here that though this reaction is doubtless conditioned by the intracellular accumulation of TMG, the accumulation can take place without forming any detectable traces of this substance [20]. In what follows we shall consider the experiments performed in a short enough time and with low enough bacterial concentrations that the fraction of converted TMG is negligible (less than 5 percent).

Reversibility, Equilibrium, and Saturation. The radioactive TMG previously accumulated by the bacteria can be displaced by the addition of nonradioactive TMG or released as a result of the additon of azide or 2,4-dinitrophenol (Table 2). The accumulation is thus reversible, as is also shown by the fact that an equilibrium is established between the external and the internal concentrations of TMG. This equilibrium state is reached in less than 5 min at 34°C. The same equilibrium is reached

Table 2. Inhibition and Reversal by 2,4-dinitrophenol and sodium azide of the intracellular accumulation of TMG

Conditions[a]	Counts min^{-1} for 100 μg bacteria[b]	TMG accumulated in	
		% dry wt. bacteria	relative values
TMG at 0 min, sample taken at 5 min	343	2.5	100
TMG at 0 min, sample taken at 10 min	345	2.5	100
TMG + NaN$_3$ at 0 min, sample taken at 5 min	27	0.20	8
TMG at 0 min, NaN$_3$ at 5 min, sample taken at 10 min	13	0.16	6
TMG + 2,4,-DNP at 0 min, sample taken at 5 min	31	0.22	9
TMG at 0 min, 2,4-DNP at 5 min, sample taken at 10 min	55	0.39	16

[a]Experiment performed according to technique given on p. 97. Radioactive TMG $5 \times 10^{-4} M$; 2,4-dinitrophenol (DNP) $10^{-3} M$; NaN$_3$, $0.02 M$.
[b]Passive transport subtracted (see p. 99).

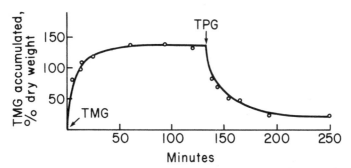

Figure 2. Accumulations of TMG by a bacterial suspension at 0°C. Conditions: radioactive TMG $5 \times 10^{-4} M$, added at time 0. Triophenyl-β-D-galactoside (TPG), nonradioactive, added at instant indicated by arrow. We see that at this temperature the accumulation equilibrium is reached in about 50 min. The addition of TPG causes a displacement of accumulated TMG.

at 0°C, but much more slowly (Fig. 2). The internal concentration at equilibrium varies as a function of the external concentration in accordance with an adsorption isotherm, within experimental error (Fig. 3). Clearly, the reaction proceeds as if there were an overall equilibrium:

$$\text{(bacteria)} + \text{(TMG)} \underset{k_2}{\overset{k_1}{\rightleftharpoons}} \text{(TMG-bacteria)} \qquad \text{(A)}$$

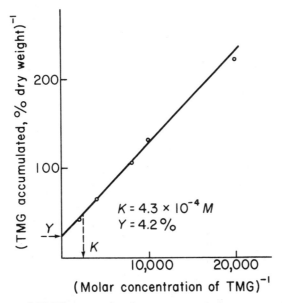

Figure 3. Amount of TMG accumulated versus external concentration of TMG (inverse coordinates). The linearity of the relationship shows that TMG accumulated at equilibrium varies as a function of external concentration of TMG, in accordance with an adsorption isotherm. K and Y correspond to the graphically determined values of equilibrium constants (1). (See text, p. 102.)

By calling G_{ex} the external concentration of TMG and $^{eq}G_{in}$ the intracellular amount of TMG at equilibrium, we can write

$$^{eq}G_{in} = Y \frac{G_{ex}}{G_{ex} + K} \tag{1}$$

where K is the dissociation constant of the bacteria–TMG "complex" and Y is a constant which we call "capacity." The "specific capacity" of the system will be the total capacity based on a given weight of bacteria (100 μg). For the time being, this equation must be considered only an empirical description (although very exact) of the phenomenon.

In general, the capacities cannot be determined for the saturating concentrations of TMG, which would allow the elimination of the possible variations of the constant K, since at such concentrations the error in estimating passive contamination would become too large. The measurement of intracellular TMG, always performed at a nonsaturating concentration of external TMG (usually $5 \times 10^{-4} M$), reflects the possible variations of constant K, as well as of capacity Y. It should be noted here that the value of the dissociation constant K, of the order of $4 \times 10^{-4} M$, does not seem to vary appreciably, even among strains, taking into account the lack of precision in its determination (± 20 percent), provided the bacteria used are in a good physiological condition.

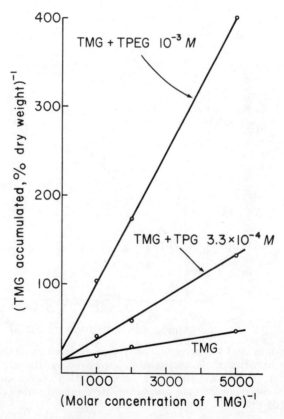

Figure 4. Inhibition of TMG accumulation by thiophenyl-β-D-galactoside (TPG) or by thiophenylethyl-β-D-galactoside (TPEG) (inverse coordinates).

Table 3. Inhibiting effect of various glycosides on the accumulation of TMG

Suspension of induced bacteria incubated 10 min in the presence of TMG $2 \times 10^{-4} M$	TMG accumulated, % dry bacterial wt.	% inhibition
Without addition	2.5	—
+ mannose $2 \times 10^{-3} M$	2.5	0
+ sucrose $2 \times 10^{-3} M$	2.6	0
+ cellobiose $2 \times 10^{-3} M$	2.4	4
+ melibiose $2 \times 10^{-3} M$	0.06	97
+ lactose $2 \times 10^{-3} M$	—	95[a]
+ phenyl-β-D-thioglucoside $2 \times 10^{-3} M$	—	0[a]
+ phenyl-β-D-thiogalactoside $2 \times 10^{-3} M$	—	95[a]

[a]Only the relative values are given here, since these results were obtained in other series of experiments with different bacterial suspensions.

Competition and Stereospecificity. The accumulation of radioactive TMG is inhibited by phenyl-β-D-thiogalactoside. It is not inhibited by its glucosidic homologue, phenyl-β-D-thioglucoside, which differs from it only in the configuration around the 4-carbon (Table 3). In general, glycosides that have an unsubstituted galactosidic radical, in an α or β linkage, inhibit the system competitively (see Fig. 4), while glycosides that do not have a galactosidic radical produce no effect, or a weak, noncompetitive effect (Table 3). The addition of the competitor *after* TMG results in displacement of the accumulated TMG (see Fig. 2). The system that concentrates TMG thus exhibits a very strict stereospecificity, comparable to that of an enzyme.

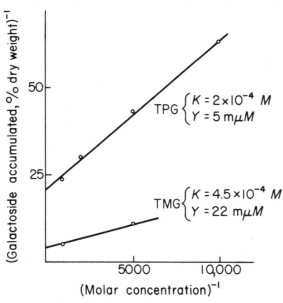

Figure 5. Comparison of accumulation of TMG and thiophenyl-β-D-galactoside (TPG) by the same bacterial suspension (inverse coordinates). The accumulation of the two galactosides is determined independently (we use radioactive TPG labeled with ^{35}S) on several samples of the same bacterial suspension. K and Y are the constants of equation (1) determined in this experiment for each of the two galactosides.

Table 4. Relative specific affinities of various thiogalactosides for galactoside permease (y) and β-galactosidase (z)[a]

Compound	Dissoc. const., molar conc. × 10^{-4}		Relative affinities[b]	
	y	z	y	z
Methyl-β-D-thiogalactoside	4	120	1	1
Phenyl-β-D-thiogalactoside	4	10	2	12
Phenylethyl-β-D-thiogalactoside	2	0.15	2	800
Galactoside-β-D-thiogalactoside	0.16	160	25	0.8

[a]Concerning the galactoside permease, the dissociation constants were determined: for TMG directly, by measuring accumulation at various external concentrations (see p. 97); for the other galactosides, by competitive displacement of TMG (see p. 102 and Fig. 4).

For β-galactosidase, the dissociation constants were determined from competitive inhibition and hydrolysis of o-nitrophenyl-β-D-galactoside. A partially purified enzymatic extract was used, in medium 56, pH 7.0.

The error of dissociation constant determinations is of the order of ±20 percent.
[b]TMG = 1.

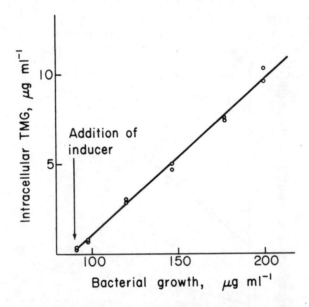

Figure 6. Increase in accumulation capacity with regard to TMG in a growing culture in the presence of TMG. Culture in stage of exponential growth in synthetic medium containing succinate. At point indicated by the arrow, addition of the nonradioactive TMG, 5×10^{-4} M. Growth followed by measurement of optical density, expressed here on the coordinates as mass increase per ml. At suitable intervals, a sample was removed, centrifuged, washed, and resuspended in the presence of radioactive TMG (10^{-3} M) and chloromycetin. Determination of accumulation capacity was made according to the usual technique (p. 97). The coordinates show the accumulation capacity per ml of culture (note that, as on other graphs, specific capacity is recorded).

The galactoside competitors of TMG are themselves accumulated by the induced cells, as was found for thiophenyl-β-D-galactoside (TPG) (Fig. 5) and for o-nitrophenol-β-D-galactoside. The measurement of the competitive effect (displacement of TMG) (see Fig. 4) allows us to evaluate the constant K of equation (1) for galactosides other than TMG (Table 4). In the case of TPG, the constants K and Y were evaluated directly (Fig. 5). We note that for the same bacterial suspension the capacity Y for TPG is smaller than for TMG, while the affinity ($1/K$) is stronger for TPG than for TMG.

Induction. The specific capacity of the system, zero in noninduced bacteria, begins to increase immediately after the addition of TMG to a growing culture, and continues to increase thereafter for several cell generations (Fig. 6). In the course of this increase the value of the dissociation constant K of equation (1) does not vary appreciably, as shown in Fig. 7. The induction phenomenon is thus essentially expressed by an increase in the *specific capacity* Y.

The induction of the system is blocked by chloromycetin, which is known to inhibit protein synthesis without inhibiting nucleic acid synthesis or energy metabolism. The formation of the system is also inhibited by various steric analogues of amino acids, such as thienylalanine, which, without blocking protein synthesis [41], are known to inhibit the formation of certain active proteins, particularly the synthesis of β-galactosidase. Finally, the induced formation of the system does not take place in the absence of exogenous methionone in an *E. coli* mutant that does not synthesize it (Table 5).

The formation of the system is thus linked to protein synthesis. When the inducing concentration of TMG is sufficient (5×10^{-4} M or more), the total capacity increases *linearly* as a function of bacterial growth, for at least one or two divisions,

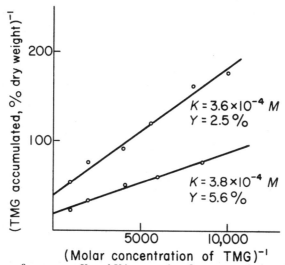

Figure 7. Variation of constants K and Y in equation for accumulation during induction of a culture. A culture in exponential growth phase in a synthetic succinate medium has TMG (2×10^{-4} M) added to it. Samples are removed after about 20 min and 1 hr of growth in the presence of TMG. The bacteria are centrifuged and washed. The accumulation of TMG is determined at several external concentrations of TMG in each of the two samples. We find that while the capacity Y has doubled between the first and second sample, the value of the constant K has not varied significantly.

Table 5. Inhibition of the induced formation of galactoside permease by chloromycetin, β-2-thienylalanine, and by a deficiency in methionine in an *E. coli* mutant auxotrophic for methionine[a]

Additions to the culture	TMG accumulated, % dry wt.
Methionine $10^{-4} M$, TMG $10^{-3} M$	1.8
Without methionine, TMG $10^{-3} M$	0.1
Methionine $10^{-4} M$, chloromycetin 50 g/cm^3, TMG $10^{-3} M$	0.0
Methionine $10^{-4} M$, β-2-thienylalanine $10^{-4} M$, TMG $10^{-3} M$	0.0

[a]*E. coli* (K12M⁻), washed suspension, in medium 56 at 0.2 percent maltose, incubated for 1 hr at 34°C.

from the moment of the addition of the inducer. Thus we can write

$$\Delta Y = p \, \Delta x$$

where x is the total bacterial mass and p is a constant depending on the strain and on the conditions of induction. This relationship, identical with that found for the induced synthesis of galactosidase [40], means that the increase in capacity is linked to the synthesis of a *new* cellular substance, i.e., one whose elements are incorporated after the addition of the inducer [40, 38, 22].

The similarity and the close link between the induction of this system and the induction of β-galactosidase appear clearly when we compare the inducing power of various thiogalactosides, for galactosidase on one hand, and for the system of TMG concentration on the other (see Table 6). With some differences, the significance of which is dubious, the relative inducing activity of the various compounds is the same for the two systems: the alkylthiogalactosides are good inducers, while the arylthiogalactosides induce weakly or not at all. Galactoside-β-D-thiogalactoside is also

Table 6. Induction of galactoside permease and of β-galactoside by various thiogalactosides[a]

Inducer added during growth	y TMG accumulated, % dry weight	z β-Galactosidase activity per μg dry bacteria
Without inducer	<0.05	2
Methyl-β-D-thiogalactoside	3.7	8,500
Propyl-β-D-thiogalactoside	3.8	8,800
Isopropyl-β-D-thiogalactoside	2.6	9,500
Hexyl-β-D-thiogalactoside	0.3	8.5
Phenyl-β-D-thiogalactoside	0.1	2
Benzyl-β-D-thiogalactoside	0.2	16
Phenylethyl-β-D-thiogalactoside	1.2	110
Galactoside-β-D-thiogalactoside	0.2	4

[a]Culture in synthetic medium at 0.2 percent succinate, in the presence of $10^{-3} M$ inducer. Accumulation measured in the presence of $10^{-3} M$ TMG.

106

inactive. It should be noted with particular interest that phenylethylthiogalactoside, whose aglycone is both aromatic and aliphatic, exhibits a weak, although significant, inducing activity for β-galactosidase and for the concentration system.

It should be added that the induced synthesis of the system is totally inhibited by glucose (10^{-3} M). We know that this inhibitory effect of glucose is encountered in the induction of a great many enzymes, one of which is β-galactosidase [33, 17, 8, 9].

Kinetic Scheme and Nature of the System. Let us see how we can understand the functioning of the system. Its strict specificity imposes the first conclusion: since only galactosides can expel the TMG and saturate the system, the latter must contain stereospecific acceptors, forming a reversible complex with the galactosides. The capacity Y depends on the number (or activity) of these acceptors. Since the value of Y increases in the course of induction, it must be expressed as an increase in the number or activity of the acceptors. Since this increase is induced under the same conditions as galactosidase synthesis, since it is proportional to the synthesis of new protein, and since it is blocked by inhibitors of protein synthesis, it is most probable that the induced formation of the system consists of the synthesis of a specific "acceptor" protein.

With this in mind, we must consider the role of the acceptor protein (which we shall designate by the symbol l) in the accumulation phenomenon. The simplest hypothesis for this role would be that of a "stoichiometric receptor," with the "internal" TMG presumably being adsorbed by these specific acceptors. This model would explain the main properties of the system (saturation, equilibrium, specificity, reversibility), but it cannot be retained for the following reasons:

1. In fully induced bacteria, the TMG at saturation corresponds to over 5 percent of the bacterial dry weight. If we assume that TMG is adsorbed on specific protein acceptors would mean, on the average, an acceptor for each bacterial protein fraction of molecular weight 2000, a clearly absurd conclusion.

2. According to the stoichiometric model, the capacity expressed in moles of galactoside adsorbed per unit bacterial mass would, for a given suspension, have to be the same for different galactosides. This is far from being the case for TMG and TPG (Fig. 7); although TPG can completely displace TMG, at saturation the cells concentrate 5 times less TPG than TMG.

The stoichiometric hypothesis is thus not likely, and we must assume that the "acceptor" protein plays a *catalytic* part in reaction (A). In order to explain the properties of the system, particularly saturation and competitive displacement, the following diagram is suggested (catalytic model):

$$G_{ex} + y \overset{\text{entry}}{\underset{k_2}{\overset{k_1}{\rightleftharpoons}}} Gy \xrightarrow{k_3} y + G_{in} \xrightarrow[k_4]{\text{exit}} G_{ex} \tag{B}$$

according to which acceptors y play the role of an enzyme catalyzing the entry reaction, while the exit takes place by an independent process, i.e., a process not involving the acceptors.

By assuming the Henri–Michaelis kinetics for the entry reaction and a rate proportional to G_{in} of intracellular TMG for the exit reaction, we can write

$$\frac{dG_{in}}{dt} = k_3 y \frac{G_{ex}}{G_{ex} \times k_2/k_1} - k_4 G_{in}.$$

By making

$$\frac{k_3}{k_4} y = Y \text{ and} \frac{k_2}{k_1} = K$$

we have, for the equilibrium state, equation (1), which, as we have seen, expresses the experimental properties of the system.

This model obviously implies that the passive penetration and exit of TMG are extremely slow compared to the rate of the reaction catalyzed by y, without which there could be no accumulation. Thus we believe that protein y is associated in some way with the cellular membrane, or at least with the osmotic barrier that functionally separates the exterior and the interior of the cell.

It can be clearly seen that the proposed model takes into account not only equilibrium but also the displacement of radioactive TMG by its analogues or by metabolic inhibitors; this displacement is the result of inhibition of the entry reaction, without inhibition of the exit reaction. The model explains why for the same suspension the capacities are not the same for different galactosides, since the rate constants k_3 and k_4 will be different in absolute and relative values for different substances. Finally, it predicts that the number of intracellular molecules of TMG at equilibrium will be proportional *but not equal* to the number of specific acceptors.

To the extent that the constants k_3 and k_4 of the entry and exit reactions remain invariant, the determination of capacity Y is a measure of the quantity of acceptor protein. The linear relationship found between the increase in intracellular TMG and the growth of bacterial mass in the presence of inducer shows that this hypothesis is valid within wide limits. We must, however, keep in mind that the amount of TMG accumulated at equilibrium results not only from the rate of the entry reaction catalyzed by system y, but also from the rate of the exit reaction. It seems most probable that the rate of this latter reaction expresses the *passive* permeability of the osmotic cellular barrier, a permeability that could vary, depending on certain physiological factors. Whatever the case, all the results indicate that under our experimental conditions the rate constant of this reaction does not vary significantly for a given substance.

It is possible that besides the acceptor protein y the entry reaction involves other constituents (coenzymes, carriers) responsible in particular for the coupling of the system with reactions donating free energy. The data in our possession, although they bring to light the existence, the catalytic role, and the specific properties of the protein, do not allow any useful discussion of the mechanism of the reaction catalyzed.

In conclusion, we attribute the reversible accumulation of galactosides in induced bacteria to cellular impermeability, compensated by the activity of a catalytic system, protein in nature, and with the kinetic properties, specificity, and inducibility of an enzyme.

In order to stress simultaneously the functional uniqueness of this system and its similarity to an enzyme system proper, we propose to designate its specific constituent (y) by the name *galactoside permease*. In the second part of this report, we shall consider the functional and genetic relationships between the galactoside permease and the β-galactosidase. We shall see that these data fully confirm our interpretation by demonstrating both the independence of the permease from β-galactosidase and the functional association between the two systems.

Functional Relationships between Galactoside Permease and β-Galactosidase

The conclusions we have just summarized suggest that the in vivo metabolism of a galactoside in *E. coli* normally involves the intervention of permease (y) *before* hydrolysis of the substrate by the β-galactosidase (z). This implies the following sequence:

$$\text{galactose-}R \xrightarrow{(y)} \text{galactose-}R \xrightarrow{(z)} \text{galactose} + R$$
$$\text{(external)} \qquad \text{(internal)}$$

This scheme assumes: (a) that the permease and hydrolase are two distinct systems; and (b) that the galactosidase is "internal" with respect to the permease.

We know that various mutants characterized by inability to metabolize lactose have been isolated in *E. coli* (strains K12 and ML) [28, 37, 7]. If the above scheme

Table 7. In vivo hydrolysis of orthonitrophenyl-β-D-galactoside by normal bacteria and by bacteria deficient in galactoside permease[a]

Strain	Physiological type[b]	Genotype (K12)	Galactoside permease		β-galactosidase		Elementary phenotype
			induced	non-induced	induced	non-induced	
ML30	Normal	—					
K12		wild	+	0	+	0	$y^+z^+i^+$
ML3	Crypt. neg.	—					
K12W2241		Lac⁻₁	trace	0	+	0	$y^-z^+i^+$
K12W2244	Abs. neg.	Lac⁻₄					
K12W2242		Lac⁻₂	+	0	0	0	$y^+z^-i^+$
ML308	Constit. normal	—					
K12S4		Lac⁻ᵢ	+	+	+	+	$y^+z^+i^-$
ML35	Constit. crypt.	—					
K12W13011		Lac⁻ᵢ; Lac⁻ᵢ	0	0	+	+	$y^-z^+i^-$
ML3088	Abs. neg.	—	+	+	0	0	$y^+z^-i^-$
ML3080	Abs. neg.	—	0	0	0	0	y^-z^- ?
K12W2247	—	Lac⁻₇	trace	0	trace	0	?

[a]Galactosidase and galactoside permease tests were performed on each strain after growth in a medium with 0.2 percent maltose in the presence (induced bacteria) or in absence (noninduced bacteria) of TMG ($10^{-3} M$). The + sign indicates an equal or higher activity than the 20 percent activity of normal induced bacteria. The sign 0 indicates an activity lower than 1 percent. "Trace" indicates an activity of about 1–5 percent.
[b]Galactoside metabolism.

and the assumptions it implies are correct, there should in principle exist among these mutants at least two different biochemical types, one corresponding to the loss of permease and the other to the loss of β-galactosidase. The study of a series of lactose-negative mutants shows that these two predicted types really do exist (Table 7), and that there are also several other types which we shall soon consider.

As can be seen, in the presence of TMG some mutants ("absolute negatives" K12W2244 and K12W2242) synthesize normal amounts of galactoside permease but no detectable traces of galactosidase. It is possible using these organisms to determine the intracellular concentration of true galactosides, such as ONPG or lactose, while it is impossible to do this with wild-type bacteria, since the substrate is hydrolyzed as it is concentrated.

The second type of mutant ("cryptic"; ML3 and K12W2241) is of particular interest. In the presence of sufficiently high concentrations of TMG these organisms synthesize only traces of galactoside permease but normal amounts of β-galactosidase. These bacteria, as long as they are physiologically intact, do not hydrolyze ONPG and do not metabolize lactose except very slowly (as compared with a wild strain having the same β-galactosidasic activity) (Table 8).

These properties of cryptic mutants, which up to now have seemed paradoxical, can be explained easily if β-galactosidase can be found effectively *inside* an osmotic cellular enclosure which is highly impermeable (to lactose, and even more to ONPG) and whose permeability to galactosides is ensured by the permease, which thus governs the penetration and the metabolism of galactosides.

The existence of these two types of lactose-negative mutants, in addition to their properties, thus entirely confirms the hypotheses concerning the role of permease and proves that this system is distinct from the β-galactosidase.

Study of the hydrolysis of galactosides by intact cells of wild-type bacteria also permits us to domonstrate the functional role of permease.

Table 8. Mutations affecting the β-galactosidase and galactoside permease in *E. coli* (strains ML and K12)

Induced bacteria[a]	Galactoside permease, TMG accumulated, % dry weight	β-Galactosidase, ONPG hydrolyzed, nmole min^{-1} mg^{-1}	In vivo hydrolysis, ONPG hydrolyzed, nmole min^{-1} mg^{-1}	Crypticity factor[b]
Normal (ML30)	3.1	7200	255	28
Cryptic (ML3)	>0.1	6700	25	270

[a]Bacteria cultivated in the presence of $10^{-3}\,M$ TMG, washed, resuspended in medium 56. Activities of β-galactosidase and galactoside permease determined under normal conditions (p. 97). Hydrolysis in vivo measured in the presence of $M/400$ ONPG and 0.2 percent succinate in agitated medium.
[b]Note that the conditions of in vivo hydrolysis are not strictly comparable to the conditions of the determination of the β-galactosidase activity (presence of Na$^+$ ions in the second case, but not in the first), which can affect the absolute value of the crypticity factor (ratio of in vitro hydrolysis, column 2, to in vivo hydrolysis, column 3), but still permits comparison of two strains.

Let us remember, first, that the hydrolytic activity of intact bacteria with respect to o-nitrophenyl-β-D-galactoside is lower than the activity of extractable β-galactosidase of the same bacteria [14, 26, 45, 43]. This indicates the isolation of β-galactosidase with respect to the external medium. The experiment shown in Fig. 8 indicates that in induced wild-type bacteria, the activity of permease, rather than that of β-galactosidase, limits hydrolytic activity. In fact, we can see that in induced bacteria incubated at 34°C for several hours, the activity of permease drops sharply, while that of β-galactosidase remains constant. However, the hydrolytic activity of intact bacteria diminishes together with the loss of permease activity.

Another example demonstrates certain differences in the specific affinity of permease and hydrolase. We can see (Table 4) that, even if only relative values are considered, the list of affinities of the two systems for a series of thiogalactosides show very marked and characteristic differences. The galactosido-β-D-thiogalactoside (TDG) shows a weak affinity for the β-galactosidase, similar to that of the TMG. On the other hand, the affinity of TDG for permease is much higher than that of TMG. The hydrolysis of galactosides in vivo by induced bacteria is inhibited much more strongly by TDG than by TMG. This can be easily explained if permease governs penetration of galactosides in vivo, and, consequently, governs the functioning of the β-galactosidase.

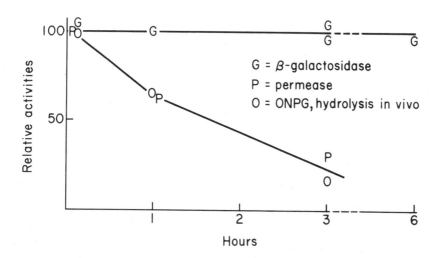

Figure 8. Relative variation of the activity of β-galactosidase, galactoside permease, and the hydrolysis of ONPG in vivo in a culture deficient in carbon source. A culture of *E. coli* ML308 (constitutive mutant) in exponential growth on maltose is centrifuged, resuspended in medium 56 without maltose, and stirred at 34°C. At 0, 1, 3, and 6 hr, the activity of β-galactosidase is determined (toluene added to suspension, see p. 97), as well as that of the galactoside permease (accumulation of TMG, see p. 97), and the rate of hydrolysis in vivo of o-nitrophenyl-β-D-galactoside (ONPG) (see p. 97). For the sake of comparison the value 100 has been attributed to the initial activity at time 0. (For activity of β-galactosidase the value 100 corresponds to the average of all determinations.) We can see that the rate of hydrolysis in vivo diminishes in direct proportion to the loss of permease activity, while the variations in β-galactosidase activity are not significant, even after 6 hr.

We shall briefly mention a very important aspect of the relationship between galactoside permease and β-galactosidase, namely, the role of permease in the induction of β-galactosidase. This problem is discussed at length elsewhere [34, 8]. We shall emphasize only two essential points.

1. The concentration of TMG needed to saturate the inducer system of cryptic mutants Lac⁻$_1$ (low in permease) is about 100 times higher than the concentration sufficient to saturate normal bacteria. Thus it is clear that in normal bacteria it is the TMG concentrated by the permease which induces the β-galactosidase. In the cryptics, the induction could be due either to the passive permeation of TMG or to traces of permease activity.

2. The inducibility (i.e., the differential rate of the β-galactosidase synthesis), at a low concentration of TMG (10^{-5} M), is much higher in the already induced bacteria than in noninduced bacteria. This may be related to the fact that induced bacteria, with a strong permease activity, achieve a high intracellular concentration of the inducer, which is not achieved by noninduced bacteria [34, 8].

The properties of lactose-negative mutants, like those of normal bacteria show, therefore, that β-galactosidase and galactoside permease are different but closely related functionally. We shall now see that the two systems are also very closely related genetically.

Genetic Relationships of β-Galactosidase and Galactoside Permease

These relationships are of two kinds:

(a) On one hand, certain mutations seem to affect the two systems simultaneously.

(b) On the other hand, several mutations that affect the two systems either *separately* or *equally* form a closely linked group in *E. coli* K12.

We know that mutants that are distinguished from the wild type in that their synthesis of β-galactosidase is *constitutive* and no longer inducible have been isolated in *E. coli* ML and K12 [27, 7]. Furthermore, it can be seen from Table 7 that in these organisms (ML308, K12S4) not only β-galactosidase but also permease is formed in the absence of an external inducer. This poses the question as to whether these organisms are double mutants in which the different determinants, corresponding respectively to the permease and to the galactosidase, would each have mutated *independently* toward a "constitutive" allele, or whether we are dealing with a single mutation, affecting simultaneously the inducibility of two enzymes. The second hypothesis appears to be correct. In fact, according to the first hypothesis there could be strains inducible for permease and constitutive for galactosidase and vice versa. No such strains were found. True, the methods of selection used to isolate constitutive mutants are only effective when the two systems are constitutive *at the same time*. In order to eliminate this ambiguity, we have isolated from the wild type a certain number of spontaneous mutants constitutive for β-galactosidase, using the direct technique of Cohen-Bazire and Jolit [7], without the intervention of any method of selection. In the four mutants isolated in this manner the permease was also constitutive; thus the constitutive mutation affects the two systems simultaneously.

We reach the same conclusion by the following observation: the lactose-negative strain ML3, inducible for galactosidase and cryptic, i.e., low in permease, gives a spontaneous and relatively frequent lactose-positive mutation: this is the classic mutation of *E. coli mutabile* [32, 31, 35, 46]. This mutation is expressed here by the acquisition of an *inducible* permease (normal *E. coli* type). From strain ML3 we have directly isolated a mutant constitutive for galactosidase (ML35). These organisms are still low in permease, hence cryptic, and form large quantities of galactosidase, but do not grow in the lactose media. This "constitutive cryptic" strain, however, gives lactose-positive mutants having permease with the same frequency as does the ML3 strain; but in these mutants the permease is invariably *constitutive*. The constitutive nature of the permease in these mutants is evidently related to the *previous* mutation, which, in this first stage, was only manifested for galactosidase, the permease being absent.

All in all, the mutations involved can be classified into three elementary types, and defined and symbolized as follows:

(a) Capacity (y^+) or incapacity (y^-) to synthesize permease;

(b) Capacity (z^+) or incapacity (z^-) to synthesize β-galactosidase;

(c) Inducible (i^+) or constitutive (i^-) character for the synthesis of the two systems. This last phenotype can only be manifested when at least one of the preceding is positive.

In the last column of Table 7, we noted the complex phenotype corresponding to each strain. We can see that all the combinations predicted by the above hypothesis are represented.

Let us see now to what extent these different elementary mutant types are associated with definite loci, allelic or not. We shall do this by using the results of J. and E. Lederberg et al. [28, 25], as well as those of Cohen-Bazire and Jolit [7], concerning the allelism relationships of different lactose-negative and "constitutive" mutants of *E. coli* K12, and we shall interpret the data of Table 7 in this light.

The following conclusions result from this comparison:

(a) The mutations expressed by the loss of permease alone ($y^+ \rightarrow y^-$) have taken place at the locus Lac_1. It does not seem that any mutation involving galactosidase took place in this area.

(b) The mutations expressed by the loss of galactosidase alone ($z^+ \rightarrow z^-$), are associated with at least two different areas (Lac_2 and Lac_4), one of which (Lac_4) is *very closely linked* to Lac_1.

(c) The constitutive mutation ($i^+ \rightarrow i^-$) must be connected to an area (Lac_i), distinct from Lac_1 (and probably also from Lac_4) but *very closely linked* to Lac_1.

It should be noted that a mutation distinct from the preceding ones (Lac_7^-), not closely linked to Lac_1, seems to have an effect on the two systems, annulling the permease and reducing considerably the capacity for galactosidase synthesis.

It can be seen that the situation is not simple, since in at least some cases (Lac_2 and Lac_4) mutations which are little related and not in allelic areas are expressed by phenotypes that cannot be distinguished [28]. However, an important fact emerges: the existence of a complex area (Lac$_1$, Lac$_1$, Lac$_4$) where closely adjacent mutations control, either selectively or simultaneously, the inducibility and synthesis of each of these two systems, which as we have previously seen are closely associated functionally. This situation is reminiscent of one revealed by observations of Hartman et al. [19] concerning mutations affecting the capacity for synthesis of certain metabolites in *E. coli*, as well as by comparable observations of Pontecorvo and co-workers [42] with *Aspergillus*.

Discussion and Conclusions

Galactoside Permease

The main observations and the conclusions resulting from them may be summarized as follows.

There is a system in *E. coli* (galactoside permease) which has the property of accumulating galactosides inside the cell. This system has the kinetic properties and steric specificity of an enzyme. Its functioning is coupled to reactions that give metabolic energy. Normally, in wild-type strains, permease is inducible. Its formation, inhibited by agents which inhibit synthesis of proteins, takes place under the same conditions, with the same kinetic characteristics, and in the presence of the same specific inducers, as the synthesis of β-galactosidase. *Distinct* mutations affect the capacity to synthesize either β-galactosidase or permease. The mutants without galactosidase accumulate galactosides but do not hydrolyze them. In the mutants without permease, the galactosidase is practically inactive in vivo (we say that it is cryptic); this indicates that in living intact bacteria the osmotic cellular enclosure is impermeable to glycosides; the galactosidase is inside this enclosure, so that the galactoside permease governs the in vivo accessibility of the β-galactosidase, and consequently governs the metabolism of galactosides as well as the induction of the β-galactosidase. These functional relationships between galactoside permease and the β-galactosidase are mirrored by close genetic relationships; several loci governing the synthesis and inducibility of galactosidase and permease, respectively, are closely linked in a complex locus in *E. coli* K12.

Other Permease Systems

The hypothesis that the use of exogenous substrates by the microorganisms is determined, in the first place, by the specific permeability properties of the cellular membrane has often been advanced by microbiologists. In view of the lack of adequate experimental data, it has generally caused more embarrassment and confusion than interest, and its validity has often been denied [44]. However, very recently it has been revived and defended with vigor and precision by B. Davis [13]. In contrast, much research over many years concerning the phenomena of adsorption and secretion in the tissues and cells of higher animals made it difficult to doubt the selectivity of certain cellular mechanisms. Rosenberg and Wilbrandt [48] have shown that the modes of transport of various organic, nonionic substances across the

114

renal epithelium and the intestine are difficult to explain without this hypothesis. The *selective* permeability of human erythrocytes with respect to certain glycosides has been thoroughly established [48, 24, 29]. Since 1934 Danielli [10–12] has developed and illustrated the hypothesis that membranes in both animal and vegetable cells contain, as a rule, a lipoid layer or constitutent, and are, because of this fact, only very slightly permeable to polar substances such as glycosides. The rapid and selective permeation of certain glycosides into certain cells, such as human erythrocytes, would be due, according to Danielli, to specific and specialized constituents, which he believed to be proteins and for which he hypothesized various possible mechanisms of action.

To go back to microorganisms, it now seems certain that specific systems analogous to galactoside permease govern the permeation of many substrates in *E. coli*, as well as in other microorganisms. This generalization is justified by the numerous observations made.

First, it is easy to see that the properties which we have had to assign to galactoside permease imply that other *specific* systems ensure the transfer of *nongalactosidic* glycosides which *E. coli* can metabolize at a rate equal to or faster than lactose. If, in fact, the osmotic cellular barrier is impermeable to lactose, it must also be equally impermeable to, e.g., maltose, which, however, is metabolized rapidly by *E. coli*. But this rapid permeation of maltose cannot be the effect of a nonspecific transfer system, since such a system would allow lactose to penetrate as well, and thus cells which can metabolize maltose rapidly could not be cryptic with respect to lactose. If we do not want to assume that *different* osmotic enclosures enclose different intracellular enzymes, we must assume that the in vivo metabolism of maltose in *E. coli* implies a specific transfer system, or, at any rate, one specific enough to be inactive toward lactose.

Furthermore, we know that the paradox of specific crypticity has often been encountered in microorganisms, for various substrates. This has raised various speculations, the discussion of which is of no interest now. The example of galactoside permease allows us to assume that specific crypticity is due, as a general rule, to the absence of a specific permease. Various examples of crypticity, involving mainly the disaccharides, have been noted in yeast [30, 21]. In *E. coli*, Doudoroff et al. [16] studied a mutant incapable of metabolizing free glucose; the mutant, however, possessed hexokinase and metabolized free glucose hydrolyzed from maltose by *intracellular* amylomaltase. Emphasizing the paradox, Doudoroff [15] showed that it could not be resolved unless one assumed a *specific* transfer mechanism present in normal bacteria but absent in the mutant.

Recently, Monod, Halvorson, and Jacob [39] resumed the comparative study of this mutant and normal bacteria. They found that the latter have a constitutive system which reversibly accumulates α-methylglucoside (a substance that is not metabolized by *E. coli*). The accumulation is strongly inhibited by glucose, which completely displaces the α-methylglucoside already accumulated. This system, comparable to galactoside permease in its kinetic properties, is totally absent in the glucose-negative mutant, but is restored, along with the capacity to utilize the glucose, by transduction with a phage from the wild-type strain. These results seem to confirm the Doudoroff hypothesis.

In *Pseudomonas*, Kogut and Podoski [23], as well as Barrett et al. [1], have shown that the Krebs cycle is cryptic with regard to exogenous citrate in organisms

115

grown in the absence of citrate, but that this crypticity disappears after a period of growth in the presence of citrate. This "adaptation" was inhibited by uv-radiation [23] or by amino acid analogues [1], which indicated that the disappearance of crypticity was due to the synthesis of an inducible "enzymatic" system. Green and Davis [18], after discovering a very similar situation in *Aerobacter aerogenes*, found that the formation of a decryptifying system was inhibited by glucose, a property characteristic of many inducible proteins, as we have noted in connection with galactoside permease. These authors were able to confirm that the osmotic enclosure impermeable to citrate corresponded closely to the total volume of the cell, which allows us to assume that the decryptifying system is effectively associated with the cellular membrane.

Finally, Cohen and Rickenberg [5, 6] recently discovered in *E. coli* systems which accumulate various exogenous amino acids. Complemented by the observations of Britten et al. [2], their findings suggest that there exists, in fact, a *distinct* system for each amino acid or type of natural amino acid. In all their properties (equilibrium, saturation, specificity, sensitivity to metabolic inhibitors) these systems are very similar to galactoside permease. Isotopic competition experiments show immediately that these systems control the entry of exogenous amino acids into metabolism, and hence their incorporation into proteins, but that they are not involved in the metabolism and incorporation of endogenous amino acids. These "amino acid permeases" of *E. coli* are not inducible, but Cohen [3] was able to show that their formation is related to protein synthesis.

Definition of Permeases

Permeases and Active Transport. Comparison of these observations seems to justify the hypothesis that in microorganisms the cellular membrane, or at least the osmotic barrier which delineates the internal metabolic space, is in general only slightly permeable to water-soluble organic substances, and that the penetration of exogenous organic substrates is ensured mainly by specific systems analogous to that which concentrates the galactosides in *E. coli*. In order to designate these systems we propose the generic term "permease." But the word and the idea will not be useful unless they are used in a limiting fashion. We shall define a permease as a system of proteinaceous nature which ensures the catalytic transfer of a substrate through a cellular osmotic barrier, and which has the steric specificity properties and the kinetic activity of an enzyme, but is distinct and independent from enzymes ensuring the actual metabolism of the substrate. This definition does not prejudge the mechanism of action of permeases, but it implies two important hypotheses:

(a) That transfer by a permease includes the transitory formation of a *specific complex* between the protein of the permease and the substrate:

(b) That the permease is a *functionally specialized* system and is not involved in the intracellular metabolism itself.

We deliberately did not include in this definition the condition that permease should catalyze an "active transport," i.e., a transfer reaction against a concentration or activity gradient. In fact, even in the cases of galactoside permease and of amino acid permease in *E. coli,* we cannot affirm that the "concentrated" substrates are really completely "free," and that they are in solution in the internal medium and in the same phase in the external medium. Thus we cannot speak with certainty of a concentration gradient. This applies a fortiori to the citrate permease system, which is detectable only through "decryptification." It is possible a priori that the activity of certain permease systems is not expressed by a perceptible accumulation of substrate, even when the latter is not metabolized. Conversely, the active accumulation of a substance could obviously take place by mechanisms that do not imply permeases, as they are defined above. It must, therefore, be clear that permease does not necessarily presuppose active transfer, or vice versa.

Permease and Enzyme. The study of only in vivo activity obviously does not permit the unambiguous identification of an enzymatic system. However, in the case of galactoside permease several distinct criteria, including specific affinity, specific inducibility, and specific mutations, lead to converging definitions and pinpoint this system very clearly. But these criteria only demonstrate the existence of a specific constituent of the system and give no indication of its mode of action. Thus there is no point at this time in discussing possible mechanisms of action or in speculating on whether or not permease is in fact an enzyme. On the enzyme side, we already know that permease has the required steric specificity, kinetic action, and inducibility. We know that certain mutations affect it at the same time and in the same way as they do β-galactosidase. We have indirect evidence of its protein nature. In order for it to be an enzyme in the accepted meaning of the word, permease must also catalyze a substrate reaction, i.e., the formation or rupture of a covalent bond. This is quite possible and even probable, but not certain, and mechanisms for its action can be imagined which do not imply the formation or rupture of covalent bonds involving the substrate [11]. This alternative will be the first possibility considered in a study of the mechanism of action of the permease.

Summary

There exists in *E. coli* a system (galactoside permease), the activity of which is expressed by the intracellular accumulation of exogenous galactosides. The formation of this system is specifically induced by certain galactosides; it is linked to protein synthesis. The activity of galactoside permease follows the Michaelis–Henri law and is inhibited by sodium azide and 2,4-dinitrophenol. Galactoside permease controls the penetration of galactosides into the cells and their hydroloysis by β-galactosidase. Specific and distinct mutations affect the capacity to synthesize galactoside permease and β-galactosidase, respectively. A single mutation determines the inducible versus constitutive character of both systems.

On the basis of these and of many other observations, we propose the hypothesis that intracellular penetration of organic substrates (and especially of highly polar substances) is, in general, catalyzed in microorganisms by specific permeases similar to galactoside permease.

Acknowledgments

We wish to thank Prof. Helferich, Director of the Chemical Institute at the University of Bonn, and Mr. D. Turk, who have studied and achieved the synthesis of many thiogalactosides. We also express our gratitude to Prof. J. Lederberg of the University of Wisconsin, who made available to us a number of mutant strains of *E. coli* K12.

References

1. Barrett, J. J., A. D. Larson, and R. E. Kallio (1953). *J. Bacteriol.*, **65**, 187–192.
2. Britten, R. J., R. B. Roberts, and E. F. French (1955). *Proc. Natl. Acad. Sci.*, **41**, 863.
3. Cohen, G. N. Unpublished results.
4. Cohen, G. N., and H. V. Rickenberg (1955). *C.R. Acad. Sci. (Paris)*, **240**, 466–468.
5. Cohen, G. N., and H. V. Rickenberg (1955). *C.R. Acad. Sci. (Paris)*, **240**, 2086–2088.
6. Cohen, G. N., and H. V. Rickenberg (1956). *Ann. Inst. Pasteur*, **91**, 693.
7. Cohen-Bazire, G., and M. Jolit (1953). *Ann. Inst. Pasteur*, **84**, 937–945.
8. Cohn, M. (1956). *Enzymes: Units of Biological Structure and Function* (Henry Ford Hospital Intern. Symp.). Academic Press, New York, pp. 41–46.
9. Cohn, M., and J. Monod (1953). *Adaptation in Microorganisms*. Cambridge University Press., pp. 132–149.
10. Danielli, J. F. (1952). Structural aspects of cell physiology. *Symp. Soc. Exptl. Biol.*, **6**, 1–15.
11. Danielli, J. F. (1954). Active transport and secretion. *Symp. Soc. Exptl. Biol.*, **8**, 502–516.
12. Danielli, J. F. (1954). *Colston Papers* (7th Symp. Colston Res. Soc.). Butterworths, London, 14 pp.
13. Davis, B. D. (1956). *Enzymes: Units of Biological Structure and Function* (Henry Ford Hospital Intern. Symp.). Academic Press, New York, pp. 509–522.
14. Deere, C. J., A. D. Dulaney, and I. D. Michelson (1939). *J. Bacteriol.*, **37**, 355.
15. Doudoroff, M. (1951). *Phosphorus Metabolism*, **1**, 42–48.
16. Doudoroff, M., W. Z. Hassid, E. W. Putnam, A. L. Potter, and J. Lederberg (1949). *J. Biol. Chem.*, **179**, 921–933.
17. Gale, E. F. (1943). *Bacteriol Revs.*, **7**, 139–173.
18. Green, H., and B. D. Davis. In B. D. Davis, Henry Ford Hospital Intern. Symp. (Cf. [13].)
19. Hartman, P. E., et al. In M. Demerec, Henry Ford Hospital Intern. Symp. (Cf. [13], pp. 131–134.)
20. Herzenberg, L. In preparation.
21. Hestrin, S., and C. C. Lindegren (1950). *Arch. Biochem.*, **29**, 315–333.
22. Hogness, D. S., M. Cohn, and J. Monod (1955). *Biochim. Biophys. Acta*, **16**, 99–116.
23. Kogut, M., and E. P. Podoski (1953). *Biochem. J.*, **55**, 800–811.
24. Kozawa (1914). *Biochem. Z.*, **60**, 231.
25. Lederberg, E. M. (1952). *Genetics*, **37**, 469–483.
26. Lederberg, J. (1950). *J. Bacteriol*, **60**, 381–392.
27. Lederberg, J. (1951). In *Genetics in the 20th Century*. Macmillan, New York, pp. 263–289.
28. Lederberg, J., E. M. Lederberg, N. D. Zinder, and E. R. Lively (1951). *Cold Spring Harbor Symp. on Quantitative Biol.*, **16**, 413–441.
29. LeFevre, P. G. (1954). *Active Transport and Secretion* (Symp. Soc. Exptl. Biol.), Academic Press, New York, pp. 118–135.
30. Leibowitz, J., and S. Hestrin (1945). *Advan. Enzymol.*, **5**, 87–127.
31. Lewis, I. M. (1934). *J. Bacteriol*, **28**, 619.

32. Massini, R. (1907). *Arch. Ilyg.*, **61**, 250–292.
33. Monod, J. (1942). *Studies on the Growth of Bacterial Cultures.* Hermann, Paris.
34. Monod, J. (1956). *Enzymes: Units of Biological Structure and Function* (Henry Ford Hospital Intern. Symp.). Academic Press, New York, pp. 7–28.
35. Monod, J., and A. Audureau (1946). *Ann. Inst. Pasteur,* **72**, 868–878.
36. Monod, J., G. Cohen-Bazire, and M. Cohn (1951). *Biochim. Biophys. Acta,* **7**, 585–599.
37. Monod, J., and M. Cohn (1953). *Advan. Enzymol.,* **13**, 67–119.
38. Monod, J., and M. Cohn (1953). *Congress of Microbiology Symposium Bacterial Metabolism, Rome,* pp. 42–62.
39. Monod, J., H. O. Halvorson, and F. Jacob. Unpublished results.
40. Monod, J., A. M. Pappenheimer, Jr., and G. Cohen-Bazire (1952). *Biochim. Biophys. Acta,* **9**, 647–660.
41. Munier, R., and G. N. Cohen (1956). *Biochim. Biophys. Acta,* **21**, 592–593.
42. Pontecorvo, G., L. M. Roper, K. D. Hemmons, A. W. MacDonald, and A. W. J. Buffon (1950). *Advan. Genetics,* **3**, 73–115.
43. Rickenberg, H. V. (1954). Ph.D. Thesis, Yale University.
44. Roberts, R. B., P. H. Abelson, D. B. Cowie, E. T. Bolton, and R. J. Britten (1955). *Studies of Biosynthesis in* Escherichia coli. Carnegie Institute, Washington, D.C., 521 pp.
45. Rotman, B. (1955). *Bacteriol Proc.,* 133.
46. Ryan, F. J. (1952). *J. Gen. Microbiol.,* **7**, 69–88.
47. Turk, D. (1955). Thesis, Chemisches Institut der Universität, Bonn.
48. Wilbrandt, W. (1954). *Active Transport and Secretion* (Symp. Soc. Exptl. Biol.). Academic Press, New York, **8**, 136–162.

119

Reprinted from *Bact. Rev.*, **21**, 169–194 (1957)

BACTERIAL PERMEASES [1]

8

GEORGES N. COHEN AND JACQUES MONOD

Service de Biochimie Cellulaire, Institut Pasteur, Paris, France

I. INTRODUCTION

The selective permeation of certain molecular species across certain tissues, or into certain cells, has been recognized for a long time as a phenomenon of fundamental importance in animal physiology. The situation is, or was up to quite recently, different in the field of microbiology. Although the importance of recognizing and studying selective permeability effects had been frequently emphasized, particularly in recent years by Doudoroff (22) and by Davis (18), the available evidence appeared ambiguous, and the very concept of selective permeation was looked upon with suspicion by many microbiologists, who believed that, in the absence of direct proof, it served mostly as a verbal "explanation" of certain results.

During the past few years, however, definite proof of the existence, in bacteria, of stereospecific[2] permeation systems, functionally specialized and distinct from metabolic enzymes, has been obtained. It now appears extremely likely that the entry into a given type of bacterial cell of most of the organic nutrilites which it is able to metabolize is, in fact, mediated by such specific permeation systems. None of these systems has been isolated or analyzed into its components. But the stereospecific component of certain of these systems has been indirectly identified as a protein, and defined by a combination of highly characteristic properties. The generic name "permeases" has been suggested for these systems. Although this designation may be criticized, it has the overwhelming advantage that its general meaning and scope are immediately understood.

The object of the present review is to discuss critically the recent evidence from different labo-

ratories concerning a few systems where the properties and identity of the stereospecific component can best be studied and where the physiological significance of "permeases" as connecting links between the intracellular and the external worlds is most clearly in evidence. We wish to emphasize that this is not a review of the literature on osmotic properties of bacteria, or on active transport. We shall be primarily interested in the specificity of permeation, and only secondarily in its thermodynamic aspects. Actually a certain amount of confusion has been entertained in this field because the question of the selectivity (stereospecificity) of permeation processes has not always been clearly distinguished from the problem of energetics of active transport of molecules across cellular membranes. Selective permeation need not necessarily be thermodynamically active. Conversely, active transport may be nonstereospecific. The fact that these two aspects of permeation processes are often, as we shall see, very closely associated renders the distinction even more important.

We shall therefore limit the discussion to the permeation of organic molecules, and exclude the problem of the penetration of inorganic ions such as phosphates about which the excellent review of Mitchell (55) may be consulted.

II. ACCUMULATION, CRYPTICITY, AND SELECTIVE PERMEABILITY

That the entry of organic substrates into bacterial cells may be mediated by more or less selective permeation systems has been suggested primarily by two kinds of observations concerning, respectively: (a) the capacity of certain cells to accumulate internally certain nutrilites; (b) the state of "crypticity" of certain cells toward certain substrates, *i.e.*, their incapacity to metabolize a given substrate, even though they possess the relevant enzyme system.

Let us see why both accumulation and crypticity phenomena were strongly suggestive, yet inconclusive, as evidence of the operation of selective permeation systems.

The classical work of Gale on the uptake of

[1] The work performed in the Service de Biochimie Cellulaire of the Institut Pasteur, has been supported by grants from the Rockefeller Foundation, the Jane Coffin Childs Memorial Fund for Medical Research and the Commissariat à l'Energie Atomique.

[2] A stereospecific system is one whose activity is primarily dependent upon the spacial configuration of the reacting molecules.

amino acids in staphylococcal cells posed the problem of accumulation mechanisms 10 years ago (27, 28). As is well known, Gale and his associates found that staphylococcal cells grown on casein hydrolysate contain large amounts of glutamic acid, lysine, and other amino acids, which could be extracted by water from crushed, but not from intact, cells. These observations appeared to indicate that the cells were very highly impermeable to the amino acids. If this were true, then the entry of the amino acids could not occur by simple diffusion, since simple diffusion is by definition a reversible process: It had to be mediated by some special, unidirectional transfer mechanism. This conclusion was also suggested by the fact that glutamic acid enters the cells only in the presence of glucose. However, lysine, which is accumulated to a similar extent as glutamic acid, and is equally retained by intact cells, does not require glucose for its entry. An alternative mechanism therefore has to be considered, namely that the amino acids are retained within staphylococcal cells by intermolecular forces, for instance by some kind of macromolecular receptors. If so, no permeable barrier, nor any permeation mechanism need be assumed to account for the accumulation (28). As we shall see again later, these two alternative interpretations must both be considered and weighed against each other, whenever attempting to interpret the mechanism of accumulation of a compound by a cell. A choice between them is always difficult; all the more so since they are not mutually exclusive: proving a contribution to the accumulation process by one of these mechanisms does not in itself disprove contribution of the other.

The paradoxical finding that enzymes active against a given substrate may, in some cases, be extracted from cells which, when intact, are inert toward the same substrate, has been noted many times and diversely interpreted by puzzled microbiologists. There is, of course, no paradox when the "cryptic" state of the cells concerns a whole class of chemical compounds, since there is no difficulty in assuming that the solubility and/or electrical properties of a class of compounds may forbid their passage through the cell membrane. The phosphorylated metabolites (nucleotides, hexose phosphates) provide classical examples.

The paradox arises when interpretations in terms of nonspecific forces or properties become inadequate; that is to say, when crypticity is highly *stereospecific*. The study of the metabolism of disaccharides by yeasts has furnished several of the earliest described cases of specific crypticity. For instance, intact baker's yeast does not ferment maltose although autolyzates of the same yeasts contain maltase (α-glucosidase). Analogous observations have been made with other yeasts for cellobiose, and cellobiase (β-glucosidase), sucrose and sucrase, etc. (52). Similarly, Deere *et al.* (19) described, in 1939, a strain of *Escherichia coli* which did not ferment lactose, although lactase (β-galactosidase) was present in dried preparations of these cells.

An essential point is that the cells which are cryptic towards a given carbohydrate, nevertheless behave as a rule quite normally towards other carbohydrates. For instance, an *E. coli* strain which is cryptic towards lactose metabolizes glucose, maltose, and other carbohydrates at a high rate. Now, if crypticity to a particular carbohydrate is attributed to the impermeability of the cell membrane, then the membrane must be impermeable to all compounds presenting similar solubility properties and molecular weight; that is to say virtually all carbohydrates. Therefore, those carbohydrates that do enter the cell and are metabolized at a high rate must be supposed to use some highly specific stratagem for getting through the barrier.

Several particularly striking examples of selective crypticity have been revealed by the studies of Doudoroff *et al.* (22–24). For instance, a mutant of *E. coli* was incapable of metabolizing glucose, although it metabolized maltose via the enzyme amylomaltase (66) which catalyzes the reversible reaction:

$$n(\text{glucose-}\alpha\text{-1-4 glucose}) \xrightleftharpoons{\text{(amylomaltase)}}$$

(maltose)

$$(\text{glucose})_n + n \text{ glucose}$$

(amylose)

Although free glucose is liberated in this reaction, the organisms were found to metabolize quantitatively both moieties of the maltose molecule. Therefore, it appeared that glucose could be used when liberated intracellularly by amylomaltase, while free glucose from the external medium could not be used by these cells. Later observations showed, moreover, that hexokinase could be extracted from these paradoxical organisms. The conclusion that the cells of this mutant strain were impermeable to glucose seemed inescapable. But a membrane impermeable to glucose could not possibly be permeable to maltose, except via a stereospecific permeation system.

This type of interpretation of specific crypticity effects, although quite logical, often appeared arbitrary and unreasonable since, to account for a metabolic paradox concerning a single compound, one had to assume the existence of a multitude of specific permeation systems for which no positive evidence existed, and towards which no direct experimental approach seemed open. An alternative interpretation was therefore often preferred; namely, that where specific crypticity occurred, it was due to a state of inactivity of the intracellular enzyme concerned. The activity was supposed to be released only upon release of the enzyme from the cell (5, 52, 62). This interpretation seemed simpler and more attractive in many respects than the specific permeation hypothesis for which only negative evidence could be adduced.

III. GALACTOSIDE-PERMEASE

The actual demonstration and identification of a specific permeation system, as distinct from other similar systems and from intracellular metabolic enzymes, rests upon a sort of operational isolation *in vivo*, which requires a combination of different experimental approaches. The galactoside-permease system, which we shall now discuss, has offered remarkable opportunities in this respect (9, 60, 71).

Before introducing this system, it should be recalled that *Escherichia coli* metabolizes lactose and other galactosides via the inducible enzyme β-galactosidase. Analogs of β-galactosides where the oxygen atom of the glycosidic linkage is substituted by sulfur (34) are not split by galactosidase, nor are they used by *E. coli* as a source of energy, carbon, or sulfur (61, 38, 7, 60):

CH₂OH

OH — O — S—R
H
OH H
H — H
H OH

(R-β-D-thiogalactoside)

A. Accumulation of Galactosides in Induced Escherichia coli: Kinetics and Specificity

When a suspension of *E. coli*, previously induced by growth in the presence of a galactoside, is shaken for a few minutes with an S^{35} labeled thiogalactoside, and the cells are rapidly separated from the suspending fluid (either by centrifugation or by membrane filtration), they are found to retain an amount of radioactivity corresponding to an intracellular concentration of galactoside which may exceed by 100-fold or more its concentration in the external medium. Noninduced cells (*i.e.*, cells grown in the absence of a galactoside) do not accumulate any significant amounts of radioactivity.

The accumulated radioactivity is quantitatively extracted by boiling water. Chromatographic analysis of extracts shows a major spot, which by all criteria corresponds to the free unchanged thiogalactoside. A minor component (which may consist of an acetylated form of the galactoside) is also evident when accumulation has taken place in the presence of an external source of energy. This compound does not seem to be a product or an intermediate of the accumulation reaction, and its formation may be disregarded in discussing the kinetics of accumulation.

The accumulation is reversible: when the uptake of galactoside is followed as a function of time, a stable maximum is seen to be reached gradually (within 5 to 20 min at 34 C, depending on the galactoside used). If at this point an unlabeled galactoside is added to the medium at a suitable concentration, the radioactivity flows out of the cells (figure 1). The amount of intracellular galactoside at equilibrium, in presence of increasing external concentrations of galactoside,

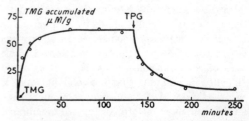

Figure 1. Accumulation of radioactive thiomethyl-β-D-galactoside (TMG) by induced *Escherichia coli* at 0 C (71). At time indicated by arrow, addition of unlabeled thiophenyl-β-D-galactoside (TPG).

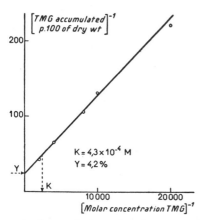

Figure 2. Accumulation of radioactive thio-methyl-β-D-galactoside (TMG) as a function of external concentration (71). Reciprocal coordinates. Y and K are the constants of equation (1).

follows quite accurately an adsorption isotherm (figure 2). Calling the external concentration of galactoside G_{ex}, and the amount taken up by the cells at equilibrium $^{eq}G_{in}$, one may write:

$$^{eq}G_{in} = Y \frac{G_{ex}}{G_{ex} + K} \qquad (1)$$

K is the dissociation constant of the "bacterium-galactoside complex," and Y is another constant, called capacity, which expresses the maximal amount of galactoside which the cells take up at saturating concentration of a given galactoside.[3]

The displacement of a labeled galactoside by another, unlabeled, galactoside also follows quite accurately the classical laws of competition for a common site (figure 3). This allows the determination of specific affinity constants for any competitive compound. The specificity of the system proves very strict: only those compounds which possess an unsubstituted galactosidic residue (in either α or β linkage) (69) present detectable affinity for the competition site. Glucosides or other carbohydrates, even though they may differ from galactosides only by the position of a single hydroxyl, do not compete with the galactosides. Moreover, all the effective competitors which have been tested have proved also to be

[3] The "total capacity" is defined as the capacity per unit volume of cell suspension. The "specific capacity" is the capacity per unit weight of organisms. It may be expressed in per cent dry weight or preferably in moles per unit dry weight.

accumulated within the cells. The affinity constant for each can then be determined either directly, by measurement of accumulation, or indirectly by displacement of another galactoside. The two values agree reasonably well.

The results show that the accumulation of galactosides within induced cells is due to, and limited by, stereospecific sites able to form a reversible complex with α and β galactosides. However, a choice must be made between two entirely different interpretations of the role of these sites.

B. Stoichiometric vs. Catalytic Model

The simplest interpretation (stoichiometric model) would be that the galactosides (G) are accumulated within the cells in stoichiometric combination with specific receptor sites (y), according to an equilibrium:

$$G + y \underset{k_2}{\overset{k_1}{\rightleftharpoons}} (Gy)$$

The constant K of equation (1) would then represent the dissociation constant of the complex Gy, while the constant y would correspond to the total number of available receptor-sites.

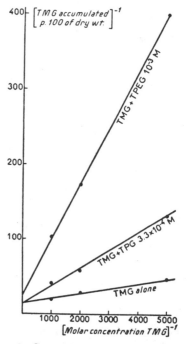

Figure 3. Competitive displacement of thio-methyl-β-D-galactoside (TMG) by thiophenyl-ethyl-β-D-galactoside (TPEG) and thiophenyl-β-D-galactoside (TPG), in *Escherichia coli* (71).

The second interpretation (catalytic or permease model) assigns to the specific "permease" sites the role of *catalyzing* the accumulation of the galactosides into the cell, rather than serving as final acceptors. In order to account for the properties of the system, one is then led to the following scheme:

$$G_{ex} \xrightarrow[entry]{(y)} G_{in} \xrightarrow[exit]{} G_{ex}$$

according to which the intracellular galactoside (G_{in}) is a steady-state intermediate between an *entry* reaction, catalyzed by the stereospecific sites, and an independent *exit* reaction. The *entry* reaction involves the transitory formation of a specific complex between the sites and the galactoside, and should follow the kinetics of enzyme reactions. The *exit* reaction is assumed *not to involve the sites*, and its rate to be proportional to the amount of intracellular galactoside G_{in}.[4] According to these assumptions, the rate of increase of the intracellular galactoside is given by:

$$\frac{dG_{in}}{dt} = y\,\frac{G_{ex}}{G_{ex} + K} - cG_{in} \qquad (2)$$

If $\frac{y}{c} = Y$, equation (2) reduces to equation (1) for equilibrium conditions, when $\frac{dG_{in}}{dt} = 0$. The constant K again corresponds to the dissociation constant of the galactoside-site complex, while the capacity constant Y is now the ratio of the permease activity, y, to the exit rate constant, c. Insofar as the latter remains constant, *the level of intracellular galactoside (G_{in}) at equilibrium is proportional to the activity of the permease.*

In choosing between the catalytic and the stoichiometric interpretations, the first argument to consider is one of common sense: in induced cells, the level of galactoside accumulation may be very high, and actually exceed 5 per cent of the dry weight of the cells. If the intracellular galactoside were adsorbed onto stereospecific sites (presumably associated with cellular proteins), there would have to exist, in highly induced cells, one such site for each fraction of cellular protein of molecular weight 2,000; an assumption which seems quite unreasonable.

The kinetics of intracellular galactoside accumulation also prove incompatible with the stoichiometric model while in good agreement with the assumptions of the permease model. The most significant facts in this respect are the following:

(a) According to the stoichiometric model, the rate of entry of galactosides should be proportional to the number of free sites and to the galactoside concentration. Actually, the initial rate of entry of a galactoside is not significantly faster than its rate of exchange during the steady state, even at saturating concentrations, when few sites remain free. Moreover, at saturating concentrations, the initial rate of entry is *independent* of galactoside concentration (figure 4) (41). Both findings are predicted by the permease model, according to which the initial and steady-state rates of entry should be equal, and proportional to the steady-state level of accumulation G_{in}. Also, in accordance with this expectation, the rate of entry at nonsaturating galactoside concentrations is proportional to the steady-state level. This means that the constant K of equation (1) corresponds effectively to the dissociation constant of the permease while the exit rate is proportional to G_{in}.

(b) According to the stoichiometric model, the capacity constant Y, *i.e.*, the saturation value for intracellular accumulation, corresponds to the number of available specific sites and should be the same, with a given cell suspension, for all galactosides. Actually, the values vary rather widely from one to another compound, the ratio being, for example, 5 to 1 for thiomethyl-galac-

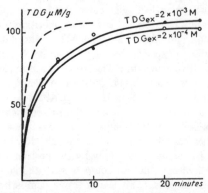

Figure 4. Uptake of radioactive thio-di-β-D-galactoside (TDG) at two different saturating concentrations (41). It is seen that the initial rate of uptake is not significantly different at the two concentrations. The dotted line is the curve of the uptake at the highest concentration as expected on the basis of the stoichiometric model.

[4] In principle, the rate of the exit reaction should probably be considered as proportional to the difference between G_{in} and G_{ex}. In practice, G_{ex} is negligible compared to G_{in}.

toside and thiophenyl-galactoside. This agrees with the catalytic model since both the activity constant of the permease, and the exit rate constant, should be expected to depend on the structure of the galactoside. Morever, the capacity constant (Y) and the affinity constant $(1/K)$ vary independently so that a galactoside endowed with relatively high affinity for the sites (e.g., thio-di-β-D-galactoside, or TDG) may be accumulated, at saturation, to a lesser extent than another one with lower affinity (e.g., thiomethyl-β-D-galactoside, or TMG). Therefore, according to the permease model, at suitable concentrations, more TMG molecules should be "displaced" from a cell by addition of TDG, than the cell takes up TDG molecules. Actually, in one experiment, addition of TDG (10^{-4} M), displaced 75 per cent of the TMG, i.e., about 100 μM/g for an uptake of 15 μM/g of TDG.

Such results are evidently incompatible with the stoichiometric site hypothesis which may be dismissed. The kinetic evidence leaves no doubt that the role of the specific sites must be catalytic.

However, the specific purpose of the catalytic model which we have considered so far is to account for the fact that the steady-state level of intracellular accumulation is proportional to the activity of the permease. The model does not imply any specific assumption regarding the nature of the forces which bind (loosely) the "internal" galactoside to the cell, thereby allowing its accumulation. Again two different interpretations of such a model might be considered. According to one, the cell membrane would be freely permeable to galactosides. The "accumulated" galactoside therefore could not be free. It would be bound loosely to some nondiffusible cell constituent. The *entry* reaction would then consist of the catalytically activated binding of the galactoside to the x constituent, while the *exit* reaction would involve the dissociation of the G-x complex:

$$G + x \xrightarrow{(y)} Gx \rightarrow G + x$$

The second interpretation (permease model, *sensu stricto*) assigns the binding essentially to a high degree of impermeability of the cell membrane (or other osmotic barrier) toward carbohydrates. The permease sites catalyzing the *entry* must then be assumed to be associated with the osmotic barrier itself. The simplest interpretation

of the *exit* reaction then is to consider it as "leakage" through the membrane, increasing in rate as the internal concentration builds up, to the point where it equilibrates the intake.[5]

It should be stressed that the kinetics of galactoside accumulation do not, by themselves, allow a choice between these two different interpretations. The first one is unlikely, however, for the same common sense reasons as the stoichiometric model: the amounts of galactoside accumulated in certain cells are so high that it would be difficult to find enough molecules or groups of any kind to account for the binding. However, the most decisive reason for adopting the second interpretation is the evidence that cells genetically or otherwise devoid of permease are *specifically cryptic toward galactosides*. This evidence will be reviewed later (see page 176).

C. Metabolic and Energy Relationships of the Permease Reaction

Even adopting the permease model as valid, it would be rash to consider the intracellular galactoside as necessarily free and in solution in a phase comparable to the external medium. The physical state of the intracellular galactoside being undetermined, the work involved in the accumulation process is unknown. That the accumulation process must involve work and that the necessary metabolic energy must be channeled via the permease system itself is evident, however, from the fact that the steady-state concentration is proportional to the rate of the permease reaction. For it were supposed that the accumulation process released, rather than consumed, energy or that another system, independent of the permease, channeled the energy for accumulation, then the equilibrium concentration would be independent of permease activity, although the *rate of entry* might remain proportional to it.

This conclusion is confirmed by direct evidence

[5] It may be useful to point out that caution should always be exercised in interpreting differences of intracellular accumulation at equilibrium as due to effects on the entry reaction. It is possible if not probable that certain conditions may affect the exit reaction and thereby alter the equilibrium, by influencing, for example, the properties of the cell membrane. Direct measurements of the rates of entry and exit are required to decide such an issue.

indicating that the accumulation of galactoside by the permease is linked to the metabolic activity of the cell.

In the first place, the accumulation process is inhibited by typical uncoupling agents such as 2,4-dinitrophenol (M/250) or azide (M/50) (9, 71). When these inhibitors are added in the steady state, the intracellular concentration decreases rapidly. An external source of energy is not required, however, although the system is somewhat more active when one is present.

In addition, it should be mentioned that, according to Kepes (40), a small, but significant, increase of respiratory activity occurs when a suitable thiogalactoside is added to cells possessing permease, while noninduced cells or cells genetically devoid of permease show no such increase. The increase is so small that it cannot be detected in the presence of an external source of energy, when the oxygen consumption is too intense. It is only observed as an increase of the endogenous respiration. This extra oxygen consumption is accompanied by an extra CO_2 production. The extra CO_2 produced in the presence of unlabeled thiogalactoside, by cells previously homogeneously labeled with C^{14}, is also labeled, showing that the extra oxidation corresponds to an extra consumption of endogenous reserves, not to an oxidation of the galactoside itself. This extra oxygen consumption could correspond to the work involved in concentrating the galactosides into the cells (40).[6]

There is, at present, no available evidence concerning the mechanism of the energy coupling. Special attention should be called to the following point: while the uncouplers NaN_3 or 2,4-DNP inhibit the *accumulation* of galactosides, they do not inhibit to a comparable extent the *in vivo* hydrolysis of galactosides by intracellular galactosidase. Since, as we shall see later, there is little doubt that the permease limits, *in vivo*, the rate of this hydrolysis, it would seem that the uncouplers do not inhibit the entry of galactosides via the permease, but only the energy coupling which allows the permease reaction to function as a pump, against a concentration gradient. When the concentration gradient is in favor of

---

[6] It remains to be seen whether it may not be linked, in part at least, with the formation of the "minor component" which was mentioned on page 171.

entry, which is so when the intracellular hydrolase splits the substrate as soon as it enters, the uncouplers appear to exert no inhibitory action.

D. The Induced Synthesis of Galactoside-Permease. Permease as Protein

The fact that galactoside-permease is an inducible system has been of particular value for its study and characterization. As we have mentioned, the system is active only in cells previously grown in the presence of a compound possessing a free unsubstituted galactosidic residue. No other carbohydrates show any inductive activity. Not even all galactosides are inducers. The specificity of induction can best be studied using *thio*galactosides which are not hydrolyzed or "transgalactosidated" in the cells. The specificity pattern of induction is strikingly parallel to that of β-galactosidase although there are some minor differences, which may be significant (table 1), in the relative inducing activity of different compounds. Since probably all galactosides are concentrated by the permease, all inducers are also "substrates" of the system. However, several compounds known to be actively concen-

TABLE 1*

Induction of galactoside-permease and β-galactosidase by various thiogalactosides

Inducer Added during Growth	Galactoside-Permease (Specific Activity), μmoles TMG/g	β-Galactosidase (Specific Activity), mμmoles ONPG Hydrolyzed $\times$ min^{-1} $\times$ mg^{-1}
None	<2	2
Methyl-β-D-galactoside	176	8,500
Propyl-β-D-galactoside	181	8,500
Isopropyl-β-D-galactoside	124	9,500
Hexyl-β-D-galactoside	14	8.5
Phenyl-β-D-galactoside	5	2
Benzyl-β-D-galactoside	10	16
Phenyl-ethyl-β-D-galactoside	59	110
Galactoside-β-D-thiogalactoside	10	4

The cultures were made on synthetic medium, in presence of inducer 10^{-3} M. The permease was measured in presence of radioactive 10^{-3} M TMG.

* From H. V. Rickenberg, G. N. Cohen, G. Buttin and J. Monod (71).

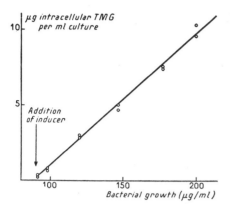

Figure 5. Induced synthesis of galactoside-permease in growing *Escherichia coli* (71). Carbon source: succinate. Inducer: thiomethyl-β-D-galactoside (TMG) 5·10⁻⁴ M. It is seen that the increase in total permease activity is proportional to the increase of bacterial mass from the time of addition of inducer.

trated (phenyl-β-D-thiogalactoside; thio-di-β-D-galactoside) show little or no inducing activity.

The induction is effective only under conditions allowing the synthesis of protein; it is blocked by chloromycetin or in the absence of a required amino acid. Perhaps more significant yet is the fact that the system is not formed in the presence of β-2-thienylalanine. This compound, it should be recalled, does not inhibit the synthesis of protein in *Escherichia coli* but it is incorporated into the proteins formed in its presence which are biologically inactive (67).

Thus, there is little doubt that the induction corresponds in effect to the synthesis of the specific protein component of the system. The kinetics of this induced synthesis follow a remarkably simple law. As measured by the total capacity, the permease increases linearly with the total cell mass, from the time of addition of the inducer (figure 5). One may write:

$$Y = p\Delta x$$

where x is the increase in cell mass after addition of inducer, and p is the differential rate of synthesis (65). This relation, it should be recalled, is typical of inducible enzymes (studied under adequate conditions) and suggests that the increase in capacity, *i.e.*, in permease, corresponds to the *de novo* synthesis of a protein (38, 65, 63). It is of interest to mention that, while the induction results in an increase of capacity, *i.e.*, permease activity, it has no effect on the affinity constant

$(1/K)$. These findings confirm that the interpretation of the two constants is correct.

If it is added that the induction of permease, like that of many enzymes, is blocked by glucose (*cf.* M. Cohn, in this issue), it will be seen that the inductive behavior of this system is in every way similar to that of the most typical inducible enzymes. Taken together with the evidence concerning the kinetics and specificity of accumulation, these findings leave no doubt that a specific, inducible, protein is responsible for the activity of this system. The complete system may, and probably does, involve also noninducible and nonspecific, or less specific, components. The term "permease" should be used primarily to designate the specific, inducible, protein component of the system, while the expression "permease system" implies all the components.

It is not excluded, of course, that the system may comprise a sequence of two (or more) inducible proteins catalyzing successive steps in the *entry* reaction. There is, at present, no necessity for this assumption.

A further characteristic of the permease may be mentioned at this point: it is an SH-dependent system, inhibited by *p*-chloro-mercuribenzoate (*p*CMB). The inhibition is partially reversed by cysteine and glutathione. Substrates, *i.e.*, thiogalactosides, protect against this inhibition in proportion to their affinity, showing that the mercurial acts directly on the specific, galactoside-binding, protein component of the system. (It may be added that β-galactosidase is also rapidly inactivated by *p*CMB, and also protected by galactosides.)

E. Functional Significance of Galactoside-Permease; Specific Crypticity

As we have already stressed, the kinetics of galactoside accumulation do not, by themselves, impose the permease hypothesis. The only assumption which is required to account for the kinetics of accumulation is that the stereospecific sites act catalytically. This could be described, as we indicated, by a model which would not involve any permeability barrier. It is only by studying the relationship of the accumulation system to other systems, in particular, β-galactosidase, that we can decide between the permease and other models. As a matter of fact, one of the problems to consider is whether the permease and the hydrolase really are distinct systems, rather than two functions of the same system.

If the permease model is correct—*i.e.*, if (a) the cells are virtually impermeable to galactosides, except for the specific activity of the permease, (b) the permease is distinct from β-galactosidase, (c) galactosidase is strictly intracellular—one should expect at least two phenotypes among mutants of *E. coli* incapable of metabolizing galactosides, one type corresponding to the loss of the permease, the other to the loss of the galactosidase. Actually, the two predicted types have been found, together with a few others which we shall discuss later, among spontaneous and induced "lactose" mutants of *E. coli*.

Cells of the first mutant type ("absolute-negatives"), grown in presence of a suitable inducer (thiomethyl-galactoside), accumulate normal amounts of galactoside, but they form no detectable trace of β-galactosidase. These organisms will, in particular, accumulate large amounts (up to 20 per cent dry weight) of lactose, while the induced, normal cells do not accumulate lactose, which is split and metabolized as soon as it is taken up.

Cells of the second mutant type (cryptics), grown in presence of sufficiently high concentrations of thiomethyl-galactoside, form normal amounts of galactosidase (as revealed by extrac-

tion), but none or only traces, of permease. So long as they are physiologically intact, these cells hydrolyze galactosides at a much slower rate than normal cells possessing equal amounts of galactosidase. Moreover, the rate of hydrolysis is a linear function of galactoside concentration instead of being hyperbolic, as in the normal type (figure 6) (36). This indicates that the rate of hydrolysis of galactosides in these cells is limited by a diffusion process rather than by a catalyst. The organisms are, in particular, almost inert towards lactose, while their metabolic behavior towards other carbohydrates is normal. In other words, these organisms are specifically cryptic towards galactosides.

These properties of the cryptic mutants, which for a long time had appeared paradoxical, as we recalled (page 170), are immediately explained if β-galactosidase is effectively inside a highly impermeable barrier which the galactosides can cross only by forming a complex with the permease. The existence of these two mutant types proves that permease and β-galactosidase are genetically and functionally distinct, and that they normally form a metabolic sequence *in vivo*.

The study of the hydrolysis of galactosides *in vivo*, in wild-type organisms, also brings out the functional role of the permease.

For instance, the hydrolysis *in vivo* of true galactosides is inhibited by thiogalactosides, al-

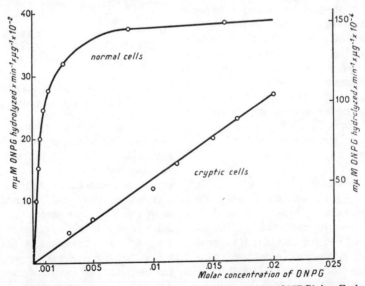

Figure 6. *In vivo* hydrolysis of ortho-nitro-phenyl-β-D-galactoside (ONPG) by *Escherichia coli* (36). Upper curve: normal type. Lower curve: cryptic (permeaseless) mutant. Ordinates on left apply to upper curve. Ordinates on right apply to lower curve.

TABLE 2

In vivo hydrolysis of o-nitrophenyl-β-D-galactoside by cells in which the syntheses of β-galactosidase and permease have been differentially inhibited

	β-Galactosidase (units/ml Culture), mμ-moles ONPG Hydrolyzed $\times$ min^{-1} $\times$ ml^{-1}	Galactoside-Permease (units/ml Culture), mμmoles TMG Accumulated/ml	Rate of *in Vivo* Hydrolysis, mμmoles ONPG Hydrolyzed $\times$ min^{-1} $\times$ ml^{-1}
Cells grown without *p*-fluorophenylalanine..	636	24	84.9
Cells grown with *p*-fluorophenylalanine.....	660	3.8	9.8
Per cent reduction of activity by growth in presence of the analog...................	0	84	88

Two cultures of *Escherichia coli* ML30 were grown respectively in the absence and in the presence of DL-*p*-fluorophenylalanine 5 $\times$ 10^{-4} M. Thiomethyl-β-D-galactoside 5 $\times$ 10^{-4} M was added 1 minute after the analog. The cell mass was allowed to increase in the two cultures from 63 μg dry weight to 252 μg dry weight/ml. The activities of extracted β-galactosidase, of the galactoside-permease, the rate of *in vivo* hydrolysis of ONPG were determined according to (71).

lowing a determination of the affinities of the inhibitors for the total system. The pattern of affinities which is thus disclosed is quite different from that of β-galactosidase studied *in vitro*, but it is very close to the pattern of affinities of the permease, suggesting that, in normal induced wild type, the permease, rather than the galactosidase, is the limiting factor for the hydrolysis of galactosides. This is probably also the explanation of the fact, which has been known for several years, that the *in vivo* hydrolysis of certain galactosides by the wild type is slower than expected on the basis of the extractible galactosidase activity of the cells (49).[7]

A striking experiment allows the conversion of normal cells into phenocopies of the cryptic, permeaseless organisms. It is based on the fact that cells grown in the presence of *p*-fluorophenylalanine incorporate the analog into their proteins, in place of tyrosine and phenylalanine (67). Now it would seem that certain proteins formed under these conditions retain their specific activity, while others do not. Actually, when normal *E. coli* is grown and induced in presence of suitable concentrations of *p*-fluorophenylalanine, the cells form normal amounts of β-galactosidase, but only traces of permease. These cells behave like cryptics, in that they hydrolyze *o*-nitrophenylgalactoside at a much lower rate than controls possessing equal amounts of galactosidase and higher levels of permease (table 2).

All these experiments, therefore, concur in

[7] This conclusion does not necessarily imply that the *in vivo* and the extracted β-galactosidase must have necessarily the same activity.

demonstrating that the permease controls *in vivo* the communications between the intracellular β-galactosidase and the external medium.

F. Functional Relationships of Permease: Induction

The control by permease of the entry of galactosides into the cells carries other consequences which are particularly interesting. Galactosides are not only substrates (or specific inhibitors) of permease and galactosidase; they are also (7, 61, 71) inducers of the two systems. Therefore, by controlling the intracellular concentration of inducers, the permease should control the kinetics of β-galactosidase *and of its own* induction. As this problem is reviewed in this same issue by M. Cohn (pp. 140–168), we shall only briefly record here the main points.

First, the induction of β-galactosidase requires much higher concentrations of inducers in cryptic mutants than it does in the wild type; actually, with thiomethyl-galactoside, more than a 100-fold difference in external inducer concentration is required for saturation of the induction system. Since the wild type effects a 100-fold or more intracellular concentration of the inducer while the cryptic does not concentrate it at all, the interpretation is immediate.

Second, at low inducer concentrations, the kinetics of galactosidase synthesis, expressed in differential rates (increase of enzyme *versus* increase in bacterial mass), is autocatalytic while it is linear at higher (saturating) concentrations. This is easily understood since, at low (nonsaturating) concentration, the differential rate of synthesis will depend on the intracellular concentration factor, *i.e.*, on the specific activity of the permease, and therefore it should increase as induction proceeds.

If this interpretation is correct, the cryptic mutants, devoid of permease, should show linear kinetics of induction at all concentrations of inducer. This actually occurs as shown in figure 7. This result, obtained by Herzenberg (36), is an important confirmation of the contention (contrary to frequently expressed views) that the induced synthesis of enzymes does not involve an increased synthesis of enzyme-forming system.

At low concentrations of inducer, however, the induction of the permease-galactosidase system is autocatalytic, since permease is a system which concentrates its own inducer. This entails some remarkable consequences. *E. coli* cells which already possess permease are induced by very low concentrations of inducer, which are ineffective on noninduced cells. As the permease is distributed between daughter cells, these will inherit this "sensitiveness" and go on producing enzyme in presence of these low inducer concentrations. Therefore, under suitable conditions, the capacity to synthesize permease may be manifested as a self-perpetuating and as a clonally distributed property (13; M. Cohn, this issue; 68). The interest of this effect as a possible model of certain types of cellular differentiation is evident.

G. Genetic Relationships of Galactosidase and Galactoside-Permease

Permease and galactosidase are genetically distinct, as we already noted, since they are affected by different specific mutations. However, there are interesting genetic relationships between the two systems.

In the first place, a fairly large number of mutations manifested by the loss of the capacity to synthesize either β-galactosidase or galactoside-permease have been isolated in *E. coli* K12 and they have been found to be all very closely linked, although no two of them were alleles (J. Lederberg, unpublished; Monod and Jacob, unpublished). (Actually, the complexity of the locus in question had been recognized several years ago by E. Lederberg (48)). This finding is interesting, although not very surprising, since the work of Hartman, Demerec, and others (20) has revealed that the genes controlling enzyme sequences in bacteria are often, if not as a rule, closely linked, and may even be arranged in the same order as the reactions themselves.

Although most mutations affect specifically either permease or galactosidase, giving rise to the "cryptic" and "absolute negatives" mentioned previously, a few appear to suppress both systems simultaneously. These might be deletions. But another mutant type deserves special mention. This is the constitutive (12, 50) in which *both* the permease and the β-galactosidase are formed in the absence of inducer (71). The mutation to constitutive is spontaneous and may occur in "cryptic" organisms devoid of permease. These "constitutive cryptics," which form very large amounts of β-galactosidase without external inducer but are unable to metabolize lactose, may in turn mutate spontaneously to a permease-positive type able to grow on lactose. When this occurs, the permease also is constitutive. Therefore, this single step, spontaneous mutation to constitutive, controls the constitutive *versus* inducible character of *both* galactosidase and permease. It should be added that the locus of the

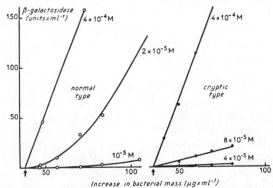

Figure 7. Kinetics of induced β-galactosidase synthesis in normal *Escherichia coli* and in the cryptic mutant type (36). Inducer: iso-propyl-thio-β-D-galactoside.

TABLE 3*

Glucuronide-permease of Escherichia coli

Bacterial Suspension	Additions	S³⁵-TPU Accumulated	Accumulation of Induced Control
		μM/g	%
Noninduced....................	None	0	0
Induced by methyl-β-glucuronide 10⁻³ M..............	None	47	100
Same........................	Phenyl-β-glucuronide 2 × 10⁻² M	0	0
Same........................	Thiomethyl-β-glucuronide 2 × 10⁻² M	6	13
Same........................	Thiomethyl-β-glucuronide, methyl ester 2 × 10⁻² M	47	100
Same........................	Methyl-β-glucoside 2 × 10⁻² M	46	97

Bacterial suspensions were shaken at 37 C for 10 minutes with S³⁵-thiophenylglucuronide (TPU) 5 × 10⁻⁴ M.

* From F. Stoeber (77).

TABLE 4

Independent synthesis and independent functioning of galactoside-permease and glucuronide-permease

Cells Previously Grown with Succinate as Main Carbon Source, in Presence of:	Permease Activities (Accumulation of Radioactive Compound, μM/g) in Presence of Following Assay Mixtures:			
	S³⁵ thiogalactoside*	S³⁵ thioglucuronide†	S³⁵ thiogalactoside + S³⁵ thioglucuronide	S³⁵ thiogalactoside + S³⁵ thioglucuronide
Isopropyl-β-D-galactoside..............	93	1	124	—
Methyl-β-D-glucuronide...............	1	57	—	56
Glucuronide + galactoside..............	100	64	85	66

* S³⁵ labeled thiomethyl-β-D-galactoside.

† S³⁵ labeled thiophenyl-β-D-glucuronide.

constitutive mutation is also closely linked to the loci controlling specifically the capacity to synthesize β-galactosidase and permease.

IV. OTHER CARBOHYDRATE PERMEASES OF ESCHERICHIA COLI

The existence, in *Escherichia coli*, of a specific permease which controls the metabolic utilization of galactosides, e.g., lactose, implies, almost necessarily, as we have already pointed out, that other equally specific permeation systems must control the permeation of the numerous other carbohydrates which *E. coli* is normally able to metabolize. However, the positive identification of a permease requires an experimental material (in particular, adequate nonmetabolizable radioactive substrates) which is not always readily available. So far, only two other carbohydrate permeation systems for specific carbohydrates have been identified in *E. coli*, one active on glucosides, the other on glucuronides.

A. Glucuronide-Permease

Stoeber (77) has recently described an inducible system which concentrates glucuronides in *E. coli*. It should be recalled that *E. coli* forms an inducible β-glucuronidase. Stoeber found that S³⁵-labeled thiophenyl-β-D-glucuronide (35), which is not hydrolyzed by glucuronidase, is accumulated, unchanged, by glucuronide-induced *E. coli*. The accumulation, which is inhibited by 2,4-DNP and NaN₃, is reversible, and the intracellular *versus* extracellular concentration follows an adsorption isotherm [equation (1) with K ca. 10⁻⁴ M]. The system is strictly stereospecific: the accumulation is competitively inhibited by free, unsubstituted glucuronides, not by glucosides, galactosides or other carbohydrates (table 3). The induced formation of the system is blocked by β-2-thienylalanine.

By all these properties, this system is closely analogous to galactoside-permease, while it is sharply defined by its strict specificity of combination and of induction. The two systems may be either separately or simultaneously induced in the same strain of *E. coli*. In the doubly induced cells, glucuronide and galactoside are simultaneously accumulated, each to a similar extent as in the singly induced cells (table 4).

The two permeases are thus synthesized independently, and function independently, without any cross interference.

B. Glucoside Accumulation in Escherichia coli (64)

E. coli does not measurably metabolize α-methyl-glucoside. However, suspensions of E. coli shaken with C^{14} labeled α-methyl-glucoside accumulate, reversibly, up to 100 μmoles/g of the compound. Radioautograms of hot water extracts from the cells show a single spot with the same R_f as free α-methyl-glucoside. The accumulation is not inhibited significantly by NaN_3 or 2,4-DNP. It is reversible and the variation of internal versus external concentration of α-methyl-glucoside follows again an adsorption isotherm [equation (1) with a K of 2×10^{-4} M].

The system is constitutive: cells grown in presence of glucose or succinate, with or without α-methyl-glucoside, are about equally active.

The specificity of the system has not been studied in detail, for lack of adequate compounds. However, the accumulation of α-methyl-glucoside is very powerfully inhibited by glucose, even at extremely low concentrations. There is little or no inhibition with fructose and galactose. Since glucose is used by the cells, it is difficult to determine whether the inhibition is competitive and whether it is due to glucose itself or to a product of its metabolism.

These observations suggest that the accumulation of α-methyl-glucoside in E. coli is due to a constitutive permease system possessing a high affinity for glucose. We have already recalled Doudoroff's studies on a mutant of E. coli K12 which appeared to be cryptic towards glucose (see page 170). As it turns out, cells of this glucose-negative strain also lack the capacity to accumulate α-methyl-glucoside although, after proper induction, they prove capable of accumulating galactosides. Moreover, 12 independent glucose-positive isolates, obtained from the glucose-negative mutant by transduction with an adequate phage (P1), were found to have also regained the capacity to accumulate α-methyl-glucoside.

These findings indicate that the α-methyl-glucoside accumulating system must be closely related to the system presumably responsible for the permeation of glucose in E. coli. Further studies are required to identify these systems.

V. THE PERMEATION OF KREBS CYCLE INTERMEDIATES AND OTHER ORGANIC ACIDS IN PSEUDOMONAS AND AEROBACTER

The operation of the permeation systems which we have studied so far can be tested directly, independently of the activity of the corresponding metabolic enzymes. The existence, in Pseudomonas and Aerobacter, of specific systems insuring the permeation of certain intermediates of the Krebs cycle and of other organic acids has been inferred from a different kind of evidence, essentially from crypticity relationships.

A. Citrate and other Intermediates of the Krebs Cycle

The interferences of permeability effects in the metabolism of intermediates of the Krebs cycle had been frequently suspected in the past, as an explanation of discrepancies observed between the metabolism of intact cells and the enzymic activities of extracts (25, 39, 78).

In 1953, Kogut and Podoski (43), and Barrett et al. (2) discovered independently that the oxidation of citrate and of other intermediates of the cycle by intact cells of Pseudomonas was adaptive, while the enzymes of the cycle itself were constitutive. The specificity of this inductive effect, as shown by Kogut's results (table 5), reveals the existence of at least five different inducible systems. It is particularly interesting to note, for instance, that cells grown on citrate, and able to use citrate without lag, are still cryptic towards iso-citrate and cis-aconitate. This illustrates strikingly the specificity of the systems involved. In sharp contrast to the strictly adaptive behavior of intact cells, extracts from cells grown on citrate, fumarate or succinate showed no significant difference in their oxidative activity towards citrate (2) and other intermediates of the cycle (43).

Both groups of authors formulated the hypothesis that the adaptive behavior of intact cells was due to specific, inducible permeation factors. They tested the effects of agents known to inhibit the synthesis of active enzymes. Barrett et al. (2), in particular, showed that the amino acid analogs, ethionine and p-fluorophenylalanine, blocked adaptation to citrate. Thus unadapted Pseudomonas cells are cryptic towards most intermediates of the Krebs cycle. Adaptation to a given intermediate suppresses crypticity towards the corresponding compound, and this

TABLE 5*

Oxidation of tricarboxylic acid cycle intermediates by washed suspensions of Pseudomonas sp.

Growth Substrate	Substrate for Oxidation									
	Succinate	Fumarate	Malate	Oxal-acetate	Pyruvate	Acetate	Citrate	cis-Aconitate	iso-Citrate	α-Keto-glutarate
Succinate	+	+	+	+	0	0	0	0	0	0
Fumarate	+	+	+	+			0			0
Malate	+	+	+	+			0			0
Acetate	0	0			+	+	0			0
Citrate	0	0	+		0	0	+	0	0	0
iso-Citrate	0	0	+				+	+	+	0
α-Ketoglutarate	0	0			0	0	0			+

+ = Substrate oxidized linearly from the moment of tipping.

0 = Substrate not linearly oxidized initially.

* From M. Kogut and E. P. Podoski (43).

"suppression of crypticity" requires protein synthesis.

An entirely similar situation has since been discovered, in *Aerobacter*, for citrate utilization, by Green and Davis (18), who showed, in addition, that the adaptation was inhibited by glucose. As we have already recalled, this "glucose effect" is typical of many, in fact of most, inducible enzymes.

Gilvarg and Davis (29) had previously found a mutant of *Escherichia coli* which lacked the condensing enzyme while it possessed all the other enzymes of the Krebs cycle. This mutant required α-ketoglutarate for growth, but it could not utilize citrate in its place. In contrast, a mutant of *Aerobacter*, which also lacked the condensing enzyme, grew equally well on either α-ketoglutarate or citrate. These findings show clearly that citrate is an obligatory intermediate of α-ketoglutarate synthesis, in both organisms, but that *E. coli* is, for all practical purposes, impermeable to it, while *Aerobacter* can be "decryptified" by proper adaptation. It may be recalled, at this point, that the operation of a "complete" Krebs cycle in microorganisms had been repeatedly denied, precisely because many microbes, such as yeast and *E. coli*, did not metabolize certain intermediates of the cycle. It is now established beyond doubt, not only that the enzymes of the cycle exist, but that the cycle actually operates *in vivo*, in *E. coli* as well as yeast (29, 44, 72, 76, 79, 80).

The findings summarized previously leave little doubt that the entry of Krebs cycle intermediates into bacteria is controlled by stereo-specific permeases and that the failure of certain organisms to metabolize certain intermediates is due to their incapacity to synthesize the appropriate permease system. It is likely that the actual operation of these systems could be tested directly if proper analogs of their substrates were available, or if proper mutants were used.

B. Tartaric Acid Permeation in Pseudomonas

Recent observations of Shilo and Stanier (75) have given clear indications of the existence of stereospecific permeation factors for different isomers of tartaric acid in *Pseudomonas*. The attack of each isomer by most strains is due to a distinct inducible dehydrase which produces oxalacetic acid (45, 47, 74). Two kinds of observations showed that, besides the specific dehydrase, an additional, inducible factor is required for the *in vivo* metabolism of tartrates.

In the first place, when cells were grown on limiting amounts of tartrate as sole source of carbon and energy, depletion of tartrate caused the population to enter the stationary phase, invariably followed by a loss of the capacity for immediate attack on tartrate. This loss could not be explained by the inactivation of any of the known intracellular enzyme systems involved in the dissimilation of tartrate, including in particular the specific dehydrase, as tested with extracts.

The dissimilated cells, in other words, were cryptic towards tartrate while they still metabolized oxalacetate. The oxidative activity of the dissimilated cells could be regained rapidly by re-exposure to tartrate. But UV irradiation blocked this readaptation, although it had no effect on the activity of the system, suggesting that readaptation involved resynthesis of a protein component which had presumably been inactivated during the dissimilation period.

Further indications that the access of tartaric acid isomers to the intracellular dehydrases is controlled by strictly stereospecific factors were given by the study of the inhibition *in vitro* and *in vivo*. *Meso*-tartaric acid is a powerful inhibitor of the *d*-dehydrase, and *d*-tartaric acid is a powerful inhibitor of *meso*-dehydrase in extracts; however, no such inhibition is observed in whole cells adapted to a single isomer. But if one uses cells able to attack the *three* isomers of tartrate, the oxidation of *meso*-tartrate can be inhibited *in vivo* by *d*-tartrate.

These facts could perhaps, like most specific crypticity effects, be explained by *ad hoc* alterations of the enzymes upon release from the cell. It is much more probable that they reveal a marked difference of stereospecific requirements, between the intracellular enzyme and a specific permeation factor. In fact, the results suggest that the permeation factors for tartrate isomers may be more exacting, in their stereospecificity, than the metabolic enzymes. The nature of the isomerism of tartaric acids makes this problem particularly intriguing. It may well be that the marvelous discriminative power possessed by the microorganisms which, about 100 years ago, helped Pasteur (70) to separate the different isomers of tartaric acid, was based largely on the properties of their permeases.

VI. AMINO ACID PERMEASES

A. The Accumulation of Exogenous Amino Acids by Escherichia coli

Many instances have been reported in the past in which the growth of auxotrophic mutants requiring amino acids was inhibited competitively by other structurally related amino acids; whereas the nonexacting wild-type was not inhibited (1, 21, 37, 42). To explain such data, various schemes were proposed, some of which (42) assumed a selective permeability barrier where the interactions were supposed to occur. However, no positive evidence was given in favor of this hypothesis.

This problem can best be studied by using labeled compounds to follow the uptake of amino acids by the cells (10, 11). In particular, the uptake of radioactive valine by *Escherichia coli* K12 has been studied in some detail. It was found that when *E. coli* K12 is shaken at 37 C, in presence of this amino acid, under conditions where protein synthesis was blocked, radioactivity was rapidly accumulated into the cells, in amounts corresponding to a concentration factor (with respect to the external medium) of up to 500.

Chromatographic analysis of the radioactive material extracted by boiling water shows that the concentrated material consists exclusively of valine. The amounts of intracellular valine vary with the external concentration, according to an adsorption isotherm; the "capacity" (see page 172) of the system at saturating external concentration (5×10^{-5} M L-valine) is of the order of 20 μmoles per g dry weight, i.e., 4×10^6 molecules of L-valine per bacterial cell. The apparent dissociation constant of the system is of the order of 3×10^{-6} M.

The accumulation is inhibited by 2,4-DNP and NaN$_3$, and is optimal in the presence of an external energy source.

The accumulation of valine is reversible: the intracellular C^{14}-valine can be displaced by nonradioactive valine, and also by the structurally related amino acids, leucine or isoleucine, while phenylalanine or proline, even at much higher concentrations, have shown no effects. Moreover, only the L-isomers are effective competitors. Substitution of the amino or carboxyl groups of the competitors, or replacement of the isopropyl group of valine by a dibutyl, diphenyl or dibenzyl group, suppresses all competitive capacity. Similarly, peptides containing valine, leucine, or isoleucine have little or no affinity for the valine accumulating system (table 6). The system responsible for accumulation is therefore strictly stereospecific. A quantitative study of the displacement shows that the antagonistic action of leucine or isoleucine is competitive (table 7).

This system is constitutive: it is present in organisms grown in the absence of valine or of any other amino acids. The total valine accumulating capacity of a growing culture increases linearly with the bacterial mass; this increase is inhibited by 5-methyl-tryptophane or thienylalanine, which do not stop protein synthesis, but are known to inhibit the synthesis of active enzymes (67).

The existence of several other systems similar to the valine one, but different in specificity, has been demonstrated in *E. coli*. One system is char-

TABLE 6*

Structural conditions required for the competitive displacement

Additions	Internal Accumulation of L-Valine
M	μ moles/g
None...........................	15.7†
L-isoleucine 5 × 10⁻⁵..............	2.8
L-leucine 5 × 10⁻⁵................	2.1
Unlabeled L-valine 10⁻³............	2.1
D-valine 10⁻³.....................	19.2
D-isoleucine 10⁻³.................	17.2
D-leucine 10⁻³....................	12.5
DL-N-monomethylvaline 10⁻³......	18.3
DL-valinamide 10⁻³................	15.9
DL-dibutylalanine 10⁻³............	12.5
DL-diphenylalanine 10⁻³...........	19.1
DL-dibenzylalanine 10⁻³...........	18.5

Radioactive DL-valine was present in all suspensions (5 millimicromoles L-valine/ml). A sample was taken after 1-minute incubation at 37 C; then the presumed competitor was added and a new sample was taken after 1 minute.

* From G. N. Cohen and H. V. Rickenberg (11).

† The controls without addition of each experiment differ at maximum of ±7% from this mean value.

acterized by the capacity to accumulate L-phenylalanine. Radioactive phenylalanine is displaced by nonradioactive L-phenylalanine, but not by the D-isomer; it is also displaced by *p*-fluorophenylalanine, but not by phenyl-lactate, phenylpyruvate or phenylserine, nor by unrelated amino acids such as isoleucine of proline.

Another system accumulates L-methionine, which is displaced by nonradioactive L-methionine or by L-norleucine, but not by the D-isomers, nor by phenylalanine or proline or other unrelated amino acids.

The three systems studied are thus stereospecific and independent one from the other, as is proved by the absence of any effect of the typical substrate of each upon the functioning of the other two.

Britten, Roberts, and French (6), using different techniques, independently found that various exogenous amino acids were accumulated by *E. coli*, independently of protein synthesis. L-proline, in particular, is highly concentrated and the results suggest that the proline concentrating system is specific for this one amino acid.

TABLE 7*

Competitive displacement of radioactive L-valine by L-isoleucine

Radioactive L-Valine	L-Isoleucine	Internal Accumulation of L-Valine
M	M	μmoles/g
5 × 10⁻⁶	0	19.1
	2.5 × 10⁻⁶	7.8
	5 × 10⁻⁶	3.1
5 × 10⁻⁵	0	34.1
	10⁻⁵	30.4
	5 × 10⁻⁵	19.0
	10⁻⁴	8.3
	5 × 10⁻⁴	5.5

* From G. N. Cohen and H. V. Rickenberg (11).

Thus, except for inducibility, the amino acid concentrating systems are very similar to the galactoside concentrating system for the interpretation of which we have already discussed the catalytic *versus* stoichiometric models. We have seen that the evidence eliminates the latter in favor of the permease theory. However, the analogy between the two classes of systems would be insufficient to decide in favor of one or the other model. The validity of the permease model for the amino acid systems must be discussed on the basis of direct evidence.

To begin with, we may remark that the "common sense argument" based on the amount of substrate accumulated is valid for amino acids as well as for galactosides. For instance, proline may, according to Bolton *et al.* (4), be accumulated to the extent of several hundred μmoles per gram, a quantity roughly equivalent to ⅓ the number of ribonucleic acid (RNA) nucleotides and ¹⁄₂₀ the total number of protein bound amino acids in the cell. Such a result would be very difficult to interpret on the hypothesis that the retention of the amino acids is due to binding to nondiffusible molecules.

B. Physiological Interactions between Amino Acids Explained by the Permease Model

The strongest arguments which favor the permease theory for amino acids is that it explains most satisfactorily the so far poorly understood physiological interactions observed in *E. coli* between structurally related amino acids.

We shall first consider valine and its effects on the growth of different strains of *E. coli*:

a. Toxic effects of valine on E. coli K12. It is known that the growth of *E. coli* K12 is inhibited by valine and that the inhibition is released by the addition of isoleucine (81) or leucine (73). The D-amino acids show no action either as inhibitors or as antagonists.

If the valine accumulating system is, in effect, a permease system which controls not only the accumulation of valine, but actually the entry of this amino acid into the cell and its access to all intracellular systems, the antagonism is immediately accounted for; the ratio isoleucine/valine necessary to displace 50 per cent of the valine is the same as the ratio of isoleucine/valine that restores 50 per cent of the normal growth. The detoxifying effect of isoleucine and leucine is thus undoubtedly linked to the inhibition of valine accumulation by the permease. The toxic effect of valine is therefore conditioned by the activity of the permease, but it is not an inherent consequence of it, since a valine-resistant mutant possesses a valine permease of the same affinity and activity. Consequently, the toxic effect of valine need not have anything to do with the intracellular metabolism and/or synthesis of isoleucine and leucine. Actually, it has been shown that the toxicity of valine for *E. coli* K12 is due to an alteration of the proteins synthesized in its presence (8).

b. The growth of auxotrophic mutants of E. coli. The growth of auxotrophic mutants of *E. coli* requiring one of the three amino acids: valine, leucine, or isoleucine is competitively inhibited by the two other members of the group. Here again, the ratio of isoleucine/valine at which the growth of a valine-requiring mutant is inhibited by 50 per cent, corresponds to the 50 per cent displacement ratio for valine accumulation through the permease in presence of isoleucine. The interpretation of these effects is again simple and straightforward in terms of the permease model while interpretations in terms of reciprocal effects upon biosynthetic pathways (37) were complex and unsatisfactory.

c. Incorporation of exogenous valine into proteins by wild-type of Escherichia coli. In normal wild-type *E. coli* (K12 excepted) neither leucine, isoleucine, nor valine exerts any positive or negative effect on the growth rate. It is, however, possible to demonstrate that the incorporation of exogenous valine is controlled by a system which

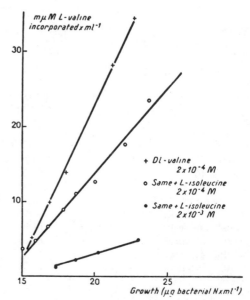

Figure 8. Incorporation of radioactive exogenous L-valine by wild-type *Escherichia coli* in absence and in presence of unlabeled L-isoleucine, at increasing concentrations (see text, this page) (11).

has the specificity of the permease. When wild-type *E. coli* ML is grown in presence of radioactive valine, this exogenous amino acid competes very efficiently with the valine which is endogenously synthesized from the carbon source of the medium: 95 per cent of the valine incorporated into the protein is radioactive. Under these conditions, the total radioactivity incorporated increases linearly with the increase in bacterial mass during growth. The slope of the straight line obtained measures the specific radioactivity of valine in the proteins synthesized from the time of addition and gives therefore the ratio of exogenous to endogenous valine in the newly formed proteins.

As is seen in figure 8, when the cultures are grown in presence of radioactive valine and unlabeled isoleucine, the specific activity of the proteins decreases as the concentration of isoleucine increases. Since, as we have mentioned, the presence or absence of valine or isoleucine is without action on the growth of these cells, this effect could hardly be attributed to a drastic alteration of the valine content of the proteins. It is evidently due to an altered ratio of incorporation of endogenous to exogenous valine into the proteins. Again, this effect is explained and predicted by the permease model. We can thus give

TABLE 8*

Interactions between valine, leucine and isoleucine in Escherichia coli

Organism	Effect of Valine on Growth	Antagonist	Effect of Antagonist on Accumulation of Valine	Effect of Antagonist on Growth
K12S	Inhibits growth	Isoleucine, leucine	Competitive displacement	Suppresses growth inhibition
ML328f	Required for growth	Isoleucine, leucine	Competitive displacement	Inhibits growth
Valine-resistant mutant of K12S or *E. coli* ML	None	Isoleucine	Competitive displacement	Nil. Inhibits incorporation of exogenous, but not of endogenous, valine

* From G. N. Cohen and H. V. Rickenberg (11).

a unitary explanation of the valine-isoleucine antagonism for *E. coli* K12, for the auxotrophs, and for the normal *E. coli* strains. The three apparently unrelated phenomena now clearly appear to be different expressions of the same initial event; namely, competition of structurally related amino acids for a structure which has the same specificity and affinity characteristics as the valine-accumulation system. All the evidence concurs in showing that, while this structure is essential for the entry of exogenous amino acids into the metabolic space of the cells, it plays no role in the synthesis or incorporation of internally synthesized amino acids. Table 8 summarizes the information for the various types of strains.

Many other examples of inhibition of the utilization of a given amino acid by another structurally related natural amino acid, or by structural analogs, have been described in the literature. Lampen and Jones (46), and Harding and Shive (33) have described the inhibition of the growth of *E. coli* by norleucine and its competitive reversal by methionine. The first authors conclude from the competitive aspect of the antagonism that norleucine inhibits utilization, rather than synthesis, of methionine. Harding and Shive apply the methods of inhibition analysis and draw the conclusion that methionine functions in the biosynthesis of leucine, isoleucine, and valine, probably in their amination, and that norleucine inhibits this function of methionine. The progress made since 1948 in the study of the biosynthesis of valine, leucine, and isoleucine by *E. coli* has revealed no such function of methionine. Actually, it has been found that norleucine is incorporated in the proteins of *E. coli* where it substitutes for methionine (R. Munier and

G. N. Cohen, unpublished); however, the synthesis of methionine is unimpaired, in the presence of norleucine, and the nonincorporated methionine is found in the culture medium. The known properties and specificity of the methionine permease account completely for the competitive suppression of norleucine inhibition by methionine.

A similar situation is encountered with thienylalanine and *p*-fluorophenylalanine which also cause the synthesis of "false proteins." The analogues are incorporated in place of phenylalanine and tyrosine in the proteins of *E. coli*, and the inhibition of growth which ensues is due to the biological inactivity of these "false proteins."

Suppression of the inhibition by phenylalanine is again explained by the directly determined displacements at the permease level. Other interpretations, based on *p*-fluorophenylalanine inhibiting the synthesis of tyrosine from phenylalanine (3) are erroneous, since such a pathway for tyrosine synthesis has been excluded in *E. coli* (17).

An antagonism which has been correctly analyzed occurs in the inhibition of the growth of *E. coli* by diamines, studied by Mandelstam (53). The wild type of *E. coli* is not inhibited by diamines, whereas the growth of a lysine-requiring mutant is inhibited by the C_5, C_6 and C_7 diamines. Direct estimation has shown that the inhibition of growth is due to the inhibition of uptake of lysine.

The essential justification of the permease model for the amino acid accumulation systems in *E. coli* is that it accounts for the observed antagonisms. The stoichiometric model or variations of it could possibly account for some of these effects, but only at the expense of additional *ad hoc* assumptions. The number and the variety of the effects which are simultaneously explained by the permease model leave no doubt that it must be valid in its essential features.

This rapid review of some examples of antagonisms shows how cautious one should be in the interpretation of antimetabolic activities of structural analogs. The possibility that the effects observed are due to specific permeases must be taken into account. However, it would probably be equally dangerous to use the permease interpretations indiscriminately in cases where no actual data concerning the specificity and relative affinities of the presumed systems are available.

C. Accumulation of Amino Acids by Other Microorganisms

As mentioned before, the experiments of Gale and his co-workers (27, 28) provided the first example of the accumulation of amino acids by bacterial cells. It is also well known that yeasts possess a "free amino acid pool" and may accumulate exogenous amino acids into this pool (32). An essential difference between these gram-positive organisms and *E. coli* is that the "free" intracellular amino acids are not in equilibrium with amino acids in the external medium. It seems probable, nevertheless, that the bulk of the amino acids are indeed free, which implies necessarily the existence of an impermeable barrier and the occurrence of active transport (28). Whether this active transport is mediated by stereospecific factors comparable to *E. coli* permeases is uncertain. Halvorson and Cohen (31) have observed that the rate of uptake of valine and phenylalanine into the pool of yeast is inhibited by structurally unrelated amino acids, including D-isomers. A similar situation had been disclosed previously in *Neurospora crassa* by Mathieson and Catcheside (54), who showed that the uptake of histidine in this organism is inhibited by a whole series of other amino acids. It should be stressed that these authors were able, on this basis, to explain the fact that histidine-requiring mutants of *Neurospora* are inhibited by other amino acids while the wild type is uninhibited.

GENERAL DISCUSSION

As an economical way of summarizing the more general conclusions which appear justified by the evidence reviewed here, it will be convenient to consider a model cell to which we shall try to attribute only the *minimal* properties required to account for the behavior of the different permeation systems studied. The discussion of this model will also give us an occasion of bringing in certain elements of evidence which we have not considered so far, because they did not pertain directly to the study of any one permeation system.

In attempting to construct such a model it might be useful to consider the information concerning selective permeation of organic compounds in tissues of higher organisms. Actually, the wealth of information is so considerable and so complex that summarizing it in a few statements is virtually impossible. It must be noted, however, that by far the largest fraction of this information concerns "transtissular" permeation, that is to say, transport from one extracellular space to another extracellular space across more or less complex cellular tissues or organs. The problem is experimentally quite different from the one we have attempted to analyze in microorganisms. Yet, Wilbrandt (84) has shown that the more likely interpretation of such "transtissular" permeation is a "membrane carrier mechanism," involving specific carriers, operating within otherwise impermeable cellular membranes.

The most significant observations directly concerning cellular permeation in cells of higher organisms have been made on erythrocytes (51, 84). The existence, in the erythrocytes of primates, of a transmembrane carrier system for hexoses, is clearly established. The system appears to function exclusively in equilibrating intracellular and extracellular concentrations; there is no evidence of active transport. Since, in this work, the rates of hexose permeation are determined by measurements of osmotic equilibration, there is no ambiguity as to the state of the intracellular compound, and to the existence of a real osmotic barrier. The intervention of a specific carrier is shown by the nonlinearity of the rates of entry with respect to concentrations, and by competition between different sugars. The stereospecificity of the carrier is not very strict, since several different hexoses and pentoses appear to compete for the same system. The existence in these cells of different independent carrier systems for different compounds or classes of compounds is likely, but they do not seem to have been clearly identified one from the other.

Let us now describe the model which we shall use as a basis of discussion.

The bacterial cell is supposed to be enclosed within an osmotic barrier (55) highly impermeable toward polar substances, such as carbohydrates, hydroxy and other organic acids, amino acids, and the like. The impermeability of the barrier is not supposed to be absolute; however, leakage in either direction may occur, slowly tending to equilibrate the inside and outside concentrations. In places within the barrier, there exist different proteins (the permeases) which are able to form stereospecific, reversible complexes with different hydrophilic compounds. Dissociation and association of the specific complex may occur either on the inside or on the outside of the osmotic barrier. The effect of a permease, therefore, is to activate catalytically the equilibration of the concentrations (activities) of the substrate on either side of the membrane, i.e., inside and outside the cell, according to the scheme:

$$G + y \underset{1}{\overset{2}{\rightleftharpoons}} Gy \underset{3}{\overset{4}{\rightleftharpoons}} y + G$$

outside   permeability barrier   inside

Many permease systems, but not necessarily all of them, are coupled to an energy donor, the net effect of which is to inhibit the "inside" association reaction (reaction 4). When this occurs, the substrate accumulates within the impermeable barrier, i.e., within the cell, until the difference of internal and external concentrations is high enough for the nonspecific leakage through the permeability barrier to equilibrate the entry via the permease.

Let us now consider the justification, the possible meanings, and the limitations of the different assumptions contained in this scheme. The first essential assumption is the existence of an osmotic barrier enclosing the whole cell and impermeable to polar compounds. The justification for this assumption is that it accounts at once (a) for the capacity of bacterial cells to accumulate and retain certain compounds in an apparently free state; (b) for the crypticity of certain cells towards certain polar compounds. Both effects have been amply documented here and we have discussed the reasons which make other interpretations of accumulation (see pages 173–174) and of specific crypticity (see page 176) virtually impossible. We need only to refer to discussion.

As final evidence of the existence of a cellular osmotic barrier in bacteria, we cite the recent work of Mitchell and Moyle (58), which allows the estimation of the actual intracellular osmotic pressures within certain bacterial cells. Using an elegant technique, they find a pressure of 20–25 atmospheres for *Staphylococcus aureus*, (*Micrococcus pyogenes* var. *aureus*) a value which indicates that most of the low molecular weight intracellular compounds must be free and in solution within the osmotic barrier (59). Mitchell and Moyle have also devised (57) new techniques for studying the rate of equilibration of internal and external osmotic pressure of bacterial cells. Using various carbohydrates and polyhydric alcohols, they find that, in general, these rates are extremely low; sometimes so low that osmotic equilibrium is never reached.

These observations illustrate the low permeability of bacterial cells toward sugars in general. However, they are not altogether free of complexities and apparent contradictions, and they cannot, we believe, be interpreted directly in terms of *rates of permeation*, since most of the compounds used are metabolized by the cells at high, variable, and unknown rates. It would be of great interest to see these techniques applied to unnatural isomers, which would not be metabolized, and for which specific permeases would presumably not be available. The tests performed with such unnatural compounds would give an estimation of the truly *nonstereospecific* permeability of the membrane.

We shall not dwell on the important problem of the physical nature of the osmotic barrier. Modern work with "protoplasts" indicates clearly that the cell wall, while responsible for the resistance of the cells to internal hydrostatic pressure (83) is not a significant element of the permeability barrier since the permeability properties of protoplasts are similar to those of intact cells. Worthy of mention here is that galactoside-permease is present and functional in *E. coli* protoplasts, as prepared by Rickenberg (personal communication). The existence of a separable membrane, limiting the protoplast (83) is indicated by the formation of "ghosts" when protoplasts burst. According to Mitchell and Moyle (56), "membrane" fractions of *Staphylococcus aureus* have a high lipid content, which may account for the impermeability of the cell to hydrophilic compounds. These facts certainly encourage the identification of this "ghost" structure with the cellular osmotic barrier and bearer of the permease proteins, but the evidence for this identification is, so far, purely circumstantial.

Before turning to another problem, note that the existence of several independent subcellular osmotic units with different permeability characteristics is not excluded, and may have to be considered in the future. For the time being, there is no necessity for this more complex picture.

The second essential assumption of the model is the existence of different, independent, stereospecific permease proteins, functionally specialized for the permeation of specific compounds and distinct from the intracellular metabolic enzymes dealing with the same compounds. The presentation of the evidence bearing on this point has been the object of the present review. We need not discuss this evidence again, but it will be useful to summarize briefly as follows the essential experimental justifications of the permease hypothesis:

1. Certain strains or mutants of bacteria are specifically cryptic, i.e., metabolically inert towards a given compound, although possessing a competent intracellular enzyme system for the metabolism of the compound, and although capable of metabolizing other closely related compounds at a high rate. This suggests that the permeation of such compounds into the cells involves a stereospecific process.

2. Certain polar compounds are (reversibly) accumulated by certain cells, and the accumulation is inhibited by steric analogs of the compound. Conversely, sterically different compounds are accumulated simultaneously without any cross-interference.

These observations show the dependence of accumulation on a stereospecific component and the functional independence of different accumulation systems.

3. The kinetics of accumulation prove that the stereospecific sites act as intermediates, not as final acceptors for the accumulated compound. Moreover, the blocking or inactivation of the stereospecific component results in the cells becoming specifically cryptic towards the corresponding compound.

Therefore, the specific sites act as catalysts for the entry of the compounds into the metabolic space of the cell.

4. The formation of the stereospecific component of certain systems is provoked by specific inducers. This induced formation is blocked by agents which prevent selectively the synthesis of biologically active proteins. These observations show that the specific components of different systems are different proteins individually and independently synthesized by the cell.

5. The capacity of the cell to synthesize a given permeation system may be suppressed by specific mutations which have no effect on the corresponding intracellular metabolic enzyme. Other specific mutations, which suppress the intracellular enzyme, do not suppress the corresponding permeation system. Similarly, certain agents inhibit the synthesis of the permeation system, without interfering with the synthesis of the corresponding metabolic enzyme. These findings prove that the permeation systems are distinct and independent of the homologous metabolic enzymes.

At the present time, at most eight different permeases have been positively identified in a single organism (*E. coli*). This is not a large number. However, as we have stressed several times, proof of the existence of a permease, and of specific crypticity relationships, for a single compound belonging to a homogeneous class, necessarily implies that other, equally specific, permeases insure the permeation of the other compounds of the class which are rapidly metabolized by the same cells. On this basis, it can be estimated that *E. coli* must possess, or be able to synthesize, at least 30 to 50 different permeases dealing with organic substrates. The entry of certain inorganic ions, in particular phosphate, is undoubtedly catalyzed, as it has been shown by the work of Mitchell (55). However, the participation of permeases, in the sense defined above, cannot be tested in this process since none of the necessary criteria (stereospecificity, independence from intracellular enzyme, specific crypticity) is applicable.

The third assumption is that the permease may be coupled to an energy-yielding reaction, and thereby act as a pump, or uncoupled and thermodynamically passive, when it functions as an equilibrator of outside and inside concentrations. This is a synthetic and abstract description of the observed relationships, rather than an assumption. The justification for assuming a coupling is, of course, in part, the effect of metabolic inhibitors. But it should be pointed out that this evidence is not so unequivocal as is generally believed. The metabolic "uncouplers" may act in different ways; they are inactive with certain systems (glucoside-permease) where work is probably performed. The effect, or absence of effect, of an external source of energy is not a good test either, since such a source is often not required. Finally, the best evidence that work is performed in the accumulation of galactosides in *E. coli* is kinetic. As we have seen (see p. 174), this evidence shows that the coupling must be at the level of the specific permeation reaction itself.

There is no reason, however, to suppose that coupling is necessary for the permease protein to act as a thermodynamically passive permeation catalyst. Moreover, as we pointed out, the uncouplers seem to inhibit only the *accumulation*, not the permeation of galactosides. Although the evidence on this last point is certainly insufficient, it is of interest in suggesting a unified interpretation of specific permeation mechanisms, whether or not active transport is performed.

In the minimal model which we have set up, no attempt has been made to represent the actual mechanism of energy coupling. It should be noted, however, that the coupling necessarily implies the participation of further components in the system, namely coenzymes or transporters of chemical potential. Even when the permease acts passively, the participation of other (nonspecific) components may be required. Obviously the actual mechanism of specific permeation must be more complex than the deliberately bare and abstract model we have set up. The object of this model has been to symbolize only the indispensable assumptions, and the identified components: the permeability barrier, the stereospecific permease protein, the dependence of permeation on the formation of a reversible permease-substrate complex, the possibility for the system to perform work or not, depending on conditions. Of the existence and nature of nonspecific components, of the detailed mechanism of action of the permease, and of the mechanism of coupling, nothing positive can be said at present. Such problems have given rise to very ingenious speculations (15, 30, 82) into which we shall not go because a whole variety of equally likely, albeit quite different, schemes would have to be considered.

We may, however, briefly consider the question of whether the permeases are proper enzymes or not, and show that it is not meaningless. As we have seen, the permeases behave exactly like enzymes in many characteristic respects.

They are proteins which form stereospecific, reversible complexes with certain compounds; certain of them are inducible under the same conditions as typical inducible enzymes; galactoside-permease is controlled by a system of mutations exactly parallel to the system which controls β-galactosidase.

However, enzymes are characterized essentially by the property of catalyzing a *chemical reaction* of their substrates, *i.e.*, the formation, or rupture, or transfer of covalent bonds. There is, at present, no direct evidence that such events occur at the substrate level, with permeases. When there is work performed in the process, it is virtually certain that covalent bonds are broken or transferred and, to that extent, the permease acts like an enzyme in activating their breakage or transfer. But these chemical events need not involve the permeating substrate itself. Whether or not the permeation process involves a *chemically altered* form of the substrate as an intermediate, is one of the first problems which will have to be considered in studying the mechanism of action of the permeases.

The hypothesis, that the rapid permeation of hydrophilic organic compounds into cells is insured by stereospecific and functionally specialized protein components of the plasma membrane, is not new or original. It has been frequently invoked in the past to account for selective permeability effects, the most precise and elaborate formulations having been given by Danielli (14–16). The recent microbiological work summarized in this review has given precise and extensive experimental support to this concept. The specific inducibility and independent mutability of certain of the bacterial permeases, allow us to individualize, identify, and study these systems under exceptionally favorable conditions.

The examples we have reviewed leave little doubt that the entry of all the main organic nutrilites (carbohydrates, organic acids, amino acids) into bacteria is controlled by specific permeases.

Thus the role of permeases as chemical connecting links between the external world and the intracellular metabolic world appears to be decisive. Enzymes are the element of choice, the Maxwell demons which channel metabolites and chemical potential into synthesis, growth and eventually cellular multiplication. Occurring first in this sequence of chemical decisions, the permeases assume a unique importance; not only do they control the functioning of intracellular enzymes, but also, eventually, their induced synthesis. Moreover, since the pattern of intermediary metabolism appears more and more to be fundamentally similar in all cells, the characteristic, differential, chemical properties of different cells should depend largely on the properties of their permeases.

ADDENDUM

Recent experiments performed by Sistrom at the Pasteur Institute answer the critical question whether galactosides intracellularly accumulated by the action of galactoside permease in *Escherichia coli* are free or bound (see page 174). These experiments depend on the fact that protoplasts of *E. coli* retain the permeability properties of intact cells, including in particular galactoside permease activity (see pages 188–189), while their resistance to osmotic pressure differences is very greatly reduced. Differences in osmotic pressure between the intra- and extracellular phases may result from either a decrease of the external osmolality or an increase of the internal pressure resulting, for example, from intracellular accumulation of a compound in an osmotically active state, *i.e.*, in a *free* state. The bursting of protoplasts is easily measured as a decrease of the optical density of the suspension, so that protoplasts can be used as sensitive indicators of variations of their own osmotic pressure. By comparing the extent of lysis which occurs when the internal pressure increases as a result of permease activity with that caused by a known decrease in external pressure, one can determine approximately the intracellular concentration of free permease substrate at the time of lysis.

Using this experimental principle, Sistrom studied the lysis caused by accumulation of various galactosides in *E. coli* protoplasts prepared by a modification of the lysozyme-versene technique of Repaske [Biochim. et Biophys. Acta, **22**, 189–191 (1956)]. The strain used was a mutant possessing an inducible galactoside permease, and devoid of galactosidase, *i.e.*, able to accumulate, but unable to hydrolyze lactose or other galactosides. It was observed that protoplasts prepared from induced cells (*i.e.*, cells grown in presence of an inducer of galactoside permease) and suspended in 0.1 M phosphate buffer, underwent lysis upon addition of M/100

lactose, while protoplasts from uninduced cells were insensitive to addition of lactose. Comparisons with lysis provoked by decreasing the molarity of the suspending buffer showed that the induced protoplasts had accumulated *free* lactose to the extent of about 22% of their dry weight, a figure closely approximating direct estimations performed on intact cells. The addition of a rapidly metabolizable carbohydrate, namely glucose, did not result in any significant lysis.

These experiments demonstrate that the bulk, if not the totality, of the substrates of galactoside-permease are accumulated within the cells in a free form. Therefore the accumulating mechanism must be catalytic, and the retention is due to a permeability barrier, not to binding or other intermolecular forces.

REFERENCES

1. AMOS, H., AND COHEN, G. N. 1954 Amino acid utilization in bacterial growth. 2. A study of threonine-isoleucine relationships in mutants of *Escherichia coli*. Biochem. J., **57**, 338–343.
2. BARRETT, J. T., LARSON, A. D., AND KALLIO, R. E. 1953 The nature of the adaptive lag of *Pseudomonas fluorescens* toward citrate. J. Bacteriol., **65**, 187–192.
3. BERGMANN, E. D., SICHER, S., AND VOLCANI, B. E. 1953 Action of substituted phenyl-alanines on *Escherichia coli*. Biochem. J., **54**, 1–13.
4. BOLTON, E. T., BRITTEN, R. J., COWIE, D. B., CREASER, E. H., AND ROBERTS, R. B. 1956 Annual Report, Biophysics. In *Carnegie Institution of Washington year book*.
5. BONNER, D. M. 1955 Aspects of enzyme formation. In *Symposium on amino acid metabolism*, pp. 193–197. Edited by W. D. McElroy and B. Glass, Johns Hopkins Press, Baltimore, Md.
6. BRITTEN, R. J., ROBERTS, R. B., AND FRENCH, E. F. 1955 Amino acid adsorption and protein synthesis in *Escherichia coli*. Proc. Natl. Acad. Sci. U. S., **41**, 863–870.
7. BUTTIN, G. 1955 Contribution à l'étude d'un enzyme inductible chez *Escherichia coli*. Diplôme d'Etudes Supérieures, Université de Paris.
8. COHEN, G. N. 1957 Synthèse de protéines "anormales" chez *Escherichia coli* K12 cultivé en présence de L-valine. Ann. inst. Pasteur, *in press*.
9. COHEN, G. N., AND RICKENBERG, H. V. 1955 Etude directe de la fixation d'un inducteur

de la β-galactosidase par les cellules d'*Escherichia coli*. Compt. rend., **240**, 466–468.

10. COHEN, G. N., AND RICKENBERG, H. V. 1955 Existence d'accepteurs spécifiques pour les aminoacides chez *Escherichia coli*. Compt. rend., **240**, 2086–2088.

11. COHEN, G. N., AND RICKENBERG, H. V. 1956 Concentration specifique réversible des aminoacides chez *Escherichia coli*. Ann. inst. Pasteur, **91**, 693–720.

12. COHEN-BAZIRE, G., AND JOLIT, M. 1953 Isolement par sélection de mutants d'*Escherichia coli* synthétisant spontanément l'amylomaltase et la β-galactosidase. Ann. inst. Pasteur, **84**, 937–945.

13. COHN, M. 1956 On the inhibition by glucose of the induced synthesis of β-galactosidase in *Escherichia coli*. In reference 26, pp. 41–46.

14. DANIELLI, J. F. 1952 Structural factors in cell permeability and secretion. In *Structural aspects of cell physiology*, Symposia of the Society for Experimental Biology, VI, pp. 1–15, Academic Press, Inc., New York.

15. DANIELLI, J. F. 1954 Morphological and molecular aspects of active transport. In *Active transport and secretion*, Symposia of the Society for Experimental Biology, VIII, pp. 502–516, Academic Press, Inc., New York. Butterworths Scientific Publications, London (14 pp.).

16. DANIELLI, J. F. 1954 The present position in the field of facilitated diffusion and selective active transport. In *Colston Papers*, VII.

17. DAVIS, B. D. 1951 Aromatic biosynthesis. I. The rôle of shikimic acid. J. Biol. Chem., **191**, 315–325.

18. DAVIS, B. D. 1956 Relations between enzymes and permeability (membrane transport) in bacteria. In reference 26, pp. 509–522.

19. DEERE, C. J., DULANEY, A. D., AND MICHELSON, I. D. 1939 The lactase activity of *Escherichia coli*-mutabile. J. Bacteriol., **37**, 355–363.

20. DEMEREC, M. 1956 In reference 26, pp. 131–137.

21. DOERMANN, A. H. 1944 A lysineless mutant of *Neurospora* and its inhibition by arginine. Arch. Biochem., **5**, 373–384.

22. DOUDOROFF, M. 1951 The problem of the "direct utilization" of disaccharides by certain microorganisms. In *Symposium on phosphorus metabolism*, **1**, pp. 42–48. Edited by W. D. McElroy and B. Glass, Johns Hopkins University Press, Baltimore, Md.

23. DOUDOROFF, M., HASSID, W. Z., PUTNAM, E. W., POTTER, A. L., AND LEDERBERG, J. 1949 Direct utilization of maltose by *Escherichia coli*. J. Biol. Chem., **179**, 921–934.

24. DOUDOROFF, M., PALLERONI, N. J., MACGEE, J., AND OHARA, M. 1956 Metabolism of carbohydrates by *Pseudomonas saccharophila*. I. Oxidation of fructose by intact cells and crude cell-free preparations. J. Bacteriol., **71**, 196–201.

25. FOULKES, E. C. 1951 The occurrence of the tricarboxylic acid cycle in yeast. Biochem. J., **48**, 378–383.

26. GAEBLER, O. H., Editor. 1956 *Enzymes: Units of biological structure and function*. Academic Press, Inc., New York.

27. GALE, E. F. 1947 The passage of certain amino acids across the cell wall and their concentration in the internal environment of *Streptococcus faecalis*. J. Gen. Microbiol., **1**, 53–76.

28. GALE, E. F. 1954 The accumulation of amino acids within staphylococcal cells. In *Active transport and secretion*, Symposia of the Society for Experimental Biology, VIII, pp. 242–253, Academic Press, Inc., New York.

29. GILVARG, C., AND DAVIS, B. D. 1954 Significance of the tricarboxylic acid cycle in *Escherichia coli*. Federation Proc., **13**, 217.

30. GOLDACRE, R. J. 1952 The folding and unfolding of protein molecules as a basis of osmotic work. Intern. Rev. of Cytol., **1**, 135–164.

31. HALVORSON, H. O., AND COHEN, G. N. 1957 Incorporation comparée des aminoacides endogènes et exogènes dans les protéines de la levure. Ann. inst. Pasteur, *in press*.

32. HALVORSON, H., FRY, W., AND SCHWEMMIN, D. 1955 A study of the properties of the free amino acid pool and enzyme synthesis in yeast. J. Gen. Physiol., **38**, 549–573.

33. HARDING, W. M., AND SHIVE, W. 1948 Biochemical transformations as determined by competitive analogue-metabolite growth inhibitions. VIII. An interrelationship of methionine and leucine. J. Biol. Chem., **174**, 743–756.

34. HELFERICH, B., AND TÜRK, D. 1956 Synthese einiger β-D-Thiogalactoside. Chem. Ber., **89**, 2215–2219.

35. HELFERICH, B., TÜRK, D., AND STOEBER, F. 1956 Die Synthese einiger Thioglucuronide. Chem. Ber., **89**, 2220–2224.

36. HERZENBERG, L. A. 1957 Biochim. et Biophys. Acta, *in press*.

37. HIRSCH, M.-L., AND COHEN, G. N. 1953 Peptide utilization by a leucine-requiring

mutant of *Escherichia coli*. Biochem. J., **53**, 25-30.

38. HOGNESS, D. S., COHN, M., AND MONOD, J. 1955 Studies on the induced synthesis of β-galactosidase in *Escherichia coli*: the kinetics and mechanism of sulfur incorporation. Biochim. et Biophys. Acta, **16**, 99-116.

39. KARLSSON, J. L., AND BARKER, H. A. 1948 Evidence against the occurrence of a tricarboxylic acid cycle in *Azotobacter agilis*. J. Biol. Chem., **175**, 913-921.

40. KEPES, A. 1957 Metabolisme oxydatif lié au fonctionnement de la galactoside-perméase d'*Escherichia coli*. Compt. rend., **244**, 1550-1553.

41. KEPES, A., AND MONOD, J. 1957 Etude du fonctionnement de la galactoside-perméase d'*Escherichia coli*. Compt. rend., **244**, 809-811.

42. KIHARA, H., AND SNELL, E. E. 1952 L-alanine peptides and growth of *Lactobacillus casei*. J. Biol. Chem., **197**, 791-800.

43. KOGUT, M., AND PODOSKI, E. P. 1953 Oxidative pathways in a fluorescent *Pseudomonas*. Biochem. J., **55**, 800-811.

44. KORKES, S., STERN, J. R., GUNSALUS, I. C., AND OCHOA, S. 1950 Enzymatic synthesis of citrate from pyruvate and oxaloacetate. Nature, **166**, 439-440.

45. KRAMPITZ, L. O., AND LYNEN, F. 1956 Formation of oxaloacetate from d-tartrate. Federation Proc., **15**, 292-293.

46. LAMPEN, J. O., AND JONES, M. J. 1947 Interrelations of norleucine and methionine in the nutrition of *Escherichia coli* and of a methionine-requiring mutant of *Escherichia cóli*. Arch. Biochem., **13**, 47-53.

47. LA RIVIERE, J. W. M. 1956 Specificity of whole cells and cell-free extracts of *Pseudomonas putida* towards (+), (−), and meso-tartrate. Biochim. et Biophys. Acta, **22**, 206-207.

48. LEDERBERG, E. M. 1952 Allelic relationships and reverse mutation in *Escherichia coli*. Genetics, **37**, 469-483.

49. LEDERBERG, J. 1950 The β-D-galactosidase of *Escherichia coli*, strain K-12. J. Bacteriol., **60**, 381-392.

50. LEDERBERG, J., LEDERBERG, E. M., ZINDER, N. D., AND LIVELY, E. R. 1951 Recombination analysis of bacterial heredity. Cold Spring Harbor Symposia on Quant. Biol., **16**, 413-443.

51. LE FEVRE, P. G. 1954 The evidence for active transport of monosaccharides across the red cell membrane. In *Active transport and secretion*, Symposia of the Society for Experimental Biology, VIII, pp. 118-135, Academic Press, Inc., New York.

52. LEIBOWITZ, J., AND HESTRIN, S. 1945 Alcoholic fermentation of the oligosaccharides. Advances in Enzymol., **5**, 87-127.

53. MANDELSTAM, J. 1956 Inhibition of bacterial growth by selective interference with the passage of basic amino acids into the cell. Biochim. et Biophys. Acta, **22**, 324-328.

54. MATHIESON, M. J., AND CATCHESIDE, D. G. 1955 Inhibition of histidine uptake in *Neurospora crassa*. J. Gen. Microbiol., **13**, 72-83.

55. MITCHELL, P. 1954 Transport of phosphate through an osmotic barrier. In *Active transport and secretion*, Symposia of the Society for Experimental Biology, VIII, pp. 254-261, Academic Press, Inc., New York.

56. MITCHELL, P., AND MOYLE, J. 1951 The glycerophospho-protein complex envelope of *Micrococcus pyogenes*. J. Gen. Microbiol., **5**, 981-992.

57. MITCHELL, P., AND MOYLE, J. 1956 Liberation and osmotic properties of the protoplasts of *Micrococcus lysodeikticus* and *Sarcina lutea*. J. Gen. Microbiol., **15**, 1956, 512-520.

58. MITCHELL, P., AND MOYLE, J. 1956 Osmotic function and structure in bacteria. In *Bacterial anatomy*, pp. 150-180. Edited by E. T. C. Spooner and B. A. D. Stocker. Cambridge University Press, Cambridge, England.

59. MITCHELL, P., AND MOYLE, J. 1957 Autolytic release and osmotic properties of "protoplasts" from *Staphylococcus aureus*. J. Gen. Microbiol., **16**, 184-194.

60. MONOD, J. 1956 Remarks on the mechanism of enzyme induction. In reference 26, pp. 7-28.

61. MONOD, J., COHEN-BAZIRE, G., AND COHN, M. 1951 Sur la biosynthèse de la β-galactosidase (lactase) chez *Escherichia coli*. La spécificité de l'induction. Biochim. et Biophys. Acta, **7**, 585-599.

62. MONOD, J., AND COHN, M. 1952 La biosynthèse induite des enzymes (adaptation enzymatique). Advances in Enzymol., **13**, 67-119.

63. MONOD, J., AND COHN, M. 1953 Sur le mécanisme de la synthèse d'une proteine bactérienne. *Symposium on bacterial metabolism*, pp. 42-62. 6th Intern. Congr. Microbiol., Rome, Italy.

64. MONOD, J., HALVORSON, H. O., AND JACOB, F. 1957 Compt. rend., *in preparation*.

65. MONOD, J., PAPPENHEIMER, A. M., JR., AND COHEN-BAZIRE, G. 1952 La cinétique de

la biosynthèse de la β-galactosidase chez *Escherichia coli* considérée comme fonction de la croissance. Biochim. et Biophys. Acta, **9**, 648–660.

66. MONOD, J., AND TORRIANI, A.-M. 1950 De l'amylomaltase d'*Escherichia coli*. Ann. inst. Pasteur, **78**, 65–77.

67. MUNIER, R. L., AND COHEN, G. N. 1956 Incorporation d'analogues structuraux d'aminoacides dans les protéines bactériennes. Biochim. et Biophys. Acta, **21**, 592–593.

68. NOVICK, A., AND WEINER, M. 1957 Enzyme induction, an all or none phenomenon. Proc. Natl. Acad. Sci. U. S., *in press*.

69. PARDEE, A. B. 1957 An inducible mechanism for accumulation of melibiose in *Escherichia coli*. J. Bacteriol., **73**, 376–385.

70. PASTEUR, L. 1860 Note relative au *Penicillium glaucum* et à la dissymétrie moléculaire des produits organiques naturels. Compt. rend., **51**, 298–299.

71. RICKENBERG, H. V., COHEN, G. N., BUTTIN, G., AND MONOD, J. 1956 La galactoside-perméase d'*Escherichia coli*. Ann. inst. Pasteur, **91**, 829–857.

72. ROBERTS, R. B., ABELSON, P. H., COWIE, D. B., BOLTON, E. T., AND BRITTEN, R. J. 1955 Studies of biosynthesis in *Escherichia coli*. Carnegie Inst. of Wash. Publ. 607, Washington, D. C.

73. ROWLEY, D. 1953 Interrelationships between amino-acids in the growth of coliform organisms. J. Gen. Microbiol., **9**, 37–43.

74. SHILO, M. 1957 The enzymic conversion of the tartaric acids to oxaloacetic acid. J. Gen. Microbiol., **16**, 472–481.

75. SHILO, M., AND STANIER, R. Y. 1957 The utilization of the tartaric acids by pseu-

domonads. J. Gen. Microbiol., **16**, 482–490.

76. STERN, J. R., SHAPIRO, B., AND OCHOA, S. 1950 Synthesis and breakdown of citric acid with crystalline condensing enzyme. Nature, **166**, 403–404.

77. STOEBER, F. 1957 Sur la β-glucuronide-perméase d'*Escherichia coli*. Compt. rend., **244**, 1091–1094.

78. STONE, R. W., AND WILSON, P. W. 1952 Respiratory activity of cell-free extract from *Azotobacter*. J. Bacteriol., **63**, 605–617.

79. SWIM, H. E., AND KRAMPITZ, L. O. 1954 Acetic acid oxidation by *Escherichia coli*: evidence for the occurrence of a tricarboxylic acid cycle. J. Bacteriol., **67**, 419–425.

80. SWIM, H. E., AND KRAMPITZ, L. O. 1954 Acetic acid oxidation by *Escherichia coli*: quantitative significance of the tricarboxylic acid cycle. J. Bacteriol., **67**, 426–434.

81. TATUM, E. L. 1946 Induced biochemical mutations in bacteria. Cold Spring Harbor Symposia Quant. Biol., **11**, 278–284.

82. THOMAS, C. A. 1956 New scheme for performance of osmotic work by membranes. Science, **123**, 60–61.

83. WEIBULL, C. 1956 Bacterial protoplasts; their formation and characteristics. In *Bacterial anatomy*, pp. 111–126. Edited by E. T. C. Spooner and B. A. D. Stocker, Cambridge University Press, Cambridge, England.

84. WILBRANDT, W. 1954 Secretion and transport of non-electrolytes. In *Active transport and secretion*, Symposia of the Society for Experimental Biology, VIII, pp. 136–162, Academic Press, Inc., New York.

$$9$$

Kinetic Studies of Galactoside Permease in *Escherichia coli*

A. KEPES

Abstract

Galactoside permease–positive *E. coli* accumulates radioactive thiogalactosides in the intracellular space until a steady state is reached. Measurements of the steady-state concentration level, of the affinity, and of the rate of turnover in the steady state make possible the calculation of the activity of the uptake mechanism and of the passive exit rate constant. The entry mechanism is catalytic; it is coupled with metabolic energy donors. The energy consumption corresponds to the hydrolysis of one mole of ATP per mole of TMG flowing inward. Passive exit is not by free diffusion, but can be understood as a carrier diffusion.

A model is proposed which is compatible with the observed behavior of the entry and exit parameters under various conditions.

Introduction*

A steadily increasing number of phenomena of selective permeation, active transport across multicellular biological membranes, and concentration or extrusion of substances by the cells have already been described. The importance of these phenomena in biological control is obvious. The mechanism of such phenomena has been the subject of numerous speculations; however, the basic structures for the selectivity of permeation or for energy coupling, in the case of active transport, have never been identified.

---

*Abbreviations used: TMG, methyl-β-thio-D-galactoside; TDG, galactosyl-β-thio-D-galactoside; TPG, phenyl-β-thio-D-galactoside; DNP, 2,4-dinitrophenol; pCMB, p-chloromercuribenzoate; ATP, adenosine triphosphate.

The bacterial permeases described by Rickenberg, Cohen, Buttin, and Monod [1] are particularly suited to the study of such a mechanism. The subject of this study, galactoside permease, is particularly well individualized because of its stereospecificity, its inducible synthesis, and the existence of simple mutants. Furthermore, it affords simple techniques in the use of nonmetabolizable and radioactive thiogalactosides, and because of the high rates of permeation which permit us to measure the energy requirements of the process.

It was shown [1] that the concentration of intracellular substrate in a steady state obeys the equation*

$$G_{in} = Y \frac{G_{ex}}{K_m + G_{ex}} \tag{1}$$

which could result either from the adsorption of galactosides on Y-specific sites (μmoles/g) with an affinity constant of $1/K_m$ or from a catalytic entry mechanism obeying the Michaelis law with constant K_m. The rate of this entry reaction,

$$V_{in} = V_{in}^{max} \frac{G_{ex}}{K_m + G_{ex}} \tag{2}$$

must be compensated for in a steady state by a passive exit proportional to the concentration gradient†:

$$V_{ex} = G_{in} k_{ex} \tag{3}$$

In steady state $V_{in} = V_{ex}$, hence

$$G_{in} = \frac{V_{in}^{max}}{k_{ex}} \frac{G_{ex}}{K_m + G_{ex}} \tag{4}$$

This equation is formally identical to (1) except that Y, the saturation capacity, is no longer an intrinsic parameter of the system but a quotient V_{in}^{max}/k_{ex}, which can vary.

Previous kinetic studies permitted elimination of the adsorption hypothesis and supported the hypothesis of a catalytic mechanism [2].

Study of the rates of entry, exchange, and exit under various conditions might therefore uncover the independent variations of these processes (the catalyzed entry and the passive exit) and thus throw some light on their nature.

The finding of an increased oxygen consumption related to the functioning of the permease [3] entails the following question: Is the concentration of galactosides

---

*Symbols used: G, a β-galactoside or β-thio galactoside substrate for the permease or its concentration; G_{in}, G_{ex}, concentration in the intracellular or extracellular medium; K_m, concentration of half-saturation; Y, G_{in}^{max}, intracellular concentration at equilibrium for a saturating external concentration or "specific capacity"; V, reaction rate; k, reaction rate constant; the subscripts "in" and "ex" refer to the reaction of entry and exit, respectively; $t_{1/2}$, time of half-equilibration or half-exchange.
†We neglect G_{ex} before G_{in}. This introduces an error of 1 percent or less.

147

by permease a case of active transport? Or, in more concrete terms: are the accumulated galactosides in the same molecular form as in the medium, and does the energy spent appear in the form of osmotic work, or in the form of bond energy with a cellular constituent still to be defined? Nothing at the present time seems to support this second hypothesis. On the contrary, the osmotic contribution of intracellular galactosides [4] is a strong argument in favor of active transport.

Finally, the passive penetration of galactosides in permease-less cells should offer a model of passive permeation which could be compared fruitfully with the passive exit mechanism of galactosides in cells where the permease does function.

Material and Methods

The experiments reported here have been performed, except when otherwise indicated, using strain ML-308 with a constitutive permease and ML-3, a permeaseless strain.

The microorganisms are cultivated in mineral medium 63 with potassium succinate or glucose as the source of carbon and agitated at 34°C. They are collected by centrifuging during the exponential growth phase and resuspended at a density of about 200 μg/ml in a mineral medium to which 4 g/liter of potassium succinate and 50 μg/ml of chloromycetin are added. The suspension is agitated at the experimental temperature for at least 15 min before adding the substrate.

The following substrates were used: methyl-β-D-thiogalactoside, or TMG; β-D-thiogalactoside-β-D-galactoside (thiodigalactoside) TDG; phenylthio-β-D-galactoside TPG; lactose (glucose-6-β-D-galactoside).

The radioactive thiogalactosides were labeled with ^{35}S (TDG and TPG) or with ^{14}C on the methyl group (TMG).

At various times after addition of radioactive substrate to the agitated suspension, 1-ml aliquots were removed and passed through a millipore filter. The filter on the filter holder had been coverd earlier with 5–6 ml of ice-cold medium. The suspension is mixed with this volume and instantaneously cooled. This moment, which is considered the exact sampling time, is defined to the nearest 2–3 sec. The filtration is followed by two rinses with 2 ml of ice-cold medium. The millipore is then dried and its radioactivity is measured without further handling. The intracellular lactose concentration was measured, after elution with boiling water, by the Somogyi–Nelson method.

Oxygen consumption was measured with a dropping mercury electrode at a constant voltage of 0.6 V against a saturated calomel electrode.

The emission of ^{14}CO$_2$ was measured by bubbling CO$_2$-free air through the incubation mixture and then through a sodium hydroxide solution in a microabsorber.

Results

Exchange

Figure 1 shows a kinetic experiment of accumulation (A) and exchange (B and C) at steady state of TMG at a concentration of $1 \times 10^{-3} M$. In these latter two (B

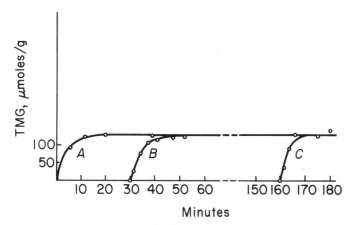

Figure 1. Kinetics of the entry of TMG (A) and exchange kinetics in steady state (B and C) for ML308 suspended in an inorganic medium with maltose. Concentration of TMG $10^{-3}M$ throughout, temperature 15°C.

and C) nonradioactive TMG is added to the microbes at time 0 and the labeled TMG is added after 30 and 160 mins, respectively, without changing the final concentration of TMG. It is observed (a) that all the TMG is exchangeable, and the plateaus of the curves A, B, and C join each other; and (b) that the exchange takes place at the same rate as the initial equilibration. The entry-reaction rate is thus practically independent of the intracellular concentration.

Determination of Parameters

Figure 2 shows an experiment on the exchange of TDG at steady state at two different concentrations.

We see that the initial slopes are proportional to the concentrations reached at equilibrium.

The shape of these curves is exponential and can be represented as

$$\log (G_{in}^t - G_{in}^\infty) = f(t).$$

This gives the straight lines in Fig. 2b. From these straight lines it is easy to determine $t_{1/2}$, the time of half-equilibration.

The initial exchange rate or the slope at the origin of the curves in Fig. 2a is as follows:

$$V_{in} = G_{in}^\infty \frac{\ln 2}{t_{1/2}} . \tag{5}$$

By comparing equations (4) and (5), we find that

$$k_{ex} = \frac{\ln 2}{t_{1/2}} . \tag{6}$$

Thus we see that if the measurement of the intracellular concentrations at steady state as a function of the extracellular concentration of galactoside permits the graphic

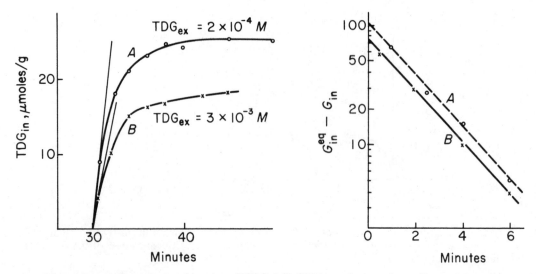

Figure 2. At left, exchange kinetics of TDG, ML-308 in an inorganic medium with maltose, TDG $2 \times 10^{-4}M$ (A) and $3 \times 10^{-5}M$ (B). Temperature 26°C. At right, the same experiment represented according to eq. (5).

determination of Y and K_m, then a kinetic experiment permits the determination of $t_{1/2}$ and hence of k_{ex}. The rate of entry at saturation is given by

$$V_{in}^{max} = Y k_{ex}. \tag{7}$$

G_{ex} and K_m are expressed in moles/liter, G_{in} and Y in μmoles/g of dry weight.

If we want to consider the ratio of the extra- and intracellular concentrations, we must take into account the available volume.

The upper limit of the concentration ratio is reached for low values of G_{ex} and G_{in} and is equal to the slope at the origin of the hyperbolic curve $G_{in} = f(G_{ex})$, [eq. (1)]. This ratio is

$$R = (Y/K_m) \times (\text{dry wt./available volume}). \tag{8}$$

Table 1 summarizes the parameters thus defined for TMG, TDG, and TPG with respect to the permease of ML-308 at 26°C. Note that no pair of basic parameters shows any clear correlations. The substrate with the strongest affinity is TDG, the substrate corresponding to the greatest capacity is TMG, and the substrate with the most rapid exit is TPG. The rate of exchange of TPG is so rapid that only its lower limit could be evaluated. R is higher for TDG than for the other two substrates (the ratio of dry weight to volume was evaluated at 0.25).

The values of Y and the parameters resulting from it are subject to large variations from one culture to another, depending on the strains, and, of course, on the degree of induction for inducible strains. In contrast, the relative values of these parameters for the three substrates, as well as the K_m's and $t_{1/2}$'s are characteristic of the system. The accumulation of lactose by strain W-2244 (which does not utilize lactose) reaches 550 μmoles/g at +4°C for a concentration of $5 \times 10^{-3} M$ in the medium, or 20 percent of the dry weight of the cells.

Paper chromatography showed that the only detectable reducing sugar was lactose.

150

Table 1. Characteristics of galactoside permease in *E. coli* ML-308 for three thiogalactosides at 26°C.

Characteristic	Substrate		
	TMG	TDG	TPG
Michaelis constant K_m (moles/liter)	5×10^{-4}	2×10^{-5}	2.5×10^{-4}
Capacity Y (μmoles/g)	160	40	32
Time of half-equilibration $t_{1/2}$ (min)	0.75	1.35	<0.25
Maximum initial rate of entry			
in state V_{in}^{max}	148	20.4	>86
k_{ex} (ml/g/min)	0.82	0.59	>2.7
Ratio R of maximum concentration	65	400	26

Energy Requirements

Since accumulation of galactosides is inhibited by inhibitors of oxidative phosphorylation such as 2,4-DNP or sodium azide, the effect of the functioning of galactoside permease on oxygen consumption was measured with a dropping mercury electrode. Chloromycetin was omitted in the course of the experiments because of its polarographic interference. Furthermore, the carbon-source was omitted in order to prevent the slight increase in the Q_{O_2} due to the functioning of the permease from being masked by the intensive basal respiration. The fluctuations in these measurements are much larger than those occurring in determination of permease activity.

As soon as thiogalactoside is added, respiration increases and then remains constant for 30–40 min.

If the concentration of thiogalactoside is not saturating for the permease, a further addition causes a new increase in Q_{O_2}.

The endogenous Q_{O_2}, which is 5–10 μmoles/g dry weight per minute, is conventionally subtracted in order to determine the Q_{O_2} due to the functioning of the permease.

All the Q_{O_2} values were extrapolated to the substrate saturating concentration by using the K_m determined for the permease. Their average for TMG is 8.9 μmoles of oxygen with a standard error of ± 3 μmoles/g/min for 16 determinations. The value for TPG is 15.0 $\pm$ 3.4 μmoles/g/min for 12 determinations.

The strains without permease, noninduced ML-30 or ML-3, show no increase in Q_{O_2} upon addition of thiogalactoside.

The direct proportion between "extra oxygen" and the functioning of permease is thus both qualitative and quantitative. In order to exclude the possibility that some impurity present in the substrate might be the cause of this oxygen consumption, we examined the CO_2 released in order to determine whether it originated from the substrate or from the microbes.

The release of $^{14}CO_2$ from a rinsed suspension of microbes which had been grown on ^{14}C-labeled fructose was measured in the absence and in the presence of nonradioactive TMG and TPG. As seen in Fig. 3, the liberation of $^{14}CO_2$ has

151

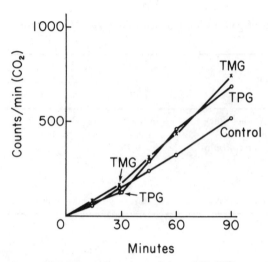

Figure 3. Liberation of $^{14}CO_2$ as a function of time. ML-308 grown on ^{14}C-labeled fructose washed and suspended in an inorganic medium without an external carbon source. Temperature 21°C. TMG $2 \times 10^{-3}M$, TPG $10^{-3}M$.

increased by a factor of 1.75 upon addition of $2 \times 10^{-3} M$ TMG, and by a factor of 2.5 by addition of $10^{-3} M$ TPG (we did not determine the absolute values of Q_{CO_2}). These relative increases are of the same order of magnitude as those of Q_{O_2} for the same two substrates.

These experiments show that the functioning of the galactoside permease stimulates cellular oxidations. It seems that this stimulation may be correlated with the energy requirements of galactoside transport. On the other hand, these requirements seem to be proportional to the rate of turnover and do not cease once steady state has been reached.

However, the possibility was not excluded that a part of the steady-state exchange may occur without calling on metabolic energy. A slight excess of oxygen consumption, limited to the first few minutes of measurement, could have escaped observation.

A kinetic experiment with 2,4-dinitrophenol as inhibitor allows us to solve this problem. Figure 4 shows such an experiment. We can see that 5 min after addition of DNP the exchange still represents 62 percent of that observed prior to addition of the inhibitor, while the rate of uptake is reduced to 28 percent of the control. If we consider that the source of the residual energy is the same in both these experiments, i.e., sufficient to ensure 28 percent of the initial rate, we then arrive at the conclusion that a fraction of the order of 34 percent of the steady-state exchange can be produced without a new supply of metabolic energy. If this result is transposed to TMG under the conditions of respirometric measurements, we find a ratio TMG/O = $60/(17.8 \pm 6) = 3.8 \pm 1.3$. If we consider that V_{in} of TMG is rather overestimated under the conditions of respirometric experiments which were made without a carbon source, then we must assign to the TMG/O ratio the more probable value of 3, which could reflect the utilization of high-energy pyrophosphate bonds in the transport process.

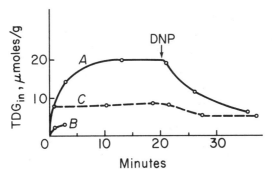

Figure 4. Entry kinetics of the TDG and exit kinetics after addition of DNP $2 \times 10^{-2} M$ (A). Entry kinetics of TDG 5 min after addition of DNP (B). Exchange in 1 min under conditions identical to A (C). ML-308 in medium 62 plus inorganic succinate. TDG $10^{-4} M$. Temperature, 26°C.

Temperature Coefficient

Figure 5 represents two exchange experiments, with TMG on one hand and with TDG on the other, at three temperatures, 34°C, 24°C, and 14°C. It is striking to see that although the entry rate of the two substrates varies, with a Q_{10} of the order of 2, the concentration reached at the steady state is practically independent of temperature for TDG, while for TMG it varies inversely with temperature. From this we can deduce that the constant of the exit rate for TDG has a Q_{10} which is about equal to that of the entry rate, while the exit of TMG has a Q_{10} of the order of 4.

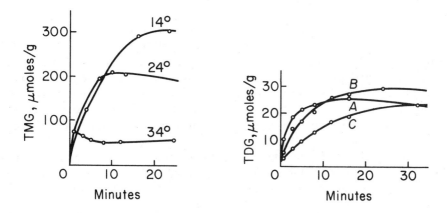

Figure 5. Exchange kinetics of the TMG ($10^{-3} M$) and TDG ($10^{-4} M$) at three different temperatures. ML-308 in an inorganic succinate medium.

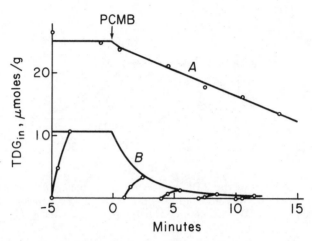

Figure 6. Kinetics of exit of previously equilibrated TDG upon addition of pCMB $10^{-4}M$ (A). Exchange after 45 sec and 90 sec under the same conditions (B), ML-308, in mineral medium with succinate, TDG $10^{-4}M$. Temperature, 26°C.

Inhibitors

We have seen that in the presence of 2,4-DNP (and the same is true for sodium azide) the accumulation of thiogalactosides is inhibited. If the inhibitor is added after accumulation, the intracellular concentration drops slowly. These agents apparently act through the inhibition of energy coupling and not directly on the specific catalyst. The cellular reserves of energy-rich pyrophosphate compounds, as well as a possible exchange with conservation of energy, may explain the slowness of the observed effect. A direct inhibitor of the mechanism catalyzing entry should cause an exponential exit with the same time constant as the steady-state exchange.

The substance p-chloromercuribenzoate (pCMB) is an inhibitor for SH groups and as such inhibits many metabolic enzymes. However, in its presence, the turnover due to permease decreases immediately and exponentially.

Inactivation by pCMB is total and irreversible after a sufficient time of action. The rate of inactivation depends on pCMB concentration, but for the same concentration of inhibitor it depends on whether a thiogalactoside is present. The presence of a thiogalactoside exerts a protective effect in a more or less competitive fashion. This protection by a specific substrate which does not react directly with pCMB suggests that the inhibitor acts directly on the molecule that carries the specific sites, i.e., on the permease itself.*

Figure 6 shows an experiment of permease inhibition by pCMB at a concentration of $10^{-4}M$. The substrate is TDG at a concentration of $10^{-4}M$. We can see the exponential decrease in the rate of exchange, which contrasts with the slowness and quasi-

---

*This observation provides the origin of the specific labeling technique of Fox and Kennedy (Paper 12). [Note added by A. Kepes, 1973.]

linear course of TDG exit. At the moment the exchange rate has dropped to a negligible value, exit of TDG is about five times slower than steady-state exchange.

Competition

We have seen in the first paragraph that the stationary level is proportional to the rate of turnover for a given substrate at a specific temperature, and eq. (4) shows that the coefficient of proportionality is $1/k_{ex}$.

Figure 7 shows, using the method of Lineweaver and Burk, the inhibition of the rate of entry and of the stationary level of TDG by TMG at various concentrations. Figure 8 shows the inhibition of the rate of entry and of the stationary level of TMG by TDG under the same conditions.

We can see that TMG inhibits the stationary level more than the rate of entry of TDG, and, conversely, that TDG inhibits the rate of entry of TMG more than its steady-state level.

This implies that the coefficient of proportionality between the steady-state level and rate of entry $1/k_{ex}$ of each substrate is modified by the presence of another substrate added as a competitor. The acceleration of the turnover of TDG in the presence of TMG when measured directly confirms this interference.

Another anomaly appears on examination of the curves in Fig. 7. This is that at very low concentrations of TMG the rate of turnover of TDG is not inhibited; on the contrary it is stimulated. In our discussion we shall attempt to interpret this phenomenon.

Another atypical inhibition, that exerted by glucose on the galactoside permease, has been observed by Horecker [6]. The addition of glucose depresses the accumulation level of thiogalactosides. A kinetic study has shown that glucose acts essentially

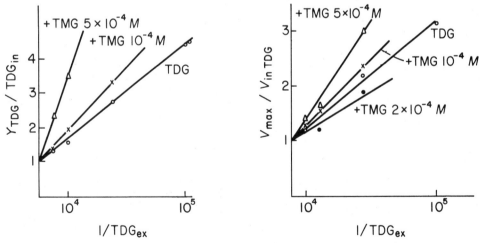

Figure 7. Representation according to Lineweaver and Burk of intracellular concentrations in steady state (at left) and rates of entry (at right) of TDG in the presence and absence of TMG. Temperature, 26°C.

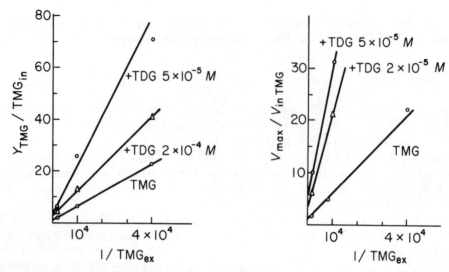

Figure 8. Representation according to Lineweaver and Burk of intracellular concentrations in steady state (at left) and of the rates of entry of TMG in the absence and presence of TDG. Temperature, 26°C.

by accelerating the exit of thiogalactoside, while its effect on the rate of entry is slight or nil.

Passive Permeation in Permease-less Strains

ML-3, lacking galactoside permease, does not concentrate the galactosides, so the experimental method had to be slightly modified: suspensions about 10 times more concentrated were used with 0.1-ml test samples, so that contamination of the filter was diminished by a factor of 10. We were able to estimate it to be less than 10 percent of the final plateau.

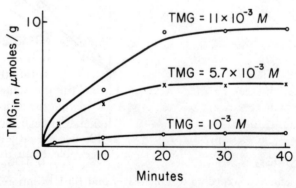

Figure 9. Kinetics of entry of TMG in the permease-deficient strain ML-3, at three concentrations. Temperature, 26°C.

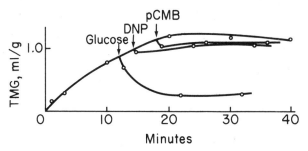

Figure 10. Kinetics of entry of TMG in the permease-less strain ML-3 (A). Displacement by addition of glucose $2 \times 10^{-2}M$ (B). Effect of DNP $2 \times 10^{-3}M$ (C) and of pCMB $10^{-4}M$ (D).

Figure 9 shows the kinetics of entry of TMG at three different concentrations. Three facts may be deduced from this experiment. First, the final level of intracellular concentration is proportional to the concentration in the medium. Second, the equilibrium time is of the order of 10 times longer than that observed with permease. Finally, the final concentration at equilibrium corresponds to an available volume of the order of 1 μl/mg of dry cells, clearly a lower value than the aqueous space usually assumed, i.e., about 4 μl.

As could have been predicted by the proportionality of the extracellular and intracellular concentrations, the entry of a galactoside in ML-3 is not inhibited by competition with another galactoside, and similarly there is no displacement after the establishment of equilibrium. However, the intracellular galactoside is exchangeable with the galactoside in the medium roughly at the same rate as that of the initial entry.

Metabolic inhibitors such as DNP, as well as pCMB, have no effect on the permeation of TMG, as shown in Fig. 10. This same graph shows that the addition of glucose rapidly and markedly diminishes the intracellular level of galactoside.* We are dealing here with a true active expulsion, which is parallel to the effect of glucose on actively accumulated galactosides in strains with galactoside permease.

Discussion

It follows from these data that the specific accumulation of galactoside cannot be due to adsorption on specific intracellular sites. It is enough to consider Table 1, which shows the variability of the accumulation capacity at saturation for the different substrates, or to consider Fig. 5, which shows that for the same substrate the capacity varies widely with temperature. We can thus state that the action of permease is catalytic, and, further, that the specific catalytic sites that govern the phenomenon are essentially accessible to extracellular substrate and hence are largely independent of the intracellular concentration. It is for a site accessible from the outside that

*ML-3 was grown with glucose as carbon source. With different carbon sources, this effect is not observed. [Note added by A. Kepes, 1973.]

157

the different galactosides enter into competition, since it is their extracellular concentration which controls the competition. This accessibility from the outside stresses the role of these catalytic sites in transport from outside to inside.

The possible nature of a transport catalyst is at the moment in the realm of pure speculation. However, in the case of galactoside permease we may infer that the catalysis is of enzymatic type, due to energy coupling. In fact, the use of metabolic energy implies the cleavage of an energy-rich bond, a rupture that takes place only in the presence of galactoside.

The respiratory acceleration caused by the galactoside in the presence of permease can be the manifestation of the respiratory control described by Lardy and Wellman [7] in mitochondria; the latter is released by the addition of ADP, a phosphate acceptor. It would suffice to assume that the action of permease leads directly or indirectly to the hydrolysis of an ATP molecule with the appearance of an ADP molecule, which would then relieve the respiratory inhibition. The number of TMG molecules transported by an atom of oxygen consumed is perfectly compatible with the use of ATP or a compound derived molecule for molecule from ATP for the transport of one TMG molecule.

The thermodynamic work needed for the transport of 1 mole of TDG at a concentration ratio of about 400 is $\Delta F = 3600$ cal/mole, so the energy from the hydrolysis of ATP is more than sufficient.

Energy units smaller than ATP would be more economical, but if one molecule of ATP had to give more than one of these energy-donor molecules, the ratio of TMG to O would have to be equal to at least 6, which is incompatible with our results.

The active-transport hypothesis is practically inevitable in view of the presently available data on the problem. The osmotic contribution of intracellular galactosides cannot be explained by any other hypothesis except by assuming that the intracellular galactosides are chemically modified. However, the only modified form, i.e., the acetylated derivative of thiogalactosides, which is observed when the permease is functional, represents a very small fraction of the accumulated galactosides. This fraction varies considerably from one substrate to another. This derivative is not detectable when the permease functions without a carbon source. At any rate, the galactoside molecules which had been subjected to one or more cycles of entry and exit from the cells are chemically unchanged, while the acetylated derivative never gives back a detectable amount of the initial galactoside under these conditions.

The model which seems to take into account all our observations without any unnecessary complications is diagrammed in Fig. 11. It shows permease as an enzyme P, which, acting on an energy-donor molecule $R{\sim}T$ and one molecule of galactoside G_{ex}, would activate the latter. This activation could be effected by the combination of galactoside with T (for transporter), a carrier molecule (or radical), TG having physical and chemical characteristics that would allow it to pass the barrier that prevents the native galactoside from crossing. The reaction $R{\sim}T + G \rightarrow GT + R$ would be practically irreversible, and the rate of this reaction is the limiting step in the whole set of phenomena. This rate itself depends mainly on the concentration of G_{ex} and not of $R{\sim}T$, which is assumed always to be in excess. It would suffice if this molecule GT, when inside the cell, would deactivate itself spontaneously, or if it would dissociate

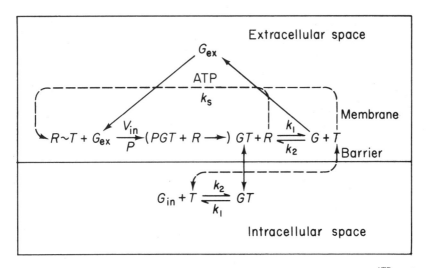

Figure 11. Proposed model of mechanism. (The reaction $T + R \xrightarrow[k_3]{ATP} R \sim T$ is supposed to take place in the cytoplasm. The arrow, as drawn, may be misleading in this regard. [A. Kepes, 1973].)

(it is not necessary to postulate a second enzyme). The existence of the activated molecule is short lived and in catalytic quantities; this explains why its existence could not be demonstrated.

There are, however, two facts supporting the existence of an activated complex of the substrate. The first is the possibility of exchange with conservation of energy, or more exactly the difference in the degree of inhibition of entry and exchange of TDG by the same dose of DNP (Fig. 4). This exchange can take place at the level of the active complex, either bound to the enzyme or not. The second argument can be derived from the paradoxical phenomenon of acceleration of entry of TDG in the presence of a very low concentration of TMG (Fig. 7). Since it is known that the activation of TMG is eight times more rapid than that of TDG (Table 1), it suffices that a large excess of TDG could displace the activated TMG from its complex in order to be able to benefit partly from its rate of activation and to overcompensate for the competition at the permease level.

The mechanism of passive exit under the influence of a concentration gradient poses no special problem in itself. However, the difference of the constants k_{ex} for strains with and without permease, the increase in k_{ex} in the presence of glucose, its reduction by inhibitors of entry, and the change in k_{ex} of TDG when it is in competition with TMG seem to exclude a simple diffusion exit. In effect, we see no plausible reason why the permeability barrier should modify its physical characteristics so that the diffusion of TDG should be accelerated without changing that of the TMG.

We propose the hypothesis that the exit of galactosides takes place by a "facilitated diffusion" process, i.e., in a complexed state with a carrier that enables the substrate to cross the diffusion barrier. This complex would be dissociable and would release the native galactoside outside the barrier. In Fig. 11 this complex is represented as identical to GT, which permits entry. We assume for this purpose that the reaction $GT \rightleftharpoons G + T$ is reversible and spontaneous and that it can take place on the outside surface of the barrier, as well as on its inner surface. The equilibrium would by far favor the dissociation. This reversible reaction represents a passive exit mechanism which depends on the concentration gradient.

The steady state is characterized by the fact that all the intermediates are at a constant concentration.

On the external surface:

$$\frac{d(GT)}{dt} = V_{in} + k_2 G_{ex} T - k_1 GT = 0 \tag{9}$$

In the intracellular space

$$\frac{d(GT)}{dt} = k_1 GT - k_2 G_{in} T = 0 \tag{10}$$

from which

$$GT = \frac{k_2}{k_1} G_{in} T \tag{11}$$

Substituting in (9),

$$V_{in} + k_2 G_{ex} T - k_2 G_{in} T = 0 \qquad \text{or} \qquad G_{in} = G_{ex} + \frac{V_{in}}{k_2 T} \tag{12}$$

If we consider that G_{ex} is negligible compared with G_{in} and that

$$V_{in} = V_{in}^{max} \frac{G_{ex}}{K_m + G_{ex}} \tag{2}$$

we have

$$G_{in} = \frac{V_{in}^{max}}{k_2 T} \frac{G_{ex}}{K_m + G_{ex}} \tag{13}$$

formally identical to eq. (4) with $k_{ex} = k_2 T_{in}$.

The exit constant is thus a transport constant proportional to the gradient as long as T_{in} is constant itself.

The constancy of the level of concentration of T must be adjusted in steady state. At the onset of uptake, T_{in} appears in the intracellular space in a stoichiometric amount with G_{in}, whereas in steady state its concentration must be considerably below that of G_{in} to explain the slowness of exit in the case of inhibition of permease.

Furthermore, in order to explain the stimulation of TDG efflux by TMG and the constancy of k_{ex} for a given substrate, T_{in} must depend more on the nature of the substrate than on the degree of saturation of permease. One of the possible mechanisms for such an adjustment consists in assuming that the total amount of T, the sum of $R{\sim}T$, GT, and free T, is limited.

When the permease does not function, it is all in the form $R{\sim}T$.

In steady state, $R \sim T$ is regenerated from R and free T with a rate equal to V_{in}:

$$R \times T \times k_3 = V_{in} \qquad (14)$$

Under these conditions,

$$R = T + GT$$

but according to eqs. (11) and (13),

$$GT = \frac{V_{in}}{k_1}$$

Substituting in eq. (14), we have

$$T(T + \frac{V_{in}}{k_1}) = \frac{V_{in}}{k_3}$$

If $T << V_{in}/k_1$, or $T << GT$, then we have $T = k_1/k_3$. T is thus constant for a given substrate, whatever the activity of the permease, and varies with the nature of the substrate since k_1 is the dissociation constant of GT. If the affinity of T is stronger for TDG than for TMG, the level of free T will be lower for the former than for the latter. It follows that the rate of exit of TDG will be accelerated in the presence of TMG. The same mechanism would explain the slow exit in case of inhibition of the permease (Fig. 6), since in this case T_{in} is not regenerated.

The effect of glucose can also be explained if it enters the cell by a mechanism different in its stereospecificity, but which liberates the same carrier in the cell. Since glucose is consumed as soon as it enters, T is liberated in excess with respect to its steady-state level and it contributes to the exit of galactosides.

This last mechanism functions also in the absence of galactoside permease, and causes the active expulsion of galactosides by glucose. This is an example of low-specificity active transport of galactosides, coupled with an active enzymatic transport of glucose and of carrier T.

The number of facts needing explanation compelled us to work out a detailed model, and, of course, it is only a possible model; only the identification of the constituents of the system can lead to a solution of the mechanism of permeases and perhaps of active transport in general.

References

1. Rickenberg, H. W., G. N. Cohen, G. Buttin, and J. Monod (1956). *Ann. Inst. Pasteur,* **91,** 829.
2. Kepes, A., and J. Monod. (1957). *C. R. Acad. Sci. (Paris),* **244,** 809.
3. Kepes, A. (1957). *C. R. Acad. Sci. (Paris),* **244,** 1550.
4. Sistrom, W. R. (1958). *Biochim. Biophys Acta,* **29,** 579.
5. Herzenberg, L. Personal communication.
6. Horecker, B. Personal communication.
7. Lardy, H., and H. Wellman (1952). *J. Biol. Chem.,* **195,** 215.

THE ROLE OF PERMEASE IN TRANSPORT

ARTHUR L. KOCH[*]

Service de biochimie cellulaire, Institut Pasteur, Paris (France)

(Received May 15th, 1963)

SUMMARY

Unidirectional fluxes of galactosides into and out of *Escherichia coli* strains were measured. Lowering the temperature to 0° inhibited these processes very strongly. It also inhibited the influx in cryptic cells. Efflux of labeled galactosides was increased during influx of other substrates of the galactoside permease. The efflux was also increased by other carbohydrates if the cells had been induced to have permeases for them. Only under these circumstances would these carbohydrates inhibit influx.

Influx was almost unaffected by energy substrates or poisons, while efflux could be greatly increased by energy poisons and decreased by energy sources.

Agents that inhibited galactoside permease had a much stronger influence on influx than on efflux.

A model to account for these observations is presented.

INTRODUCTION

The galactoside transport system of *E. coli*[1–3] is a particularly favorable object for the study of transport for a number of reasons: transport may be specifical'y enhanced by induction, mutants are available which lack or constitutively form the permease, labeled substances are available which are accumulated but not further metabolized, and also available are competitive inhibitors of the transport. Important for the present study is the availability of a chromogenic substrate (ONPG)[4] which is further metabolized through the action of β-galactosidase (EC 3.2.1.23). Transport of this substance in cells containing the hydrolyzing enzyme is almost independent of energy metabolism as compared to accumulating systems[2]. Measurements of this hydrolysis *in vivo* can thus be considered to reflect the transport *per se* and not its coupling to intermediary metabolism or of intermediary metabolism itself. Thus, by choice of substrate one is able to effectively convert an "active" into a "facilitated" transport system.

In addition, we have employed measurements of the rate of escape of a labeled non-metabolized substrate from cells into unlabeled media. Together these measure-

Abbreviations: ONPG, *o*-nitrophenyl-β-D-galactoside; TMG, thiomethylgalactoside; IPTG, isopropyl-β-D-thiogalactoside; TDG, thiodigalactoside; TPG, thiophenylgalactoside; DNP, dinitrophenol; and the symbols for the proposed model defined in Scheme I.

[*] Permanent address: Departments of Biochemistry and Microbiology, University of Florida, Gainesville, Fla. (U.S.A.). This investigation was carried out during the tenure of a special fellowship from the Division of General Medical Sciences, United States Public Health Service.

ments allow a clear separation of the total active process into three distinct phases:
(1) externally, the highly stereospecific binding or permease step; (2) the less specific
carrier or transport mechanism which bridges the impermeable barrier of the cell;
(3) internally, the mechanism which couples transport to energy metabolism.

MATERIALS AND METHODS

Conditions and techniques were those previously employed in this laboratory[1–3].
Most of the experiments reported in this paper were carried out with derivatives
of *E. coli* ML 30. This strain is inducible for β-galactosidase and β-galactoside permease. Most of the experiments were carried out with the constitutive strain ML 308
which produces both at high level in the absence of inducer. Also employed were
the cryptic (*i.e.*, permeaseless) strains ML 3 which inducibly forms β-galactosidase
and ML 35 which constitutively forms the enzyme. Finally, ML 3088 which forms
no enzyme and a lower level of permease was employed to study the accumulation
of ONPG under conditions where this substance is not hydrolyzed secondarily.

Hydrolysis *in vivo* was followed by the increase in absorbancy at 420 mμ in
a thermostatted Beckman spectrophotometer. The reaction was generally carried out
by the addition of 0.2 ml of cells to the blank and to 2.5 ml of $2·10^{-3}$ M ONPG
at 28° in Medium 63. This medium is a potassium-based medium which supports
the growth of the organisms with a variety of carbon sources.

The enzyme content of the cells was measured by ONPG hydrolysis on cell
suspensions after toluene and deoxycholate treatment. The assay was carried out
under the same conditions as employed for study *in vivo* even though enzyme action
could be greatly stimulated by the addition of Na^+ or of thiols. For the measurement
of hydrolysis *in vivo* in cryptic cells, 10^{-2} M ONPG was employed routinely. This
concentration was employed for genetic cryptics as well as cells rendered cryptic by
the action of formaldehyde or mercurials.

To measure the exit of label of a non-utilizable, or gratuitous galactoside, use
of the high temperature coefficient of the transport carrier (see below) was employed.
Typically 200–300 ml of a growing culture near the end of exponential growth (2.5 mg
dry wt. per ml) were harvested and concentrated to 10 ml in Medium 63. Methyl
carbon labeled TMG to a final concentration of 10^{-3} M was added. This concentration
is approx. twice the effective K_m for accumulation. The thick suspension was vigorously aerated for 10 min at 28°. A carbon source and chloramphenicol were usually
added. Then the culture, while still being aerated, was placed in an ice bath and
thoroughly chilled. It was centrifuged quickly in the cold, washed once, resuspended
in the cold, and kept cold until 1.0-ml portions were suddenly diluted into a total
volume of 8.8 ml of media at the chosen temperature and appropriate composition. At suitable times 1.0-ml portions were filtered on 1-in millipore HA filters
in an ice-cold stainless-steel filter apparatus containing the cold media and washed
with two portions of cold medium. The filters were then mounted, dried, and counted.
For the ONPG accumulation experiments, the bacteria from the millipore were
suspended in 1.0 ml of 0.1 N NaOH, and the filter was removed. The ONPG was
hydrolyzed by 6.0-min immersion in a boiling-water bath. The cells particulate
matter was removed by centrifugation and the supernatants read in the spectrophotometer in microcuvettes.

RESULTS AND DISCUSSION

Temperature dependence

Although many "active" transport systems have high temperature coefficients, this is not surprising since energy metabolism is itself highly temperature dependent. The temperature dependence of the hydrolysis *in vivo* is of interest because it seems to be a transport process which has a reduced or absent dependence on the energy metabolism of the cell. It was found that the rate of hydrolysis was slowed 70–90 fold. on lowering the temperature from 28° to 0°. Most of this large change in rate is near 0°, being 11 fold between 0° and 10°. The measurements at low temperatures required an extended period of time; however, it could readily be shown that hydrolysis proceeded at a constant rate at 0° and returned to the rate characteristic of 28° when the reaction mixture was removed from the ice bath after as long as 22 h.

It should be noted that the subsequent step of hydrolysis by β-galactosidase is not limiting at 0°. At 28° ONPG is hydrolyzed 13 times faster by cells whose permeability barrier has been broken by toluene and deoxycholate than by intact cells. The discrepancy is probably greater if the internal environment of the cell resembles more closely the optimum conditions for the enzyme action than does this particular growth medium. Additional support for the assumption that the enzyme is in excess comes from the fact that the steady-state rate of hydrolysis is reached within a fraction of a second after the addition of the cells to the cuvette[5]. This would not be the case if high internal levels of substrate had to be accumulated before the enzyme could keep pace with the transport process. At 0° the excess of enzyme is about 12 times greater than at 28° since hydrolysis by the enzyme is only decreased 6.3 fold by this temperature change while the transport rate has been decreased 70–90 fold.

The hydrolysis by cryptic cells is also characterized by a rather high temperature coefficient. However, here it is difficult to properly assess the numerical value since the permeation is slow and can be easily obscured, either by a small amount of cell lysis or damage to the permeability barrier. This is especially true at the lower temperatures since hydrolysis resulting from either of the latter two processes would have a low temperature coefficient. However, if cells forming enzyme are kept in the exponential phase of growth for long periods of time, or if cells induced to form enzyme for only a short period of time before harvest are employed, ratios of 30 fold and occasionally 70 fold for the activities at 28° to 0° are found. These high values appear to preclude leakage through pores of the intact cells since hydrolysis limited by diffusion through pores should have a very low temperature coefficient except under very exceptional circumstances. This supports the contention[3] that permeation in cryptic cells is not a passive process. Further support is given by the observation that hydrolysis *in vivo* of permease-containing cells at a concentration equal to 50 times the effective K_m is only a little bit larger than the value at 10 times K_m. The change observed can be entirely attributed to the hyperbolic kinetics and does not increase further by the amount characteristic of the corresponding differences in cryptic cells (Table I). Thus, penetration by the process present in the cryptic cells is either missing in permease-containing cells or present but not independent of the permease-linked process.

The observations of the effect of temperature appear to fit with the observations of the Carnegie group[6]. These workers had found that an amino acid accumulated at ordinary temperature was retained for long periods at 0° even though accumulation could not take place at this temperature. This was put to test for our system, and it was found that TMG accumulated at 28° and retained by the cells during the chilling and centrifugation in the cold was lost very slowly to the suspension medium. The cells lost radioactivity with a half life of 126 min; the loss continues at this rate for at least 20 h at 0°. The retention of label has been qualitatively confirmed in each of thirty odd "exit" experiments (such as is illustrated in Figs. 1 and 2) with cells of several genotypes and grown with a number of carbon sources. In all experiments the extra-

TABLE I

HYDROLYSIS *in vivo* AT HIGH ONPG CONCENTRATION SUCCINATE-GROWN ML 308 AT 28°

	$A\Delta$/min
$2 \cdot 10^{-3}$ M ONPG	0.098
$1 \cdot 10^{-2}$ M ONPG	0.138, 0.132
$5 \cdot 10^{-2}$ M ONPG	0.146, 0.145
$5 \cdot 10^{-2}$ M ONPG + formaldehyde	0.042
Calculated $K_m = 1.02 \cdot 10^{-3}$ M	
Calculated effect to increase in saturation between $1 \cdot 10^{-2}$ M and $5 \cdot 10^{-2}$ M	0.013

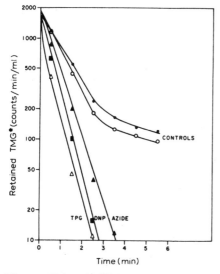

Fig. 1. Exit of [^{14}C]TMG from succinate-grown ML 308 at 28°. Final concentration of TPG, $1.2 \cdot 10^{-4}$ M; NaN$_3$, $2.7 \cdot 10^{-2}$ M; DNP, $1.2 \cdot 10^{-3}$ M.

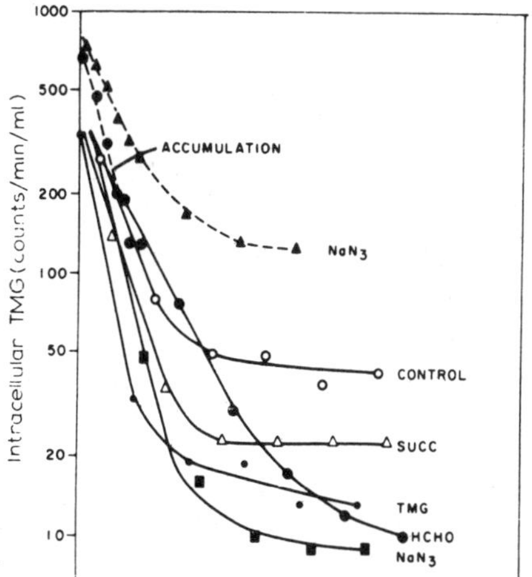

Fig. 2. Exit of [^{14}C]TMG from succinate-grown partially induced ML 30. ML-30 cells were induced for 10 min with $2 \cdot 10^{-4}$ M IPTG, chloramphenicol was added, and the cells were washed. Solid lines: exit experiments with the indicated additions. Half times of the initial straight portion: control, 2.8 min; succinate, 2.5 min; TMG, 1.5 min; formaldehyde, 3.5 min; NaN$_3$, 1.6 min. Dotted lines: experiments with high external radioactive TMG present. Curve labeled accumulation was prepared by subtracting the radioactivity of each sample from the average plateau value of 820 counts/min/ml. Curve labeled NaN$_3$ is the radioactivity from the moment of addition of NaN$_3$ to the accumulated culture: half time for accumulation, 2.0 min; half time for decay in NaN$_3$, 3.1 min. The cells hydrolyzed ONPG under the standard conditions at 48 μmoles/g·min.

Biochim. Biophys. Acta, 79 (1964) 177–200

polated activity at the time of dilution of the cold cells into warmed medium dropped very slowly throughout the course of the experiments, each of which lasted for approx. 1 h.

It may thus be generally concluded that cells capable of retaining accumulated thiogalactoside for the short period required for washing are negligibly damaged in the sense that pores or cracks, etc., have been formed in the cell membrane to allow an appreciable loss of internal thiogalactoside.

The temperature coefficient for the exit process is indeed high: the half time at 0° is 126 min, at 15.5° it is 3.2 min, and at 28° it is 0.87 min. These are averages for five measurements at 15.5° and seven measurements at 28° obtained on different days with the same strain, ML 308, grown on the same source of carbon, potassium succinate. One of the data is the control in Fig. 1. From these numbers, one can calculate a Q_{10} of 11.2 for the decade from 0° to 10° and 2.4 for that from 18° to 28°. Since, as we shall see, the exit is influenced by a large number of factors, it does not appear profitable to calculate energies of activation. The salient conclusion to be drawn is that the three processes—the permease-mediated hydrolysis *in vivo*, the cryptic hydrolysis *in vivo*, and the exit of accumulated thiogalactosides—all are retarded to large and similar degrees by lowering the temperature to 0°.

The influence of competitors on exit

If other substrates for the β-galactoside permease are added, they cause an inhibition of ONPG hydrolysis of varying degrees and a depression of the level of radioactivity in the accumulation-type of experiment[1-3]. This second type of finding could have been simply the result of inhibition and/or competition for entry. However, the exit type of experiment shows definitely (Table II) an acceleration of exit caused by the entry of certain galactosides.

TABLE II

THE INFLUENCE OF COMPETITORS ON EXIT

	Half lives (min)						
	TDG	*Control*	*TMG* $1.2 \cdot 10^{-4} M$	*TPG* $1.2 \cdot 10^{-4} M$	*ONPG* $4 \cdot 10^{-4} M$	*ONPG* $2 \cdot 10^{-3} M$	*ONPG* $10^{-2} M$
ML 308							
28°	2.5, 1.5, 1.5	0.87*	0.4, 0.2	0.33			
15.5°		3.2**	1.45		1.1		0.34, 0.3
*ML 3088****							
28°		1.9	0.8			0.7	
15.5°		10, 10.7	4.8		4		3, 4.8

* Average of 7 measurements.
** Average of 5 measurements.
*** ML 3088 is devoid of β-galactosidase and can not hydrolyze ONPG.

First, we shall consider the exception, TDG. This substance has been characterized in previous studies[3] by its high affinity for permease but low degree of accumulation. Thus, binding with the permease slows exit which thus presumably must take place through or be dependent upon the permease. On the other hand, a substance like TMG which is accumulated to a high degree induces a more rapid exit of internal [^{14}C]TMG. Both of these substances inhibit ONPG hydrolysis and displace other galactosides.

It is to be noted that the accumulation-type of experiment employed as an assay for permease is usually carried out at roughly this concentration of TMG. Under these conditions the exit in the steady state is very much faster than would be the case in the absence of entry. It is therefore not surprising that the previously published value[3] for the half time of entry (0.75 min) at 26° yields, when extrapolated to 28°, a value intermediary between the exit half times with and without TMG. A second point to be made is that roughly two-thirds of the flux of TMG at the plateau steady state is supported by the counterflow and thus need not be supported by energy from intermediary metabolism.

Also shown in Table II is the fact that TPG greatly speeds up exit. This substance is characterized by its speed in attaining equilibrium accumulation (the half time is less than 0.25 min); it had been hoped that we would be able to distinguish the effects of net entry from the steady-state exchange with this compound. The initial rate of TMG exit in the first few seconds is faster than is the subsequent rate, as apparent from Fig. 1. However, the effect is not pronounced enough to be definitive.

The use of ONPG as a competitor appeared initially to be more fruitful in distinguishing the effects of net inward flux and the steady-state exchange. In the presence of β-galactosidase, ONPG would enter but would be hydrolyzed before it could leave. In the absence of enzyme, it should soon reach a steady value where entry balances exit. ONPG does rapidly extrude labeled TMG, and the influence of the concentration of ONPG is consonant with the K_m of 10^{-3} M for its hydrolysis *in vivo*. The increased rate of exit was maintained throughout the observable period of the decay of internal TMG. This was also true in two of the three experiments with ML 3088 which lacks the β-galactosidase. Unfortunately, the control values are 2–3 times smaller with this organism than with its parent, ML 308. This is readily interpreted by the hypothesis that this organism is defective in permease as well as devoid of β-galactosidase, produced by the neighboring gene of the same operon[7]. It is known from studies from MELVIN COHN's laboratory[8] that this organism has a steady-state accumulation level that is 2.7 times smaller than does the parental type*. At 15.5°, 10^{-2} M ONPG speeds exit by a factor of 10 times in the normal and only 2 or 3 times in the mutant, a highly significant difference. One would not expect a larger difference since in both cases a counterflow will develop. In the absence of β-galactosidase, ONPG will eventually escape; and in the presence of β-galactosid-

---

* Although ML 3088 has a reported plateau level 2.7 times lower than its parental strain and has a 3.2 times slower exit, it should not be concluded that the permease level is 8.6 times less than ML 308. This is because the plateau measurements are at 37° and the exit values quoted are for 15.5° in the absence of external TMG. At 28° the ratio of exit rates in the two strains is lower (about 2), and it should be lower yet at 37°. The presence of external TMG does not appear to affect the ratio significantly. ML 3088 probably possesses a permease with the same kinetic properties as does ML 308. The accumulation of ONPG in ML 3088 as a function of substrate increases with increased substrate and does not plateau. This is because the accumulation is low so that the background due to simple equilibration is important. This can be corrected for simply by adding formaldehyde to the accumulated suspensions, and waiting until only the equilibrium amount of ONPG is present. These measurements then are used to correct the previous plateau measurements. When this is done, the half-saturation values of about 3·10^{-4} M are obtained. This is one-third that for ONPG hydrolysis by ML 308. A similar result has been obtained with mutants and parental types of the K_{12} strain of *E. coli* (A. KEPES, unpublished).

Biochim. Biophys. Acta, 79 (1964) 177–200

ase, galactose produced by enzyme action must leave the cell. This is because galactose cannot be utilized in these succinate-grown organisms since the galactokinase is an inducible enzyme and these cells have not been previously exposed to this sugar. Galactose interacts with the galactoside permease[9],[10] (see also below) so that a steady state of flow and an interacting counterflow is eventually established in both the normal and deficient organism.

In any case, the conclusion can be drawn that exit is accelerated by a flow inwards whether or not the internal level of galactoside becomes high and is inhibited by binding to the specific permease.

The influence of induction of permease on exit

The results of experiments on induction are shown in Table III. It is seen that the rate of exit increases progressively with induction from a basal level. This is true both in the absence of counterflow and when TMG or ONPG are present in the external medium. It is thus clear that exit is "induced" along with the entry. This is possibly correlated to the findings of HORECKER et al.[10] with the distinct but interrelated galactose system where an "exitase" is induced by galactose although there is no increase in the rate of galactose entry. Whether the exit in our experiments is through the β-galactoside permease or because of it is discussed below, but here we must consider the clear discrepancy between this finding and the implications of the previously reported proportionality between the degree of growth under inducing conditions and the level of steady-state accumulation of non-metabolized thiogalactoside. This proportionality should only be found if the number of permease sites on the membrane were proportional to growth and if the exit mechanism were constant per unit amount of protoplasm, independent of the state of induction.

TABLE III

VARIATION OF PERMEASE LEVEL BY INDUCTION OF ML 30

Induction* (%)	ONPG hydrolysis in vivo (μmoles/g · min)	Half time (min)				
		No addition	TMG $1.2 \cdot 10^{-3}$ M	ONPG 10^{-2} M	NaN₃	Formaldehyde
				$15.5°$		
18		13.6	3.7	2.6		9.1**
35		5	2.3	1.6		9.1**
87		3.6				
ML 308		3.2	1.45	0.35, 0.3		10**
				$28°$		
7	48	2.8	1.5		1.6	3.5
20	195	2.0	0.95		1.0	
ML 308	580–700	0.87	0.4, 0.2		0.35, 0.5	2.9***

* Induction (%) = per cent growth under inducing conditions. Growth determined turbidimetrically.
** The values at 15.5° are believed to be comparable, but faster than those found in other experiments under similar conditions and faster than the 13.6 min found in the absence of formaldehyde in the least induced culture. This is discussed in the text and is attributable to acid in the reagent and to paraformaldehyde.
*** Average of 6 measurements, range 2.6–3.3.

Biochim. Biophys. Acta, 79 (1964) 177–200

The increase in the rate of exit caused by induction is documented further in Fig. 2. The curves show the influence of energy poisons and sources and co-substrates for the permease on exit on a culture which has been induced to form the permease for a very brief period of time. Also shown are the kinetics of entry and azide-induced loss of accumulated TMG by these cells. The straight portion of each line is 2–3 times less steep than for a permease-positive ML 308, as indicated in various places in this paper and in KEPES[3]. There is one exception—the slope in the presence of formaldehyde. This value is the same as that obtained with cells containing the maximum amount of permease. Thus the influence of permease is manifested by its interaction with a number of processes, but not that of permeation by the pathway remaining active in the presence of formaldehyde.

It would logically follow that a plot of the increase of mass *versus* plateau level would be curvilinear for small mass increment. New experiments of this kind are shown in Fig. 3 and are in accord with this prediction. The results of this experiment can be stated in another way; namely, the half time for the saturation of the plateau is 10–15 min instead of the observed 80-min doubling time of the culture. The plot of hydrolysis *in vivo* of ONPG approaches the doubling time of the organisms more closely. It still, however, yields a curved line. This is because in the initial phase β-galactosidase is limiting relative to the cryptic transport process and only in the second phase does permease become limiting. Subtraction of the cryptic rate leads a line extrapolating to the time of addition of inducer.

Presumably, the difference in the results of the present experiment and those previously reported from this laboratory results in part from two improvements in technique. The first is the use of the millipore technique which allows the rapid separation of the cells from the medium instead of the laborious and time-consuming centrifugation and washing by centrifugation. In this process there will inevitably be considerable redistribution. The second improvement is the use of $3 \cdot 10^{-4}$ M IPTG as an inducer instead of the relatively less efficient TMG.

Another possible reason for the discrepancy from this and the previous results is the temperature of assay, which was 34° in the original and 28° in the present one. It is shown below that the difference in exit rate in cells with and without functioning permease decreases with increasing temperature. Thus, induction affects the exit rate more markedly at lower temperature and thus may have escaped previous detection.

Although the exit rate alters during the induction process by a factor of 3 or more, the dependence of accumulation on substrate concentration does not alter (Fig. 4), in confirmation of previous results[1].

Variations in permease level brought about by growth on different carbon sources

Comparison of the first and third columns in Table IV shows a clear correlation of the rate of hydrolysis *in vivo* and the rate of exit of the accumulated non-metabolizable thiogalactoside of cultures grown on various substrates. It is also clear that the effect is not simply "the" glucose effect since maltose- and galactose-grown cells have lowered hydrolysis *in vivo* and, as shown below, interact with glucose to a smaller extent than do glucose-grown cells. Nor is it simply a matter of growth rate since glycerol supports almost as rapid proliferation as does glucose, and yet permease levels are high when the cells are actively growing.

Biochim. Biophys. Acta, 79 (1964) 177–200

Thus, from both the effect of induction and of variation in nutrient, one comes to a correlation, though clearly not a linear relationship, between the two observed quantities of exit rate and entrance rate.

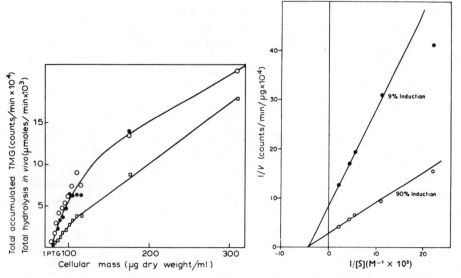

Fig. 3. Induction of ability to accumulate TMG and to hydrolyze ONPG by intact cells. $3 \cdot 10^{-4}$ M IPTG was added to ML 30 growing in succinate. At various times, chloramphenicol was added to a final concentration of 50 μg/ml and the cells centrifuged. TMG accumulation was measured in the presence of succinate and chloramphenicol at $28°$. ●—●, measurements after 12 min; ○—○, measurements after 25 min. Hydrolysis *in vivo* was measured at $2 \cdot 10^{-3}$ M ONPG (□—□). The initial slope was 1100 μmoles/g·min. The half times mentioned in the text were computed from the corresponding graphs for the activity per unit amount of bacteria.

Fig. 4. Determination of K_m for a partial and almost completely induced culture. ML 30 growing in succinate with 74-min doubling time was made $3 \cdot 10^{-4}$ in IPTG. 10 min later, chloramphenicol was added to a portion of the culture which was then harvested. Calculated per cent induction, 9%. Another portion was diluted in the same IPTG concentration and incubated for 4 h. Calculated per cent induction, 90%. Standard assays *in vivo* gave 79 and 766 μmoles/g·min, respectively. Accumulation measurements at $28°$ were made at 12, 24, and 36 min of incubation. The average values are plotted. The K_m value for both lines is $2.4 \cdot 10^{-4}$ M. The ratio of maximal incorporation is 3.0.

The influence of formaldehyde

In searching for an agent that would rapidly seal the cell and thus allow prolonged washing on the millipore filter, we found that formaldehyde did rapidly stop the permease action as measured both by a profound lowering of hydrolysis *in vivo* and a slowing of the rate of exit. It apparently has a slight secondary effect of breaking down the permeability barrier. Since the permeation is itself stopped by chilling, the results of including formaldehyde in the ice-cold wash medium were the reverse of those desired since the secondary effect outweighed the primary one in these circumstances. But nonetheless, this agent has served a very useful role in dissection of the steps of the transport process. Formaldehyde decreases hydrolysis *in vivo* to the level exhibited by genetic cryptic cells (ML 3 induced or ML 35). Thus, under the usual conditions of $2 \cdot 10^{-3}$ M ONPG, the rate is reduced to 3–5% of its original value in the presence of formaldehyde. This is the same level that is produced by *p*-chloro-

Biochim. Biophys. Acta, 79 (1964) 177–200

mercuriphenylsulfonic acid. Together there is no more inhibition. Thus, the two agents clearly influence the permease; however, they do so in distinctly different ways. It has previously been reported[11] that the mercurial reaction can be prevented in the presence of a substrate for the permease. Such studies are performed at concentrations where the reaction is slow enough to study the kinetics. Formaldehyde inhibits very rapidly. The half time is much less than 1 min at concentrations that give 50% inhibition. The half inhibitory concentration is about $5 \cdot 10^{-4}$ M and is independent of the ONPG concentration employed in the measurement, and thus is noncompetitive.

A concentration of $5 \cdot 10^{-4}$ M also decreases the rate of TMG accumulation to about half and gives rise to significant loss from culture that has previously accumulated TMG.

In the experiment reported here, we employed the relatively high concentration of $7 \cdot 10^{-2}$ M. This concentration reacts instantaneously for all practical purposes. If the formic acid in the formalin preparation is properly neutralized, it does not inhibit the assay of β-galactosidase within the experimental limit and does not alter the turbidity of cell suspension.

As mentioned above, it does break down the permeability barrier to a small extent. This was concluded from the finding that the hydrolysis of cryptic cells increased about 20% at 28° and about 100% at 0°. Large amounts of formaldehyde give more leakage and smaller amounts give less. The amount of leakage appears to be decreased if the paraformaldehyde in the formalin is hydrolyzed.

If 20% is deducted from the hydrolysis rates observed in the presence of formaldehyde, agreement is found with the rate of hydrolysis of cryptic cells. Thus ML 3, partially or fully induced, or ML 35 grown in various carbohydrates show rates of hydrolysis *in vivo* in the range of 30–60 μmoles/g·min in 10^{-2} M ONPG.

When exit is measured in the presence of formaldehyde, half lives observed do not seem to be affected by the genotype or cultural conditions (see Tables III and IV). The increase in half time caused by formaldehyde for a fully adapted succinate-grown ML 308 culture is 4–5 fold at 15.5° and 3 fold at 28°. This is much less than the 20–30 fold change produced by formaldehyde in rate of hydrolysis of ONPG of the same cultures.

TABLE IV

INFLUENCE OF CARBON SOURCE FOR GROWTH OF ML 308 ON TRANSPORT IN THE ABSENCE OF SOURCE

	Exit TMG at 15.5°		Hydrolysis in vivo of ONPG at 28°	
	No addition (min)	Formaldehyde (min)	$2 \cdot 10^{-3}$ M (μmoles/g·min)	10^{-2} M + formaldehyde (μmoles/g·min)
Succinate	2.0–4.4[*]	12–15	580–700[*]	46, 68
Glycerol	3.2[**]	18	470[**]	83
	1.9	18	750	89
Glucose	5.0–8.6[*]	10, 15	126–216[*]	50
Maltose	6.8	20, 16	265	70, 44
Galactose	3.2, 3.2	15, 11	260, 390	35, 63, 89
Broth			160	50

[*] Range of at least 4 independent measurements.
[**] Culture employed had overgrown and its growth rate had slowed at the time of harvest.

Biochim. Biophys. Acta, 79 (1964) 177–200

Seemingly then, this cryptic pathway is constant and does not depend on conditions which cause major influence in the level of the permease. This cryptic path appears to serve as a mechanism of entry in the hydrolysis *in vivo* in the presence of formaldehyde and as a basal mechanism of exit which can be further stimulated by a functioning permease.

In passing, it should be noted that a glucose-grown cell has a much less effective permease system per unit amount of bacteria and also has a lowered level compared to the basal level of the cryptic mechanism of the same cells. At first, it was thought that cells lysed to a larger extent during growth on this carbon source than on succinate. However, it was concluded that this does not reflect lysis or leakage of cells since, in addition to the quantitative data, it was also found that the temperature coefficient for cryptic hydrolysis is as high for glucose- as it is for succinate-grown cells.

Like the case of partially induced cells, cells grown in a medium that yields lowered permease levels do not give a proportional lowering in the plateau accumulation levels since the exit is also slowed, though not to the same degree. The published ratio of plateau accumulation values for succinate-grown cells to glucose-grown cells is 2.3 (ref. 8), while the ONPG values indicate a factor closer to 4–5 for the decrease in permease site concentration.

Influence of carbohydrates on transport

In addition to their negative adaptive influence, it was found that growth on glucose and maltose altered the cells so that hydrolysis *in vivo* of ONPG was inhibited and TMG exit accelerated in the presence of these carbohydrates (Table V). The effect was specific in that glucose-grown cells fail to respond to maltose; though in the reverse case maltose cells do respond to glucose, presumably because glucose is

TABLE V

THE INFLUENCE OF VARIOUS CARBOHYDRATES ON CELLS GROWN ON DIFFERENT SOURCES[*]

		Hydrolysis in vivo (%)	Exit half life (min)			Hydrolysis in vivo (%)	Exit half life (min)
Succinate-grown	No addition	(100)	3.2	Galactose-grown	No addition	(100)	3.2, 3.2
	Succinate	100	5		Glucose	77, 99	4.5
	Glycerol	100	5		Maltose	77	3.8
	Glucose	86–100	4		Galactose	Inhibits	1.8, 2.0
	Maltose	86	4				
	Galactose	Inhibits	0.7				
Glucose-grown	No addition	(100)	7.2, 7.0, 7.6, 8.6, 5.0	Glycerol-grown	No addition	(100)	3.2[**], 1.9
					Succinate	100	3.6
	Glucose	33–38	3.3, 4.8, 3.6, 3.1		Glycerol	99	20[**], 6.4
					Glucose	95	
	Maltose	100	6.3		Maltose	92	5.1
	Galactose	Inhibits	4		Galactose	Inhibits	1.1
Maltose	No addition	(100)	6.8, 4.5				
	Glucose	58	4.3				
	Maltose	43	3.8, 4.1				
	Galactose	Inhibits	2.4				

[*] All sources tested at 0.25 %. Hydrolysis measurements *in vivo* at 28°; exit measurements at 15.5°.
[**] This culture had entered the stationary phase of growth.

Biochim. Biophys. Acta, 79 (1964) 177–200

produced by the cells growing on maltose. Although it had been believed that the glucose permease is constitutive, these results strongly suggest that growth on these carbohydrates results in the formation of new specific permease that through some competition, possibly for sites on the membrane, depresses the level of β-galactoside permease. Similarly, growth on maltose induces maltose permease[12] at the expense of galactoside permease. When these new permeases function, they inhibit entry and facilitate exit of galactosides in the same way that the usual substrates for the galactoside permease do. This is most readily explained by the assumption that the several different permeases function through a common, less specific element in the transport process. In principle, this could be a common carrier or a common energy source. The common carrier is the simplest explanation for both phenomena, and support for the notion will be given below.

Galactose inhibits entry and speeds exit in all cases. No detailed inhibition data are given in the table because the inhibition is not constant during the experimental run and is not of constant form, varying from noncompetitive behavior at low concentrations of galactose to competitive at high concentrations. At high concentration of galactose, the apparent K_m for ONPG hydrolysis appears very high. Further work is needed to detail these kinetics.

The kinetics of glucose inhibition appears to be noncompetitive of the special type that does not yield zero inhibition even at very high concentrations of the inhibitor. In different batches of glucose-grown cells, the level of inhibition may vary a little; but with the same batch the per cent inhibition is constant, independent of substrate concentration or inhibitor concentration over a wide range. At very low glucose concentration, however, the inhibition is not constant because the glucose is consumed by the microorganisms and the rate of hydrolysis then abruptly rises to the control value. Using 5-cm cells instead of the usual 1-cm cells and employing the Cary or Zeiss spectrophotometer, we could establish that the inhibition was not all or none but decreased progressively. The K_m for inhibition is about $1 \cdot 10^{-5}$ M (KOCH AND STEEGER, unpublished). This is quite comparable to the K_m for glucose uptake of these cells as adduced from growth-rate measurements in our laboratory and those reported previously[13]. Therefore, it is likely that the functioning of the permease for glucose uptake inhibits galactoside uptake.

Cells grown for a portion of a generation on glucose have a lower degree of inhibition of hydrolysis *in vivo* of ONPG by glucose than do cells grown for many generations on glucose, but the inhibition appears to be of the same noncompetitive kind. Maltose-grown cells are also inhibited noncompetitively by maltose and to a smaller extent by glucose.

The effect of these carbohydrates and succinate and glycerol as energy sources is discussed in the following section.

The influence of energy metabolism

As mentioned above, hydrolysis *in vivo* is much less energy dependent than is accumulation. $3 \cdot 10^{-2}$ M azide partially inhibits hydrolysis (Table VI). Massive amounts of azide (0.15 M) cause little further inhibition. On the other hand, these levels of azide immediately cause the net efflux of TMG previously accumulated from cells and eventually yield a stationary level of TMG not greatly different from that in the medium (see Fig. 2).

Biochim. Biophys. Acta, 79 (1964) 177–200

It might be considered that this lack of inhibition of ONPG hydrolysis *in vivo* by azide results from some property that is unique to ONPG. That this is not so can be shown in several ways. First, it was found that ONPG accumulation in the ML 3088, which lacks the β-galactosidase and therefore accumulates ONPG, is as sensitive to azide as is TMG accumulation. Secondly, studies were made with a mutant that constitutively forms permease and a very modified β-galactosidase that has very low activity and very high K_m (10^{-2} M for ONPG; D. PERRIN, personal communication). Effective hydrolysis by cells of this strain* should require the active accumulation by the permease when the external substrate concentration is low. Indeed, it was found that KCN blocked virtually 100 % the hydrolysis at $4 \cdot 10^{-4}$ M ONPG while inhibition was 66 % at 10^{-2} M.

TABLE VI

INFLUENCE OF 0.03 M SODIUM AZIDE ON HYDROLYSIS *in vivo*

	ONPG (M)	Per cent control rate
Succinate-grown cells		
	10^{-2}	81
	$2 \cdot 10^{-3}$	72
	$4 \cdot 10^{-4}$	62
	$8 \cdot 10^{-5}$*	57
Glucose-grown cells		
	10^{-2}	77
	$2 \cdot 10^{-3}$	59

* With this concentration of substrate the hydrolysis is first order, and the data have been plotted suitably on logarithmic paper to measure the inhibition of the first-order rate constants.

A similar but less dramatic effect could be shown with minimally induced ML-30 cultures. The same suspension of cells used for the experiments of Fig. 2 were 75 % inhibited by azide under the standard conditions where a fully induced culture would be 28 % inhibited.

Other experiments were carried out with succinate-grown ML-308 cells which were washed and studied in concentrated suspension in Thunberg tubes. Oxygen removal with an oil pump under these conditions easily proved capable of lowering the rate of ONPG hydrolysis to 60 % of the original value, but it could not be lowered further by extended pumping or incubation to exhaust residual O_2 or by the addition of azide.

ONPG hydrolysis *in vivo* is not stimulated at all by the addition of succinate to washed suspensions of succinate-grown cells. If such cells are incubated with aeration at 37° in the absence of carbon source, the permease is lost[1,14]. The rate of loss can be decreased by the presence of chloramphenicol during the incubation, but the salient point is that hydrolysis *in vivo* does not even then become stimulated by the addition of a carbon source. (As noted above, galactose formed is not the source of energy in

* This mutant, 13PO, is a derivative of K_{12} and was kindly obtained from Dr. D. PERRIN. The necessary controls showed that K_{12} strains were similar in azide sensitivity to that of the ML strains. It must be further noted that a stimulation instead of an inhibition is obtained with sodium azide. This is attributable to the sodium ion activation of the enzyme. Sodium azide penetrates the cell much more rapidly than does sodium chloride and, therefore, we have no suitable reference to test with NaN_3 as used in the other studies.

Biochim. Biophys. Acta, 79 (1964) 177–200

these experiments because the cells lack the enzyme for its utilization. Formation of galactokinase is precluded in some of the experiments by the addition of chloramphenicol to the assay system *in vivo*.)

On the other hand, accumulation initially takes place in the absence of a carbon source; but when the immediate energy supplies are exhausted, the internal concentration is rapidly depleted, although not to zero, and may be raised by the subsequent addition of an energy source[14].

It is necessary to conclude that conditions may be obtained where accumulation responds to the addition of an additional energy source while hydrolysis *in vivo* does not.

It might be argued that hydrolysis *in vivo* is doubly supported by the electrochemical gradient of ONPG from outside to inside and of galactose from inside to outside. Indeed, the entry of galactose speeds the exit of TMG, so it is quite likely that galactose exit speeds ONPG entry. It is, however, unlikely that a one for one exchange accounts completely for the greatly decreased sensitivity to energy poisons of the hydrolysis *in vivo* compared to accumulation. This follows from the fact that hydrolysis in the presence of azide continues at a constant rate and does not slow even after 1 h at 28°. If exchange be the mechanism in the presence of azide, then it need be extremely efficient because a small degree of uncoupling of exit to entry should lead to a progressive decrease in rate. This argument becomes much stronger as the results of the observation (Table VIII) that azide is no more effective at speeding exit in galactose-grown cells than in succinate-grown cells. Such cells have a 4-fold increased "exitase" for galactose[10] and therefore an increased chance of exit without a forced exchange. Further, these organisms now possess galactokinase, which should serve as a further drain on both phosphate energy and galactose. It may be noted as well that azide is also no more effective here than for the hydrolysis *in vivo* of ONPG in succinate-grown cells. Washing cells in 0.1 M buffered potassium arsenate and carrying out the hydrolysis in this medium inhibits hydrolysis to the same degree as does azide. It might be argued that internal energy sources are not completely removed in that phosphate is more efficiently used than is arsenate and that the cell barrier may prevent the entry of arsenate to the internal sites for energy production or coupling to the transport. While this is possible, it is rendered less likely by the fact that the same small effect of arsenate could be observed starting with cells grown in a medium low in phosphate. The necessary controls in the presence of formaldehyde showed that the cell permeability barrier was not damaged by the arsenate treatment. Potassium fluoride initially has no influence on ONPG hydrolysis, but after 20 min the rate lowers to about half and continues at this rate.

Together these observations form a convincing picture that transport in the direction of the thermodynamic gradient may proceed without energy expenditure by the cell. We would now like to present complementary observations that the exit process is strongly energy dependent, *i.e.*, metabolic energy decreases and metabolic poison increases the rate of exit. These observations are presented in Tables VII and VIII. In succinate-grown cells metabolic inhibitors increase the rate of exit by a factor of 2–4. This is comparable to speeding exit of galactose observed in the elegant experiments of ROTMAN AND GUZMAN[15]. The inhibitors also cause the loss of TMG from accumulated cultures. In this case the loss is half as fast since the exchange process can take place in the presence of azide. This exchange in the presence of

Biochim. Biophys. Acta, 79 (1964) 177–200

metabolic poison has been shown for the galactose system by HORECKER *et al.*[16], and we have made similar experiments (Fig. 5) to show that preloading cells with unlabeled TMG causes a temporary influx of labeled TMG in the presence of azide. Previous results of exchange or incorporation experiments in the presence of DNP from this laboratory can be interpreted in this same way[3]. The difference in half life in the exit and decay of accumulation would imply, as was concluded above, that much of the flux of galactoside in the steady state of accumulation represents an energy-free exchange.

TABLE VII

THE EFFECT OF ENERGY POISONS ON EXIT HALF TIME OF ML 208 GROWN IN SUCCINATE*

	Control (min)	0.03 M NaN₃ (min)	10⁻³ M DNP (min)
Formaldehyde	0.87 3.0	0.35, 0.5 3.0	0.39, 0.5 3.0
Agent added to accumulated culture	—	0.70	0.75

*Tests performed at 28°.

TABLE VIII

THE EFFECT OF ENERGY POISONS ON EXIT HALF TIME OF ML 308 GROWN IN THE CARBON SOURCE INDICATED*

	Exit half time (min)	0.03 M NaN₃
Succinate	2.90	
	4.4	0.7
Glucose	6.3	1.1
	7.6	0.7, 6.4**
Maltose	6.8	4, 1.9**
	4.1	0.7, 0.7
Galactose	3.2	0.7
Glycerol	3.25	0.7
	1.9	0.3

*Tests performed at 15.5°.
**Straight-line semi-logarithmic plots were not obtained. Numbers given are initial, and subsequent half times.

Energy reserves of the cell for the transport process are not very great since the effect of azide on the rate of exit over the control and on the rate of loss of TMG from accumulated cultures is almost instantaneous (less than 15 sec). In the presence of azide the loss of label continues in many cases until less than 1 % of the original label remains.

With cells grown on other carbon sources we have had less reproducible results. For example, the two cultures of maltose-grown cells gave a 4-fold different relative influence of azide. We believe this was due to the presence of amylomaltose in one suspension and less in the other. This could serve as a source of anaerobic energy and/or a source of maltose that would decrease exit of TMG by competition for the common transporter mechanism. When such source is exhausted, exit should speed,

Biochim. Biophys. Acta, 79 (1964) 177–200

as was observed. Support for this idea was given by an experiment (not shown) with maltose-grown cells which temporarily incorporated TMG in the presence of azide. Succinate-grown cells would not do this as shown in other experiments, nor did this particular batch of cells when the azide had been added previously to the TMG to allow the depletion of the amylomaltose.

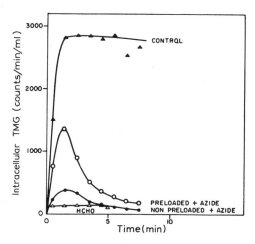

Fig. 5. The effect of preloading on TMG uptake. ML-308 succinate-grown cells were washed and resuspended in succinate plus chloramphenicol containing media. One portion was made 10⁻³ M in TMG, aerated for 10 min at 38°, chilled, centrifuged, and washed as in an exit experiment. ▲—▲, uptake of TMG by preloaded cells; ○—○, uptake of TMG by preloaded cells in presence of azide; △—△, uptake of TMG by preloaded cells in presence of formaldehyde; ●—●, uptake of TMG by non-preloaded cells in presence of azide.

In the case of glucose-grown cells, the cells may develop a more or less effective anaerobic metabolism and become less sensitive then to azide. Such anaerobic energy sources take about 1 min to become mobilized. The difference between the two glycerol-grown cultures was commented on before; the culture with a slow exit rate was a culture almost in the stationary phase of growth and had a decreased permease as measured by hydrolysis *in vivo*.

The addition of a carbon source can, under certain conditions, lower the rate of exit (Table V). This is most clearly evidenced by the addition of glycerol to glycerol-grown cells. Probably succinate also has this effect; but it is utilized much more slowly[14], so the effect becomes evident only in later phases of an exit experiment as a decrease in exit rate after the majority of the TMG has escaped. The effect is less marked at 28°, possibly because most of the label has escaped before the energy from succinate becomes available.

This raises the point that the effect of glucose and maltose in cells grown on these sources is composite. Their entry speeds exit, but also slows exit to the extent that they serve as energy sources. Thus, these influences in forcing exit are greater than indicated by the data of Table V. In fact, their influence is less marked or even reversed at 28°. However, glucose speeds exit as effectively at 15.5° as at 0°.

Finally, we note that the effect of azide is duplicated by DNP and by KCN, whereas KF has no immediate action on either ONPG hydrolysis or on exit.

Biochim. Biophys. Acta, 79 (1964) 177–200

Interactions of various factors

With the various conditions employed above, a large number of possible interactions can be studied. In this section we hope to detail the pertinent ones. First,
we will discuss results of exit-type experiments.

The action of formaldehyde in causing a slowing of the exit is completely dominant over the influence of energy poisons in speeding exit. This is shown in Table VII.
p-Chloromercuriphenylsulfonic acid slows exit to the same extent, and together
there is no further action. Formaldehyde is also dominant over the speeding caused
by an unlabeled substrate for β-galactoside permease and over the effect of glucose
in glucose-grown organisms. The only exception to this dominance is found with high
(10^{-2} M) ONPG in which there is an appreciable flux of ONPG through the cryptic
mechanism driven by the internal cleavage.

TABLE IX

THE INTERACTION OF GLUCOSE AND TMG ON EXIT HALF TIME OF GLUCOSE-GROWN ML 308[*]

	Half time (min)
Control[**]	5.0
0.25 % glucose	3.1
$1.2 \cdot 10^{-3}$ M TMG	2.5
$1.2 \cdot 10^{-4}$ M TMG	4.6
0.25 % glucose + $1.2 \cdot 10^{-3}$ M TMG	2.55
0.25 % glucose + $1.2 \cdot 10^{-4}$ M TMG	2.65

[*] Exit half times measured at 15.5°.
[**] This control value is the lowest value for these conditions seen in 5 experiments with this
type of cell at this temperature.

The influence of glucose in speeding exit in glucose-grown cells does not further
speed exit over that caused by a large ($1.2 \cdot 10^{-3}$ M) concentration of TMG but does
at lower concentration (Table IX).

There appears to be little influence of decreased temperature on the effect of
glucose on glucose-grown organisms. At both 15.5° and at 0°, glucose speeds exit
by a factor of about 2. Similarly, the speeding due to TMG is nearly the same at both
temperatures. It follows that movement of both the transporter and the loaded
transporter are inhibited similarly by lowering the temperature.

It is possible to speed the exit rate further by a combination of agents. Thus,
with the actively growing glycerol culture, the following data were obtained at 15.5°:
control, 1.9 min; NaN$_3$, 0.3; 10^{-2} M ONPG, 0.35; NaN$_3$ plus 10^{-2} M ONPG, ~ 0.1.
The last figure is necessarily approximate because of the rapidity of the exit. It is
interesting to note that in this experiment the fastest exit observed is of the order
of 180 times the rate of the formaldehyde-resistant exit process.

Turning to the hydrolysis experiment *in vivo*, we will first discuss results with
cryptic cells. Hydrolysis by glucose-grown constitutive cryptic (ML 35) cells is slightly
inhibited by glucose (27 %) at $2 \cdot 10^{-3}$ M ONPG and a little more (36%) at 10^{-2} M
ONPG. This inhibition is somewhat more than that exhibited by the corresponding
succinate-grown cells (11%) at $2 \cdot 10^{-3}$ M ONPG. Although the glucose effect is slight,
it is observed in the presence or absence of azide. Azide alone does virtually nothing.

Biochim. Biophys. Acta, 79 (1964) 177–200

Likewise, the cryptic hydrolysis is uninfluenced by $1.2 \cdot 10^{-3}$ M TMG and 0.25 % galactose. This latter finding serves as a control for the drastic inhibition by galactose of hydrolysis *in vivo* by cells containing permease and grown in a variety of carbon sources.

In permease-positive, glucose-grown cells the action of glucose and of azide appears to be independent. Thus, high azide, 0.15 M, decreases the rate to 46 %; 0.25 % glucose decreases it to 38%. Together hydrolysis is decreased to 10%. Since 46% × 38 % = 18 %, there may be a small synergism.

The effect of variation of substrate concentration is very straightforward. The hydrolysis *in vivo* obeys simple Michaelis–Menten kinetics, and K_m values of $10.3 \cdot 10^{-4}$ M and $10.8 \cdot 10^{-4}$ M were obtained in measurements made 6 months apart on succinate-grown ML 308. With glucose-grown ML 308, a lower value of $7.9 \cdot 10^{-4}$ M was obtained. All of these results are uncorrected for formaldehyde-resistant hydrolysis. The data of Table I require that this correction be omitted for the succinate-grown cells, but it must be employed certainly for either partially induced cells or cells grown on carbon sources where the permease mechanism does not saturate the transporteur or cryptic part of the pathway. Such correction brings the result with glucose-grown cells in accord with the succinate one. Such a correction would also partially account for the lower values for K_m reported by BUTTIN[17].

Finally, we would note that cryptic hydrolysis is initially directly proportional to ONPG concentration. However, at high ONPG concentrations, $5 \cdot 10^{-2}$ M, the rate slowly, over 10–15 min, increases by a factor of 2–2.5. Presumably, this results from the build up of galactose and its exit via the common transporteur mechanism. This response can not be an adaptive process because it takes place in the presence of chloramphenicol.

CONCLUSIONS

A model which accounts for the data obtained in this paper and in previous studies is given in Scheme I. It is an elaboration of those previously given[1-3]. Like the previous model[3], the permease serves a role of speeding and rendering more specific the transport process, but does not itself bridge the permeability barrier. The existence of a separate transporter or carrier, which bridges this barrier, was postulated previously on the basis of glucose-caused expulsion of TMG in cryptic cells. This part of the model is strongly supported in the present work by the specific interaction of certain carbohydrates on both entry and on exit and by the finding that an element with a high temperature coefficient is involved in each of three processes: (1) hydrolysis *in vivo* of ONPG in the permease-containing cells; (2) hydrolysis *in vivo* in cryptic cells; (3) the exit process. In the present model the energy coupling is placed inside the permeability membrane instead of outside as in the previous model. This is not only much more reasonable because it does not require the transport of an energy-rich compound outwards, but also accounts for the many observations indicating that transport in this system can take place in the absence of metabolic energy if no thermodynamic work need be performed. It also readily explains the influence of energy sources and poisons on the exit process. If this model prove correct, it will be one of the relatively rare instances of a process "pulled" by an exergonic process instead of being "pushed" by it. Usually nature chooses to drive reactions by "pushing" because it leads to higher concentrations of intermediates

Biochim. Biophys. Acta, 79 (1964) 177–200

and consequently the possibility of more rapid reaction. Transport inwards would appear to be a reasonable case to be the exception because, of course, the metabolic machinery is internal to the barrier.

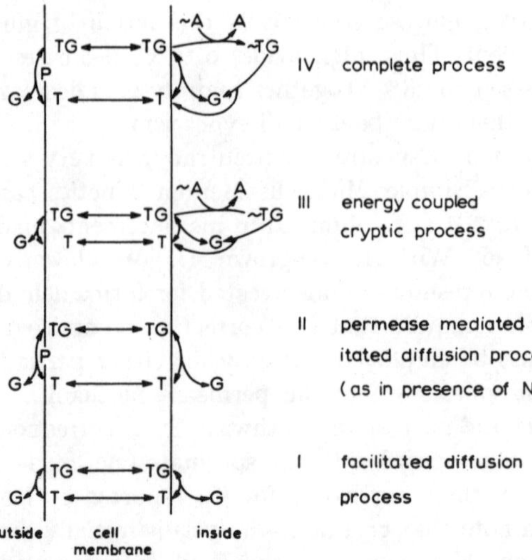

Scheme I. Model for sugar transport in *E. coli*. G = sugar; galactoside, galactose, glucose, or maltose and perhaps others. P = permease—inducible, stereospecific, fixed in position (several P's of same or different specificity may react with same T). T = transporter or carrier element which crosses the barrier in unspecified manner with or without sugar. A = immediate energy source, reserves of ∼ A are not large. ∼ TG = "activated" transporter sugar. Chemical nature unspecified (although depicted as separate from membrane, must be on or in membrane). There are none, or at most very few, pores, cracks, fissures, etc. allowing completely passive passage.

The model is also reasonable from the point of view of evolution. For at the moment when protoplasm becomes cytoplasm a barrier must appear which is generally impermeable to the intermediates involved in energy metabolism and those needed for the formation of the macromolecules. It was simultaneously necessary then to have a mechanism to allow passage of certain classes of substances. Later on in evolution the permeases were added to speed the specific entry of certain sugars, etc., under certain conditions. Independently, the energy-coupling step was added; this can both speed up the transport process and permit the accumulation of the substance to drive subsequent metabolism. An important feature of the model is that both the permease and the energy coupling are in addition to the simpler processes of the primitive mechanism.

In comparing various experimental transport systems, the possibility of the presence or absence of these various processes should be borne in mind. For example, some of the less specific amino acid transport systems in mammalian systems may simply reflect the lack of permeases and, therefore, exhibit the properties of the transporter element alone. It may be that permeases may be restricted to the rapidly growing unicellular forms. We thus feel, as does CHRISTENSEN[18], that the concept of the permease may increase "parochialism" in biology; but we also feel that it may aid in increasing our understanding of transport in general.

The new model proposed here is consistent with models proposed in other

Biochim. Biophys. Acta, 79 (1964) 177–200

laboratories[5,18,19]. In particular, the present model goes a long way toward a compromise between the points of view of this laboratory and those expressed by MITCHELL[19].

At ice-bath temperature, movement of the transporter is limiting. The movement of both T and TG in this system appears to be slowed to about the same degree. This system is different in this respect from amino acid transport system in *E. coli*[6]. At 28° in the presence of energy, ONPG hydrolysis *in vivo* is most probably not limited by the crossing of the barrier but is limited by the binding affinity of the permease. In the absence of energy coupling, the internal level of TG would rise to a value to give the same steady-state dissociation rate as the rate of the permease step. The energy-coupled step would also do this, though now there would be a lower steady-state concentration of TG at the inside of the barrier. The influence on rate of energy metabolism would then be indirect (and minor) through the increased concentration of free T to react with external G through the mediation of the permease.

Glucose and maltose via their specific permeases compete very successfully for T. The fact that the competition is nearly independent of substrate probably means that certain regions of the bacterial membrane have T accessible to permease molecules of only single specificities and others to permease molecules of the varying specificities*.

The complementary fact pointing to the conclusion that a given molecule of T can react with at least two different permease molecules is that exit is increased by glucose and maltose in cells grown on these sources.

Agents which block the permease and agents which interrupt the energy coupling inhibit both exit and hydrolysis *in vivo*. However, the two classes of agents do not alter the two types of flux to the same degree. This discord is, of course, a reflection that the mechanism is not symmetrical about the membrane. Consequently, even though the same steps are involved in both entry and exit, different steps can be rate limiting. The permease step is rate controlling on entry and energy linkage of minor importance, and the reverse is true on exit. The external concentration is so low that external TG dissociates with sufficient rapidity so that this step is not limiting and permease increases the overall rate to a relatively small extent (3 fold at 28°, 4–5 fold at 15.5°). Presumably, this results from decreasing external TG and increasing, therefore, internal T available to react with internal G.

Nonradioactive galactosides, galactose, and glucose and maltose also speed exit of radioactive TMG but in a different way than simply altering the energy supply. It must be assumed that passage of TG in both directions across the membrane is a much more rapid process than the passage of free T. In such a case, entry of one molecule in the fashion of a turnstile would permit the exit of one molecule. Flux measurements of all components will be required to further investigate this aspect.

The rate of either entry or exit depends on the concentration gradient of the species that actually traverses the membrane, *i.e.*, TG in our model. In the exit experiments the external concentration of TG is virtually zero (certainly in the presence of permease) and, therefore, the rate varies with the internal concentration

---

* It also would be presumed that the T can react with G at the external membrane nonenzymically even if the permease molecules are present. This follows from the fact that exit half times are observed with formaldehyde-treated and PCMS-treated cells, similar to the extrapolated value to zero degree of induction of cells with functioning permease.

of TG. TG at the inner surface of the membrane is lost by three processes: migration, dissociation, and activation. It is formed by only one, *i.e.*, the uncatalyzed combination of T and G. If the activated state is relatively short-lived and if we assume, in the absence of better information, that $\sim A$ activates in a bimolecular reaction, we may then diagram this part of the model as follows:

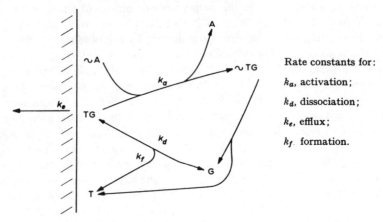

Rate constants for:

k_a, activation;

k_d, dissociation;

k_e, efflux;

k_f formation.

The corresponding kinetic equation is

$$\frac{d(TG)}{dt} = k_f(G)\,(T) - (k_e + k_a(\sim A) + k_d)\,(TG)$$

Since the amount of T is presumed small with respect to the internal concentration of G, we may make the standard Bodenstein steady-state approximation by setting $d(TG)/dt$ to zero and solving for TG:

$$(TG) = \frac{k_f(G)\,(T)}{k_e + k_a(\sim A) + k_d}$$

Since exit is proportional to (TG) and the above equation shows that (TG) is proportional to (G), we can readily understand why exit is a first-order process and does not show saturation kinetics. In particular, it does not reflect the kinetic properties of the permease. The rate constant and the corresponding half-life change as $k_a(\sim A)$ is altered. Exit is slow when $k_a(\sim A)$ is large and fast when it is small or zero. The fastest exit observed in the presence of energy poisons can be further increased by increasing T on the internal side of the membrane as the result of ONPG flux inwards.

Throughout the duration of almost all of the exit experiments $k_a(\sim A)$ appeared to be constant. Thus, it is lowered abruptly by azide, etc., very rapidly compared to the smallest experimental time of 30 sec employed. On the other hand, glycerol raised it rapidly, so that a slow exit from the start was observed. The two exceptions are: succinate addition, which takes several minutes at 28° to slow exit, and azide addition to one batch of glucose-grown cells, where the rapid azide exit rate was briefly observed followed by a slower exit, presumably resulting from the mustering of anaerobic energy reserves.

Formaldehyde slows exit to a much smaller extent than it does entry. This is then the mirror situation to the energy poisons. Its slowing of exit results from an indirect effect, that in its presence a higher external TG concentration is required to maintain the exit rate; and this reduces internal T and slows the rate of migration of TG outward.

Thus, the model can account for the simpler situations of entry alone or exit alone. We now must consider the more complicated case of accumulation of a non-metabolized substrate. In the β-galactoside system it is very clear that thermodynamic work can be performed and a concentration gradient of free osmotically active galactoside[20] can be created. The present model acts as a pump by actively dissociating the GT complex so that the slower spontaneous reaction of G and T can not reach equilibrium. Thus, the ratio of internal (G) to (TG) is maintained at a much higher than equilibrium value. As fast as T and G recombine, they are actively redissociated. In this sense, the pump works by actively preventing exit as well as by actively "pulling" entry. At the steady state the increased leakage due to increased internal (G) just compensates entry.

In terms of the original model[1,2], this raises one grave difficulty. The present model as delineated so far imagines that exit proceeds through the same pathway as does entry—consequently, at the steady-state plateau of accumulation the permease catalyzes a reaction that is at chemical equilibrium. Obviously, one does not need a catalyst to speed up a reaction that is at chemical equilibrium; and consequently, permease would not be needed to maintain accumulation contrary to experiment. Cryptic cells should accumulate against a concentration gradient, although perhaps much more slowly, and level off at a much smaller concentration. This also is contrary to experiment*. The original model overcomes these difficulties by postulating that exit is through an independent passive leak. Consequently at the steady state, the permease carries out a unidirectional reaction inward, compensating the leak. Thus, the concentration dependence of the accumulation process mirrors that of the permease itself; and the cryptic cell should simply and slowly equilibrate with the external milieu.

An entirely different type of model could be proposed in which the saturation type of kinetics observed with accumulation represents a limitation on energy reserves. Thus, accumulation would take place with external concentrations until (TG) and (T) were the same on both sides of the barrier. It is then apparent that the ratio of internal to external G would be a function of rate constants and the steady-state value of ($\sim$ A) and independent of the other constituents of the model, including permease. The observed steady-state plateau values would then be a measure of the steady-state concentration of $\sim$ A at different external concentrations.

Originally this concept was rejected because partially induced cultures showed the same half saturation value as did fully induced cultures, even though the energy utilization must be less. This question we felt was important, so we repeated and confirmed (Fig. 4) the original observations on this point using the newer "millipore" technique.

It is possible to resuscitate the hypothesis of energy limitation by insisting that the limitation is either on the local formation of $\sim$A or on the rate of delivery of

---

* KEPES[3] reported that the plateau level of accumulation of the cryptic cells corresponded to one-fourth the concentration in cell water as in the environment. RICKENBERG (personal communication) finds equal concentrations under his experimental conditions. So far, no one has found conditions where the internal concentration exceeds that of the surroundings with cryptic cells.

~A to the sites on the membrane. For either of these possibilities, the total capability of the cells to produce ~A is not limiting; yet accumulation will show saturation kinetics with kinetic parameters independent of the degree of induction. We do not like this hypothesis because we believe that our hyperbolas are too beautiful to be approximations of the more complicated saturation curves required by this modification of hypothesis. Also, we do not favor it because the half saturation for ONPG accumulation is found to be so near that for the energy less-dependent hydrolysis *in vivo* of the same substrate.

A more reasonable explanation for these several problems is the original one, *i.e.*, in the steady-state permease molecules catalyze a reaction that causes a net flux inwards counterbalanced by a flux outwards somewhere else. However, we can not presume that the quantitatively important path of efflux is a simple passive leak because both the cryptic hydrolysis and the exit measurements have high temperature coefficients. Also in accord with this idea is the fact that the cryptic hydrolysis pathway can be saturated (Table I) by the permease process in cells containing a maximal amount of permease.

But, if most of the efflux takes place through the same pathway as does influx and if only a small fraction proceeds independently, a net influx through permease site would take place and the observed identity of K_m of partially and fully induced cultures for accumulation would be explained.

A very small number of pores could thus account for the experimental facts. We feel that it is more probable that this exit takes place through unenergized sites of Type I or Type II of the model. Whether these are merely defective or lacking in some activating enzyme or temporarily inactive is, of course, immaterial for the present considerations.

This hypothesis could account as well for the finding that the amount of TMG accumulated by cryptic cells is highly variable*. Clearly, even a small inward flux of the transporter carrying another substance would tend to pump TMG back out of the cell. One could only hope even with the best of circumstances to obtain an accumulation of galactoside to a level slightly higher than that existing in the environment.

Other alternatives must be considered. If ~A is an energy-rich, but not a very energy-rich, compound, one could then consider mechanisms where counterflow is induced by the generation of ~A pushed or backed up by the gradient of external compound flowing inward. This ~A could then force the flow in the other direction at a later instant of time or at another site. In the latter case, each site might be specific for one sugar, and it would not be necessary to postulate a common transporter or access of the same transporter to two different permease sites.

ACKNOWLEDGEMENTS

The author wishes to thank Dr. J. MONOD and Dr. A. KEPES for their hospitality while he was a visitor in their laboratory. The author also wishes to acknowledge the able assistance of Mrs. F. PETTIS during the preparation of the manuscript.

A little of the experimental research was done in the author's own laboratory

---

* See footnote on p. 198.

in Gainesville under the support of Grant C-3255 from the National Cancer Institute, U.S. Public Health Service, and Grant G-8851 from the National Science Foundation.

REFERENCES

1 H. V. RICKENBERG, G. N. COHEN, G. BUTTIN AND J. MONOD, *Ann. Inst. Pasteur*, 91 (1956) 829.
2 G. N. COHEN AND J. MONOD, *Bacteriol. Rev.*, 21 (1957) 169.
3 A. KEPES, *Biochim. Biophys. Acta*, 40 (1960) 70.
4 J. LEDERBERG, *J. Bacteriol.*, 60 (1950) 381.
5 B. ROTMAN, *J. Bacteriol.*, 76 (1958) 1.
6 R. J. BRITTEN AND F. T. McCLURE, *Bacteriol. Rev.*, 26 (1962) 292.
7 F. JACOB AND J. MONOD, *Cold Spring Harbor Symp. Quant. Biol.*, 26 (1961) 193.
8 M. COHN AND K. HORIBATA, *J. Bacteriol.*, 78 (1959) 624.
9 B. L. HORECKER, J. THOMAS AND J. MONOD, *J. Biol. Chem.*, 235 (1960) 1580.
10 B. L. HORECKER, J. THOMAS AND J. MONOD, *J. Biol. Chem.*, 235 (1960) 1586.
11 A. KEPES, 12 *Colloq. Ges. Physiol. Chem.*, (1961) 100.
12 H. WIESMEYER AND M. COHN, *Biochim. Biophys. Acta*, 39 (1960) 440.
13 J. MONOD, *La Croissance des Cultures bactériennes*, Herman et Cie, Paris, 1942, p. 70.
14 A. L. KOCH, *Ann. N.Y. Acad. Sci.*, 102 (1963) 602.
15 B. ROTMAN AND R. GUZMAN, *Pathologie-Biologie*, 9 (1961) 806.
16 B. L. HORECKER, M. J. OSBORN, W. L. McLELLAN, G. AVIGAD AND C. ASENSINO, in A. KLEIN-ZELLER AND A. KOTYK, *Membrane Transport and Metabolism*, Academic Press, London, 1961, p. 378.
17 G. BUTTIN, *Thesis*, to the Faculté de Science de Paris for Diplome d'Étude Superior (Science Naturelles), 1955.
18 H. N. CHRISTENSEN, *Advan. Protein Chem.*, 15 (1960) 267.
19 P. MITCHELL, in A. KLEINZELLER AND A. KOTYK, *Membrane Transport and Metabolism*, Academic Press, London, 1961, pp. 22–34.
20 W. R. SISTROM, *Biochim. Biophys. Acta*, 29 (1958) 575.

Copyright © 1966 by the American Society of Biological Chemists, Inc.
Reprinted from *J. Biol. Chem.*, **241**, 2200–2211 (1966)

The Role of Energy Coupling in the Transport of β-Galactosides by *Escherichia coli*[*]

(Received for publication, November 17, 1965)

Herbert H. Winkler‡ and T. Hastings Wilson§

With the technical assistance of Angela DeCarlo

From the Department of Physiology, Harvard Medical School, Boston, Massachusetts 02115

11

SUMMARY

Lactose and *o*-nitrophenylgalactoside (NPG) were transported against considerable concentration gradients by ML 308-225, a mutant of *Escherichia coli* which lacks β-galactosidase but possesses a constitutive transport system for β-galactosides. The kinetics of influx of NPG during this accumulation were found to be the same as those found for this galactoside moving *down* a concentration gradient into ML 308, *i.e.* transport into the parent cell containing β-galactosidase. When cells of ML 308-225 were exposed to azide, azide plus iodoacetate, or dinitrophenol transport of these galactosides against a concentration gradient was completely abolished. However, the presence of functional membrane carriers in such inhibited cells was indicated by (*a*) very rapid equilibration of external and internal concentrations of galactoside, (*b*) inhibition of the rate of this equilibration by chemical analogues, and (*c*) transient accumulation of substrate by washed cells preloaded with galactoside. All the evidence was consistent with the hypothesis that the same membrane carriers were involved in active transport by control cells and facilitated diffusion by poisoned cells.

The most striking finding was that the addition of metabolic inhibitors reduced the K_t of exit about two orders of magnitude, whereas the K_t of entrance remained constant. It was inferred from these studies that energy coupling reduced the affinity of the carrier for its substrate on the inner surface of the plasma membrane.

The possible physiological control of energy coupling of transport was discussed in the light of the present observations.

All the available evidence is consistent with the hypothesis that active transport across a biological membrane consists of at least two distinct components: first, a substrate-specific membrane factor or "carrier" which facilitates movement across the permeability barrier; and second, a mechanism which couples metabolic energy to the carrier and produces a net movement of substrate from the extracellular environment into the cell against an electrochemical gradient. The glucose transport systems of yeast (1) and erythrocyte (2, 3) have been shown to possess only the first component and are designated carrier-mediated transport or facilitated diffusion. In this type of system, the membrane carriers facilitate the equilibration of intracellular and extracellular substrate concentrations, without the expenditure of energy by the cell. Simple passive diffusion is not involved, however, as the transport shows substrate specificity, saturation kinetics, and competitive inhibition by chemical analogues. Strong evidence for a mobile carrier for sugars in the erythrocyte was provided by the "counterflow" experiments of Park *et al.* (4) and by Rosenberg and Wilbrandt (5) after the prediction of the phenomena by Widdas (6). Park *et al.* incubated erythrocytes in a balanced salt solution containing xylose until the intracellular and extracellular concentrations of sugar had equilibrated. They then added a high concentration of glucose, a substrate with a high affinity for the carrier. They observed a transient net movement of xylose out of the cell against a concentration gradient. This behavior, according to the model shown in Fig. 1, is the result of the inhibition of the influx of xylose by the presence of glucose, with little or no effect on efflux. This model is consistent with all the experimental data for facilitated diffusion in the erythrocyte (6–9).

One attractive hypothesis is that this same model may be applied to active transport with the simple modification that the affinity for uptake must be greater than the affinity for exit. If we accept the model shown in Fig. 1, a cell capable of accumulating a substance intracellularly to a concentration 100 times in the medium must possess a K_t of entrance 100-fold less than the K_t of exit. Various workers have shown in both animal and bacterial cells that when an active transport system is poisoned with compounds that abolish the production of adenosine triphosphate, the membrane carrier remains intact although movement against a gradient is no longer possible (10–17). It is assumed that the effect of the energy uncoupling of the active transport system has been to change either the K_t of entrance or the K_t of exit (or both) so that the affinities of the two processes are equal.

* This research was supported by grants from the United States Public Health Service (AM-05736). A preliminary report of this work was presented at the 49th meeting of the American Society of Biological Chemists, Atlantic City, April 9 to 14, 1965.

‡ Supported by a Predoctoral Fellowship from the National Science Foundation. This work represents a portion of a thesis to be submitted to Harvard University in partial fulfillment of the requirements for the degree of Doctor of Philosophy.

§ Supported by a Career Development Award of the United States Public Health Service (5-K3-GM-15,303).

2200

This paper describes a study of the role of energy coupling in membrane transport of β-galactosides in *Escherichia coli*. Metabolic inhibitors converted the active transport system to a facilitated diffusion system in which the affinities of entrance and exit were equal. Energy coupling increased the K_t of exit without greatly affecting the K_t of entrance.

EXPERIMENTAL PROCEDURE

Bacteria—Three strains of *E. coli* ML were used in this study: ML 308 ($i^-z^+y^+$) which is constitutive for both the transport system and β-galactosidase (EC 3.2.1.23); ML 35 ($i^-z^+y^-$), which lacks the transport system but possesses β-galactosidase; and ML 308-225 ($i^-z^-y^+$), which has the same transport activity as ML 308 but has less than 0.01% of the β-galactosidase. The first two organisms were isolated by Monod and his collaborators at the Pasteur Institute.

ML 308-225 was obtained by first mutagenizing ML 308 with ethyl methanesulfonate (18); second, growing the mutagenized culture in liquid medium with 0.2% glucose to eliminate auxotrophs; third, treating the mutants twice with medium containing penicillin G (Lilly) (19, 20) and 0.25 mM lactose to reduce the number of cells which were lactose-positive, and then plating the surviving cells on agar plates containing peptone, the indicators neutral red and crystal violet, and 0.2% lactose. White colonies from these plates were grown in casein hydrolysate medium to exponential growth and tested for transport activity as measured by the uptake of the nonmetabolizable substrate thiomethyl-β-galactoside, and β-galactosidase as measured by o-nitrophenyl-β-galactoside hydrolysis in the sonically disrupted or toluenized cells. Fig. 2 shows the transport activity of both the parent and ML 308-225.

The mineral medium used in this study was Medium 63 (21) which contains KH_2PO_4 (13.6 g), $(NH_4)_2SO_4$ (2.0 g), $MgSO_4 \cdot 7H_2O$ (0.2 g), $FeSO_4 \cdot 7H_2O$ (0.005 g), and H_2O (1 liter) adjusted to pH 7.0 with KOH. Sodium chloride (50 mM) was added in

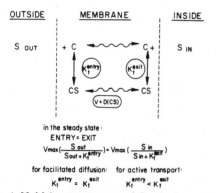

FIG. 1. Model for an energy-uncoupled active transport system. S, substrate (designated as inside or outside of the cell); C, membrane carrier; CS, membrane carrier-substrate complex; D, diffusion constant of carrier or carrier-substrate complex through the membrane; K_t, equilibrium constant for the reaction $S + C = CS$. It is assumed that: (*a*) the chemical reactions at each interface are much more rapid than the diffusion of carrier or carrier-substrate complex; (*b*) a linear concentration gradient of both C and CS exists in the membrane; (*c*) the diffusion constants for C and CS are the same.

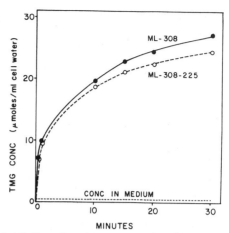

FIG. 2. Comparison of transport capacity of the parent (ML 308) and the mutant (ML 308-225). Cells were incubated in media containing 0.5 mM TMG-^{14}C at 10°.

some experiments but had no effect on transport processes. In all experiments the carbon source for growth was 1% casein hydrolysate (Nutritional Biochemicals). Cells were diluted 1:100 from the maintenance culture and grown overnight. In the morning, 1 ml of cells was transferred to a side-arm flask containing 30 ml of fresh media and grown on a rotary shaker at 37° until late logarithmic growth phase was attained. These cells were then centrifuged, washed with cold medium, recentrifuged, and resuspended in medium containing chloramphenicol (50 μg per ml) in the absence of carbon source.

Identification of Accumulated Material—Washed cells of ML 308-225 were incubated with radioactive lactose or NPG[1] for 10 min at 23°, centrifuged, and washed once at 0°. The pellet was then extracted with 5% trichloracetic acid for 10 min at 0°. The suspension was centrifuged and the supernatant solution was extracted four times with ether and then neutralized. Chromatograms of the neutralized cell extracts were run on Whatman No. 1 filter paper in the descending direction in each of the following solvent systems: A, 1-butanol-acetic acid-water (2:1:1); B, 1-propanol-water (3:1); C, pyridine-1-butanol-water (4:6:3); and D, pyridine-ethyl acetate-water (4:10:3). Each solvent gave clear separation between the galactoside (NPG or lactose) and galactose. Autoradiograms were prepared by exposure of chromatograms to x-ray film (Kodak "No Screen" medical film) for 1 to 7 days. In addition, the radioactivity on the chromatograms was determined semiquantitatively in a Nuclear-Chicago strip scanner. After exposure of ML 308-225 to NPG-^{14}C, 95% of the radioactivity accumulated within the cell had an R_F corresponding to NPG. In the lactose experiments, 98% of the cell radioactivity corresponded to the R_F of lactose. Three separate experiments were performed with NPG and two with lactose.

Determination of Cell Water—A known volume of washed ML 308-225, grown in casein hydrolysate, was resuspended at

[1] The abbreviations used are: NPG, o-nitrophenyl-β-galactoside; TMG, thiomethyl-β-galactoside; CMB, p-chloromercuribenzoate; TDG, thio-β-digalactoside.

a known optical density in a medium containing inulin-^{3}H. A 35-ml aliquot of this suspension was pipetted into a tared centrifuge tube and centrifuged, the tube was drained, and the weight of the pellet was determined. The inulin space and *percentage* of dry weight were measured on this pellet. The inulin space was 10% of the wet weight of the pellet under these conditions. The dry weight was 27% of the wet weight of the cells (corrected for inulin space). The wet weight of the cells minus the dry weight was taken as the weight of the cell water. There were 2.7 μl of cell water per mg of dry weight. In 1.0 ml of a cell suspension with an optical density of 100 Klett units (No. 42 filter), there was 0.6 μl of cell water.

In a number of experiments, the cell volume was determined by the microhematocrit of a dense suspension. Microhematocrit tubes (1.4-mm diameter; 75-mm length) were filled with a dense cell suspension (greater than 10%) of known optical density and centrifuged, and the *percentage* of packed cells was measured. On assumption that the extracellular volume in this packed cell mass was 10% and the cell specific gravity was 1.1, the cell water was calculated. This latter method required only a small number of cells and hence could conveniently be carried out on the same batch of cells used for transport experiments when a precise measurement of cell water was required.

Assay of NPG Transport in ML 308—The transport activity in ML308 was determined by measuring the rate of *o*-nitrophenol formation when intact cells were incubated in the presence of NPG. This process (hydrolysis *in vivo*) has been shown to be proportional to the transport rate, since there is a large excess of β-galactosidase which catalyzes immediate hydrolysis of NPG after it enters the cell (22). The rate of yellow color formation was determined with various concentrations of NPG at both 23° and 10° in poisoned and unpoisoned cells by two methods. In the first method, an aliquot of the incubation mixture containing cells, NPG, and the indicated additions was taken at various time intervals, and the reaction was stopped by addition of 2 volumes of 0.7 M Na$_2$CO$_3$ at 0°. The cells were then removed by centrifugation at 4° and the absorbance at 420 mμ was determined. In the second method, the rate of hydrolysis was followed continuously in a cuvette in a Gilford spectrophotometer with temperature control.

Measurement of Galactoside Uptake—Washed cells were suspended at a final concentration of 300 μg of dry weight per ml in medium with the indicated additions. The cells were incubated for 30 min at room temperature with or without the indicated metabolic inhibitors prior to the addition of labeled substrate. After temperature equilibration at 10°, labeled substrate was added to give the desired substrate concentration. At various time intervals, 0.5-ml samples were pipetted onto the center of a Millipore filter (0.65-μ pore size) which had been precooled with 10 ml of medium at 0°. Care was taken to prevent the cell suspension from touching the outer region of the filter which was in contact with the chimney as higher values were obtained in these cases. The cells were then quickly washed on the filter with 10 ml of ice-cold medium. As additional washing of the cells did not further reduce the counts, it was presumed that all extracellular radioactivity was removed and that no loss from the cells occurred under these conditions. The filter was placed in a vial and 15 ml of liquid scintillation fluid, a toluene-ethanol mixture (12:7) containing 0.4% 2,5-diphenyloxazole and 0.01% 1,4-bis-2'(5'-phenyloxazolyl)benzene, were added. The vials were shaken vigorously and counted in a liquid scintillation

counter (either Packard or Nuclear-Chicago). The presence of the filter did not significantly alter the counting efficiency. In some of the experiments with poisoned cells, 5 times as many cells (0.5 ml of 1500 μg of dry weight per ml) were pipetted on Millipore filters with a 1.2-μ pore size. Less than 5% of the cells were lost during filtration under the latter conditions.

For measurement of the initial uptake rate into preloaded cells, this technique was modified. First, the cells were filled with a high concentration of nonradioactive galactoside by incubation for 30 min at 23° in medium containing 25 mM lactose. These preloaded cells were diluted 10-fold by addition of ice-cold medium, and centrifuged. The pellet was resuspended in cold medium and again centrifuged. The supernatant fluid was decanted, the centrifuge tube was carefully wiped, and the pellet was taken up in a small volume of ice-cold medium, the temperature being maintained at 0° to minimize exit of preloaded material. At zero time, 50 μl of this cell suspension were added to a test tube with 0.45 ml of medium containing the radioactive sugar at 10° with a microliter syringe cooled to 0°. As quickly as possible, most of this 0.5-ml sample was taken up in a Pasteur pipette (cooled to 10°), delivered onto a precooled filter, and washed with 10 ml of ice-cold medium. The volume delivered by the pipette was calculated from the difference between the counts in the initial 0.5 ml and the counts remaining in the pipette and test tube. With this method, incubation periods of 5 to 6 sec could be achieved.

Determination of Non-Carrier-mediated Component—In some experiments, particularly with NPG as substrate, the non-carrier-mediated component of the entrance or exit rate was appreciable and appropriate corrections were necessary. This non-carrier-mediated transfer was presumed to be due to simple passive diffusion since the rate was directly proportional to the concentration difference across the membrane and larger compounds such as lactose shown very much slower movement than NPG or TMG. The non-carrier-mediated transport rate was determined by a number of different procedures. One method involved the measurement of the rate of NPG entrance into the transport-negative mutant ML 35 (average value about 0.2 μmole per g of wet weight per min per 1 mM concentration gradient at 10°. Another method was with the use of *p*-chloromercuribenzoate (10^{-4} M) to block the carrier-mediated component. This non-carrier-mediated value was subtracted from the observed rate of entry into ML 308 to obtain the true carrier-mediated rate. The diffusion correction for NPG uptake into energy-uncoupled ML 308-225 was taken as the rate of entry in the presence of the competitive inhibitor thio-β-digalactoside (10^{-2} M). The exit rate of NPG from ML 308-225 due to passive diffusion was taken as the rate in the presence of CMB (10^{-4} M). These values averaged 0.05 μmole per min per g of wet weight per 1 mM concentration gradient at 10°. The explanation for the somewhat different values with different methods is not clear. The diffusion component for lactose exit was found to be extremely small (0.002 μmole per g of wet weight per min per 1 mM concentration gradient) in comparison with the carrier-mediated rate and therefore corrections were unnecessary.

Measurement of Galactoside Exit from ML 308-225—Either control cells or poisoned cells (1 ml at a density of 2 mg of dry weight per ml) were incubated for 20 min at room temperature with various concentrations of radioactive galactoside to obtain cells containing the desired levels of intracellular galactoside.

During this preincubation period, the sugar concentration in the poisoned cells equilibrated with that in the medium; control cells accumulated substrate to high intracellular levels. The cells were then centrifuged at 0° for 10 min. The supernatant fluid was carefully decanted, and the inside of the tube was wiped free of adhering drops of radioactive medium (the volume of contaminating medium in the tube after this procedure varied from 20 to 60 μl). The pellet was then quickly taken up in 30 ml of medium at 10° (with or without poisons, as required) so that a dilute suspension of about 70 μg of dry wt per ml was obtained. At various intervals, 5-ml samples were removed and filtered on precooled Millipore filters. In these experiments, the cells were not washed on the filter since the medium contained so little radioactivity and the volume of medium retained by the filter was only 20 μl (as measured by inulin-^{3}H).

Chemicals—Lactose-1-^{14}C was obtained from Nuclear-Chicago, and thiomethyl-^{14}C-β-galactoside and inulin-methoxy-^{3}H was from New England Nuclear. The *o*-nitrophenyl-β-galactoside-1-^{14}C was prepared by Dr. Bernard Pitt according to the method of Seidman and Link (23). Chloramphenicol was a gift of Parke, Davis, and methyl-β-galactoside was a gift from Corn Products. Thiomethyl-β-galactoside, thiophenyl-β-galactoside, *o*-nitrophenyl-β-galactoside, and *p*-chloromercuribenzoate were obtained from Calbiochem; thio-β-digalactoside and isopropylthio-β-galactoside were obtained from Mann. α-D-Lactose (glucose-free) was obtained from Sigma; sodium iodoacetate, potassium azide, and ethyl methane sulfonate were from Eastman Kodak.

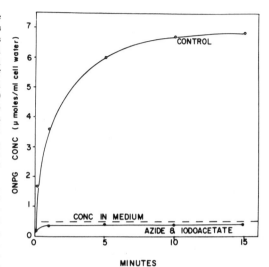

FIG. 4. Effect of azide plus iodoacetate on the active transport of NPG by ML 308-225. Inhibited cells were preincubated with 30 mM azide plus 1 mM iodoacetate for 30 min at 10° before the addition of 0.5 mM NPG-^{14}C.

RESULTS

Active Transport of NPG and Lactose by β-Galactosidase-negative Mutant—When a washed cell suspension of *E. coli* ML 308-225 (β-galactoside transport-positive, β-galactosidase-negative) was incubated with radioactive *o*-nitrophenylgalacto-side the galactoside was accumulated within the cell to concentrations which depended on the external level of NPG (Fig. 3). The maximum concentration ratio (ratio of intracellular concentration to extracellular concentration) observed in these experiments was about 150-fold. At the end of similar experiments, cells were extracted with cold trichloracetic acid, and over 95% of the radioactivity of such an extract was found in NPG (see "Experimental Procedure" for details). Thus, the intracellular radioactivity in such experiments was due to NPG and not to galactose or other metabolic products.

Lactose, the natural substrate for this transport system, was also accumulated against large concentration gradients by ML 308-225. When cells were incubated at 10° with lactose the largest concentration ratio (ratio of intracellular concentration to extracellular concentration) was of the order of 200-fold (Fig. 3). When chromatography was performed on extracts of these cells, over 98% of the accumulated material was found to be lactose.

Effect of Metabolic Poisons on Active Transport—The effect of a variety of metabolic poisons were tested on active transport of galactosides by ML 308-225. Fig. 4 illustrates the effect of preincubating cells for 30 min in azide (30 mM) plus iodoacetate (1 mM) on the subsequent intake of radioactive NPG. Transport against a concentration gradient was completely abolished under these conditions. The sugar, however, rapidly entered the cell until its intracellular concentration was the same as that in the incubation medium.

Fig. 5 shows that when poisoned cells were incubated in solu-

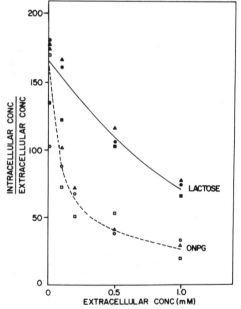

FIG. 3. Effect of external concentration on steady state concentration ratio in ML 308-225. Cells were incubated with the indicated concentration of NPG-^{14}C or lactose-^{14}C at 10° for 10 min, at which time the concentration in the cells had reached a steady state. Results of three experiments with each sugar are shown. *ONPG*, *o*-nitrophenyl-β-galactoside.

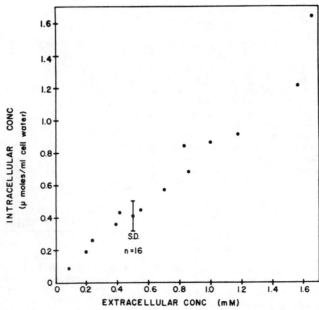

Fig. 5. Equilibration of intracellular NPG with that in the medium at various extracellular concentrations. ML 308-225, preincubated with 30 mM azide plus 1 mM iodoacetate, were incubated with NPG-¹⁴C (plus the two metabolic inhibitors) at 10° for 22 min, at which time the intracellular concentration had reached a steady state.

TABLE I

Effect of metabolic inhibitors on active transport of β-galactosides

Cells were preincubated for 30 min at 23° with the indicated metabolic inhibitors before the addition of radioactive substrate. Uptake of radioactivity was measured after 10 and 20 min of incubation at 23°. Similar values were obtained at these two time intervals. Each value represents the mean of three experiments. An inhibition of 100% indicates that the steady state concentration (intracellular) did not exceed that in the medium.

Substrate for transport	Concentration	Strain	Temperature	Inhibition of active transport		
				Azide (10 mM)	Azide (30 mM) + iodoacetate (1 mM)	DNP (1 mM)
	mM			%	%	%
TMG....	0.5	ML 308	23°	91	100	100
TMG....	0.1	ML 308	23	97	100	100
NPG....	0.5	ML 308–225	10	99	100	99
Lactose..	1.0	ML 308–225	10	97	100	94

tions containing NPG at concentrations varying from 0.1 mM to 1.6 mM equilibration of the sugar between cell and medium was regularly observed. Although active transport was completely inhibited under these conditions, rapid entrance of the sugar into the cell was still observed. Sodium azide (10 mM) and 2,4-dinitrophenol (1 mM) were also extremely potent inhibitors of active transport (Table I). The three metabolic inhibitors affected the active transport of lactose and thiomethylgalactoside in the same manner as that for NPG giving 91 to 100% inhibition (Table I).

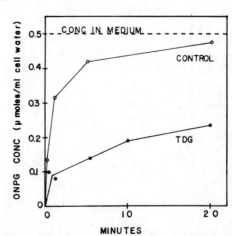

Fig. 6. Effect of TDG on the equilibration of intracellular and extracellular concentrations of NPG in energy-uncoupled ML 308-225. Cells preincubated with 30 mM azide plus 1 mM iodoacetate were added to 0.5 mM NPG-¹⁴C with or without 10 mM TDG at 10°.

Evidence for Presence of Membrane Carriers in Poisoned Cells—
A variety of experiments suggested that entry of sugar into the energy-uncoupled cells was by a carrier-mediated process and not by simple passive diffusion. Although accumulation of sugar did not occur in poisoned transport-positive organisms,

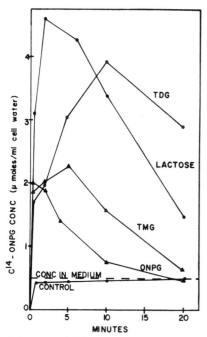

FIG. 7. Uptake of ONPG in energy-uncoupled ML 308-225 preloaded with galactoside. Cells were first incubated with 30 mM azide plus 1 mM iodoacetate and 25 mM nonradioactive galactoside for 30 min at 23°. These cells were then centrifuged and washed once at 0°. The washed pellet was resuspended in a solution containing 0.5 mM NPG-¹⁴C plus metabolic inhibitors at 10°.

the rate of entry was much faster than that found in the transport-negative mutant.

The rate of equilibration of the NPG in the cell with that in the medium was strongly inhibited by the addition of the chemical analogue, thiodigalactoside (Fig. 6). Further evidence that the membrane carriers were intact is provided by the experiments shown in Fig. 7 which are analogous to the "counterflow" experiments performed in erythrocytes (4, 5). Washed cells of ML 308-225 were poisoned with azide plus iodoacetate and preloaded with high concentrations of various nonradioactive galactosides. After the cells were washed at 0°, they were resuspended in a medium containing radioactive NPG. As predicted by the hypothesis (Fig. 1), a transient accumulation of NPG-¹⁴C was observed in the poisoned cell due to inhibition of efflux of NPG-¹⁴C by the preloaded galactoside. Loss of the preloaded sugar via the carrier would be expected to relieve the inhibition of efflux and diffusion equilibrium could be established. Fig. 8 shows a similar experiment in which lactose-¹⁴C was the substrate for uptake; preloading resulted in large transient gradients.

In a test of whether these preloading effects were due to transfer via a mechanism independent of the transport system determined by the "y" gene, a similar experiment was performed with the transport-negative organism ML 35 (i⁻z⁺y⁻). Since β-galactosidase was present in these cells, the nonmetabolizable substrate thiomethylgalactoside was utilized. TMG-¹⁴C entered the preloaded poisoned ML 308-225 at a rapid rate and a large transient concentration gradient was observed. When the preloaded cells of ML 35, either poisoned or unpoisoned, were incubated in TMG-¹⁴C, there was only slow entrance of the radioactive galactoside and *no* detectable effect from preloading (Fig. 9).

Kinetics of NPG Uptake in ML 308—In ML 308, which possesses both β-galactosidase and the transport system, transport of NPG has been shown to be rate-limiting for the over-all process resulting in its hydrolysis to o-nitrophenol and galactose by the intact cell (22). Because of the very large excess of β-galactosidase, NPG is hydrolyzed immediately after entering

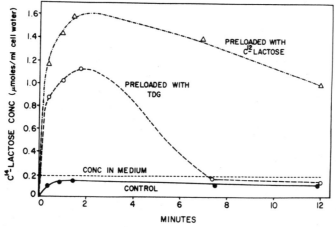

FIG. 8. Uptake of lactose-¹⁴C in energy-uncoupled ML 308-225 preloaded with lactose-¹²C or TDG. See legend of Fig. 7 for experimental details.

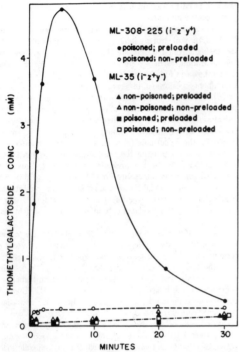

ML-308-225 (i⁻z⁻y⁺)

● poisoned; preloaded
○ poisoned; non-preloaded

ML-35 (i⁻z⁺y⁻)

▲ non-poisoned; preloaded
△ non-poisoned; non-preloaded
■ poisoned; preloaded
□ poisoned; non-preloaded

FIG. 9. Effect of preloading on TMG uptake in ML 308-225 and ML 35. For preloading, cells were incubated for 30 min in the presence of 30 mM TMG-^{14}C plus 30 mM azide and 1 mM iodoacetate. At the end of this preincubation period, the intracellular TMG concentration was 9 mM in ML 35 and 6 mM in ML 308-225. Cells were then washed and centrifuged at 0°. The washed pellet was resuspended in 0.5 mM TMG-^{14}C plus azide and iodoacetate at 10°.

the cell; neither accumulation nor exit of NPG can occur. Table II shows that at 23° the K_t of entrance of NPG was not markedly changed in cells which had been incubated with either 10 mM azide or 30 mM azide plus 1 mM iodoacetate. In experiments performed at 10°, there was a relatively small increase of K_t after poisoning.

Kinetics of Galactoside Uptake in ML 308-225—The K_t and V_{max} of NPG and lactose entrance into poisoned and unpoisoned cells of ML 308-225 (β-galactosidase-negative) were measured (Table III). The K_t of entrance in unpoisoned cells was also calculated from the steady state concentrations obtained at various extracellular concentrations of lactose or NPG as described by Rickenberg *et al.* (22). This calculation is valid in actively transporting cells since the K_t of entrance is the only saturable parameter determining the steady state level. (It will be shown in a later section the K_t of exit in these cells is extremely high.) These values compare favorably with those obtained from measurement of initial rates. Table III shows

TABLE III

Effect of metabolic inhibitors on kinetics of entrance into ML 308-225

Actively transporting cells were sampled either between 5 and 15 sec (initial rate) or at 10 min (steady state). Poisoned cells were sampled between 5 and 10 sec. As shown in Figs. 11 and 12, preloading had no effect on unpoisoned cells. Results are expressed ±S.D. See "Experimental Procedure" for details of sampling and preloading methods. K_t and V_{max} (initial rate) were calculated by plotting the reciprocal of the rate with respect to the reciprocal of the substrate concentration in the manner of Lineweaver and Burk.

Substrate and inhibitor	Preloading	K_t	V_{max}
		mM	μmoles/min/g wet weight
Lactose			
None[a]	None[a]	0.8[a] ± 0.1 (3)	
None	None	0.6 ± 0.5 (4)	39 ± 28 (5)
Azide (30 mM) + iodoacetate (1 mM)	None	0.9 ± 0.3 (3)	3.4 ± 2.6 (4)
Azide (30 mM) + iodoacetate (1 mM)	^{12}C-Lactose	1.0 ± 0.4 (3)	37 ± 23 (4)
NPG			
None[a]	None[a]	0.3[a] ± 0.1 (3)	
None	None	0.5 ± 0.2 (4)	15 ± 3 (4)
Azide (30 mM) + iodoacetate (1 mM)	None	0.5 ± 0.1 (3)	1.0 ± 0.4 (3)
Azide (30 mM) + iodoacetate (1 mM)	^{12}C-Lactose	1.3 ± 0.1 (3)	22 ± 2 (3)

[a] Values determined from plot of intracellular steady state concentration with respect to extracellular concentration according to Rickenberg *et al.* (22), K_t is the extracellular concentration at which the intracellular steady state concentration is half-maximal.

TABLE II

Effect of metabolic inhibitors on uptake of NPG by ML 308

Inhibitor	Temperature	K_t[a]	V_{max}[a]
		mM	μmoles/min/g wet weight
None	23°	0.9 ± 0.2 (8)	69 ± 16 (8)
Azide (10 mM)	23	1.1 ± 0.4 (4)	58 ± 4 (4)
Azide (30 mM) + iodoacetate (1 mM)	23	1.5 ± 0.2 (5)	19 ± 5 (5)
None	10	0.3 ± 0.1 (4)	8.5 ± 2.3 (4)
Azide (30 mM) + iodoacetate (1 mM)	10	0.6 ± 0.2 (4)	1.6 ± 0.4 (4)

[a] Values are expressed as mean values ± S.D.; number of experiments is given in parentheses. These values were obtained by plotting the reciprocal of the rate of NPG uptake (rate of hydrolysis *in vivo*) with respect to the reciprocal of the substrate concentration in the manner of Lineweaver and Burk.

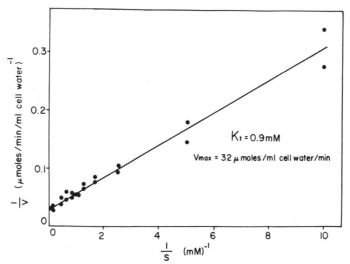

Fig. 10. Effect of concentration on uptake of lactose-^{14}C into energy-uncoupled ML 308-225 preloaded with lactose-^{12}C. Cells were first incubated for 30 min with lactose-^{12}C plus 30 mM azide and 1 mM iodoacetate. After washing, the cells were incubated at 10° in the presence of lactose-^{14}C plus metabolic inhibitors for an average time of 5.7 sec with the rapid sampling technique described in "Experimental Procedure."

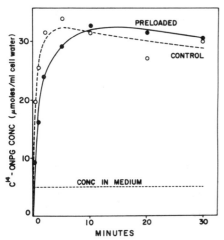

Fig. 11. Effect of preloading ML 308-225 with NPG-^{12}C on the subsequent uptake of NPG-^{14}C. Uptake of NPG-^{14}C in control cells was measured in the usual manner. Preloaded cells were incubated with 5 mM NPG-^{12}C for 30 min; then a small volume of NPG-^{14}C was added without changing the concentration appreciably. The uptake of counts was then measured and plotted on the same scale as that for the control cells.

that the K_t of entrance was only affected slightly by metabolic poisons. The V_{max} appeared to be markedly reduced in inhibited cells which were not preloaded (see below).

Since poisoned cells were found to contain an appreciable intracellular concentration of substrate even at the 15-sec point,

it was obvious that *initial* rates were not being measured but rather a composite of entrance plus an additional efflux component. To block this efflux of radioactive material, the cells were preloaded with galactoside and then the 5- to 6-sec rate of uptake measured (Fig. 10). Under these conditions, K_t and V_{max} were the same in both actively transporting cells and energy-uncoupled cells (Table III).

Effect of Preloading on Uptake in Unpoisoned Cells—An alternative explanation for results obtained in the preloaded experiments must be considered. It has been suggested that in *E. coli* the flux of a sugar in one direction could accelerate the flux of another sugar in the opposite direction by facilitating the return of the carrier (17). This assumes that the diffusion constant of the carrier-substrate complex is greater than that of the unloaded carrier. This acceleration of the influx of a compound by the efflux of another has been termed "forced exchange" or "exchange diffusion." To test this possibility, we incubated actively transporting cells with nonradioactive NPG until a steady state was reached and then added radioactive NPG at this point to measure the influx under steady state conditions (Fig. 11). The rate of uptake by the preloaded cell was not greater than the control. Similar results were obtained with lactose as substrate. This provides confirmation of similar experiments performed by Kepes (14) on TMG uptake into preloaded ML 308. In a further test of the possibility of forced exchange, the initial (6-sec) rate of uptake of lactose was determined at several concentrations in control cells and in cells which had been preloaded with lactose (Fig. 12). The figure shows that the rates were the same for both preloaded and nonloaded cells. Thus, forced exchange does not seem to play a role in the β-galactoside transport system in *E. coli*, and one may use preloading to block efflux in the poisoned cell.

Exit of Galactosides—It is known from the studies of Ricken-

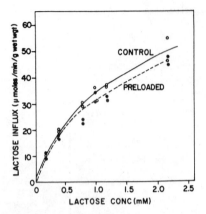

FIG. 12. Effect of concentration of lactose on the rate of uptake by preloaded and control cells. Control cells or cells preloaded with lactose-^{12}C were incubated in the presence of lactose-^{14}C and samples taken at an average time of 5.7 sec. Conditions were similar to those in experiment shown in Fig. 8.

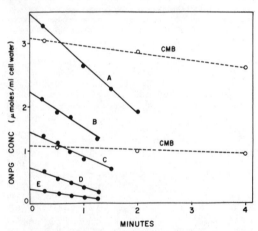

FIG. 13. Effect of intracellular concentration on exit of NPG from energy-uncoupled ML 308-225. Cells were preincubated for 30 min with various concentrations of NPG-^{14}C plus 30 mM azide and 1 mM iodoacetate. After centrifugation, the supernatant fluid was poured off and the inside of the tube was wiped free of all fluid. The pellet was resuspended in sugar-free medium containing the two metabolic inhibitors and the rate of loss of radioactivity from the cells was measured at intervals. To some preloaded cells, CMB (10^{-4} M) was added before centrifugation and was present in the external medium during the exit portion of the experiment. The CMB-insensitive exit rate was proportional to the intracellular concentration and was taken as the non-carrier-mediated exit rate.

berg *et al.* (22) that the exit of accumulated β-galactoside from *E. coli* is a first order process. However, from the work of Kepes (14) and Koch (17), it is clear that exit is not simple passive diffusion but is mediated by the same membrane carriers responsible for entrance. It must be inferred that the K_t of exit is very high.

An attempt was made to measure the K_t of exit of NPG from ML 308-225 by preloading the cells with substrate, washing the cells and measuring the initial rate of loss from the cell. The rate of exit was proportional to the internal concentration under the experimental conditions tested (Fig. 14). The exit of lactose from unpoisoned cells also exhibited a high K_t, an approximate value of 16 mM was obtained (Fig. 15). When exit measurements were made on poisoned cells, the K_t of exit was found to be much lower than found in the actively transporting cells. Fig. 13 shows the exit rates of NPG from poisoned cells at various intracellular concentrations. The simple passive diffusion component of the exit process, as measured by the exit rate in the presence of CMB, was only a small fraction of the total exit rate. These rates corrected for the diffusion component, did not increase linearly with concentration but showed saturation kinetics. When exit rates such as these were plotted in a Lineweaver-Burk fashion a K_t of 1.3 mM was obtained (Fig. 14). The mean value for the K_t of NPG exit from poisoned cells in six experiments was 1.8 mM ± 1.1 (S.D.). When similar experiments were performed with lactose as the substrate a K_t of 0.7 mM was obtained (Fig. 15). Thus, energy uncoupling caused an increase in affinity of the carrier for substrate in the exit process. Two exit experiments were performed with and without 2,4-dinitrophenol (1 mM). The results were entirely similar to those obtained above with azide plus iodoacetate.

The observed V_{max} for exit of NPG was found to be 0.9 μmole per min per g of wet weight (mean of five experiments) for the

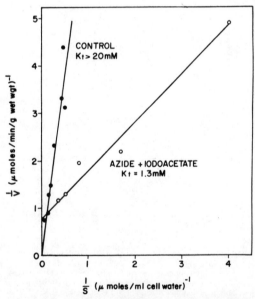

FIG. 14. The effect of intracellular concentration on the exit rate of NPG from control and energy-uncoupled ML 308-225. Data for control cells came from three experiments while those for inhibited cells came from Fig. 13; all data graphed by the method of Lineweaver and Burk. Exit rate, V, was taken as the initial rate of exit corrected for the non-carrier-mediated determined with CMB. The intracellular concentration, S, was taken as the concentration at the time exit measurements began (zero time in the experiment shown in Fig. 13).

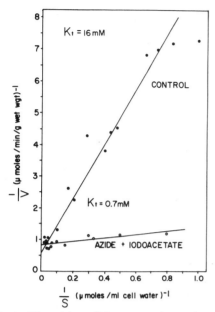

FIG. 15. Effect of intracellular concentration on the exit of lactose from control and energy-uncoupled ML 308-225. Data were graphed by the method of Lineweaver and Burk. Control cells and cells previously exposed to 30 mM azide and 1 mM iodoacetate were preloaded with lactose-^{14}C and centrifuged, and the exit rate was measured by the general method given in Fig. 13. The *control curve* is a composite of four experiments; the *inhibited curve* is a composite of three experiments.

poisoned ML 308-225 which was approximately ⅓ the value for the V_{max} of entrance. The exit V_{max} for lactose was 1.2 μmoles per min per g of wet weight (mean of three experiments) for the poisoned cell and 1.8 μmoles per min per g of wet weight (mean of three experiments) for the control cells, values that are much lower than the corresponding values for entrance. A possible explanation for the low V_{max} of exit will be discussed in the next section.

DISCUSSION

Existence of Membrane Carriers in Poisoned Cells—It has been known for many years that inhibition of energy-yielding reactions of a cell results in loss of the capacity to transport substances actively across the cell membrane. More recently, however, it has been recognized that membrane carriers persist in these poisoned cells and facilitate the equilibration of the intracellular and extracellular concentrations of a substance to which the cell is otherwise impermeable. Crane and Krane (10), for example, found that while *o*-nitrocresol abolished active transport of galactose into kidney slices the intracellular sugar concentration rapidly equilibrated with that in the incubation medium. The rate of entry into these poisoned cells was markedly inhibited, however, by the addition of the specific inhibitor, phlorizin. Experiments leading to similar conclusions have been performed on sugar transport in the small intestine (11) and *E. coli* (15–17); on amino acid transport in nucleated erythrocytes (24) and bac-

teria (25); on phosphate transport in *Micrococcus pyogenes* (26); and weak acid transport in the kidney (13).

Observations in this paper provide additional experimental support for the view that membrane carriers remain functional in the cell deprived of energy coupling. The experiments with NPG entry into ML 308 ($i^-z^+y^+$) suggest that when a substrate enters the cell from a high external concentration to a low internal one addition of a metabolic poison has only a minor effect, in confirmation of the observations of previous workers (16, 17). Furthermore, the rate of entry of 1 mM NPG into ML 308 poisoned with azide and iodoacetate was 20 times faster than that into the corresponding transport-negative organism (ML 35). The carrier-mediated rate of entry of lactose into poisoned cells at an external concentration of 1 mM (Table III) was from 1,000 to 10,000 times the corresponding rate for simple passive diffusion (See "Experimental Procedure"). In poisoned cells, the membrane carriers are also available to facilitate exit of galactosides from the cell. Exit rates of galactosides are far greater in poisoned ML 308-225 than in similarly treated ML 35.

The lack of a major change in V_{max} of entrance after poisoning either ML 308 or ML 308-225 (Table III) suggests that most, if not all, of the carriers remain intact in the absence of a source of energy. The membrane carriers for β-galactosides found in poisoned *E. coli* behave in a manner entirely analogous to those responsible for sugar transport in the human erythrocyte. Competition inhibitors such as thiodigalactoside inhibit the rate of entry of NPG into the poisoned cell. Furthermore, when cells preloaded with galactoside are placed into a solution of a radioactive galactoside a transient accumulation of the radioactive compound occurs due to inhibition of efflux by the preloaded sugar. These "counterflow" experiments require the activity of carriers for both entrance and exit.

No evidence was found for the type of nonspecific *transporteur* postulated by Kepes (14) and Koch (17). The cryptic organism ML 35 permitted only very slow entry of any galactoside and this entrance was not inhibited by TDG. Furthermore, preloading either poisoned or unpoisoned cells of ML 35 with galactoside never resulted in the transient gradients when the loaded cells were added to a radioactive substrate (Fig. 9). The entry process into ML 35 possessed all the attributes of simple passive diffusion.

Preloading of poisoned cells with a nonradioactive galactoside was used extensively throughout this study to inhibit efflux of radioactive sugar molecules competitively. There is one possible objection to such a procedure which must be considered. If loaded carriers cross the plasma membrane more rapidly than empty ones, then the efflux of sugar molecules from the cell may speed the rate of influx of radioactive substrate molecules by presenting the outer face of the plasma membrane with more carriers. Evidence for such "exchange diffusion" or "forced exchange" has been presented for ion transport in muscle (27) and for amino acid transport into ascites tumor cells (12). In the latter case, preloading of unpoisoned tumor cells with glycine stimulated the subsequent influx of a radioactive neutral amino acid, although it could be shown that the affinity for exit was so low that internal glycine could not have been effectively prevented exit of entering radioactive molecules. When a similar experiment was performed with the galactoside transport system of *E. coli* by Kepes (14) and in the present work (Figs. 11 and 12), no stimulation of influx was observed when cells were preloaded with galactoside. It should be emphasized that in the case of exchange diffusion preloading produces a *stimulation of influx;* whereas, in the case of the simple model there is no stimulation of influx, but only an *inhibition of efflux;* both mechanisms result in a stimulation of *net uptake.* Table III and Fig. 12 show that preloading bacteria with nonradioactive lactose does not stimulate influx. The stimulation of exit of a galactoside by adding another galactoside to the external medium has been taken by some (17) as evidence for exchange diffusion. Our view is that the external substrate competes for the "recapture" of the molecules leaving the cell, as will be described in a later section. All of the available data on this transport system is consistent with the assumption, made in developing the model (Fig. 1), that empty carriers and filled carriers diffuse across the membrane at equal rates.

Effect of Energy Uncoupling on Affinity of Influx and Efflux— The model in Fig. 1 predicts that in the steady state

$$\frac{\text{Concentration in cell } (S_i)}{\text{Concentration in medium } (S_0)} = \frac{K_t \text{ of exit}}{K_t \text{ of entrance}}$$

In the poisoned cell, no active transport is possible and consequently the affinity for entrance must equal the affinity for exit. For conversion of this simple equilibrating system to an active transport system, there must be either (*a*) reduction in the K_t of entrance, (*b*) increase in K_t of exit, or (*c*) some combination of the two. Since the maximum concentration ratio for galactosides observed in this study was 100- to 200-fold (Fig. 2), the ratio of K_t values must increase two orders of magnitude. Metabolic poisons were found to have little or no effect on the K_t of entrance of NPG into either ML 308 or ML 308-225. Similar results were obtained with lactose as substrate. On the other hand, a dramatic fall in the K_t of exit was observed when cells were exposed to metabolic inhibitors. The exit K_t for NPG in unpoisoned cells was too high to measure experimentally (above 20 mM). With a K_t of entrance (at 10°) of 0.5 mM and a maximum steady state ratio of cell to medium of 120-fold the K_t of exit may be calculated as 60 mM, according to the model under discussion. With lactose as substrate, the experimental K_t of entrance was 0.6 mM while the K_t of exit was 16 mM. The predicted ratio of K_t values from the steady state ratios (Fig. 2) is 120 mM, rather larger than observed experimentally. The data, however, conform at least qualitatively to the pattern predicted by the model. It is inferred from these experiments that the energy coupling results in a marked reduction in affinity of the membrane carrier for its substrate at the inner surface of the membrane.

A number of observations on poisoned cells might be reexamined in the light of these observations. The rapid net loss of accumulated substrates following poisoning may be caused at least in part by accelerated efflux from the cell. The conclusion of Osborn *et al.* (28) and Koch (17) that metabolic poisons stimulate efflux is, of course, entirely consistent with these findings.

An entirely different mode of energy coupling appears to operate for α-glucoside transport in yeast. Okada and Halvorson (29, 30) have shown that *Saccharomyces cerevisiae* containing the MG2 gene possesses a constitutive facilitated diffusion system for α-thioethylglucoside with an entrance K_t of 50 mM. When such organisms are induced by growth on α-methylglucoside or certain other α-glucosides, they develop the capacity to transport α-thioethylglucoside actively with an entrance K_t of about 2 mM. Thus, the energy coupling in the cell appears to reduce the K_t of entrance.

In certain animal cells, the immediate source of energy for the active transport of amino acids and sugars appears to be the sodium ion concentration gradient. Vidavar (24) has shown that erythrocyte ghosts, free of ATP, can actively transport glycine into the cell provided the sodium ion concentration within the cell was considerably below that in the incubation medium. It was inferred that the K_t of the carrier is reduced by high sodium ion concentration. Similar conclusions have been reached by Crane with regards to active sugar transport by the columnar absorptive cell of the small intestine (31). In *E. coli* strain ML, the rate of uptake and intracellular plateau level of galactosides did not appear to be affected by the external sodium ion concentration. In most of the experiments reported in this paper cells were suspended in a medium free of sodium ion. The addition of 50 mM sodium ion produced little or no effect on transport of NPG. A sodium ion gradient does not appear to be essential for sugar transport in these organisms, although gradients of other ions have not been excluded.

V_max of Entrance—Uncertainties in the experimentally obtained V_{max} values are largely caused by the difficulty in measuring unidirectional flux rates across the cell membrane. The observed rate of uptake of a substance into a cell is clearly the difference between the influx and efflux, the latter being a much more serious problem when the affinity for exit is high. Thus, reliable influx values may be obtained with comparative ease in actively transporting cells since efflux does not become appreciable until high intracellular concentrations are obtained; while serious experimental problems arise in poisoned cells since efflux may become significant at intracellular concentrations much lower than that in the incubation medium. A back reaction (efflux) undoubtedly accounts for the low V_{max} value obtained by 15-sec sampling of NPG uptake into poisoned ML 308-225 (Table II). Under these conditions, the intracellular concentration was often 50% of that in the incubation medium and rapid efflux reduced the net entrance rate. Shortening the sampling time to 5 sec did not reduce the efflux sufficiently to obtain reliable results.

Since sampling times shorter than 5 sec were not found feasible, attempts were made to block efflux by preloading the cell with nonradioactive substrate. As discussed earlier, the available evidence on galactoside transport in *E. coli* suggests that such preloading blocks exit without affecting influx. When the efflux from metabolically poisoned cells was blocked by preloading with galactoside the V_{max}, as well as K_t, was similar to that obtained in control cells (Table III). This observation is consistent with the view that energy coupling does not alter the number or activity of the membrane carriers.

V_max of Exit—An unexpected finding during the course of these experiments was the observation that the measured V_{max} for exit from either the poisoned or unpoisoned cell was much less than the V_{max} of entry. Koch (17) and Okada and Halvorson (30) have previously made similar observations but without special comment. The possibility that the measured exit rate was much lower than the true exit rate was suggested by the following consideration. When cells were exposed to a relatively high concentration of galactoside (3 times the K_t), the intracellular concentration rose until a plateau level was reached. At the steady state, the efflux rate must exactly equal the influx rate, the latter being readily measurable (and approaching the V_{max} of uptake). When efflux was measured directly from such cells after separation from the incubation medium and resuspended in sugar-free medium, the efflux was found to be about 10% of the previously measured influx. The V_{max} of exit extrapolated from Lineweaver-Burk plots is similarly about 10% of the corresponding value for influx. A possible explanation for this anomalous behavior is that many of the molecules leaving the cell are recaptured from the outer surface of the plasma membrane and reenter the cell. More specifically, it is postulated that there exists a small compartment, probably between the plasma membrane and the cell wall, which exchanges relatively slowly with the external medium and from which recapture can occur. According to this view, 8 or 9 molecules out of 10 which leave the cell interior are recaptured again, thus reducing the net loss from the whole cell to a low level. If this hypothesis were correct, then addition of galactosides to the outer surface of the cell should greatly accelerate the net loss of labeled molecules from the cell by competitively inhibiting recapture. The exit of radioactive lactose from ML 308-225 was stimulated many fold by the presence of 1 mM nonradioactive lactose in the incubation medium although the K_t was relatively unaffected. This marked stimulation of the exit of radioactive molecules by external nonradioactive substrate is believed to be due to blockage of recapture rather than stimulation of exchange diffusion which has been suggested by others (17). The direct experimental test of the hypothesis with cells devoid of cell wall (protoplasts) has not yet been successful since such protoplasts of the ML strain with either the penicillin or lysozyme methods were too leaky to retain appreciable amounts of transported galactosides. Perhaps this experiment may be feasible in other organisms or other strains of *E. coli*.

Physiological Control of Energy Coupling—Some type of mechanism for reversibly uncoupling energy supplies from membrane transport would be most useful to organisms under conditions in which energy supplies are limited. Under anaerobic conditions, for example, a cell might be forced to expend up to 50% of its ATP production for active glucose transport (on the assumption that 1 mole of ATP was required for 1 mole of glucose translocated (14)). The capacity of a cell to spare its energy reserves when external substrate concentration is high and yet actively accumulate substrate when the external concentration is low would provide the cell with a powerful selective advantage.

Hoffee, Engelsberg, and Lamy (32) have recently presented data to show that when washed cells are exposed to the carbon source previously used for growth, the K_t of exit of α-methylglucoside is reduced. These authors interpret their findings in the light of their hypothesis that the energy source stimulates a specific exit transport system, separate from the entry mechanism. An alternative view is that rapid flow of oxidizable substrate into the cell produces a high level of some intermediate which uncouples energy from transport (33) and reduces the K_t of exit. According to the data presented in this paper, uncoupling does not affect either the affinity or the V_{max} of entry. Consequently, the entry rate may continue at an adequate rate in the uncoupled cell provided that excessive efflux is prevented by metabolism of the intracellular substrate. When the external concentration falls considerably below the K_m of the first enzyme, however, active transport would be necessary to provide a

sufficiently high intracellular concentration for an adequate rate of metabolism. Although these views on energy uncoupling as a means of conservation of metabolic energy and control of over-all metabolism are rather speculative, they provide a working hypothesis for further experimentation.

Acknowledgment—We are indebted to Dr. Edmund C. C. Lin for advice in many aspects of this work, especially the techniques of chemical mutagenesis.

REFERENCES

1. BURGER, M., HEJMOVÁ, L., AND KLEINZELLER, A., *Biochem. J.*, **71**, 233 (1959).
2. WILBRANDT, W., AND ROSENBERG, T., *Pharmacol. Rev.*, **13**, 109 (1961).
3. LeFevre, P. G., *Pharmacol. Rev.*, **13**, 39 (1961).
4. PARK, C. R., POST, R. L., KALMAN, C. F., WRIGHT, J. H., JR., JOHNSON, L. H., AND MORGAN, H. E., *Ciba Colloquia Endocrinol.*, **9**, 240 (1956).
5. ROSENBERG, T., AND WILBRANDT, W., *J. Gen. Physiol.*, **41**, 289 (1957).
6. WIDDAS, W. F., *J. Physiol.*, **118**, 23 (1952).
7. WIDDAS, W. F., *J. Physiol.*, **125**, 163 (1954).
8. ROSENBERG, T., AND WILBRANDT, W., *Exptl. Cell Res.*, **9**, 49 (1955).
9. LeFevre, P. G., AND McGINNISS, G. F., *J. Gen. Physiol.*, **44**, 87 (1960).
10. KRANE, S. M., AND CRANE, R. K., *J. Biol. Chem.*, **234**, 211 (1959).
11. BIHLER, I., HAWKINS, K. A., AND CRANE, R. K., *Biochim. Biophys. Acta*, **59**, 94 (1962).
12. HEINZ, E., AND MARIANI, H. A., *J. Biol. Chem.*, **228**, 97 (1957).
13. KINTER, W. B., AND CLINE, A. L., *Am. J. Physiol.*, **201**, 309 (1961).
14. KEPES, A., *Biochim. Biophys. Acta*, **40**, 70 (1960).
15. HORECKER, B. L., OSBORN, M. J., McLELLAN, W. L., AVIGAD, G., AND ASENSIO, C., in A. KLEINZELLER AND A. KOTYK (Editors), *Membrane transport and metabolism*, Academic Press, Inc., New York, 1960, p. 378.
16. COHEN, G. N., AND MONOD, J., *Bacteriol. Rev.*, **21**, 169 (1957).
17. KOCH, A. L., *Biochim. Biophys. Acta*, **79**, 177 (1964).
18. LIN, E. C. C., LERNER, S. A., AND JORGENSEN, S. E., *Biochim. Biophys. Acta*, **60**, 422 (1962).
19. LUBIN, M., *J. Bacteriol.*, **83**, 696 (1962).
20. GORINI, L., AND KAUFMAN, H., *Science*, **131**, 604 (1960).
21. HERZENBERG, L. A., *Biochim. Biophys. Acta*, **31**, 525 (1959).
22. RICKENBERG, H. V., COHEN, G. N., BUTTIN, G., AND MONOD, J., *Ann. Inst. Pasteur*, **91**, 829 (1956).
23. SEIDMAN, M., AND LINK, K. P., *J. Am. Chem. Soc.*, **72**, 4324 (1950).
24. VIDAVER, G. A., *Biochemistry*, **3**, 662, 795, 799, 803 (1964).
25. KESSEL, D., AND LUBIN, M., *Biochim. Biophys. Acta*, **57**, 32 (1962).
26. MITCHELL, P., *J. Gen. Microbiol.*, **11**, 73 (1954).
27. USSING, H. H., *Nature*, **160**, 262 (1947).
28. OSBORN, M. J., McLELLAN, W. L., JR., AND HORECKER, B. L., *J. Biol. Chem.*, **236**, 2585 (1961).
29. OKADA, H., AND HALVORSON, H. O., *J. Bacteriol.*, **86**, 966 (1963); *Biochim. Biophys. Acta*, **82**, 538 (1964).
30. OKADA, H., AND HALVORSON, H. O., *Biochim. Biophys. Acta*, **82**, 547 (1964).
31. CRANE, R. K., *Federation Proc.*, **24**, 1000 (1965).
32. HOFFEE, P., ENGLESBERG, E., AND LAMY, F., *Biochim. Biophys. Acta*, **79**, 337 (1964).
33. KEPES, A., in J. HOFFMAN (Editor), *The cellular functions of membrane transport*, Prentice-Hall, Inc., Englewood Cliffs, N. J., 1964, p. 155.

Reprinted from *Proc. Natl. Acad. Sci. U.S.A.*, **54**, 891–899 (1965)

12

SPECIFIC LABELING AND PARTIAL PURIFICATION OF THE M PROTEIN, A COMPONENT OF THE β-GALACTOSIDE TRANSPORT SYSTEM OF ESCHERICHIA COLI*

BY C. FRED FOX† AND EUGENE P. KENNEDY

DEPARTMENT OF BIOLOGICAL CHEMISTRY, HARVARD MEDICAL SCHOOL

Communicated July 21, 1965

The specific permeability of the membranes of living cells is a property of obvious interest and importance. However, the elucidation of the chemical events occurring during the transport process, often leading to the concentration of a given substance within the cell against a high concentration gradient, has proved to be a singularly refractory problem.

Work in this laboratory on the lipid metabolism of *Escherichia coli* and its relation to the transport of β-galactosides[1] has led us to a more general investigation of this system, which offers many advantages for study, since much is known about its genetic control, specificity, and kinetics, largely as a result of the work of Monod and Kepes and their collaborators.[2–5]

198

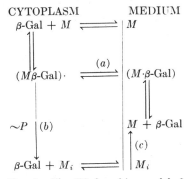

CYTOPLASM   MEDIUM

FIG. 1.—Simplified working model of the β-galactoside transport system.

We have found the highly simplified model of the β-galactoside transport system shown in Figure 1 to be useful in planning the experiments described in this report. In the model, two processes are sharply distinguished: (a) the facilitated entrance of β-galactosides into the cell, and (b) the accumulation of galactosides after entry against a concentration gradient. Process (b) requires metabolic energy and is abolished by certain metabolic poisons such as dinitrophenol and azide, which essentially do not affect process (a).

In this model, a molecule of β-galactoside in the medium combines with a component M on the exterior surface of the membrane, and passes through the membrane as a β-galactoside-M complex [process (a)]. On the inner surface of the membrane the complex may simply dissociate. If the enzyme β-galactosidase is present in the cell, and if the β-galactoside is a substrate which can be hydrolyzed by this enzyme, the continuous removal of galactoside by this reaction may lead to a substantial flux of galactoside from the medium into the cell. This is the process, independent of metabolic energy, which takes place when the hydrolysis of o-nitrophenyl-β-galactoside is measured in intact cells poisoned with azide, as in the experiments of Koch.[6] Alternatively, if sources of metabolic energy are available, component M may be converted to M_i, a form that has a greatly reduced affinity for β-galactosides. M_i moves back through the membrane to the exterior surface, where it is reconverted to the active form M, in a reaction not requiring coupled metabolic energy [process (c)]. The expenditure of metabolic energy takes place at the inner surface of the membrane, and reduces the affinity of the carrier for substrate, a view consistent with the experimental findings of Koch,[6] Horecker et al.,[5] Winkler and Wilson,[7] and others. The maintenance of a concentration gradient requires the continuous expenditure of energy.

In this paper, we wish to describe some experiments on the nature of process (a) and to present evidence that component M in Figure 1 is a protein, distinct from β-galactosidase and thiogalactoside transacetylase,[8] the two previously characterized proteins of the lactose system in E. coli, but like them, controlled by the i gene. This component is localized in the membrane-containing particulate fraction and has a high affinity for certain β-galactosides. Methods have been devised for the specific labeling of this protein, and its isolation in partially purified form as the labeled derivative.

We have obtained no evidence for an enzymatic role of the newly identified protein. Indeed, it is our working hypothesis that it may function not as an enzyme but as a substrate in process (b) (Fig. 1). Until more evidence is available as to its biochemical role in the transport process, it appears preferable to designate the newly discovered protein simply as the M protein.

Materials and Methods.—Cultures of E. coli strain ML 30 (genotype $i^+z^+y^+x^+$) and of strain ML 308 ($i^-z^+y^+x^+$), originally isolated in Monod's laboratory, were the gift of Dr. Arthur Koch. Thiodigalactoside (β-D-galactosyl-1-thio-β-D-galactopyranoside), o-nitrophenyl-β-D-galactopyranoside, and isopropyl-1-thio β-D-galactopyranoside were purchased from the Mann Chemical Co. N-ethylmaleimide labeled with tritium was synthesized from ethylamine-H[3] and maleic anhydride by a microadaptation of the method of Piutti and Giustiniani,[9] which was also used for the preparation of C[14]-labeled N-ethylmaleimide from maleic anhydride-1-C[14] and ethylamine. Purified β-galactosidase from *E. coli* was the generous gift of Drs. C. B. Anfinsen and G. Craven.

Cells were grown at 37° in synthetic medium 63 (Rickenberg *et al.*[10]) with succinate as a carbon source and were aerated on a rotary shaker.

The rate of hydrolysis of o-nitrophenyl-β-galactoside in intact cells was measured essentially by the method of Koch.[6] The cells were incubated at 28° in 0.1 M potassium phosphate at pH 7.0 containing 0.01 M sodium azide. After 3 min, o-nitrophenylgalactoside (final concentration 0.002 M) was added, and the incubation, in a final volume of 3.0 ml, was continued for another 15 min. The reaction was stopped by the addition of 5 ml of 1 M potassium carbonate. The optical density at 420 mμ was determined in a Coleman spectrometer. The values given are corrected for the slight hydrolysis exhibited by control tubes to which formaldehyde was added to block the facilitated entrance of the substrate.[6]

Thiogalactoside transacetylase was measured by a highly sensitive procedure utilizing tritiated acetyl CoA, the details of which will be published elsewhere.

Protein-bound radioactivity was determined by the following procedure. To 1.0-ml samples containing labeled protein, 0.3 ml of unfractionated sonicate of *E. coli* ML 308 (about 4 mg protein per ml) was added as carrier. Ethanol (4.0 ml) was then added and the sample was chilled in an ice-bath for 30 min. The precipitated protein was removed by centrifugation, and was washed with ethanol (1.0 ml) and with 0.3 N trichloroacetic acid (1.0 ml). The precipitate was then thoroughly dispersed in 1.0 ml of ethanol, to which 0.5 ml of a methanolic solution of Hyamine (1.0 M) was added. The contents of the tube were then thoroughly agitated and warmed to 57° for a few minutes. The precipitate dissolved completely, yielding a slightly opalescent solution, which was taken up in 10 ml of Buhler's scintillation fluid[11] for counting in a three-channel Packard scintillation spectrometer equipped with an external standardization device to detect variation in quenching. Protein was determined by the method of Lowry.[12]

The modification of the *t*-test described by Hogben[13] was used to evaluate the significance of differences between the means of experimental results.

Results.—Effects of N-ethylmaleimide on the facilitated entry of β-galactosides: When intact cells of *E. coli* of constitutive strain ML 308 are incubated with o-nitrophenylgalactoside, in the presence of sodium azide, the accumulation process is blocked, but the facilitated entry of the galactoside into the cell is not impaired.[6] Since a large excess of β-galactosidase is present, the over-all rate of splitting of the chromogenic substrate is thus limited by process (a). In an effort to identify and isolate component M, we decided to search for inhibitors of process (a) of a type which act by irreversible combination with protein, thus opening the way to the labeling of the M protein. The over-all transport process is known to be inhibited

TABLE 1

EFFECT OF N-ETHYLMALEIMIDE ON THE RATE OF HYDROLYSIS OF o-NITROPHENYLGALACTOSIDE BY INTACT CELLS

Pretreatment		Activity
None		(100)
N-ethylmaleimide	$1 \times 10^{-5} M$	89
	$5 \times 10^{-5} M$	54
	$1 \times 10^{-4} M$	7
	$5 \times 10^{-4} M$	0

Cells of strain ML 308 were harvested in early log phase, washed, and suspended in 0.1 M phosphate buffer pH 7.0 at a concentration equivalent to 0.3 mg protein/ml. Portions of the cells were treated as indicated with N-ethylmaleimide, at the final concentration shown, for 15 min at 28°. The N-ethylmaleimide was destroyed by the addition of mercaptoethanol (0.01 M), and the intact cells were assayed for the ability to catalyze the hydrolysis of o-nitrophenylgalactoside in the presence of 0.01 M sodium azide as described under *Materials and Methods*. The aliquots of untreated cells in the control tubes, corresponding to 0.06 mg of total protein, cleaved 0.34 μmoles of o-nitrophenylgalactoside in 15 min. This value is arbitrarily set at 100%.

TABLE 2

Effect of Treatment with N-Ethylmaleimide on
β-Galactosidase and Thiogalactoside Transacetylase

	Activity	
Treatment	β-Galactosidase	Thiogalactoside transacetylase
10^{-4} M N-ethylmaleimide	353	141
Control	344	145

Cells of ML 308 were treated with 10^{-4} M N-ethylmaleimide as described under Table 1. The reaction was terminated by the addition of 0.001 M β-mercaptoethanol, and identical aliquots of treated and untreated cells were disrupted in a Branson sonifier. The aliquots of sonicate used for the transacetylase assay were heated to 65° for 20 min prior to assay, while β-galactosidase was assayed directly, as described under *Materials and Methods*. The results are given in arbitrary units.

by sulfhydryl reagents.[14]   We have found N-ethylmaleimide, a reagent that reacts under mild conditions with sulfhydryl groups with high although not complete specificity,[15, 16] to be very useful for our purposes.

When intact cells were treated with N-ethylmaleimide for 15 min at 28°, upon subsequent assay the rate of hydrolysis of o-nitrophenyl-β-galactoside was strikingly inhibited (Table 1).   This effect could not be due to an inhibition of energy metabolism, since this was already effectively blocked by azide, nor was it due to inhibition of β-galactosidase or thiogalactoside transacetylase, since the activity of these two enzymes was not impaired by treatment of the intact cells with N-ethylmaleimide under the experimental conditions used here (Table 2).   We postulate that the component of the system inhibited by N-ethylmaleimide is the M protein (Fig. 1).

Protection by thiodigalactoside against inhibition by N-ethylmaleimide: It is often possible to protect enzymes against the effects of inhibitors by the addition of substrate.   If N-ethylmaleimide inactivates the transport process by reacting with the galactoside-binding component responsible for carrier-mediated diffusion (the M protein in our model), it might be possible to protect this component by the addition of thiodigalactoside, for which the transport system has a high affinity.   When intact cells of ML 308 were treated with 5×10^{-5} M N-ethylmaleimide, the presence of 0.001 M thiodigalactoside greatly reduced the inhibitory effects of this reagent (Table 3).   Since a distinct protective effect was exerted even at 10^{-5} M, the N-ethylmaleimide-sensitive component must have a high affinity for thiodigalactoside, and combination with the galactoside must reduce the rate of reaction of the inhibitor with some essential residue (presumably cysteine) of this component.

TABLE 3

Protection by Thiodigalactoside against Inhibitory Effects of N-Ethylmaleimide

	Per cent inhibition
10^{-6} M Thiodigalactoside	53
10^{-5} M "	43
10^{-4} M "	25
10^{-3} M "	19

Cells of ML 308 were treated with 5×10^{-5} M N-ethylmaleimide for 15 min at 28° as described in Table 1. The concentration of thiodigalactoside present during the treatment was varied as shown. The reaction was terminated by the addition of mercaptoethanol (0.01 M), and the intact cells were assayed for the ability to hydrolyze o-nitrophenylgalactoside in the presence of sodium azide. The values shown indicate the degree of inhibition observed in comparison with controls carried through the entire procedure, but not treated with N-ethylmaleimide.

In 1960, Kepes[14] reported a protective effect of thiodigalactoside against the inhibition of certain reactions of the transport system by p-chloromercuribenzoate, another reagent that attacks sulfhydryl groups.   Although Kepes did not analyze the inhibition or protection in detail, it seems likely that p-chloromercuribenzoate and N-ethylmaleimide attack the same component of the system.

Labeling of the M protein: The M protein in cells treated with radioactive N-ethylmaleimide under the conditions described in Table 2 should become labeled, since the reaction of this inhibitor with the sulfhydryl groups of proteins is virtually irreversible. Treatment of the cells with unlabeled N-ethylmaleimide in the presence of thiodigalactoside prior to exposure of the cells to radioactive N-ethylmaleimide in the absence of this protective agent should greatly increase the specificity of this labeling. Moreover, proteins of the lactose system are present in inducible strains such as ML 30 in significant amounts only after induction, which makes possible the following experiment on the labeling and intracellular localization of the M protein.

A culture of ML 30 growing in log phase in medium 63 at 37° with succinate as carbon source was divided into two parts. Half of the cells were induced by further growth for 4 hr in the presence of $5 \times 10^{-4} M$ isopropylthiogalactoside, a gratuitous inducer, while the other half of the culture was allowed to grow to the same cell density (equivalent to 0.15 mg protein/ml or about 10^9 cells/ml) in the absence of inducer. Both induced and uninduced cells were chilled, harvested separately by centrifugation, washed with 0.1 M potassium phosphate buffer of pH 7.0, and suspended separately in the buffer at a cell density equivalent to 0.3 mg protein/ml. To both suspensions, warmed to 28°, chloramphenicol (20 μg/ml) and thiodigalactoside (0.001 M) were added, followed by unlabeled N-ethylmaleimide ($5 \times 10^{-5} M$). After 20 min, the N-ethylmaleimide was destroyed by the addition of a 10-fold excess of β-mercaptoethanol. This treatment leads to the reaction of N-ethylmaleimide with sulfhydryl groups in both induced and uninduced cells *except those specifically protected by thiodigalactoside*. (This step leads to a large amplification of the effects observed on later treatment with labeled N-ethylmaleimide.)

The two lots of cells were next chilled and washed twice with cold 0.1 M phosphate buffer of pH 7.0. The induced and the uninduced cells were then divided into three portions of 10 ml each and suspended in the buffer at a density corresponding to 0.6 mg of protein per ml. All samples were treated for 15 min at 28° with labeled N-ethylmaleimide (final concentration $5 \times 10^{-5} M$). C^{14}-labeled N-ethylmaleimide was used with induced cells, while the uninduced cells were labeled with N-ethylmaleimide-H^3. The reaction in each case was terminated by the addition of β-mercaptoethanol (0.01 M). The cells were harvested by centrifugation and resuspended in the phosphate buffer.

Induced cells (labeled with C^{14}) were now mixed with exactly equivalent amounts of uninduced cells (labeled with H^3) and fractionated as described in Table 4. All proteins of the uninduced cells that can react with radioactive N-ethylmaleimide under the conditions described are labeled with H^3. The induced cells contain all these proteins and, in addition, the proteins controlled by the i gene. In the mixture of cells, all proteins are therefore doubly labeled, except those of the lactose system, which are labeled only with C^{14}. *The presence of labeled proteins of the lactose system in fractions derived from the mixture of cells should therefore be revealed by an increase in the ratio of protein-bound C^{14}/H^3.*

Such an experiment (Table 4) revealed that the ratio of protein-bound C^{14}/H^3 in the particulate fraction, which contains the cell wall membrane complex, was slightly but significantly higher than that of the total cellular protein, or that of the soluble supernatant fraction, which contains about 80 per cent of the total protein. Since β-galactosidase and thiogalactoside transacetylase are recovered largely in

TABLE 4

LOCALIZATION OF LABELED M PROTEIN

Fraction	Protein-bound Radioactivity (mμmoles)		Mean ratio C^{14}/H^3
	C^{14}	H^3	
Total cell protein	5.22	5.41	
	5.48	5.62	0.96 ± 0.02
	5.47	5.77	
Supernatant	4.38	4.62	
	4.28	4.58	0.94 ± 0.01
	4.15	4.36	
Particulate fraction	0.67	0.62	
	0.83	0.84	1.02 ± 0.05
	0.91	0.91	

Mixtures of induced and uninduced cells of strain ML 30 (about 4×10^{10} of each type), labeled with C^{14} and tritium, respectively, as described in the text, were disrupted by treatment in the Branson sonifier (3–5 min). The particulate fraction was collected by centrifugation at 100,000 × g for 30 min, and washed once with 0.1 M phosphate buffer of pH 7.0. Aliquots of the total protein, supernatant fraction, and washed particulate fraction were analyzed for protein-bound radioactivity as described under *Materials and Methods*. The entire procedure was carried out in triplicate.

the supernatant fraction, the failure to find a significant enrichment of C^{14} in this fraction indicates that these proteins, coded for by the *lac* operon, do not bind substantial amounts of radioactive N-ethylmaleimide under the conditions described, a finding in accord with the results of the experiment described in Table 2.

The difference between the mean ratios of protein-bound C^{14}/H^3 of the particulate and supernatant fractions is statistically significant[13] when replicate analyses are carried out ($p < 0.05$) and is reproducible from one batch of cells to the next. The possibility was considered that the result shown in Table 4 might be caused by an impurity in the C^{14}-labeled N-ethylmaleimide, or might be caused by preferential quenching of tritium in the samples derived from the particulate fraction. Such sources of artifact were eliminated by "criss-cross" control experiments, in which the induced cells were labeled with tritiated N-ethylmaleimide and the uninduced cells with N-ethylmaleimide-C^{14}. These experiments led to the same conclusion, namely, that the particulate fraction was slightly enriched in labeled protein derived solely from the induced cells. This labeled protein has the following properties expected of the M protein: (1) its synthesis is controlled by the i gene, (2) it is distinct from β-galactosidase and thiogalactoside transacetylase, (3) it reacts with N-ethylmaleimide, and (4) it is localized in the membrane-containing fraction.

The specific reactivity of the M protein in cell-free systems: In the experiment of Table 4, the labeling of the M protein was accomplished by treatment of the intact cells with radioactive N-ethylmaleimide. We have found that the M protein retains its affinity for thiodigalactoside in cell-free fractions, as shown by the following experiment.

Induced and uninduced cells of strain ML 30 were pretreated with unlabeled N-ethylmaleimide in the presence of 0.001 M thiodigalactoside, collected by centrifugation, and washed essentially as described in the experiment of Table 4. The cells were then disrupted in a Branson sonifier, the particulate fraction was collected by centrifugation at 100,000 × g for 30 min, and washed with 0.1 M phosphate buffer of pH 7.0. Aliquots of the particulate fractions from both types of cells were then tested for the ability to bind radioactive N-ethylmaleimide in the presence and in the absence of thiodigalactoside.

The results (Table 5) show that the particulate fraction from the induced cells

TABLE 5

LABELING OF M PROTEIN IN A CELL-FREE SYSTEM

Source of particulate fraction	Additions	Protein-bound radioactivity (total counts)
1. Induced cells ML 30	None	2958 ± 31
2. " " " "	0.001 M TDG	2297 ± 270
3. Uninduced cells ML 30	None	2514 ± 43
4. " " " "	0.001 M TDG	2519 ± 20

Triplicate or quadruplicate aliquots (2.0 ml containing about 1.2 mg protein) of particulate fractions from induced or uninduced cells, previously treated as described in the text, were treated with H^3-labeled N-ethylmaleimide (final concentration 2.5×10^{-5} M) for 15 min at 28°, with 0.001 M thiodigalactoside (TDG) either added or omitted as indicated. The reaction was stopped by the addition of β-mercaptoethanol (0.01 M). The assay for protein-bound radioactivity was carried out essentially as described under *Materials and Methods*. The specific activity of the N-ethylmaleimide was 2.9 million counts per minute/μmole.

contains the M protein, as revealed by the decrease in the binding of N-ethylmaleimide caused by the addition of thiodigalactoside. The difference is statistically highly significant ($p < 0.001$). No such effect is noted when the particulate fraction from uninduced cells is tested in identical fashion. It is clear that the M protein in these cell-free fractions retains its affinity for thiodigalactoside and its reactivity toward N-ethylmaleimide, making possible a simple and direct assay for this protein.

Extraction of the labeled M protein with nonionic surfactant: Metabolically active membranes throughout nature are made up principally of lipid and protein. If the M protein is a constituent of the membrane of *E. coli*, it might be possible to extract it from the particulate fraction after treatment with surfactants such as those of the "Triton" series (Rohm and Haas, Philadelphia, Pa.), known to be effective in extracting enzymes from the lipid-rich endoplasmic reticulum fraction of mammalian tissues.

This hypothesis was tested in the experiment shown in Table 6. Here the particulate fraction from cells of the constitutive strain ML 308 was assayed for its content of M protein as in the experiment of Table 5, except that after treatment with the labeled N-ethylmaleimide, the particulate fraction was extracted with the nonionic detergent Triton X-100 (t-octylphenoxypolyethoxyethanol). It can be seen that the difference in protein-bound N-ethylmaleimide caused by the addition of thiodigalactoside is entirely in the fraction extracted by surfactant; the residue shows no significant difference, indicating that essentially all of the labeled M

TABLE 6

EXTRACTION OF LABELED M PROTEIN

Additions during reaction with labeled N-ethylmaleimide	Protein-Bound Radioactivity (Total Counts)	
	Extract	Residue
1. None	1691 ± 163	373 ± 39
2. 0.001 M thiodigalactoside	1386 ± 30	370 ± 7

Cells of constitutive strain ML 308 were treated with unlabeled N-ethylmaleimide in the presence of 0.001 M thiodigalactoside essentially as described in the experiment of Table 4. The cells were then harvested by centrifugation and suspended in 0.1 M phosphate buffer of pH 7.0 containing 0.01 M MgCl₂ and 0.001 M thiodigalactoside at a cell density corresponding to 3 mg of protein/ml and disrupted in the Branson Sonifier (3 min). The sonicate was centrifuged at $3000 \times g$ for 15 min and the precipitate was discarded. The supernatant solution was then centrifuged at $40,000 \times g$ for 1 hr. The particulate fraction so obtained was washed with 0.1 M phosphate of pH 7.0 containing 0.01 M MgCl₂ and resuspended in buffer of the same composition at a protein concentration of 4 mg/ml.

Quadruplicate aliquots (0.6 ml) of the particulate fraction were assayed for the ability to bind radioactive N-ethylmaleimide in the presence and the absence of thiodigalactoside, essentially as in the experiment of Table 6, except that MgCl₂ (0.003 M) was also present during the reaction.

The reaction was stopped with mercaptoethanol (0.01 M), and the particulate fraction was recovered by centrifugation at $40,000 \times g$ for 1 hr. The supernatant solution was discarded.

The labeled particulate fraction was then extracted with 0.02 M Tris buffer of pH 7.4 containing 2% of Triton X-100 for 15 min at 28°, followed by centrifugation at $40,000 \times g$ for 1 hr. The extract and the residue were assayed for protein-bound radioactivity.

TABLE 7

FRACTIONATION OF LABELED M PROTEIN ON ECTEOLA COLUMN

Fraction	Ratio protein-bound C^{14}/H^3
Residue after extraction	0.88
Triton X-100 extract	1.20
Fraction #9 from ECTEOLA column	3.0

The extract (6.0 ml containing about 1 mg protein/ml) prepared as described in the text in a 2% solution of Triton X-100 in 0.02 Tris buffer of pH 7.4 was chromatographed on ECTEOLA (Cellex-E; California Biochemical Corp.) in the chloride form in a column 17 cm high by 0.8 cm in diameter, made up in Triton-Tris buffer of the same composition as the extract.

A band of protein-bound radioactivity was eluted with a solution containing 2% Triton X-100, 0.02 M Tris of pH 7.4, and 0.1 M NaCl. Fraction #9 was located at the leading edge of this band.

protein has been extracted under these conditions. The experiment of Table 6 also indicates the presence of the M protein constitutively in this strain.

Fractionation of labeled M protein on ECTEOLA column: Particles from induced cells of ML 30 were labeled with C^{14} and particles from uninduced cells were labeled with H^3, as in the experiment of Table 4. The labeled particles were then mixed and extracted with 0.02 M Tris buffer of pH 7.4 containing 1 per cent Triton X-100. The extract was chromatographed on a column of ECTEOLA.

The results are summarized in Table 7. It can be seen that the labeled M protein was extracted by the detergent solution, as indicated by the increased ratio of protein-bound C^{14}/H^3 in the extract as compared to the residue. Nearly all of the labeled protein was adsorbed on the column. Upon elution with buffer containing 0.1 M sodium chloride, the first radioactive fractions to emerge from the column were found to be greatly enriched in the labeled M protein.

The highest ratio of protein-bound C^{14}/protein-bound H^3 observed in fractions eluted from the column in this experiment was 3.0, indicating a radiochemical purity of about 66 per cent.

Discussion.—The particulate fraction in which the M protein is localized contains the cell wall membrane complex as well as some ribosomal material. The amount of ribosomal material can be reduced if the fraction is collected and washed by centrifugation at lower speeds (as in the experiment of Table 6) without affecting the content of M protein. Furthermore, it is known that much of the cell wall of *E. coli* may be removed by treatment with lysozyme-EDTA without loss of the β-galactoside transport system. These facts, together with the apparent lipo-protein nature of the M protein as shown by its extractability with Triton X-100, support the view that it is a constituent of the spheroplast membrane.

Further characterization of the M protein awaits the availability of larger amounts of material. The experiment described in Table 7 offers hope that the labeled derivative of this protein may be purified by conventional methods of protein fractionation. The isolation of the active protein may present greater difficulties, but should be facilitated by the assay (Table 5) now available for it.

It will be of interest to determine whether the M protein is present in various y^- mutants, and, if so, whether its properties are different from those of the protein found in wild-type cells.

We are indebted to Miss Marilynn Rumley for valuable technical assistance.

* This work has been supported by grants from the U.S. Public Health Service, National Institute of Neurological Diseases and Blindness, NB-02946, and the Life Insurance Medical Research Foundation.

† Fellow of the National Science Foundation.

[1] Tarlov, A., and E. P. Kennedy, *J. Biol. Chem.*, **240**, 49 (1965).

[2] Cohen, G. N., and J. Monod, *Bacteriol. Revs.*, **21**, 169 (1957).

[3] Kepes, A., and G. N. Cohen, in *The Bacteria*, ed. I. C. Gunsalus and R. Stanier (New York: Academic Press, Inc., 1962), vol. 4, p. 179.

[4] Kepes, A., in *The Cellular Functions of Membrane Transport*, ed. J. F. Hoffman (Englewood Cliffs, N. J.: Prentice-Hall, 1962), p. 155.

[5] Horecker, B. L., M. J. Osborn, W. L. McLellan, G. Avigad, and C. Asensio, in *Membrane Transport and Metabolism*, ed. A. Kleinzeller and A. Kotyk (New York: Academic Press, Inc., 1960), p. 378.

[6] Koch, A., *Biochim. Biophys. Acta*, **79**, 177 (1964).

[7] Winkler, H. H., and T. H. Wilson, *Federation Proc.*, **24**, 352 (1965).

[8] Zabin, I., in *Cold Spring Harbor Symposia on Quantitative Biology*, vol. 28 (1963), p. 431.

[9] Piutti, A., and E. Giustiniani, *Gazz. Chim. Ital.*, **26** (part I), 431 (1896).

[10] Rickenberg, H. V., G. N. Cohen, and G. Buttin, *Ann. Inst. Pasteur*, **91**, 829 (1956).

[11] Buhler, D. R., *Anal. Biochem.*, **4**, 413 (1962).

[12] Lowry, O. H., N. J. Rosebrough, A. L. Farr, and R. J. Randall, *J. Biol. Chem.*, **193**, 265 (1951).

[13] Hogben, C. A., *J. Lab. Clin. Med.*, **64**, 815 (1964).

[14] Kepes, A., *Biochim. Biophys. Acta*, **40**, 70 (1960).

[15] Gregory, J. D., *J. Am. Chem. Soc.*, **77**, 3922 (1955).

[16] Smith, D. G., A. Nagamatsu, and J. S. Fruton, *J. Am. Chem. Soc.*, **82**, 4600 (1960).

Reprinted from *Proc. Natl. Acad. Sci. U.S.A.*, **66**, 1190–1198 (1970)

13

β-Galactoside Transport in Bacterial Membrane Preparations: Energy Coupling via Membrane-Bound D-Lactic Dehydrogenase

Eugene M. Barnes, Jr.* and H. R. Kaback†

NATIONAL HEART AND LUNG INSTITUTE, NATIONAL INSTITUTES OF HEALTH,
BETHESDA, MARYLAND

Communicated by E. R. Stadtman, May 19, 1970

Abstract. The transport of β-galactosides by isolated membrane preparations from *Escherichia coli* strains containing a functional *y* gene is markedly stimulated by the conversion of D-lactate to pyruvate. The addition of D-lactate to these membrane preparations produces a 19-fold increase in the initial rate of uptake and a 10-fold stimulation of the steady-state level of intramembranal lactose or thiomethylgalactoside. Succinate, DL-α-hydroxybutyrate, and L-lactate partially replace D-lactate, but are much less effective; ATP and P-enolpyruvate, in addition to a number of other metabolites and cofactors, do not stimulate lactose transport by the vesicles. Lactose uptake by the membrane preparations in the presence of D-lactate requires oxygen, and is blocked by electron transport inhibitors and proton conductors; however, uptake is not significantly inhibited by high concentrations of arsenate or oligomycin. Furthermore, the P-enolpyruvate-P-transferase system is not involved in β-galactoside transport by the *E. coli* membrane vesicles. The findings indicate that the β-galactoside uptake system is coupled to the membrane-bound D-lactic dehydrogenase via an electron transport chain but does not involve oxidative phosphorylation.

Although the β-galactoside transport system of *Escherichia coli* has been examined in very great detail, the mechanism of the coupling of metabolic energy to active β-galactoside transport remains poorly understood. Scarborough, Rumley, and Kennedy[1] suggested an involvement of ATP in the lactose transport system of *E. coli*. However, recent studies by Pavlasova and Harold[2] on anaerobic methyl-1-thio-β-D-galactoside (TMG) uptake indicate that uncouplers of oxidative phosphorylation block TMG accumulation but do not alter ATP levels.

Fox and Kennedy demonstrated the existence of a "permease" protein (the *M* protein) which was a product of the *y* gene.[3] The subsequent suggestion of a role for the P-enolpyruvate-P-transferase system in TMG uptake[4] in *E. coli* raised the possibility that the *M* protein might be an inactivated Enzyme II. This topic has been discussed in detail in a recent review.[5]

Recent studies in this laboratory have described the coupling of a membrane-bound D-lactic dehydrogenase to amino acid transport in isolated membrane preparations from *E. coli*.[6] This paper reports a similar coupling of the β-galac-

1190

toside transport system to the membrane-bound D-lactic dehydrogenase and describes the general properties of the β-galactoside accumulation system in isolated bacterial membranes. Evidence is presented which indicates that electron transport, in the absence of oxidative phosphorylation, is required for β-galactoside uptake by the membrane vesicles. Furthermore, evidence is also presented which demonstrates that the P-enolpyruvate-P-transferase system is not involved in this transport mechanism.

Methods. Whole cells: *E. coli* ML 308-225 ($i^-z^-y^+a^+$) was grown on medium A[7] containing 0.4% succinate or glycerol. The ML 30 strain ($i^+z^+y^+a^+$) was grown on the same glycerol minimal medium and where indicated 0.5 mM isopropyl-β-D-thiogalactopyranoside (IPTG) was added as inducer. *E. coli* GN-2 ($i^-z^+y^+a^+$, Enzyme I$^-$) was grown as described previously.[8]

Membrane preparations: Membranes were prepared by methods already described.[8-10]

Uptake studies: The assay methods for methyl-α-D-glucopyranoside (α-MG) uptake by isolated membrane preparations have been reported.[8] The assay for β-galactoside uptake was identical to that described for amino acids.[6]

Enzyme assays: D-Lactic dehydrogenase and succinic dehydrogenase activities in the membrane preparations were assayed by a procedure reported earlier.[6]

Identification of accumulated lactose or TMG: Membrane samples that had accumulated lactose-^{14}C or TMG-^{14}C were extracted[8] or applied directly to silica gel G thin-layer plates (Mann Biochemicals, N.Y.) which were then developed with chloroform:methanol:water (60:70:26, v/v/v). The plates were radioautographed and the R_f values of the radioactive spots compared to those of known standards.

Estimation of concentration ratio: The ratio of intravesicular to external lactose concentrations was estimated by a method already described.[11]

Materials. Lactose-1-^{14}C was obtained from Amersham/Searle and β-methyl-^{14}C-thiogalactoside was a product of New England Nuclear Corp. All chemicals used in these experiments were of reagent grade and were obtained from the usual commercial sources.

Results. Effect of D-lactate on lactose uptake: The effect of D-lactate on the initial rate of lactose uptake and the steady-state level of accumulation by isolated membranes from *E. coli* ML 308-225 is shown in Figure 1. The addition of D-lactate stimulated the initial rate of transport 19-fold over controls incubated without D-lactate. The steady-state level of lactose accumulation was increased 10-fold by D-lactate. In 5 min membranes in the presence of D-lactate accumulated lactose to an intravesicular concentration about 30 times higher than that of the medium. It can also be seen that the addition of DNP results in the rapid loss of approximately 90% of the accumulated radioactivity. Over the time course indicated in Figure 1 more than 95% of the radioactivity accumulated in the membranes was recovered as unchanged lactose, and there was no detectible lactose-P at any of the times sampled. The rapid exit of lactose-1-^{14}C when excess unlabeled lactose was added is consistent with this observation.

Conversion of D-lactate to pyruvate: As indicated in Figure 2, membrane preparations of ML 308-225 rapidly converted D-lactate to pyruvate in a nearly stoichiometric fashion. In these experiments, D-lactate-U-^{14}C was employed as substrate and pyruvate was the only detectible radioactive product on thin-layer chromatograms. The loss of about 20% of the total carbon after 20 min may have been caused by partial decarboxylation of the pyruvate formed.

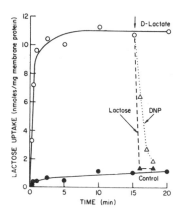 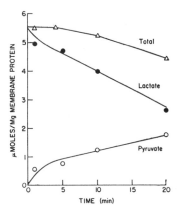

(*Left*) FIG. 1.—Effect of D-lactate on lactose uptake by *E. coli* ML 308-225 membrane preparations. Aliquots (50 μl) of membranes from ML 308-225 (grown on succinate medium) containing 0.41 mg protein were diluted to a final volume of 100 μl containing, in final concentrations, 50 mM potassium phosphate (pH 6.6), and 10 mM magnesium sulfate. After a 2-min incubation at 25°C, lithium D-lactate was added (where indicated) at a final concentration of 20 mM, and immediately thereafter lactose-1-^{14}C (10.2 mCi/mmol) was added to yield a final concentration of 0.2 mM. The incubations were continued at 25°C for the indicated times and then terminated and assayed as described previously.[6] The arrow indicates the time of addition of 10 mM ^{12}C-lactose (—▲—) or 1 mM dinitrophenol (·Δ·). Control incubations contained no D-lactate.

(*Right*) FIG. 2.—Metabolism of D-lactate by ML 308-225 membrane preparations. The experiment shown was conducted as described in *Methods*.

Energy source specificity for lactose uptake : The effect of various metabolites and cofactors on the steady-state level of lactose accumulation in membrane vesicles is shown in Table 1. Of the compounds tested, only D-lactate (line 2), DL-α-hydroxybutyrate (line 3) (a known substrate for D-lactic dehydrogenase), succinate (line 4), and L-lactate (line 5) increased lactose transport above endog-

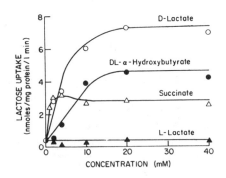

FIG. 3.—Effect of energy sources on the rate of lactose uptake by ML 308-225 membranes. Incubation mixtures were the same as described in the legend of Fig. 1 except that the energy sources indicated were added at the final concentrations shown and lactose uptake was terminated after 1 min.

enous levels. As shown above (Fig. 2), the membrane preparations converted D-lactate to pyruvate. Although the data are not shown, the same membrane preparations also stoichiometrically converted L-lactate to pyruvate and succinate to fumarate. Furthermore, pyruvate (line 19), fumarate (line 26), and α-ketobutyrate (line 31) failed to support transport.

The initial rates of lactose uptake as a function of increasing concentrations of energy sources is illustrated in Figure 3. As indicated, at concentrations which yield maximal initial rates of uptake, D-lactate is clearly the most effective compound. DL-α-hydroxybutyrate and succinate were relatively less stimulatory. At limiting concentrations, however, suc-

TABLE 1.   *Effect of various energy sources on lactose uptake by isolated membrane preparations from E. coli ML 308-225.*

Energy source (20 mM)	Lactose uptake (nmoles/mg membrane protein/10 min)	Energy source (20 mM)	Lactose uptake (nmoles/mg membrane protein/10 min)
1 None	1.87	20 Acetate	0.76
2 D-Lactate	13.0	21 Acetyl-CoA	1.59
3 DL-α-Hydroxybutyrate	10.0	22 Citrate	0.65
4 Succinate	7.95	23 Isocitrate	1.36
5 L-Lactate	5.50	24 cis-Aconitate	0.66
		25 α-Ketoglutarate	1.29
6 Glucose	1.03	26 Fumarate	2.26
7 6-P-Gluconate	1.21	27 Malate	1.87
8 Glucose-6-P	1.10	28 Oxaloacetate	0.48
9 Glucose-1-P	1.59	29 γ-Hydroxybutyrate	1.02
10 Fructose-6-P	1.29	30 β-Hydroxybutyrate	1.38
11 Fructose-1-P	1.97	31 α-Ketobutyrate	1.24
12 Fructose-1,6-P$_2$	1.08	32 ATP	1.22
13 α-Glycerol-P	1.58	33 CTP	1.23
14 Dihydroxyacetone-P	1.36	34 3',5'-AMP	1.13
15 3-P-Glycerate	1.82	35 UDP-Glucose	1.21
16 1,2-P$_2$-Glycerate	1.00	36 NADH	1.80
17 2-P-Glycerate	1.43	37 NADPH	0.79
18 P-enolpyruvate	1.20	38 Acetyl-P	1.29
19 Pyruvate	0.72	39 Carbamyl-P	1.32

Effect of various energy sources on lactose uptake by isolated membrane preparations from *E. coli* ML 308-225. Incubations were identical to those described in the legend of Fig. 1 except the energy sources listed replaced D-lactate and lactose uptake was terminated after 10 min.

cinate was slightly more effective than D-lactate. L-Lactate did not produce any significant stimulation. Although not shown, the presence of NAD or NADP in addition to D-lactate did not cause any additional stimulation of lactose uptake.

Induction of D-lactate-coupled β-galactoside uptake: As shown in Figure 4, membrane preparations from uninduced *E. coli* ML 30 took up very little lactose, nor was lactose uptake by these membranes stimulated by the addition of D-lactate. Membranes from ML 308-225 (*y* constitutive) grown on the same glycerol medium rapidly concentrated lactose in the presence of D-lactate. Membranes prepared from IPTG-induced ML 30 rapidly concentrated TMG in the presence of D-lactate as shown in Figure 5. Again, as with lactose uptake (Fig. 4), membranes from uninduced ML 30 took up very little TMG and D-lactate had no effect. The absence of D-lactate-coupled lactose or TMG uptake in uninduced ML 30 membranes is not caused by a defect in D-lactic dehydrogenase since induced and uninduced ML 30 membranes had the same D-lactic dehydrogenase activity and concentrated proline to the same extent in the presence of D-lactate.

Dependence of uptake rates on external β-galactoside concentration: Initial rates of lactose uptake by isolated membranes from ML 308-225 as a function of increasing external lactose concentration is shown in Figure 6. The initial rates of lactose uptake in the presence of D-lactate exhibit saturation kinetics and a Michaelis constant (K_m) of 0.19 mM. Similar data for TMG uptake were also obtained (Fig. 7), yielding a K_m of 0.51 mM.

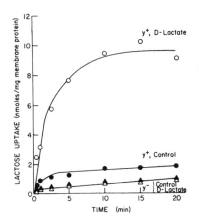
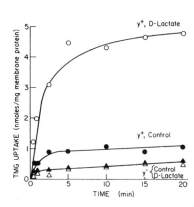

(*Left*) Fig. 4.—Effect of D-lactate on lactose uptake by membranes prepared from ML 308-225 and uninduced ML 30 grown on glycerol medium. Incubation mixtures were the same as described in the legend of Fig. 1 except that the incubations contained 0.43 mg membrane protein from ML 30 (—△—, —▲—) or 0.52 mg membrane protein from ML 308-225 (—○—, —●—) grown on glycerol medium. Closed symbols indicate omission of D-lactate.

(*Right*) Fig. 5.—Effect of D-lactate on TMG uptake by IPTG-induced and uninduced ML 30 membranes. Incubations were the same as described in the legend of Fig. 1 except that 0.2 mM ^{14}C-TMG (8.7 mCi/mmol) replaced lactose-1-^{14}C and 0.43 mg membrane protein from uninduced ML 30 (—△—, —▲—) or 0.48 mg from IPTG-induced ML 30 (—○—, —●—) was present. Closed symbols indicate omission of D-lactate.

Competition by sugars for lactose entry and exchange : The data presented in Table 2 indicate the relative capacity of various sugars to compete for the lactose uptake system of the membrane vesicles. When unlabeled sugars (0.2 or 2.0 mM) were added at the same time as lactose-1-^{14}C (0.2 mM), only β-galactosides, melibiose, and galactose caused significant inhibition of lactose entry (columns 2 and 3). Similarly, only β-galactosides, melibiose, and galactose were effective in displacing lactose from preloaded membranes (column 4).

TMG uptake in GN-2 membranes : Previous work from this laboratory[8] demonstrated that membrane preparations from *E. coli* GN-2, a mutant lacking enzyme I of the P-transferase system,[12] were unable to vectorially phosphorylate α-MG even in the presence of high concentrations of P-enolpyruvate. The data in Figure 8 indicate that GN-2 membranes, despite their inability to transport α-MG, rapidly concentrated TMG in the presence of D-lactate. As shown, GN-2 membranes exhibited a slightly higher intial rate of TMG uptake than induced ML 30 membranes (Fig. 5). The latter vectorially phosphorylated α-MG normally (data not shown). It is especially noteworthy that D-lactate did not stimulate α-MG uptake by any of the isolated membranes. Concentrations of P-enolpyruvate up to 0.1 M (which gave optimal rates of α-MG uptake) did not stimulate lactose or TMG uptake, nor was lactose-P or TMG-P detected in these experiments. Finally, membranes prepared from *E. coli* ML 308-225 failed to exhibit phosphatase activity towards TMG-P, and the addition of lactose to ML 308-225 membranes incubated in the presence of ^{32}P-enolpyruvate did not accelerate the appearance of ^{32}P$_i$ as might be expected if a lactose-P P-hydrolase were involved in this system.

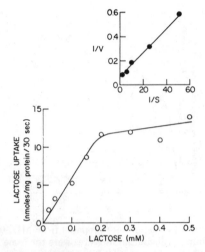

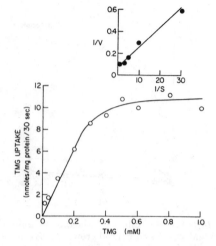

FIG. 6.—Effect of external lactose concentration on initial rates of lactose uptake by ML 308-225 membranes. Incubations were the same as indicated in the legend of Fig. 1 except that lactose-1-^{14}C was present at the concentrations shown and all incubations contained 20 mM D-lactate. Lactose uptake was terminated after 30 sec. The inset is a double reciprocal plot of the data shown.

FIG. 7.—Effect of external TMG concentration on initial rates of TMG uptake by ML 308-225 membranes. Incubations were the same as described in the legend of Fig. 6 except that TMG-^{14}C (8.7 mCi/mmol) replaced lactose.

Effect of metabolic state on lactose uptake : As indicated in Table 3, D-lactate-coupled lactose uptake by ML 308-225 membranes was inhibited 94% by exclusion of oxygen (line 2). Furthermore, the electron transport inhibitors azide

TABLE 2. *Competition by various sugars for lactose uptake and exchange by ML 308-225 membranes.*

	Lactose Accumulated (% control)		
	—Competitive Entry—		Competitive displacement
	0.2 mM	2.0 mM	2.0 mM
^{12}C-Sugar	^{12}C-sugar	^{12}C-sugar	^{12}C-sugar
TDG	55	6	37
Melibiose	75	18	23
ONPG	68	19	14
IPTG	72	24	20
TMG	99	34	28
Galactose	95	50	34
Maltose	95	89	91
Sucrose	110	98	96
Glucose	99	102	98
Mannose	99	97	85
Mannitol	113	102	93
Fructose	97	99	105

Competition by various sugars for lactose uptake and exchange by ML 308-225 membranes. Incubations were the same as described in the legend of Fig. 1 except that the ^{12}C-sugar was added at the same time as lactose-1-^{14}C in the competitive entry experiments and uptake was for 1 min. For the competitive displacement experiments, the membranes were preloaded for 5 min with lactose-1-^{14}C, then the ^{12}C-sugar shown was added, and incubations were continued for 2 min. All incubations contained 20 mM D-lactate. The values for lactose accumulated in the membranes are expressed as percentages of control values with no added ^{12}C-sugar.

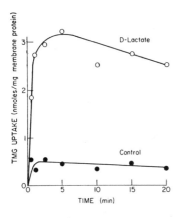

Fig. 8.—Effect of D-lactate on TMG uptake by GN-2 membranes. Incubations were the same as described in the legend of Fig. 5 except that 0.31 mg of membrane protein from GN-2 was employed.

(line 3), amytal (line 4), cyanide (line 5), antimycin A (line 6), and 2-heptyl-4-hydroxyquinoline-*N*-oxide (line 7) also effectively blocked lactose accumulation. The inhibition by cyanide and azide is consistent with the observed anaerobic inhibition and indicates the involvement of oxygen as a terminal electron acceptor. Although not shown, anaerobic incubations inhibited the membrane D-lactate dehydrogenase activity by 80%. Finally, DNP (line 8), carbonyl cyanide trifluoromethoxyphenylhydrozone (line 9), and valinomycin (line 10) were all effective inhibitors of lactose transport. These compounds, well-known uncouplers of oxidative phosphorylation, are all effective proton conductors.[13]

As also shown in Table 3, arsenate inhibited D-lactate-coupled lactose uptake by only 30% despite concentrations as high as 50 mM (line 11). Experiments performed in phosphate-free media produced a similar degree of inhibition. The transport system was similarly insensitive to oligomycin (line 12).

Lactose transport by the membranes was inhibited by the sulfhydryl reagents *N*-ethylmaleimide and *p*-chloromercuribenzoate (lines 13 and 14). This is consistent with similar findings with whole cells and the *M* protein.[14]

Discussion. The data presented here demonstrate the β-galactoside transport in *E. coli* membrane preparations is coupled to a membrane-bound D-lactic dehydrogenase via a respiratory chain. A similar system mediating the transport of a wide variety of amino acids has recently been reported.[6]

The effect of D-lactate, DL-α-hydroxybutyrate, and succinate on lactose or TMG uptake by the membrane preparations clearly does not involve the P-enolpyruvate-P-transferase system. This conclusion is supported by the observation that P-enolpyruvate failed to stimulate lactose transport in membrane preparations which readily utilized P-enolpyruvate for the transport of α-MG— nor did D-lactate substitute for P-enolpyruvate in this system. Furthermore, membranes prepared from *E. coli* GN-2 (defective in enzyme I of the P-transferase system) which were completely unable to catalyze the vectorial phosphorylation of α-MG were fully capable of supporting D-lactate-coupled TMG accumulation.

Although the precise nature of the coupling of D-lactic dehydrogenase and succinic dehydrogenase to β-galactoside transport is uncertain, the inhibitor studies presented in Table 3 provide a strong indication that an electron transfer chain mediates this coupling. Furthermore, the effect of D-lactate or succinate on β-galactoside transport is apparently not exerted through the production of stable high-energy phosphate compounds. The lack of sensitivity of β-galactoside transport to arsenate or oligomycin, the failure of added ATP to stimulate

TABLE 3. *Effect of anaerobic incubation and metabolic inhibitors on lactose uptake by ML 308-225 membranes.*

	Condition during incubation	Inhibitor concentration (M)	Inhibition of lactose uptake (%)
1	Aerobic		...
2	Anaerobic		94
3	Sodium azide	10^{-3}	83
		10^{-2}	94
4	Sodium amytal	10^{-3}	37
		10^{-2}	87
5	Sodium cyanide	10^{-3}	76
		10^{-2}	98
6	Antimycin A	2×10^{-5}	18
		2×10^{-4}	93
7	HOQNO	2×10^{-6}	29
		2×10^{-5}	70
8	Dinitrophenol	10^{-4}	84
		10^{-3}	97
9	CCCP	10^{-6}	70
		10^{-5}	98
10	Valinomycin	2×10^{-7}	62
		2×10^{-6}	81
11	Sodium arsenate	10^{-2}	17
		5×10^{-2}	29
12	Oligomycin	2×10^{-5}	4
		10^{-4}	10
13	N-ethylmaleimide	10^{-4}	32
		10^{-3}	100
14	p-Chloromercuribenzoate	10^{-4}	36
		10^{-3}	100

Effect of anaerobic incubation and metabolic inhibitors on lactose uptake by ML 308-225 membranes. All incubations were performed as indicated in the legend of Fig. 1 except that the membrane vesicles were exposed to the condition or inhibitor shown for 15 min prior to addition of D-lactate and ^{14}C-lactose. The membranes were then allowed to take up lactose for 10 min. Results are expressed as percentage inhibition of the lactose concentrating ability of a 25°C, aerobic control (12.5 nmoles/mg membrane protein). All incubations contained 20 mM D-lactate. The anaerobic experiments were performed in a nitrogen atmosphere containing less than 50 ppm oxygen. Antimycin A, HOQNO, valinomycin, and oligomycin were added as aliquots of dimethylsulfoxide solutions.

lactose transport, and the observation that similar membrane preparations cannot conduct oxidative phosphorylation[15] argue strongly against the involvement of respiration-linked phosphorylation in β-galactoside transport. In light of these observations, a possible coupling of electron transfer to active β-galactoside transport via oxidation-reduction of a membrane "carrier" protein (e.g., the M protein) with resultant conformation changes is an attractive hypothesis. However, further experimentation is necessary before such a theory can be seriously considered.

Abbreviations: DNP, 2,4-dinitrophenol; IPTG, isopropyl-β-D-thiogalactopyranoside; α-MG, methyl-α-D-glucopyranoside; ONPG, o-nitrophenyl-β-D-galactopyranoside; TDG, β-D-galactosyl-1-thio-β-D-galactopyranoside; HOQNO, 2-heptyl-4-hydroxyquinoline-N-oxide; TMG, methyl-1-thio-β-D-galactoside.

* Postdoctoral fellow of the American Cancer Society (PF-545).

† Associate member of The Roche Institute of Molecular Biology, Nutley, N.J. 07110, and a visiting scientist in the National Heart and Lung Institute until July 1, 1970. Requests for reprints should be sent to Dr. H. R. Kaback at the New Jersey address.

[1] Scarborough, G. A., M. K. Rumley, and E. P. Kennedy, *Proc. Nat. Acad. Sci. USA*, **60**, 951 (1968).

[2] Pavlasova, E., and F. M. Harold, *J. Bacteriol.*, **98**, 198 (1969).

[3] Fox, C. F., J. R. Carter, E. P. Kennedy, *Proc. Nat. Acad. Sci. USA*, **57**, 698 (1967).

[4] Kundig, W., F. D. Kundig, B. E. Anderson, and S. Roseman, *J. Biol. Chem.*, **241**, 3243 (1966).

[5] Kaback, H. R., *Ann. Rev. Biochem.*, **39**, in press.

[6] Kaback, H. R., and L. S. Milner, *Proc. Nat. Acad. Sci. USA*, **66**, 1008 (1970).

[7] Davis, B. D., and E. S. Mingioli, *J. Bacteriol.*, **60**, 17 (1950).

[8] Kaback, H. R., *J. Biol. Chem.*, **243**, 3711 (1968).

[9] Kaback, H. R., *Proc. Nat. Acad. Sci. USA*, **63**, 724 (1969).

[10] Kaback, H. R., *Methods in Enzymology*, ed. W. B. Jakoby (New York: Academic Press, in press).

[11] Kaback, H. R., and E. R. Stadtman, *Proc. Nat. Acad. Sci. USA*, **55**, 920 (1966).

[12] Tanaka, S., Fraenkel, D. G., and Lin, E. C. C., *Biochem. Biophys. Res. Commun.*, **27**, 63 (1967).

[13] Thompson, T. E., and F. A. Henn, in *Membranes of Mitochondria and Chloroplasts*, ed. E. Racker (New York: Van Nostrand Reinhold Co., 1970), p. 30.

[14] Fox, C. F., and E. P. Kennedy, *Proc. Nat. Acad. Sci. USA*, **54**, 891 (1965).

[15] Klein, W. L., A. S. Dhams, and P. D. Boyer, *Fed. Proc.*, **29**, 341 (1970).

215

Copyright © 1971 by the American Society of Biological Chemists, Inc.
Reprinted from *J. Biol. Chem.*, **246**, 5518–5522 (1971)

Mechanisms of Active Transport in Isolated Membrane Vesicles

I. THE SITE OF ENERGY COUPLING BETWEEN D-LACTIC DEHYDROGENASE AND
β-GALACTOSIDE TRANSPORT IN *ESCHERICHIA COLI* MEMBRANE VESICLES*

(Received for publication, April 1, 1971)

Eugene M. Barnes, Jr.,‡ and H. R. Kaback

From the Roche Institute of Molecular Biology, Nutley, New Jersey 07110

SUMMARY

Transport of a wide variety of amino acids and sugars by membrane vesicles isolated from *Escherichia coli* ML 308-225 is coupled primarily to D-lactic dehydrogenase. This membrane-bound, flavin-linked primary dehydrogenase is coupled to oxygen via a cytochrome system also present in the vesicle membrane. Spectrophotometric evidence shows that D-lactic dehydrogenase, succinic dehydrogenase, L-lactic dehydrogenase, and NADH dehydrogenase all utilize the same cytochrome system. There is no relationship between rates of oxidation of electron donors by the respiratory chain (succinate > NADH = D-lactate > L-lactate) and the ability of these compounds to stimulate lactose transport (D-lactate ≫ succinate > L-lactate > NADH). Furthermore, D-lactate in combination with other electron donors is no more effective than D-lactate alone for support of lactose transport. These findings indicate that the site of energy coupling of D-lactic dehydrogenase to active transport lies between the primary dehydrogenase and cytochrome b_1.

Supportive evidence for this conclusion is obtained from experiments showing that N-ethylmaleimide and p-chloromercuribenzoate inhibit transport and D-lactate-induced respiration. However, these sulfhydryl reagents do not inhibit D-lactic dehydrogenase with dichlorophenolindophenol as an artificial acceptor, nor do they significantly block NADH-induced respiration.

Previous work from this laboratory showed that concentrative uptake of amino acids (2) and β-galactosides (3) by isolated cytoplasmic membrane vesicles from *Escherichia coli* is coupled primarily to a membrane-bound D-lactic dehydrogenase. Subsequent studies have shown that transport systems for galactose,[1] glucuronic acid,[1] arabinose,[1] glucose-6-P,[2] manganese (4), and potassium (in the presence of valinomycin) (5) are also coupled primarily to this dehydrogenase. Although it was apparent from the initial studies that generation of high energy phosphate compounds was not involved in these transport systems, a requirement for electron transport was clearly shown.

This paper reports further studies on the nature of coupling of D-lactic dehydrogenase to active transport in isolated membrane vesicles. For the purposes of this communication, the β-galactoside transport system has been selected as an example; however, it should be emphasized that the studies reported here have been carried out with each D-lactate coupled transport system with qualitatively similar results. The results of those studies will be published at a later date.

METHODS

Membrane Preparations—*E. coli* ML 308-225 (i⁻z⁻y⁺a⁺) was grown on Medium A containing 1% disodium succinate (hexahydrate). Membrane vesicles were prepared from these cells as described previously (6–8).

Sugar Uptake—The assay for lactose uptake by the vesicles has been reported (3).

D-Lactic Dehydrogenase Assays—Either the assay utilizing D-[¹⁴C]lactate described previously (2) or DCI[3] acceptor was used for D-lactic dehydrogenase. Incubation mixtures for the latter procedure contained (in 1.0 ml) 0.20 M potassium phosphate (pH 7.3), 20 mM D-lactate (lithium salt), 100 to 500 μg of membrane protein, and 0.002% dichlorophenolindophenol. Reduction of DCI was followed at 620 nm with a Gilford recording spectrophotometer.

Oxygen Uptake Measurements—Rates of oxygen uptake were measured with the Clark electrode of a YSI model 53 oxygen monitor (Yellow Springs Instrument Company, Yellow Springs, Ohio). The oxygen-monitoring system was calibrated by the method of Chappell (9) and gave a typical tension of 0.47 μg-atom of oxygen per ml of reaction mixture at 25°. Assay mixtures (3.0 ml) contained 50 mM potassium phosphate (pH 6.6), 10 mM MgSO₄, membrane protein (0.2 to 1.0 mg), and 20 mM electron donor (D-lactate, succinate, etc.) with the exception of NADH which was assayed at 5 mM. All assays were carried out at 25° and rates of oxygen consumption were linear with time for a minimum of 5 min.

Difference Spectra—An Aminco-Chance dual wave length-split

* A preliminary report of portions of this work has been submitted for publication (1).

‡ Postdoctoral fellow of the American Cancer Society (PF 545). Present address, Department of Biochemistry, Baylor College of Medicine, Houston, Texas 77045.

[1] G. K. Kerwar and H. R. Kaback, manuscripts in preparation.

[2] P. Bhattacharyya, F. J. Lombardi, and H. R. Kaback, manuscript in preparation.

[3] The abbreviations used are: DCI, dichlorophenolindophenol; HOQNO, 2-heptyl-4-hydroxyquinoline-N-oxide; NEM, N-ethylmaleimide; PCMB, p-chloromercuribenzoate.

5518

216

beam recording spectrophotometer (American Instrument Company, Silver Spring, Maryland) was used for measurement of difference spectra. Assay mixtures were the same as described for oxygen uptake measurements except that membrane protein was increased to 1 to 4 mg per ml. The spectrophotometer was operated at a slit width of 2.0 mm and a scan rate of 0.5 nm per sec. Spectra were measured at 22°.

RESULTS

Substrate Oxidation and Lactose Uptake—The effect of a series of electron donors on lactose uptake is shown in Table I. Addition of D-lactate to membrane vesicles produces a 30-fold increase in the rate of lactose transport and a 20-fold stimulation of the steady state level of lactose accumulation. Under these conditions intravesicular lactose accumulates to a level over 100 times that of the surrounding medium. As indicated, other substrates such as α-hydroxybutyrate, succinate, and L-lactate are able to stimulate both initial rates and the steady state level of lactose accumulation, but none are as effective as D-lactate. The second most effective compound, DL-α-hydroxybutyrate, is apparently utilized as substrate for D-lactate dehydrogenase (10). Only very slight stimulation of lactose uptake is observed with NADH.

The rates of oxygen uptake by membrane vesicles in the presence of these substrates are also presented in Table I. In the absence of donor, oxygen utilization by the vesicles is negligible. Addition of oxidizable substrate to membrane preparations produces an immediate and rapid uptake of oxygen from the medium. The most effective electron donors for respiration in these preparations are succinate, D-lactate, and NADH, in that order of effectiveness. The concentrations of substrates used in these studies are saturating both for oxygen uptake and lactose transport. It is clear that no apparent relationship exists between rates of oxidation of these electron donors and their ability to support lactose transport. Although succinate is the best substrate for respiration, it ranks third in stimulation of transport, being less than half as effective as D-lactate. Oxidation of NADH proceeds at nearly the rate of D-lactate, but NADH is only marginally effective in stimulating lactose transport. These findings indicate that the observed specificity of β-galactoside transport for D-lactic dehydrogenase, as opposed to other dehydrogenases, cannot be accounted for solely on the basis of rates of electron flow to oxygen.

Although not shown in Table I, vesicles prepared from cells grown on complex media (*i.e.* Difco Penassay broth or nutrient broth) have both α-glycerol-P and formate dehydrogenase activities. Moreover, in such membrane preparations, addition of α-glycerol-P or formate stimulates lactose or amino acid uptake about as well as succinate.

Effect of Inhibitors on D-Lactate Respiration—The effect of inhibitors and other agents on D-lactate-dependent oxygen uptake by membrane vesicles is shown in Table II. Amytal, HOQNO, and cyanide all inhibit respiration by the vesicles in the presence of D-lactate. Previous experiments from this laboratory have established that these inhibitors also block lactose uptake by membrane preparations with similar degrees of effectiveness (3). Recent investigations of the respiratory chain of *E. coli* by Cox *et al.* (11) have identified the amytal-sensitive site as a flavoprotein between D-lactic dehydrogenase and cytochrome b_1. In addition, these workers have indicated that HOQNO acts between cytochrome b_1 and cytochrome a_2, perhaps at a quinone-containing component, and that cyanide blocks cytochrome a_2.

TABLE I

Respiration and lactose uptake by ML 308-225 membrane vesicles

Reaction mixtures for lactose uptake assays contained (in a volume of 100 μl), in final concentrations, 50 mM potassium phosphate (pH 6.6), 10 mM $MgSO_4$, 0.21 mg of membrane protein, 0.2 mM lactose-1-^{14}C (14.9 mCi per mmole), and (as indicated) 20 mM electron donor. Incubations were carried out at 25° and terminated and assayed as described previously (3). Initial rates of lactose uptake were from 30-sec and 1-min incubations and were linear over this time span. Steady state accumulation was determined from 10-min incubations. Rates of oxygen uptake by the vesicles under these conditions were determined as described under "Methods." All values are expressed per mg of membrane protein.

Electron donor	Lactose uptake		Oxygen uptake rate
	Rate	Steady state	
	nmoles/ min/mg	*nmoles/10 min/mg*	*ng-atoms/min/mg*
None	1.2	2.8	<1
D-Lactate	37.0	53.6	330
DL-α-Hydroxybutyrate	19.8	43.2	65
Succinate	15.4	15.8	540
L-Lactate	15.6	15.8	91
NADH	3.6	4.1	270

TABLE II

Effect of various compounds on D-lactate oxidation by ML 308-225 membranes

Rates of oxygen uptake by ML 308-225 membrane vesicles were assayed as described under "Methods." Amytal, HOQNO, antimycin, and valinomycin were added as aliquots of dimethyl sulfoxide solutions. Control incubations indicated no effect of this solvent on respiration. Inhibitors were incubated with membrane protein (105 μg per ml) for 15 min at 25° in the absence of D-lactate. Reaction was then initiated by lactate addition. Rates of oxygen uptake are expressed as a percentage of the rate of a control without inhibitor.

Compound	Concentration	Relative rate of oxygen uptake
	M	*%*
Sodium amytal	1×10^{-3}	68
	5×10^{-3}	40
HOQNO	4×10^{-6}	67
	2×10^{-5}	48
KCN	2×10^{-4}	68
	1×10^{-3}	16
2,4-Dinitrophenol	1×10^{-3}	98
Carbonyl cyanide-*m*-chloro-phenylhydrazone	1×10^{-5}	90
Valinomycin	2×10^{-6}	100
Antimycin	2×10^{-4}	90
Sodium azide	1×10^{-2}	96
Lactose	2×10^{-4}	100
Oxamate	5×10^{-4}	30
	5×10^{-3}	9

It is also significant that NADH oxidation is also sensitive to HOQNO (76% inhibition at 2×10^{-5} M) and cyanide (86% inhibition at 10^{-3} M)—as is succinate oxidation (44% inhibition at 2×10^{-5} M HOQNO and 93% inhibition at 10^{-3} M cyanide)—

TABLE III

Effect of oxamate on dehydrogenase activities

Rates of DCI reduction and oxygen uptake were determined as described under "Methods." Sodium oxamate (5 mM) was incubated with membrane vesicles in the assay mixture for 2 min prior to the initiation of the reaction with substrate. Results are expressed as per cent inhibition due to oxamate relative to untreated controls.

Substrate	Assay method	Inhibition by 5 mM oxamate
		%
D-Lactate	DCI reduction	91
D-Lactate	Oxygen uptake	91
NADH	Oxygen uptake	3
Succinate	Oxygen uptake	<1

indicating that most of the NADH and succinate oxidation observed is cytochrome-linked. These findings show that each dehydrogenase studied is coupled to oxygen via a membrane-bound respiratory chain.

Dinitrophenol, carbonyl cyanide *m*-chlorophenylhydrazone, and valinomycin do not significantly affect D-lactate oxidation (Table II), despite profound inhibition of lactose transport (3). This finding is not unexpected since most bacterial electron transport systems are not subject to respiratory control. Regarding the failure of antimycin and azide to block respiration, it is noteworthy that many bacterial respiratory systems are resistant to these inhibitors (12). Moreover, antimycin inhibition of β-galactoside transport by vesicles (3) may be due to induction of passive leakage due to detergent effects of this compound (13).

Numerous experiments including those shown in Table II have established that lactose is totally without effect on D-lactate oxidation. These findings are in opposition to the experiments of Kepes (14) who found that thiomethylgalactoside stimulated oxygen uptake by whole cells.

Oxamic acid, an inhibitor of the cytoplasmic, NAD-dependent D-lactic dehydrogenase (15), also blocks oxidation of D-lactate by the membrane-bound, NAD-independent D-lactic dehydrogenase (*cf.* Table II). A similar degree of oxamate inhibition is observed for either oxygen uptake or DCI reduction in the presence of D-lactate (*cf.* Table III). Furthermore, oxamate inhibition is specific for D-lactic dehydrogenase, since neither succinate nor NADH oxidation is affected. Thus oxamate inhibits at the level of the primary dehydrogenase for D-lactate.

Utilization of Electron Donors by Cytochrome System—Reduction of the membrane-bound cytochromes by D-lactate, succinate, NADH, and dithionite is illustrated in Fig. 1. This figure shows difference spectra of substrate-reduced membrane vesicles with respect to control preparations in the oxidized state. *Line 1* is a plot of D-lactate reduced against oxidized spectrum of *E. coli* ML 308-225 membrane vesicles. The absorption bands are identified (12) as Soret (432 nm), due mainly to the δ band of cytochrome b_1, flavoprotein trough (465 nm), β band of cytochrome b_1 (533 nm), α band of cytochrome b_1 (560 nm), α band of cytochrome a_1 (595 nm), and α band of cytochrome a_2 (630 nm). These peaks account for all known classes of cytochromes of *E. coli* (12) except for cytochrome *o* which is not detectable by these methods. This observation is also supported by comparison of the D-lactate (*Line 1*), succinate (*Line 3*), and NADH (*Line 4*) reduced-oxi-

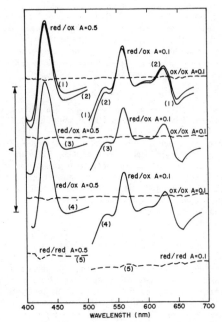

FIG. 1. Difference spectra of membrane vesicles. Absorption spectra were obtained with the spectrophotometer in the split beam mode and the oxidized-oxidized base line adjusted for aerobic membrane suspensions (3.0 ml) containing 50 mM KP$_i$ (pH 6.6), 10 mM MgSO$_4$, and 3.15 mg per ml of protein in both sample and reference beams. Then the substrate indicated was added to the suspension in the sample beam and the spectrum scanned until no further absorbance changes were observed indicating anaerobic steady state. This state was achieved in about 1 min; the spectra shown were then recorded. The electron donors were 10 mM D-lactate (*Line 1*), 0.2 mg per ml of sodium dithionite (*Line 2*), 10 mM succinate (*Line 3*), 2 mM NADH (*Line 4*). The scan in *Line 5* is a NADH reduced-D-lactate reduced difference spectrum. *red/ox*, reduced-oxidized; *ox/ox*, oxidized-oxidized; *red/red*, reduced-reduced.

dized spectra with the dithionite reduced-oxidized spectrum (*Line 2*). Reduction of the cytochrome system by each substrate studied reveals nearly complete reduction of all available cytochromes. Moreover, essentially the same absorption spectra are obtained with vesicles whose electron transport chain is reduced with D-lactate (*Line 1*), succinate (*Line 3*), and NADH (*Line 4*). Indeed, the difference spectrum between NADH reduced and D-lactate reduced vesicles (*Line 5*) shows no significant bands.

Quantitative determination of the relative amounts of each chromophore reduced by various electron donors is shown in Table IV. No significant differences are observed for reduction of Soret, cytochrome b_1, or cytochrome a_2 (cytochrome oxidase) by any of the three enzymatic electron donors. There was insufficient cytochrome a_1 to permit quantitative comparison.

These findings indicate that D-lactic, succinic, and NADH dehydrogenases utilize the same cytochrome system for electron flow in membrane vesicles. Thus the specificity of the coupling between lactose transport and D-lactic dehydrogenase (as opposed

TABLE IV

TABLE IV

Anaerobic reduction of respiratory components by dehydrogenase substrates

Values for reduction are expressed as percentages of controls reduced with dithionite and were based on the spectra shown in Fig. 1. Wave length pairs were those suggested by Jones and Redfearn (16) and values for the absorption of each component were based on the absorption difference between these wave lengths. Flavoprotein measured by this method may be subject to interference by non-heme iron (11).

Component	Wave length pair	Reduction of component		
		D-Lactate	Succinate	NADH
		%		
Soret................	432-419	105	103	101
Cytochrome b_1.......	560-575	97	81	84
Cytochrome a_2.......	630-615	99	102	113
Flavoprotein.........	465-510	63	53	61

TABLE V

Effect of thiol reagents on D-lactate and NADH oxidation

Rates of oxygen uptake by ML 308-225 membrane vesicles were measured as described under "Methods." N-Ethylmaleimide (0.5 mM) and p-chloromercuribenzoate (0.05 mM) were incubated with membrane protein (105 μg per ml) for 15 min prior to initiation of the oxidase reaction with substrate. Dithiothreitol (1.0 mM) was added to reaction mixtures (where indicated) after initiation of the oxidase reaction and the rate of oxygen uptake was measured immediately.

Reagent	Rate of oxygen uptake	
	D-Lactate	NADH
	%	
NEM..........................	27	106
PCMB.........................	18	67
PCMB + dithiothreitol..........	92	67

to succinic or NADH dehydrogenases) cannot be related either to rates of electron flow to oxygen or to a unique cytochrome system coupled to D-lactic dehydrogenase. This conclusion is supported by experiments in which lactose transport was studied in the presence of fixed concentrations of D-lactate and increasing concentrations of either succinate or NADH. There was no additional stimulation of the rate of lactose transport over that obtained with D-lactate alone under any condition studied. It seems clear therefore that the site of coupling for D-lactic dehydrogenase to transport must occur prior to entry of electrons into the cytochrome system.

Effect of Sulfhydryl Reagents on Oxidation—The effect of N-ethylmaleimide and p-chloromercuribenzoate on D-lactate oxidation is shown in Table V. Addition of NEM produces a 73% decrease in oxidation of D-lactate and PCMB reduces D-lactate-dependent oxygen uptake by 82%. These reagents are effective inhibitors of lactose transport by membranes (3). Reversal of PCMB inhibition by dithiothreitol provides further evidence that sulfhydryl groups are necessary for oxidation of D-lactate.

Inhibition of oxygen uptake by PCMB and NEM does not appear to be mediated at the level of the primary dehydrogenase for

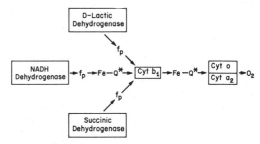

FIG. 2. Respiratory chain of *E. coli*. *Cyt.*, cytochrome.

D-lactate. Neither D-lactate-DCI reductase activity in intact vesicles (125 nmoles of DCI reduced per min per mg of protein) nor a solubilized, partially purified preparation of this enzyme[4] (300 nmoles of DCI reduced per min per mg of protein) is sensitive to NEM or PCMB inhibition. It is important to note that NADH oxidation (*cf.* Table V) is not sensitive to NEM, and the small amount of inhibition due to PCMB is not reversed by dithiothreitol. Thus, neither the primary D-lactic dehydrogenase itself nor the cytochrome system contains a sulfhydryl-sensitive site. Therefore, the site of inhibition of D-lactate oxidation by NEM and PCMB must lie between D-lactic dehydrogenase and the cytochromes.

DISCUSSION

A diagram of some features of the *E. coli* respiratory chain is shown in Fig. 2. This figure illustrates the order of the components involved in electron flow as suggested by Cox *et al.* (11). Electrons from D-lactic dehydrogenase flow to cytochrome b_1 via flavoproteins (in perhaps several stages). Electrons are then transferred to a ubiquinone component perhaps involving non-heme iron (11), which in turn transfers electrons to the cytochrome oxidases, a_2 and o. Succinic and NADH dehydrogenases also feed electrons via flavins to cytochrome b_1 and in the latter case a second quinone site appears to be involved (11). This scheme is not intended to represent complete pathways for electron transport but merely the order of events thought to occur. With reference to the findings presented here indicate that the site of energy coupling for lactose transport is localized primarily between D-lactic dehydrogenase and cytochrome b_1. It is re-emphasized that the same observations and thus the same conclusions can be extended to the coupling of D-lactic dehydrogenase to the other transport systems mentioned previously.

Although current studies do not provide details of the nature of coupling between D-lactic dehydrogenase and active transport, several observations have led to a working hypothesis. Possibly, membrane "carrier" proteins for transport substrates are themselves electron transfer intermediates localized in the membrane

[4] In collaboration with Dr. Leonard D. Kohn of the Laboratory of Biochemical Pharmacology, National Institute of Arthritis and Metabolic Diseases, Bethesda, Maryland, the membrane-bound D-lactic dehydrogenase has been solubilized and purified approximately 250- to 300-fold from whole cells of *E. coli* ML 308-225. At this stage, the preparation is approximately 80% pure as judged by disc gel electrophoresis, and contains cytochrome b_1 but no phospholipid or dehydrogenase activity toward succinate or NADH. The purification and properties of this enzyme will be published at a later date.

between primary dehydrogenases (principally D-lactic dehydrogenase in *E. coli*) and cytochrome b_1. A cycle of D-lactate-dependent reduction of the carriers followed by oxidation through cytochrome b_1 (perhaps by sulfhydryl-disulfhydryl interconversion) could account for reversible conformation changes. Decreased affinity of the reduced form of a carrier protein for its transport substrate at the interior face of the membrane could be a result of this conformational change.

There are several observations which are consistent with such a model.

1. Oxidation-reduction coupling of carriers accounts directly for utilization of electron flow for active transport.

2. Association of carrier molecules with dehydrogenase systems branched before cytochrome b_1 explains electron donor specificity.

3. Conformational changes in the carrier model do not require participation of high energy phosphate compounds. Participation of such energy sources has been suggested in an earlier model (17) but such compounds are not required for D-lactic dehydrogenase-coupled sugar or amino acid transport in vesicles (2, 3).

The model proposed here is intended only to summarize some of our working hypotheses. Another possible mechanism for energy coupling of D-lactic dehydrogenase to active transport involving participation of proton or potential gradients has been suggested recently (18, 19). However, this type of mechanism appears to be ruled out by studies presented in the following paper (20).

Acknowledgment—The authors are indebted to Dr. R. W. Hendler of the National Heart and Lung Institute who assisted us with assays for oxygen uptake and difference spectra in early stages of this investigation.

REFERENCES

1. KABACK, H. R., BARNES, E. M., JR., AND MILNER, L. S. *Proceedings of the Xth International Congress of Microbiology Mexico City, 1970*, in press.
2. KABACK, H. R., AND MILNER, L. S., *Proc. Nat. Acad. Sci. U. S. A.*, **66**, 1008 (1970).
3. BARNES, E. M., JR., AND KABACK, H. R., *Proc. Nat. Acad. Sci. U. S. A.*, **66**, 1190 (1970).
4. BHATTACHARYYA, P., *J. Bacteriol.*, **104**, 1307 (1970).
5. BHATTACHARYYA, P., EPSTEIN, W., AND SILVER, S., *Proc. Nat. Acad. Sci. U. S. A.*, in press.
6. KABACK, H. R., *J. Biol. Chem.*, **243**, 3711 (1968).
7. KABACK, H. R., *Proc. Nat. Acad. Sci. U. S. A.*, **63**, 724 (1969).
8. KABACK, H. R., in W. B. JAKOBY (Editor), *Methods in enzymology*, Vol. XXII, Academic Press, New York, 1971, p. 99.
9. CHAPPELL, J. B., *Biochem. J.*, **90**, 225 (1964).
10. SINGER, T. P., AND CREMONA, T., in W. A. WOOD (Editor) *Methods in enzymology*, Vol. IX, Academic Press, New York 1966, p. 302.
11. COX, G. B., NEWTON, N. A., GIBSON, F., SNOSWFLL, A. M. AND HAMILTON, J. A., *Biochem. J.*, **117**, 551 (1970).
12. SMITH, L., in I. C. GUNSALUS AND R. Y. STANIER (Editors) *The bacteria*, Vol. II, Academic Press, New York, 1961, p 365.
13. MARQUIS, R. E., *J. Bacteriol.*, **89**, 1453 (1965).
14. KEPES, A., *C. R. Acad. Sci.*, **244**, 1550 (1957).
15. TARMY, E. M., AND KAPLAN, N. O., *J. Biol. Chem.*, **243**, 2587 (1968).
16. JONES, C. W., AND REDFEARN, E. R., *Biochim. Biophys. Acta* **113**, 467 (1966).
17. SCARBOROUGH, G. A., RUMLEY, M. K., AND KENNEDY, E. P. *Proc. Nat. Acad. Sci. U. S. A.*, **60**, 951 (1968).
18. PAVLASOVA, E., AND HAROLD, F. M., *J. Bacteriol.*, **98**, 198 (1969).
19. WEST, I. C., *Biochem. Biophys. Res. Commun.*, **41**, 655 (1970).
20. KABACK, H. R., AND BARNES, E. M., JR., *J. Biol. Chem.*, **246**, 5523 (1971).

Reprinted from *J. Biol. Chem.*, **246**, 5523–5531 (1971)

Mechanisms of Active Transport in Isolated Membrane Vesicles

II. THE MECHANISM OF ENERGY COUPLING BETWEEN D-LACTIC DEHYDROGENASE AND
β-GALACTOSIDE TRANSPORT IN MEMBRANE PREPARATIONS FROM *ESCHERICHIA COLI*[*]

(Received for publication, April 1, 1971)

H. R. KABACK AND EUGENE M. BARNES, JR.[‡]

From the Roche Institute of Molecular Biology, Nutley, New Jersey 07110

15

SUMMARY

The results presented in this paper provide preliminary evidence for the concept that the "carriers" of the D-lactic dehydrogenase-coupled transport systems in isolated membrane vesicles from *Escherichia coli* may be electron transfer intermediates.

Initial rates of lactose transport and D-lactic dehydrogenase activity respond identically to temperature and both processes have the same activation energy of 8400 cal per mole. The steady state levels of lactose accumulation at a variety of temperatures represent equilibrium states in which there is a balance between influx and efflux. This balance can be easily influenced by raising or lowering the temperature. Temperature-induced efflux is a saturable process with an apparent affinity constant that is approximately 60 times higher than the affinity constant for influx determined under the same experimental conditions. The apparent maximum velocity of temperature-induced efflux, on the other hand, is the same as that of influx. Potassium cyanide also induces a saturable efflux phenomenon which has an apparent K_m that is much higher than that of the influx process.

p-Chloromercuribenzoate inhibits D-lactic dehydrogenase-coupled transport of lactose, galactose, arabinose, glucuronate, glucose-6-P, proline, glutamic acid, serine, alanine, tyrosine, lysine, and tryptophan, and inhibition of each system by *p*-chloromercuribenzoate is reversed by dithiothreitol. Furthermore, *p*-chloromercuribenzoate inhibits temperature-induced efflux of intramembranal lactose, exchange of external lactose with [^{14}C]lactose in the intramembranal pool, and lactose efflux induced by 2,4-dinitrophenol. Inhibition of these experimental parameters and of D-lactic dehydrogenase by *p*-chloromercuribenzoate is reversed by dithiothreitol.

Reduction of the respiratory chain between D-lactic dehydrogenase and cytochrome b_1 is responsible for carrier-mediated efflux of lactose. Anaerobiosis, cyanide, and 2-heptyl-4-hydroxyquinoline-*N*-oxide, each of which inhibits electron transfer after cytochrome b_1, cause marked efflux. Amytal causes slow efflux, and oxamate and *p*-chloromercuri-benzoate do not cause efflux despite marked inhibition of D-lactic dehydrogenase and the initial rate of lactose transport.

Addition of amino acids which are also transported by D-lactic dehydrogenase-coupled transport mechanisms results in little or no inhibition of lactose transport.

These findings are discussed in terms of a conceptual working model in which the carriers are depicted as electron transfer intermediates between D-lactic dehydrogenase and cytochrome b_1.

The data presented in the first paper in this series (2) indicate that the site of energy coupling of D-lactic dehydrogenase to active transport lies between the primary dehydrogenase and cytochrome b_1, the first cytochrome in the respiratory chain of *Escherichia coli*. In addition, the possibility that the "carriers" may be electron transfer intermediates between D-lactic dehydrogenase and cytochrome b_1 was suggested.

The experiments presented in this paper provide further preliminary evidence for a possible electron transfer nature of the transport-specific components of the D-lactic dehydrogenase-coupled transport systems.

METHODS

Membrane Preparations—*E. coli* ML 308-225[1] (i$^-$z$^-$y$^+$a$^+$) was grown on Medium A (3) containing 1% disodium succinate (hexahydrate); *E. coli* ML 30[1] (i$^+$z$^+$y$^+$a$^+$) on Medium A containing 0.5% glucuronic acid or arabinose as indicated; *E. coli* ML 3[1] (i$^+$z$^+$y$^-$a$^-$) on Medium A containing 0.5% galactose; and *E. coli* GN-2[2] (i$^-$z$^+$y$^+$a$^+$; enzyme I$^-$) on Medium 63 (4) containing 0.2% glucose-6-P. Membrane vesicles were prepared from these cells as described previously (5–7).

Transport Studies—Assays for lactose and amino acid uptake were carried out as reported previously (7–9). Glucuronate, arabinose, galactose, and glucose-6-P uptake studies were performed exactly as described for lactose and amino acid uptake with final concentrations of glucuronate, arabinose, galactose, and glucose-6-P of 0.1, 0.3, 0.2, and 0.02 mM, respectively.

[*] A preliminary report of portions of this work has been submitted for publication (1).

[‡] Postdoctoral fellow of the American Cancer Society (PF 545). Present address, Department of Biochemistry, Baylor College of Medicine, Houston, Texas 77045.

[1] These strains were obtained from Dr. T. H. Wilson.

[2] This mutant was reisolated from a culture provided by Dr. L. Heppel by selecting a glucose-negative colony from a MacConkey plate (*Difco Manual*, Difco Laboratories, Detroit, 1953, p. 131).

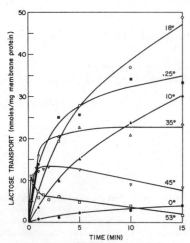

FIG. 1. Effect of temperature on the time course of lactose uptake. Aliquots (25 μl) of membranes prepared from *E. coli* ML 308-225 containing 0.158 mg of membrane protein were diluted to a final volume of 50 μl, containing, in final concentrations, 50 mM potassium phosphate (pH 6.6) and 10 mM magnesium sulfate. Lithium D-lactate and [1-¹⁴C]lactose (14.9 mCi per mmole) at final concentrations of 20 and 0.4 mM, respectively, were added to the reaction mixtures at 0° and they were immediately transferred to water baths at the temperatures shown. The incubations were carried out for the times indicated, the reactions were terminated, and the samples were assayed as described previously (8, 9). Each experimental point was corrected for a control sample obtained as described previously (5). Incubations were carried out at the following temperatures: 0° (●), 10° (▲), 18° (○), 25° (■), 35° (△), 45° (▽), and 53° (□).

D-Lactic Dehydrogenase Assays—The chromatographic assay utilizing D-[¹⁴C]lactate has been reported (8).

Oxygen Uptake Measurements—Rates of D-lactate-dependent oxygen uptake were measured with a Clark oxygen electrode as described previously (2).

Materials—In addition to [1-¹⁴C]lactose and the [¹⁴C]amino acids used, the sources of which have been reported (8, 9), potassium D-[U-¹⁴C]glucuronate (25 mCi per mmole) and D-[U-¹⁴C]glucose-6-P (127 mCi per mmole) were obtained from Amersham-Searle, and D-[1-¹⁴C]arabinose (10 mCi per mmole) and D-[U-¹⁴C]galactose (25 mCi per mmole) were obtained from New England Nuclear.

All other materials used in these experiments were of reagent grade and were obtained from commercial sources.

RESULTS

Effect of Temperature on D-Lactic Dehydrogenase Activity and β-Galactoside Transport—Fig. 1 represents time courses of lactose uptake by ML 308-225 membrane vesicles at a variety of temperatures. As shown, initial rates of uptake increase with temperature up to 53°, whereas the steady state level of lactose accumulation at 15 min increases from 0–18° and then decreases above 18°. At 53°, membranes take up lactose very rapidly for 15 sec and then lose radioactivity such that, by 1 min, approximately 50% of the radioactive lactose that had been accumulated in 15 sec is lost. After 1 min, loss of radioactivity continues, but

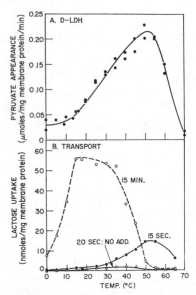

FIG. 2. *A*, effect of temperature on the initial rate of D-lactic dehydrogenase (*D-LDH*). Aliquots (25 μl) of ML 308-225 membranes containing 0.21 mg of membrane protein were diluted to a final volume of 50 μl, containing, in final concentrations, 50 mM potassium phosphate (pH 6.6), and 10 mM magnesium sulfate. Lithium D-[1-¹⁴C]lactate (0.1 mCi per mmole) at a final concentration of 20 mM was added to the reaction mixtures at 0° and they were immediately transferred to water baths at the temperatures shown. Samples assayed at 0–20° were incubated for 30 min; those assayed at temperatures above 20° were incubated for 10 min. The reaction rates were linear at the time each sample was assayed as determined by independent experiments. The reactions were terminated by the addition of 5 μl of concentrated formic acid. Each sample was then assayed chromatographically for [¹⁴C]pyruvate as described previously (8). *B*, effect of temperature on the initial rate of lactose transport and on the steady state level of lactose accumulation. Assays were carried out with ML 308-225 membrane vesicles as described in the legend to Fig. 1 with the exception that 1 mM [1-¹⁴C]lactose (14.0 mCi per mmole) was used. *15 MIN.* (○- - -○), lactose accumulation by samples incubated at the temperatures given for 15 min. *15 SEC.* (●——●), rate of lactose uptake by samples incubated at the temperatures given for 15 sec. The initial rates of uptake by samples incubated in the absence of D-lactate (*20 SEC: NO ADD.*, △) were also assayed.

at a much slower rate. These data are similar to studies carried out with whole cells (10).

The initial rate of D-lactic dehydrogenase activity increases very slightly as the temperature is raised from 0° to approximately 15°, and then increases more markedly above 15° (Fig. 2A). From 15° to approximately 50°, the reaction rate increases essentially linearly, and then decreases abruptly at temperatures exceeding 55°. In Fig. 2B, initial rates of lactose transport (*15 SEC.*) and steady state levels of lactose accumulation (*15 MIN.*) in the presence of D-lactate are plotted as a function of temperature. The steady state level of lactose accumulated in 15-min incubations exhibits a broad peak from 15–35° with a maximum at approximately 18°. The initial rate of lactose uptake (*15 SEC.*) is optimal at 50–55°, and the initial rates of lactose trans-

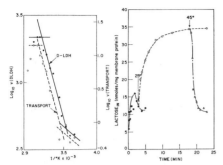

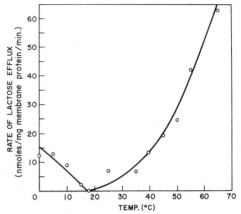

Fig. 3 (left). Arrhenius plots (11) of the initial rates of lactose transport and D-lactic dehydrogenase (D-LDH) activities. The data shown were obtained from Fig. 2, A and B. TRANSPORT (O- - -O) corresponds to 15 SEC. (●——●) in Fig. 2B, and D-LDH (●——●) to D-LDH (●——●) in Fig. 2A.

Fig. 4 (right). Effect of temperature shifts on the steady state intramembranal level of lactose accumulation. Reaction mixtures containing ML 308-225 membrane vesicles were prepared as described in the legend to Fig. 1. The samples were incubated at 45° in the presence of lithium D-lactate and [1-14C]lactose (14.9 mCi per mmole) at final concentrations of 20 and 0.4 mM, respectively. Samples were assayed at the times shown by methods described previously (8, 9). At 3 min (indicated by arrow), samples were transferred to a water bath at 25°, and assayed at the times shown. At 18 min (indicated by arrow), the remaining samples were transferred back to the 45° water bath and assayed at the times shown. ●——●, initial uptake at 45°; O- - -O, uptake at 25° after incubation at 45°; △—·—△, uptake at 45° after final shift to 45°.

port and D-lactic dehydrogenase activity have similar, if not identical, temperature profiles. Membranes incubated in the absence of D-lactate (20 SEC: NO ADD.) have almost negligible rates of uptake.

The similarity of the effect of temperature on both D-lactic dehydrogenase activity and the initial rate of lactose transport in the presence of D-lactate is further emphasized in Fig. 3. Clearly, the initial rates of transport and D-lactic dehydrogenase activity respond almost identically to temperature. Increments in the $\log_{10}$ of the velocities of both activities from approximately 20–45° are identical, yielding an activation energy of 8400 cal per mole. Furthermore, there are discontinuities in both activities at 10–15° and at 45–50°. The small discrepancies in the temperatures at which the discontinuities occur (from 3–5°) are within experimental error.

Data given in Fig. 4 show that the steady state levels of lactose accumulated at 25° or 45° represent equilibrium states which can be shifted by changing temperature. At 45°, [14C]lactose is taken up very rapidly by the vesicles which achieve a steady state level of lactose accumulation of approximately 12 nmoles per mg of membrane protein (corresponding to an intramembranal concentration of about 5.5 mM) within about 3 min. When the temperature is lowered to 25°, the membranes accumulate lactose to a steady state level of 31 to 34 nmoles per mg of membrane protein (an intramembranal concentration of 14 to 16 mM) approximately 5 min after the temperature shift. If the temperature is again raised to 45°, the vesicles re-equilibrate at the same steady state level as that observed during the initial incubation at 45° within 2 min.

The temperature dependence of efflux is shown in Fig. 5.

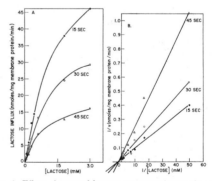

Fig. 5. Rate of lactose efflux as a function of temperature. Reaction mixtures containing ML 308-225 membranes were prepared as described in the legend to Fig. 1. The samples were incubated at 18° in the presence of lithium D-lactate and [1-14C]-lactose (14.9 mCi per mmole) at final concentrations of 20 and 1.0 mM, respectively. After 15 min, a number of control samples were assayed (60.1 nmoles per mg of membrane protein; average of five determinations), and the remaining samples were transferred to water baths at the temperatures given. Incubations were continued for 1 min, and the samples were assayed as described previously (8, 9). The differences between the control samples and the samples incubated at the temperatures given (expressed as nanomoles per mg of membrane protein) are presented as a function of temperature.

Fig. 6. Effect of external lactose concentration on the rates of lactose uptake at 45°. A, reaction mixtures containing ML 308-225 membrane vesicles prepared as described in the legend to Fig. 1 were incubated at 45° in the presence of lithium D-lactate at 20 mM final concentration and [1-14C]lactose (14.9 mCi per mmole) in the concentrations indicated. The reactions were terminated at the times shown and the samples were assayed as described previously (8, 9). 15 SEC (●——●), samples assayed at 15 sec; 30 SEC (O——O), samples assayed at 30 sec; 45 SEC (△——△), samples assayed at 45 sec. B, data from A plotted by the method of Lineweaver and Burk (12).

Membranes were first loaded with [14C]lactose by incubation at 18° for 15 min in the presence of D-lactate. The reaction mixtures were then shifted to the temperatures shown, and loss of radioactivity in 1 min was measured. Initial rates of efflux are

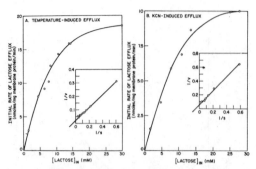

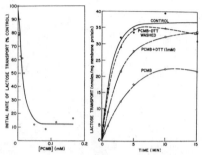

FIG. 7. Kinetics of lactose efflux. *A*, effect of internal lactose concentration on the rates of lactose efflux at 45°. Reaction mixtures containing ML 308-225 membranes were prepared as described in the legend to Fig. 1. The samples were incubated at 20° in the presence of lithium D-lactate at 20 mM final concentration and the external [1-¹⁴C]lactose (14.9 mCi per mmole) concentrations given in Fig. 6*A*. After 15 min, one set of control samples was assayed and another set of identical samples was transferred to a water bath at 45°. After 1 min, the samples were assayed as described previously (8, 9). The differences between the control samples and the samples incubated at 45° for 1 min (expressed as nanomoles per mg of membrane protein) are presented as a function of the intramembranal lactose concentration prior to the shift to 45°. By methods described previously (6), it was determined that each milligram of membrane protein contained approximately 2.2 μl of intramembranal fluid. It was assumed that all of the intramembranal lactose was in free solution. *Inset*, data plotted by the method of Lineweaver and Burk (12). *B*, effect of internal lactose concentration on the rates of lactose efflux in the presence of potassium cyanide at 20°. Reaction mixtures containing ML 308-225 membrane vesicles were prepared as described in the legend to Fig. 1. The samples were incubated at 20° in the presence of lithium D-lactate at 20 mM final concentration and the external [1-¹⁴C]lactose (14.9 mCi per mmole) concentrations given in Fig. 6*A*. After 15 min, one set of control samples was assayed. Potassium cyanide (0.01 M, final concentration) was then added to another set of identical samples and the incubations were continued at 20° for 1 min. The reactions were then terminated and the samples were assayed as described previously (8, 9). The differences between the control samples and the samples incubated for another minute in the presence of cyanide (expressed as nanomoles per mg of membrane protein) are presented as a function of the intramembranal lactose concentration prior to the addition of cyanide. *Inset*, Lineweaver-Burk plot.

FIG. 8 (*left*). Effect of PCMB on the initial rate of lactose transport. Reaction mixtures containing ML 308-225 membranes and PCMB in the concentrations shown were prepared as described in the legend to Fig. 1. Incubations were carried out at 20° for 1.5 min in the presence of lithium D-lactate and [1-¹⁴C]lactose (14.9 mCi per mmole) at 20 and 0.4 mM final concentrations, respectively, after a 3-min preliminary incubation with the concentrations of PCMB shown.

FIG. 9 (*right*). Effect of dithiothreitol on PCMB inhibition of lactose transport. Three aliquots (0.5 ml) of ML 308-225 membranes containing 4.2 mg of membrane protein were diluted to a final volume of 1 ml containing, in final concentrations, 50 mM potassium phosphate (pH 6.6) and 10 mM magnesium sulfate. PCMB (8.3 × 10⁻⁵ M, final concentration) was added to two of the samples. After a 5-min incubation at room temperature (approximately 20°), the samples were centrifuged in the cold for 15 min at approximately 20,000 × *g*. The supernatants were discarded, and the pellets were washed twice in the following solutions: control (●——●), sample incubated in the absence of PCMB, washed in 0.1 M potassium phosphate (pH 6.6); *PCMB-DTT WASHED* (△---△), sample incubated with PCMB, washed in 0.1 M potassium phosphate (pH 6.6) containing 1 mM dithiothreitol; *PCMB* (○---○), sample incubated with PCMB, washed in 0.1 M potassium phosphate (pH 6.6). After the second wash, each pellet was resuspended to 0.5 ml in 0.1 M potassium phosphate (pH 6.6) and 25-μl aliquots were assayed at 20° for the times given. The reactions were carried out as described previously (8, 9) in the presence of lithium D-lactate and [1-¹⁴C]-lactose (14.9 mCi per mmole) at final concentrations of 20 and 0.4 mM, respectively. The sample treated with PCMB and washed in the absence of PCMB was also assayed in the presence of 1 mM dithiothreitol (final concentration) (*PCMB+DTT* [*1 mM*], □---□).

linear for at least 1 min at each temperature studied (see temperature shift from 25–45° in Fig. 4). There is a sharp efflux minimum at 18°. Thus the relationship between the rate of efflux and temperature is approximately the inverse of the relationship between the steady state level of lactose accumulation and temperature (see Figs. 1 and 2*B*). Membranes are not irreversibly damaged by 1-min incubations at temperatures up to and including 50° (data not shown). Above 55°, there is rapid and irreversible loss of both D-lactic dehydrogenase and transport activities.

Kinetics of Efflux—Before presenting kinetic data on temperature-induced efflux, it is necessary to characterize the kinetics of uptake at 45° as a basis for comparison. The initial rate of lactose uptake at 45° shows saturation kinetics at each time point studied (Fig. 6, *A* and *B*). Since uptake at 45° is so rapid (see Fig. 1), it is technically difficult to obtain time points during the linear phase of uptake. For this reason, rates of uptake were

measured at 15, 30, and 45 sec as a function of external lactose concentration. In Fig. 6*B*, the data have been plotted by the method of Lineweaver and Burk (12). Each function yields a K_m of 0.4 mM. With the data obtained from experiments carried out for 15 sec, a V_{max} of 50 nmoles per mg of membrane protein per min is obtained. This value is in good agreement with previous kinetic studies carried out at 25° (9). The V_{max} at 45° is higher than that obtained at 25° (approximately 20 nmoles per mg of membrane protein per min at 25° (9)).

The rate of lactose efflux at 45° also exhibits saturation kinetics when studied as a function of intramembranal lactose concentration (Fig. 7*A*). Vesicles were first loaded with lactose by incubation at 20° in the presence of D-lactate and various concentrations of [¹⁴C]lactose and subsequently transferred to 45°. The data in the *inset* yield a K_m of 25 mM and a V_{max} of 50 nmoles per mg of membrane protein per min. Compared to influx at 45°, the K_m of efflux is approximately 60 times higher, but the V_{max} is almost identical.

Addition of energy poisons to either membrane vesicles (8) or

TABLE I

Inhibition of sugar and amino acid transport by PCMB and reversal
by dithiothreitol

Membranes prepared from E. coli ML 308-225 were assayed for
the uptake of lactose, proline, glutamic acid, serine, alanine,
tyrosine, lysine, and tryptophan. For glucuronate and arabinose
uptake, membranes were prepared from E. coli ML 30 grown on
glucuronate or arabinose, respectively, as described under
"Methods." In order to avoid complications due to galactose
transport via the β-galactoside transport system, membranes prepared from E. coli ML 3 (y⁻) grown on galactose as described
under "Methods" were used for galactose uptake. These membrane preparations did not transport either lactose or methyl-β-
thio-D-galactopyranoside in the presence of D-lactate. Studies on
glucose-6-P transport were performed with membrane preparations from E. coli GN-2 grown on glucose-6-P as described under
"Methods." Membrane vesicles prepared from the organisms
described above were treated exactly as described in the figure
legend to Fig. 9: I. Control—membranes were washed twice with
0.1 M potassium phosphate buffer (pH 6.6); II. PCMB—membranes treated with 8.3×10^{-5} M PCMB as described in Fig. 9 were
washed twice with potassium phosphate buffer (pH 6.6); III.
PCMB-dithiothreitol—membranes treated with 8.3×10^{-5} M
PCMB were washed twice with 0.1 M potassium phosphate buffer
(pH 6.6) containing 1 mM dithiothreitol. After washing, the
membrane pellets were treated as described in Fig. 9. Uptake
studies were carried out for 1 min at 20° as described previously
(8, 9) with the sugars and amino acids given at the concentrations
and specific activities given under "Methods" or reported previously (8, 9).

| Transport substrate | Initial rate of transport | | | Inhibition by PCMB (I − II/ I × 100) | Reversal of PCMB inhibition by dithiothreitol (III/I × 100) |
	I. Control	II. PCMB	III. PCMB-dithio-threitol		
	nmoles/mg membrane protein/min			%	%
Lactose	16.2	4.5	15.0	72.2	93.0
Glucuronate	3.78	0.82	3.95	78.0	104.0
Galactose	2.25	0.75	2.03	65.0	94.0
Arabinose	2.96	0.81	1.81	63.0	61.0
Glucose-6-P	0.49	0.073	0.398	85.0	81.0
Proline	2.32	0.394	1.830	73.0	80.0
Glutamic acid	1.12	0.60	1.09	46.5	97.0
Serine	5.84	1.40	5.64	76.0	97.0
Alanine	0.83	0.21	0.44	75.0	53.0
Tyrosine	2.48	0.576	2.44	76.8	98.5
Lysine	1.06	0.89	1.10	16.0	104.0
Tryptophan	5.42	2.30	5.1	57.5	94.0

whole cells (10, 13, 14) which have been loaded first with lactose
causes rapid efflux. In the experiment described in Fig. 7B,
membranes were first loaded with [¹⁴C]lactose at 20°, and potassium cyanide was then added. Rapid efflux is induced (see Fig.
13 in addition to Fig. 7B), and, furthermore, the rate of efflux
exhibits saturation kinetics when plotted as a function of intramembranal lactose concentration. The data yield a reciprocal
plot (see inset) from which a K_m of 25 mM is obtained. The
V_{max} of cyanide-induced efflux at 25° is 20 nmoles per mg of
membrane protein per min.

Effect of Sulfhydryl Reagents on Carrier-mediated Transport—
The initial rate of lactose transport by ML 308-225 membrane

TABLE II

Inhibition of D-lactate oxidation by ML 308-225 membranes and
reversal by dithiothreitol

The samples of membrane vesicles prepared from E. coli ML
308-225 used in Table I for the transport of lactose, proline,
glutamic acid, serine, alanine, tyrosine, lysine, and tryptophan
were assayed for D-lactate oxidation with the oxygen electrode as
described previously (2).

Sample and treatment	D-Lactate oxidation	Control
	ng atom O₂/min/ mg membrane protein	%
I. Control	183	
II. 8.3×10^{-5} M PCMB Washed with KPO₄	57.7	31.5
II. +1 mM DTTᵃ	217	118.5
III. 8.3×10^{-5} M PCMB Washed with KPO₄ + DTT	211	115
III. +1 mM DTT	217	118

ᵃ DTT, dithiothreitol.

vesicles in the presence of D-lactate is sensitive to PCMB[3] (Fig. 8),
and decreases markedly with PCMB concentrations up to approximately 0.05 mM. NEM is also an effective inhibitor of
galactoside transport by membrane vesicles (9) and PCMB inhibits β-galactoside uptake into whole cells (15).

The inhibitory effect of PCMB on D-lactate-dependent concentrative uptake of lactose is reversed by dithiothreitol. As described in Fig. 9, membranes were treated with PCMB and then
washed with solutions to which dithiothreitol either had or had
not been added. The time course of lactose uptake by membranes which had been treated with PCMB and subsequently
with dithiothreitol is insignificantly different from the control
preparation. On the other hand, the sample which had been
treated with PCMB but not dithiothreitol shows marked inhibition of the initial rate of uptake and significant inhibition of the
steady state level of lactose accumulation. On addition of
dithiothreitol, the rate and extent of lactose uptake by this preparation are almost doubled.

The inhibitory effect of PCMB on β-galactoside transport is
general for all D-lactic dehydrogenase-coupled transport systems,
and, in each case, PCMB inhibition is virtually completely reversed by treating vesicles with dithiothreitol (Table I). Initial
rates of transport of lactose, galactose, arabinose, glucuronate,
glucose-6-P, proline, glutamic acid, serine, alanine, tryptophan, and tyrosine are inhibited by PCMB to varying extents.
As shown in the last column of Table I, after membrane preparations are washed in phosphate buffer containing dithiothreitol,
the inhibition observed is almost completely abolished. Detailed
studies on the effects of PCMB on each of these transport systems
will be published at a later date.[4]

The data presented in Table II represent measurements of
D-lactate oxidation by the same membrane preparations used for
most of the transport studies presented in Table I. PCMB
produces approximately 70% inhibition of D-lactate-induced
respiration; this inhibition is completely reversed by subsequent
treatment with dithiothreitol. Moreover, when dithiothreitol is
added to membranes which had been treated with PCMB but not

[3] The abbreviations used are: PCMB, p-chloromercuribenzoate;
NEM, N-ethylmaleimide.
[4] F. J. Lombardi and H. R. Kaback, manuscripts in preparation.

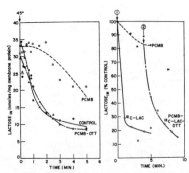

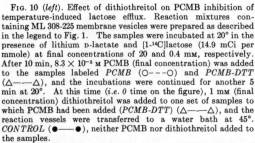

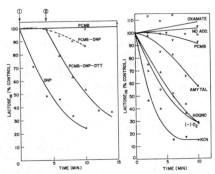

Fig. 10 (*left*). Effect of dithiothreitol on PCMB inhibition of temperature-induced lactose efflux. Reaction mixtures containing ML 308-225 membrane vesicles were prepared as described in the legend to Fig. 1. The samples were incubated at 20° in the presence of lithium D-lactate and [1-¹⁴C]lactose (14.9 mCi per mmole) at final concentrations of 20 and 0.4 mM, respectively. After 10 min, 8.3×10^{-5} M PCMB (final concentration) was added to the samples labeled *PCMB* (O---O) and *PCMB-DTT* (△—△), and the incubations were continued for another 5 min at 20°. At this time (*i.e.* 0 time on the figure), 1 mM (final concentration) dithiothreitol was added to one set of samples to which PCMB had been added (*PCMB-DTT*) (△—△), and the reaction vessels were transferred to a water bath at 45°. *CONTROL* (●—●), neither PCMB nor dithiothreitol added to the samples.

Fig. 11 (*right*). Effect of dithiothreitol on PCMB inhibition of lactose exchange. Reaction mixtures containing ML 308-225 membranes were prepared as described in the legend to Fig. 1. The samples were incubated at 20° in the presence of lithium D-lactate and [1-¹⁴C]lactose 14.9 mCi per mmole) at final concentrations of 20 and 0.4 mM, respectively. After 10 min, 8.3×10^{-5} M (final concentration) PCMB was added to the samples labeled *PCMB* (●---●) and *PCMB-¹²C-LAC-DTT* (O—O), and the incubations were continued for another 5 min at 20°. Nonradioactive lactose was then added to all of the samples at a final concentration of 2 mM (*Arrow 1*), and the incubations were continued at 20° for the times shown. After 5 min, 1 mM dithiothreitol (final concentration) was added to the appropriate samples containing PCMB (*Arrow 2*). Reactions were terminated and samples were assayed as described previously (8, 9). ¹²C-LAC (O—O), [¹²C]lactose added at zero time (*Arrow 1*) in the absence of either PCMB or dithiothreitol. The data are presented as a percentage of the control samples after prior loading for 15 min.

Fig. 12 (*left*). Effect of dithiothreitol on PCMB inhibition of 2,4-dinitrophenol (*DNP*)-induced lactose efflux. Reaction mixtures containing ML 308-225 membranes were prepared as described in the legend to Fig. 1. The samples were incubated at 20° in the presence of lithium D-lactate and [1-¹⁴C]lactose (14.9 mCi per mmole) at final concentrations of 20 and 0.4 mM, respectively. After 10 min, 8.3×10^{-5} M PCMB (final concentration) was added to the samples labeled *PCMB*, *PCMB-DNP*, and *PCMB-DNP-DTT*, and the incubations were continued for another 5 min at 20°. 2,4-Dinitrophenol (1 mM, final concentration) was then added to the appropriate samples (*Arrow 1*), and the incubations were continued for the times shown. After 5 min, 1 mM dithiothreitol (*DTT*) (final concentration) was added to the samples containing PCMB and 2,4-dinitrophenol as indicated (*Arrow 2*). The reactions were terminated and the samples were assayed as described previously (8, 9). *DNP* (●—●), DNP added at zero time (*Arrow 1*) in the absence of either PCMB or dithiothreitol; *PCMB* (O—O), PCMB added at −5 min, no further additions; *PCMB-DNP* (O---O), PCMB added at −5 min and 2,4-dinitrophenol added at zero time (*Arrow 1*); *PCMB-DNP-DTT* (△—△), PCMB added at −5 min, 2,4-dinitrophenol added at zero time (*Arrow 1*), and dithiothreitol added at 5 min (*Arrow 2*).

Fig. 13 (*right*). Effect of various electron transport inhibitors on lactose efflux. Reaction mixtures containing ML 308-225 membrane vesicles were prepared as described in the legend to Fig. 1. The samples were incubated at 20° in the presence of lithium D-lactate and [1-¹⁴C]lactose (14.9 mCi per mmole) at final concentrations of 20 and 0.4 mM, respectively. After 15 min, one of the inhibitors shown was added at the following final concentrations: potassium cyanide (■—■), 0.01 M; 2-heptyl-4-hydroxyquinoline-*N*-oxide (*HOQNO*, △—△), 8×10^{-5} M; sodium amytal (▲—▲), 0.01 M; PCMB (▽—▽), 8.3×10^{-5} M; potassium oxamate (□—□), 6 mM. For the samples labeled (−)O₂ (O—O), the incubations were carried out in tubes fitted with rubber stoppers through which a needle was inserted. During the initial 15-min incubation, the tubes were gassed with air, and, at 15 min, the gas mixture was changed to argon which was continued for the remainder of the incubation. Control samples (*NO ADD.*, ●—●) were incubated under identical conditions with the exception that no inhibitor was added after the initial 15-min incubation. Samples were assayed at the times indicated by methods described previously (8, 9).

dithiothreitol, there is complete restoration of activity. Each transport system studied in Table I and D-lactate oxidation is also sensitive to NEM; however, with this sulfhydryl reagent, the effects are not reversed with dithiothreitol.

Addition of PCMB to membrane vesicles first loaded at 20° causes marked inhibition of lactose efflux at 45° (Fig. 10). When dithiothreitol is added just before the temperature shift, however, the rate of efflux is indistinguishable from the control.

PCMB also inhibits exchange of external lactose with [¹⁴C]lactose present in the intramembranal pool (Fig. 11) and lactose efflux induced by the addition of 2,4-dinitrophenol to previously loaded membrane vesicles (Fig. 12). Moreover, PCMB inhibition of both of these phenomena is reversed by addition of dithiothreitol (Figs. 11 and 12).

D-Lactate-stimulated concentrative uptake of lactose and the loss of radioactive lactose evoked by raising the temperature or adding [¹²C]lactose or 2,4-dinitrophenol are also inhibited by

NEM; however, the inhibition observed is not reversed by dithiothreitol. These experiments are not presented in detail here.

Effect of Anaerobiosis and Electron Transfer Inhibitors on Lactose Transport—The effects of anaerobiosis and a variety of electron transfer inhibitors on the ability of previously loaded membrane vesicles to retain [¹⁴C]lactose are shown in Fig. 13. These inhibitors were selected because their sites of inhibition in the respiratory chain of *E. coli* have been well documented. Cyanide inhibits cytochrome a_2 (16), 2-heptyl-4-hydroxyquinoline-*N*-

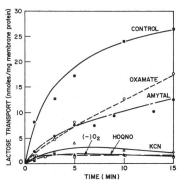

FIG. 14. Effect of various electron transport inhibitors on the time course of lactose uptake. Reaction mixtures containing ML 308-225 membranes were prepared as described in the legend to Fig. 1. In addition, where indicated, the reaction mixtures contained the inhibitor given in the figure at the concentration stated in the legend to Fig. 13. For the samples labeled $(-)O_2$ ($\square$–––$\square$), the incubations were carried out in tubes fitted with rubber stoppers through which a needle was inserted. The samples were gassed with argon for 5 min before the addition of D-lactate and [14C]-lactose by injection through the stopper. The incubations were then continued under argon. At the times indicated, the reactions were terminated and the samples were assayed by methods described previously (8, 9). The control samples ($\blacksquare$—$\blacksquare$) contained none of the inhibitors and were incubated under aerobic conditions (room air). *HOQNO*, 2-heptyl-4-hydroxyquinoline-*N*-oxide.

oxide inhibits between cytochrome b_1 and cytochrome a_2 (16), amytal inhibits at the flavin level (16), and oxamate, as shown in the previous paper (2), is an effective and specific inhibitor of the membrane-bound D-lactic dehydrogenase. Potassium cyanide, anaerobiosis, and 2-heptyl-4-hydroxyquinoline-*N*-oxide cause marked efflux of [14C]lactose from the membranes. On the other hand, amytal is only slightly effective, and PCMB and oxamate do not cause any loss of radioactivity. The concentrations of each inhibitor used here produce at least 70% inhibition of the initial rate of D-lactic dehydrogenase activity (2) and lactose uptake (Fig. 14). Removal of oxygen from the reaction mixture also blocks D-lactate oxidation and lactose transport (9).

The effects of anaerobiosis and the same electron transfer inhibitors on the time course of lactose uptake in the presence of D-lactate are shown in Fig. 14. The inhibitors were used at the same concentrations as in the experiment presented in Fig. 13. All of the inhibitors, as well as removal of oxygen, markedly inhibit the initial rate of lactose uptake (see also Fig. 9, PCMB). However, with oxamate, amytal, and PCMB (see Fig. 9), although the initial rate of uptake is markedly inhibited, the membranes accumulate significant lactose and begin to approximate the control samples by 15 min. With anaerobiosis, potassium cyanide, or 2-heptyl-4-hydroxyquinoline-*N*-oxide, inhibition is profound throughout the time course of the incubation.

Effect of Amino Acids on Time Course of Lactose Uptake—The experiments presented in Fig. 15, *A* and *B*, show time courses of lactose uptake in the presence of D-lactate and a mixture of 16 amino acids. The concentrations of amino acid shown refer to the individual concentration of each of 16 amino acids in the reaction mixtures, and, even at the lowest concentration, each amino

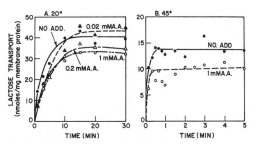

FIG. 15. Effect of amino acids on the time course of lactose uptake. Reaction mixtures containing ML 308-225 membranes were prepared as described in the legend to Fig. 1. In addition, where indicated, the reaction mixtures contained each of the following amino acids (*A.A.*) in the final individual concentrations given: proline, glutamic acid, aspartic acid, lysine, serine, glycine, alanine, threonine, tryptophan, leucine, isoleucine, valine, histidine, tyrosine, phenylalanine, and cysteine. The control samples (*NO ADD.*, $\bullet$——$\bullet$) contained no amino acids. At the times shown, the reactions were terminated and the samples were assayed by methods described previously (8, 9). *A*, lactose uptake at 20°; *B*, lactose uptake at 45°.

acid was present at a saturating concentration for its transport system (8).[4] It can be seen that, at either temperature, the presence of amino acids causes only mild inhibition of either the rate or extent of lactose uptake.

DISCUSSION

The experimental findings presented in this and the previous paper are consistent with the conceptual working model presented in Fig. 16. Detailed experiments with other D-lactic dehydrogenase-coupled transport systems which will be published at a later date show that each system behaves in a manner qualitatively similar to that shown here for the β-galactoside transport system. In the mechanism presented, the carriers (in this specific case, the M protein (17–20)) are depicted as electron transfer intermediates which undergo reversible oxidation-reduction. As shown, in the oxidized state, the carrier has a high affinity site for ligand which it binds on the exterior surface of the membrane. Electrons coming ultimately from D-lactate through one or possibly more flavoproteins reduce a critical disulfide in the carrier molecule resulting in a conformational change. With this conformational change, the affinity of the carrier for its ligand is markedly decreased and ligand is released on the interior surface of the membrane. The reduced "sulfhydryl" form of the carrier is oxidized by cytochrome b_1 and electrons then flow through the remainder of the cytochrome chain to reduce molecular oxygen to water. The reduced form of the carrier can also "vibrate" and catalyze a low affinity, carrier-mediated, non-energy-dependent transport of ligand across the membrane.

Although no direct evidence has been presented which shows unequivocally that the carriers are electron transfer intermediates or that they are the only sulfhydryl-containing components of the respiratory chain between D-lactic dehydrogenase and cytochrome b_1, this formulation is consistent with all of the experimental observations presented and is the simplest conception possible.

The concept that the carriers may be electron transfer intermediates is supported primarily by experiments in which the

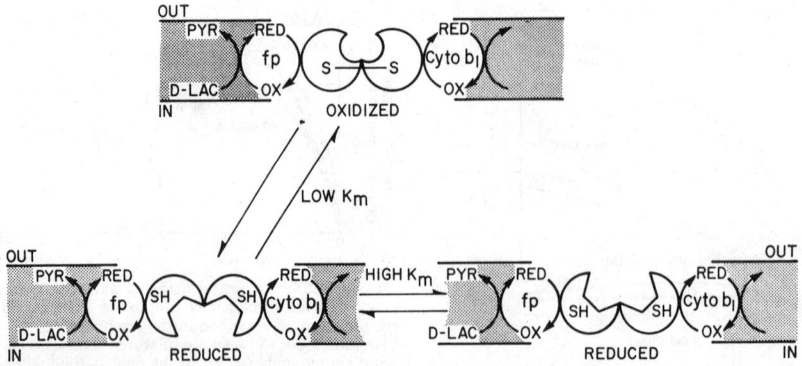

FIG. 16. Conceptual working model for D-lactic dehydrogenase-coupled transport systems. *D-LAC*, D-lactate; *PYR*, pyruvate; *fp*, flavoprotein; *Cyto b₁*, cytochrome b_1; *OX*, oxidized; *RED*, reduced. *OUT* signifies the outside surface of the membrane; *IN* signifies the inside surface. The hemispheres located between *fp* and *cyto b₁* represent the "carrier": ∪, a high affinity binding site and ∪, a low affinity binding site. The remainder of the cytochrome chain from cytochrome b_1 to oxygen has been omitted.

effects of anaerobiosis and electron transfer inhibitors on efflux and influx were studied. Since only anaerobiosis and those inhibitors which block electron transfer after the site of energy coupling induce efflux (Fig. 13), reduction of the electron transfer chain between D-lactic dehydrogenase and cytochrome b_1 must be responsible for efflux. The effect of anaerobiosis and the same inhibitors on the time course of lactose uptake (Fig. 14) is also consistent with this interpretation. Since removal of oxygen or addition of electron transfer inhibitors which inhibit after the site of energy coupling cause reduction of that site, membranes incubated under these conditions manifest profound inhibition of uptake throughout the time course of the experiment. On the other hand, inhibition before the site of energy coupling slows the rate of reduction of the energy-coupling site, but not its rate of oxidation by cytochrome b_1. Thus, vesicles incubated under these conditions manifest markedly diminished initial rates of uptake but eventually accumulate significant quantities of lactose.

As shown in this and the previous paper (2), D-lactate oxidation and the D-lactic dehydrogenase-dependent and -independent aspects of β-galactoside transport are inhibited by PCMB and NEM. Moreover, PCMB inhibition of each of these parameters is reversed by dithiothreitol. Since the site of energy coupling lies between D-lactic dehydrogenase and cytochrome b_1, and the site(s) of inhibition by sulfhydryl reagents is also between the dehydrogenase and cytochrome b_1, these data are consistent with the model presented. It must be emphasized that there may be many sulfhydryl-containing components in the respiratory chain between D-lactic dehydrogenase and cytochrome b_1. Furthermore, it is possible that the carriers are not obligatory intermediates in the respiratory chain, but reflect the oxidation-reduction potential at the site of energy coupling in an unknown manner. The mechanism presented is merely the simplest conception that accounts for the data presented.

In the model proposed, for each transport system, there should be a component between D-lactic dehydrogenase and cytochrome b_1 which has a binding site that is specific for that particular transport substrate. Evidence supporting this prediction is provided by experiments in which lactose transport was studied in the presence of a mixture of amino acids the concentrations of which were more than sufficient to saturate their respective transport systems. Little or no inhibition of either the initial rate of lactose uptake or the steady state level of lactose accumulation was observed under any of the conditions studied. Other data, the details of which will be published at a later date, indicate that the sum of the V_{max} values of all of the known D-lactic dehydrogenase-coupled transport systems in a given membrane preparation are comparable to the V_{max} of D-lactic dehydrogenase activity in that same membrane preparation.[4]

Acknowledgments—The authors would like to thank Mrs. Grace K. Kerwar for her excellent technical assistance during some of this work. They would also like to express their appreciation to Dr. T. H. Wilson for his gift of *E. coli* ML 308-225 and ML 3.

REFERENCES

1. KABACK, H. R., BARNES, E. M., JR., AND MILNER, L. S., *Proceedings of the Xth International Congress of Microbiology, Mexico City, 1970*, in press.
2. BARNES, E. M., JR., AND KABACK, H. R., *J. Biol. Chem.*, **246**, 5518 (1971).
3. DAVIS, B. D., AND MINGIOLI, E. S., *J. Bacteriol.*, **60**, 17 (1950).
4. RICKENBERG, H. V., COHEN, G. N., BUTTIN, G., AND MONOD, J., *Ann. Inst. Pasteur*, **91**, 829 (1956).
5. KABACK, H. R., *J. Biol. Chem.*, **243**, 3711 (1968).
6. KABACK, H. R., in F. BRONNER AND A. KLEINZELLER (Editors), *Current topics in membranes and transport, Vol. I*, Academic Press, New York, 1970, p. 36.
7. KABACK, H. R., in W. B. JAKOBY (Editor), *Methods in enzymology, Vol. XXII*, Academic Press, New York, 1971, p. 99.
8. KABACK, H. R., AND MILNER, L. S., *Proc. Nat. Acad. Sci. U. S. A.*, **66**, 1008 (1970).
9. BARNES, E. M., JR., AND KABACK, H. R., *Proc. Nat. Acad. Sci. U. S. A.*, **66**, 1190 (1970).
10. KEPES, A., AND COHEN, G. N., in I. C. GUNSALUS AND R. STANIER (Editors), *The bacteria, Vol. IV*, Academic Press, New York, 1962, p. 179.
11. ARRHENIUS, S. A., in J. S. FRUTON AND S. SIMMONDS, *General biochemistry*, John Wiley and Sons, Inc., New York, 1959, p. 263.

12. LINEWEAVER, H., AND BURK, D., *J. Amer. Chem. Soc.*, **56**, 658 (1934).
13. KEPES, A., in F. BRONNER AND A. KLEINZELLER (Editors), *Current topics in membranes and transport*, *Vol. I*, Academic Press, New York, 1970, p. 101.
14. KOCH, A. L., *Biochim. Biophys. Acta*, **79**, 177 (1964).
15. KEPES, A., *Biochim. Biophys. Acta*, **40**, 70 (1960).
16. COX, G. B., NEWTON, N. A., GIBSON, F., SNOSWELL, A. M., AND HAMILTON, J. A., *Biochem. J.*, **117**, 551 (1970).
17. FOX, C. F., AND KENNEDY, E. P., *Proc. Nat. Acad. Sci. U. S. A.*, **54**, 891 (1965).
18. FOX, C. F., CARTER, J. R., AND KENNEDY, E. P., *Proc. Nat. Acad. Sci. U. S. A.*, **57**, 698 (1968).
19. CARTER, J. R., FOX, C. F. AND KENNEDY, E. P., *Proc. Nat. Acad. Sci. U. S. A.*, **60**, 725 (1968).
20. JONES, T. H. D., AND KENNEDY, E. P., *J. Biol. Chem.*, **244**, 5981 (1969).

229

The Phosphotransferase System

IV

Editor's Comments on Papers 16 Through 20

16 **Kundig, Ghosh, and Roseman:** *Phosphate Bound to Histidine in a Protein as an Intermediate in a Novel Phosphotransferase System*

17 **Kundig, Kundig, Anderson, and Roseman:** *Restoration of Active Transport of Glycosides in* Escherichia coli *by a Component of a Phosphotransferase System*

18 **Tanaka and Lin:** *Two Classes of Pleiotropic Mutants of* Aerobacter aerogenes *Lacking Components of a Phosphoenolpyruvate-dependent Phosphotransferase System*

19 **Simoni, Levinthal, Kundig, Kundig, Anderson, Hartman, and Roseman:** *Genetic Evidence for the Role of a Bacterial Phosphotransferase System in Sugar Transport*

20 **Kaback:** *The Role of the Phosphoenolpyruvate-Phosphotransferase System in the Transport of Sugars by Isolated Membrane Preparations of* Escherichia coli

An important feature of the permease hypothesis was the distinction drawn between permeases and metabolic enzymes (Rickenberg et al., Paper 7; Cohen and Monod, Paper 8), a formulation that departed from earlier attempts to explain active transport in terms of enzyme activity. The phosphorylation theory, for example, pictured hexokinase or a similar enzyme as an active component in sugar transport in muscle and intestine, either "trapping" the sugars by phosphorylation as they entered the cell, or being intimately involved in the actual passage of the molecule across the membrane. Like Job, the phosphorylation theory passed through long periods of considerable difficulty without being completely forsaken (see Wilbrandt and Rosenberg, 1961). Its eventual demise was brought about by the demonstration that neither of the hydroxyl groups in positions 1 and 6 was necessary for active transport of hexoses in the intestine (Crane and Krane, 1956; Wilson and Landau, 1960).

It is ironic that after the many years of debate on the issue the participation of phosphorylation in transport was finally demonstrated through the efforts of a group that had set out to investigate a completely different problem. Kundig, Ghosh, and Roseman (Paper 16), searching for a phosphorylating enzyme involved in the metabolism of *N*-acetyl-mannosamine in bacteria, discovered the phosphotransferase system (PTS), which catalyzes the following reactions:

$$\text{phosphoenolpyruvate} + \text{HPr} \underset{\text{Mg}^{2+}}{\overset{\text{Enzyme I}}{\rightleftharpoons}} \text{phospho–HPr} + \text{pyruvate}$$

$$\text{phospho–HPr} + \text{sugar} \xrightarrow[\text{Enzyme II}]{} \text{sugar–phosphate} + \text{HPr}$$

Enzyme I, a soluble protein, catalyzes the transfer of phosphate from phosphoenolpyruvate (PEP) to a histidine residue of HPr, a soluble heat-stable protein of low molecular weight. Enzyme II is actually a group of membrane-bound enzymes, each member being specific for a particular sugar. In *Staphylococcus aureus* a fourth protein (Factor III) is required in addition to Enzyme II for the second reaction (Simoni, Smith, and Roseman, 1968). The constitutive Enzymes II of *E. coli* have been separated into two protein fractions: II-A, consisting of three separate proteins, each being required for a phosphorylation of a different sugar, and II-B, which constitutes about 10 percent of the total membrane protein (Kundig and Roseman, 1971). Enzyme II activity can be reconstituted by the appropriate combination of II-A, II-B, Mg^{2+} and phosphatidylglycerol (Kundig and Roseman, 1971).

The involvement of the phosphotransferase system in transport was strongly suggested by Kundig, Kundig, Anderson, and Roseman's (Paper 17) demonstration that osmotically shocked cells of *E. coli*, which had lost 50–80 percent of their HPr and a similar fraction of their ability to accumulate methyl-β-thiogalactoside (TMG) or α-methylglucoside, regained their transport activities when incubated with partially purified HPr. A puzzling aspect of these results is that TMG uptake was stimulated by HPr, although it is now generally accepted that galactoside uptake in *E. coli* is not mediated by the phosphotransferase system.

Tanaka, Lerner, and Lin (1967) showed that *Aerobacter aerogenes* mutants defective in the Enzyme II for mannitol had lost specifically their ability to grow on mannitol. In contrast to the limited effect of Enzyme II mutations, cells deficient in either Enzyme I or HPr lost their abilities to grow normally on seven different sugars (Tanaka and Lin, Paper 18). Simoni et al. (Paper 19) showed that Enzyme I–deficient mutants of *Salmonella typhimurium* grew slowly or not at all on 10 different sugars, the physiological defect being their inability to transport sugars into the cell. In addition, the original *car⁻* mutant of *S. aureus*, which had lost by a single mutation the ability to transport seven different carbohydrates (Egan and Morse, 1965a, 1965b, 1966), was shown to be an Enzyme I mutant by Simoni, Smith, and Roseman (1968).

The discovery of the phosphotransferase system also solved the mystery of the "missing hexokinase" (Fraenkel, Falcoz-Kelly, and Horecker, 1964). A mutant of *E. coli* (MM-6) had been isolated that grew only slowly on glucose and not at all with fructose or mannose as the sole carbon source; a second mutant (GN-2), derived from the first, had lost glucokinase activity and hence its residual ability to grow on glucose. A third strain (FR-1), isolated as a revertant of GN-2, had regained the ability to utilize all three sugars but remained glucokinase negative. Tanaka, Fraenkel, and Lin (1967) showed that Enzyme I activity was absent in the parent strain MM-6 and had been restored in the revertant FR-1, and that the unknown "nonspecific kinase" involved in the utilization of the three sugars was therefore the phosphotransferase system.

The question remained as to whether the PTS participated directly in the movement of carbohydrates across the membrane or whether it merely trapped sugars by phosphorylation as they entered the cell. The latter alternative seemed unlikely since the postulated entry of sugars into the cell might be expected to support growth even in the absence of phosphorylation by the PTS, as demonstrated for glucose in Enzyme I mutants by Simoni et al. (Paper 19). In fact, however, Enzyme I mutants were unable to grow on many carbohydrates, including glucose in low concentrations, suggesting an intimate relation between phosphorylation and the translocation of sugars across the membrane.

Conclusive evidence against the trapping mechanism was provided by Kaback (Paper 20). Membrane vesicles were shown to accumulate glucose and α-methylglucoside as their respective phosphate esters in the presence of PEP; neither uptake nor phosphorylation could be demonstrated in vesicles prepared from cells lacking Enzyme I. When the vesicles were preloaded by passive diffusion with ^{14}C-glucose and allowed to take up external ^{3}H-glucose in the presence of PEP, nearly all of the internal glucose-6-phosphate was found to be labeled with tritium. Thus the transported glucose had not entered the internal pool prior to phosphorylation. Moreover, the entry of α-methylglucoside as the unmodified sugar was found to be a nonsaturable, passive process, occurring much too slowly to account for the initial rates of uptake and phosphorylation in the presence of PEP.

Several fascinating, although mystifying, observations have recently been reported concerning the effects of PTS mutations on the synthesis of inducible catabolic enzymes. Mutants deficient in Enzyme I and HPr are unable to grow on sugars transported by the phosphotransferase system, as is to be expected. Some of these mutants, however, show little or no growth with other metabolites as well (Simoni et al., Paper 19; Fox and Wilson, 1968; Saier, Simoni, and Roseman, 1970). The "extra" growth defects are the result of the organism's inability to induce the appropriate catabolic enzymes; in certain instances, growth on the affected substances will occur if cyclic AMP is included in the growth medium (Epstein, Jewett, and Winter, 1970; Berman, Zwaig, and Lin, 1970; Saier et al., 1970). The different patterns of carbohydrate utilization appear to be related to the degree of impairment of Enzyme I or HPr; "leaky" mutants, retaining some residual Enzyme I or HPr activity, grow normally on the metabolites in question, whereas mutants that have lost all activity do not (Epstein et al., 1970; Saier et al., 1970).

These observations appear to be very closely related to the remarkable finding that β-galactosidase synthesis in Enzyme I mutants of E. coli is hypersensitive to transient repression by PTS-transported sugars (Pastan and Perlman, 1969). It is worth noting, however, if only to demonstrate the perversity of Nature's defenses against Science, that in a different strain of E. coli an Enzyme I mutant was found to be *resistant* to transient repression (Tyler and Magasanik, 1970). These observations need not be totally contradictory, since Pastan and Perlman (1969) were measuring *induced* synthesis of β-galactosidase while Tyler and Magasanik (1970) were monitoring *constitutive* synthesis. Saier and Roseman (1972) have recently reported that in Enzyme I mutants of S. *typhimurium*, the transport of inducers of certain catabolic enzyme systems is inhibited by PTS sugars. This explains, at least in part, the hypersensitivity of Enzyme I mutants of this organism to catabolite repression.

References

Berman, M., N. Zwaig, and E. C. C. Lin (1970). Suppression of a pleiotropic mutant affecting glycerol dissimilation. *Biochem. Biophys. Res. Commun.*, **38**, 272–278.

Crane, R. K., and S. M. Krane (1956). On the mechanism of the intestinal absorption of sugars. *Biochim. Biophys. Acta*, **20**, 568–569.

Egan, J. B., and M. L. Morse (1965a). Carbohydrate transport in *Staphylococcus aureus*. I. Genetic and biochemical analysis of a pleiotropic transport mutant. *Biochim. Biophys. Acta,* **97,** 310–319.

Egan, J. B., and M. L. Morse (1965b). Carbohydrate transport in *Staphylococcus aureus*. II. Characterization of the defect of a pleiotropic transport mutant. *Biochim. Biophys. Acta,* **109,** 172–183.

Egan, J. B., and M. L. Morse (1966). Carbohydrate transport in *Staphylococcus aureus*. III. Studies of the transport process. *Biochim. Biophys. Acta,* **112,** 63–73.

Epstein, W., S. Jewett, and R. H. Winter (1970). Catabolite repression as basis of pleiotropy in PEP-dependent phosphotransferase mutants of *E. coli* K-12. *Fed. Proc.,* **29,** 601 abs.

Fox, C. F., and G. Wilson (1968). The role of a phosphoenolpyruvate-dependent kinase system in β-gluoside catabolism in *Escherichia coli*. *Proc. Natl. Acad. Sci. USA,* **59,** 988–995.

Fraenkel, D. G., F. Falcoz-Kelly, and B. L. Horecker (1964). The utilization of glucose-6-phosphate by glucokinaseless and wild type strains of *Escherichia coli*. *Proc. Natl. Acad. Sci. USA,* **52,** 1207–1213.

Kundig, W., and S. Roseman (1971). Sugar transport. II. Characterization of constitutive membrane-bound enzymes II of the *Escherichia coli* phosphotransferase system. *J. Biol. Chem.,* **246,** 1407–1418.

Pastan, I., and R. L. Perlman (1969). Repression of β-galactosidase synthesis by glucose in phosphotransferase mutants of *Escherichia coli*. *J. Biol. Chem.,* **244,** 5836–5842.

Saier, M. H., Jr., and S. Roseman (1972). Inducer exclusion and repression of enzyme synthesis in mutants of *Salmonella typhimurium* defective in Enzyme I of the phosphoenolpyruvate: sugar phosphotransferase system. *J. Biol. Chem.,* **247,** 972–975.

Saier, M. H., Jr., R. D. Simoni, and S. Roseman (1970). The physiological behavior of Enzyme I and heat-stable protein mutants of a bacterial phosphotransferase system. *J. Biol. Chem.,* **245,** 5870–5873.

Simoni, R. D., M. F. Smith, and S. Roseman (1968). Resolution of a staphylococcal phosphotransferase system into four protein components and its relation to sugar transport. *Biochem. Biophys. Res. Commun.,* **31,** 804–811.

Tanaka, S., D. G. Fraenkel, and E. C. C. Lin (1967). The enzymatic lesion of strain MM-6, a pleiotropic carbohydrate-negative mutant of *Escherichia coli*. *Biochem. Biophys. Res. Commun.,* **27,** 63–67.

Tanaka, S., S. M. Lerner, and E. C. C. Lin (1967). Replacement of a phosphoenolpyruvate-dependent phosphotransferase by a nicotinamide adenine dinucleotide-linked dehydrogenase for the utilization of mannitol. *J. Bacteriol,* **93,** 642–648.

Tyler, B., and B. Magasanik (1970). Physiological basis of transient repression of catabolic enzymes in *Escherichia coli*. *J. Bacteriol.,* **102,** 411–422.

Wilbrandt, W., and T. Rosenberg (1961). The concept of carrier transport and its corollaries in pharmacology. *Pharmacol. Rev.,* **13,** 109–183.

Wilson, T. H., and B. R. Landau (1960). Specificity of sugar transport by the intestine of the hamster. *Amer. J. Physiol.,* **198,** 99–102.

Reprinted from *Proc. Natl. Acad. Sci. U.S.A.*, **52**, 1067–1074 (1964)

PHOSPHATE BOUND TO HISTIDINE IN A PROTEIN AS AN INTERMEDIATE IN A NOVEL PHOSPHO-TRANSFERASE SYSTEM*

By Werner Kundig,[†] Sudhamoy Ghosh,[‡] and Saul Roseman

RACKHAM ARTHRITIS RESEARCH UNIT AND DEPARTMENT OF BIOLOGICAL CHEMISTRY, UNIVERSITY OF MICHIGAN, ANN ARBOR

16

Communicated by Arthur Kornberg, August 10, 1964

Mammalian tissues contain a kinase involved in the intermediary metabolism of the sialic acids.[1, 2] This enzyme has been extensively purified,[3] studied in detail, and catalyzes the following reaction: N-Acyl-D-mannosamine + ATP $\xrightarrow{\text{Mg}^{++}}$ N-Acyl-D-mannosamine-6-P + ADP. To determine whether this kinase occurred in bacteria, such as *Aerobacter cloacae* and *Escherichia coli* K235,[4] that metabolize N-acetyl-D-mannosamine, extracts of these organisms were examined and found to contain a novel phospho-transferase system. The system obtained from *E. coli* K235 consisted of two enzymes, I and II, and a histidine-containing, heat-stable protein (HPr). The sequence of reactions is:

$$\text{Phosphoenolypyruvate (PEP)} + \text{HPr} \underset{\text{Mg}^{++}}{\overset{\text{I}}{\rightleftharpoons}} \text{Phospho-histidine-protein (P-HPr)} + \text{Pyruvate} \quad \text{(A)}$$

$$\text{P-HPr} + \text{Hexose} \xrightarrow[\text{Mg}^{++}]{\text{II}} \text{Hexose-6-P} + \text{HPr} \quad \text{(B)}$$

$$\text{PEP} + \text{Hexose} \xrightarrow[\text{Mg}^{++}]{\text{I + II}} \text{Hexose-6-P} + \text{Pyruvate} \quad \text{(A+B)}$$

The intermediate in the system, P-HPr, is protein-bound phosphohistidine.

Materials and Methods.—Unless otherwise specified, all materials were obtained from commercial sources. Previously published methods[1, 5] were used for the preparation, separation, and characterization of C^{12}- and C^{14}-hexosamines, N-acylhexosamines, the corresponding 6-phosphate esters, and for the periodate oxidation of the esters and the characterization of glycolaldehyde-phosphate. The following compounds were prepared as described: P-histidine,[6] N-phospho-glycine,[7] phosphoramidate,[8] and PEP.[9] An essential substrate for these experiments, P^{32}-PEP was prepared enzymatically by a published procedure[10] and with the invaluable help of Dr. M. F. Utter and Mr. Douglas Kerr, to whom we are most grateful.[11] The P^{32}-PEP (5–10 μmoles per experiment) contained 200–400 μc of P^{32} per μmole and was purified by ion-exchange chromatography; paper chromatography and electrophoresis indicated that it was homogeneous. It was diluted with unlabeled PEP prior to use.

Purification of enzymes I, and II, and HPr: The organism, *E. coli* K235, was grown to the stationary phase in Todd-Hewitt (Difco) broth supplemented with 1.5% glucose in a New Brunswick fermentor. Maximum yields of the phospho-transferase system were obtained when the culture was stirred during growth but without passage of air through the sparger. After washing with 1% KCl solution, the wet cell paste was stored at $-18°$. The cells were ruptured by sonic oscillation following suspension in 0.025 M phosphate buffer, pH 7.6 (containing 0.1% 2-mercapto-ethanol and 10^{-3} M EDTA when enzymes I and II were desired).

After centrifugation, the supernatant fluid (crude extract) was treated with charcoal to remove HPr and fractionated for I and II as outlined in Table 1. The critical step was the C_γ alumina gel treatment since I was adsorbed while II was not; after washing the gel with 0.01 and 0.05 M phosphate buffers, pH 7.6, I was eluted with 0.10 M buffer. These data suggest that both enzymes were purified approximately 300-fold. Since we have not yet determined which enzyme, I or II, was present at rate-limiting concentrations prior to their separation, the purification factor is correct for only one of these enzymes, and is not known for the other. However, the availability of the purified enzymes I and II will now permit accurate analysis for each enzyme.

TABLE 1

PURIFICATION OF ENZYMES I AND II

Fraction	Specific Activity* of			Yield (%)
	I + II	I	II	
Crude extract	0.30	—	—	100
Charcoal filtrate	0.82	—	—	62
Ammonium sulfate (30–50%)	2.2	—	—	51
Cγ alumina gel:				
Supernatant		0	16	39
Eluate		19	0	41
DEAE-cellulose:				
Cγ supernatant		0	98	24
Cγ eluate		119	0	27

* Specific activity was defined as μmoles of N-acetylmannosamine-6-P formed per mg protein per 30 min. Each incubation mixture contained the following (in μmoles) in a final volume of 0.18 ml: C^{14}-acetyl-labeled N-acetyl-D-mannosamine, 2.0 (specific activity, 5×10^5 cpm/μmole); PEP, 2.5; MgCl$_2$, 2.5; Tris-HCl buffer, pH 7.4, 10; HPr, 20 μg; enzymes I and II. Following the Cγ alumina step, either I or II was added in rate-limiting amounts, and the other enzyme was added in excess. After incubating for 30 min at 37°, the mixtures were heated at 100° for 3 min, 25-μl samples of the supernatant fluids were spotted on Whatman #3 mm paper and electrophoresed in 0.05 M pyridinium acetate "buffer," pH 6.5, at 100 v/cm for 15 min. Under these conditions, the substrate remained at the origin while the product, N-acetylmannosamine-6-P, migrated approximately 10 cm. Both areas were cut from the paper strip and counted by liquid scintillation techniques. Controls consisted of mixtures lacking I, II, or HPr or contained heat-denatured I or II. In each case, product formation was shown to be linear with time and proportional to the concentration of I or II (or to I + II in the first 3 fractions). The 0 values indicate no detectable activity (i.e., less than 0.05).

Based on analysis by a modified biuret protein method,[12] HPr was purified 8,000- to 10,000-fold as follows: the crude extract was heated for 10 min at 100°; the chilled supernatant fluid was adjusted to pH 1; the resulting precipitate was washed with 0.01 M HCl; HPr activity was extracted with 0.5 M phosphate buffer, pH 7.6, dialyzed 24 hr against 0.025 M phosphate buffer, pH 6.5, passed over a column of Ecteola-cellulose that did not adsorb HPr, and finally fractionated by adsorption and elution on two successive columns of DEAE-cellulose. The final elution was conducted with a shallow gradient of 0–0.10 M KCl in 0.01 M Tris buffer, pH 7.6. Following the elution of inactive protein, HPr activity was eluted in a single symmetrical protein peak where the specific activity of each fraction was essentially constant.[13]

Kinetic Properties and Specificity of the Complete System.—When the assay was conducted as described in Table 1, the addition of 5 μg each of purified enzymes I and II gave 0.5 μmole of N-acetylmannosamine-6-P in 30 min. Omission of I, II, HPr, or Mg^{++} gave no detectable (i.e., less than 0.005 μmole) product. Substitution of heated I or II for the active proteins also gave negative results. Each of the three proteins could be made the rate-limiting factor, and in each case (and throughout the purification steps) product formation was linear with time for at least 2 hr. The pH optimum for the complete system was 7.2–7.4, and the approximate K_m values for PEP and N-acetylmannosamine were $6 \times 10^{-4} M$. The following D-sugars could substitute for N-acetylmannosamine and exhibited the following approximate K_m values ($M \times 10^4$): glucose, 4; mannose, 20; glucosamine, 30; mannosamine, 10; N-acetylglucosamine, 9; N-glycolylmannosamine, 20. The ratios of activities obtained with glucose, mannose, N-acetylglucosamine, and N-acetylmannosamine as phosphate acceptors were approximately constant over the entire range of purification. The following D-sugars did not act as phosphate acceptors in this system: galactose, galactosamine, N-acetylgalactosamine, fructose, xylose, arabinose, ribose, glucose-1-P, and glucose-6-P.

The following compounds could not replace PEP as the phosphoryl donor: mono-, di-, and triphosphates (all 5′) of adenosine, deoxyadenosine, guanosine, deoxyguanosine, cytidine, deoxycytidine, thymidine, uridine, and inosine (alone or in mixture with other nucleotides); cyclic 3′,5′-AMP; creatine-P (with and without creatine-P transphosphorylase ± ADP or ATP); PP$_i$; P$_i$; phosphoramidate; phosphohistidine; *N*-phosphoglycine; thiamine-PP; P-glycerate; P-serine; coenzyme A and glutathione; coenzyme A + succinate ± P_i. Moreover, these compounds did not affect the rate of the reaction in the presence of PEP.

The following divalent cations could either partially or completely replace Mg^{++}: Mn^{++}, Zn^{++}, Co^{++}. In addition, Ca^{++} and Cu^{++} were highly inhibitory in this system at concentrations where the other cations were active.

Nonparticipation of Nucleotides.—As indicated above, the purified proteins exhibited activity only with PEP as the initial phosphoryl donor. By contrast, the crude extracts were fully active with P-glycerate and 10–20 per cent as active with ATP and creatine-P in place of PEP. Only PEP, however, was active in the crude system in the presence of 0.01 M KF. Additional evidence suggesting that PEP was the direct phosphoryl donor, and was not involved in a nucleotide triphosphate generating system was obtained as follows: (1) Pyruvate kinase could not be detected in fractions I, II, or HPr. In addition, pyruvate kinase in the presence or absence of varying concentrations of ADP could not substitute for any of the indicated protein fractions. (2) Creatine-P and creatine-P-ATP-transphosphorylase did not substitute for PEP in the presence or absence of ADP. Under the same conditions with ADP, crystalline yeast hexokinase readily phosphorylated glucose. (3) Addition of excessive quantities of fructose and hexokinase to PEP, I, II, and HPr gave no detectable fructose-6-P. The addition of varying quantities of ADP (10^{-6} to 10^{-2} M) to this mixture also gave negative results. The presence of fructose, hexokinase, and ADP did not affect the rate of *N*-acetylmannosamine phosphorylation when the latter was added to the mixture. (4) Various concentrations of ATP (10^{-6} to 10^{-2} M) did not replace PEP or any of the three protein components necessary for the system; in addition, the ATP did not affect the rate of reaction in the complete system. (5) The addition of purified venom 5′-nucleotidase and/or venom P-diesterase to the complete incubation mixture did not affect the rate of the reaction. (6) Proteins I, II, and HPr were each incubated in the presence and absence of PEP with C^{14}-labeled mono-, di- and triphosphates of adenosine, uridine, guanosine, and cytidine (specific activities: 28, 27, 12, and 13 $\mu c/\mu$mole, respectively; 10 μc total C^{14}-nucleotide were added to each protein fraction). Each of the three proteins retained full catalytic activity but were not radioactive after passing the fractions through Sephadex G-25 followed by pressure dialysis. Under the conditions used for the enzymatic assay (Table 1), C^{14}-nucleotide would have been detected at concentrations above 5×10^{-6} μmole per ml. Based on these experiments, particularly the C^{14} experiment, we conclude either that a nucleotide is not involved in these transfer reactions, or that such a nucleotide is firmly bound to one of the protein fractions and is not in equilibrium with nucleotide in the surrounding solution.

Characterization of Products.—As indicated above, a number of D-sugars of the gluco- and manno-configuration served as phosphate-acceptors. Three of the products have been characterized as the 6-phosphate esters. Analysis of the products isolated after phosphorylation of *N*-acetylmannosamine and *N*-acetylglucosamine showed the following molar ratios: *N*-acetylglucosamine, 1.00, P, 0.95; *N*-acetylmannosamine, 1.00, P, 0.95. These compounds gave the expected *N*-acetylhexosamines on treatment with phosphatase and yielded glycolaldehyde-P on oxidation with periodate. In addition, no P_i was liberated on treatment with

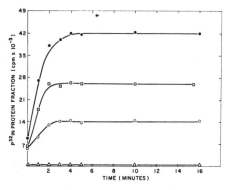

FIG. 1.—Rate of phosphate transfer from P³²-PEP to protein as a function of HPr concentration. Each incubation mixture contained the following (in μmoles) in a final volume of 0.45 ml: P³²-PEP 0.075 (specific activity, 10^7 cpm/μmole); MgCl₂, 1.5; Tris-HCl buffer, pH 7.4, 20; enzyme I, 21 μg; HPr, 50 μg (○), or 100 μg (□), or 150 μg (●). Controls (△) were mixtures lacking either I or HPr, or contained II or heat-denatured I in place of I or HPr. Aliquots of the mixtures (50 μl) were removed at the indicated times, frozen, and stored at −75°. Low-molecular-weight P³² components, like P³²-PEP, were separated from P³²-HPr by subjecting 25 μl of the samples to high-voltage electrophoresis (65 v/cm) in 0.05 M citrate buffer, pH 6.5 for 20 min. The product remained close to the origin; all radioactive areas were cut from the paper strips and counted in a gas-flow proportional counter.

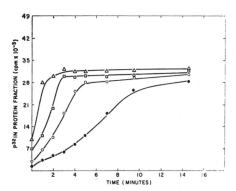

FIG. 2.—Rate of phosphate transfer from P³²-PEP to protein as a function of enzyme I concentration. Incubation mixtures were prepared as described in Fig. 1 except that complete mixtures contained 100 μg of HPr and enzyme I at one of the following levels (in μg): 4.5 (●); 9.0 (○); 21 (□); or 45 (△). Controls were the same as those shown in Fig. 1.

1 M HCl at 100° for 20 min. Finally, the product obtained from N-acetylmannosamine was fully active in the N-acetylneuraminic acid-9-P synthetase reaction.[14] The product formed from glucose was shown to be glucose-6-P by including TPN and G-6-P dehydrogenase in the phosphorylating system; TPN was immediately reduced to TPNH as determined spectrophotometrically. Phosphoglucomutase activity was not detected in this system.

Pyruvate was measured with lactic dehydrogenase and DPNH. With each of the seven sugars that served as P-acceptors in the complete system, 1.00 ± 0.03 mole of pyruvate was formed per mole of hexose-P.

Properties of HPr.—The purified material showed a typical protein ultraviolet absorption spectrum, was nondialyzable, and studies with Sephadex gels showed that it was not retarded by G-50, slightly retarded by G-100, and fully retarded by G-200. The following substances were not detected in the preparation and therefore, if present, would have to be at levels (based on protein content) of less than 0.01 per cent for phosphorus, anthrone-reactive hexose, anthrone-reactive pentose, or hexosamine, and less than 0.05 per cent thiobarbituric acid-reactive compounds (such as 2-keto-3-deoxy sugar acids). The factor was stable at 100° at neutral pH for at least 20 min, at pH 1 at room temperature for several hours, and was completely resistant to prolonged digestion and dialysis with the following nucleases: purified venom phosphodiesterase, polynucleotide phosphorylase, pancreatic RNase and DNase, and venom 5'-nucleotidase.

All of the protein[12] and catalytic activity in purified HPr was adsorbed by charcoal but not by mixed-bed ion-exchange resins. It was precipitated with protamine sulfate and lost activity on treatment

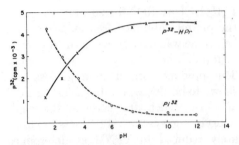

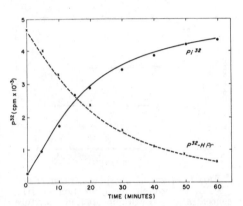

FIG. 3.—Hydrolysis of P^{32}-HPr as a function of pH. Each sample contained 150 μg of P^{32}-HPr (6100 cpm; specific activity 2 $\times$ 10^7 cpm/μmole) in a final volume of 0.25 ml adjusted to the indicated pH. After 30 min at 23°, a 50-μl aliquot was removed to measure the pH, and 0.20 ml was neutralized and spotted on S and S 589 (green) paper. The chromatograms were developed with 0.1 M Na$_2$CO$_3$, 95% ethanol, 3.5:6.5. P^{32}-HPr remained at the origin while P^{32}-inorganic phosphate migrated close to the solvent front. Radioactivity measurements were performed as described in Fig. 2; the results obtained with the 0.20-ml aliquots are given above.

FIG. 4.—Hydrolysis of P^{32}-HPr: pH 3.9 at 40°. The sample, 1.5 mg P^{32}-HPr (47,500 cpm; 1.6 $\times$ 10^7 cpm/μmole), was maintained at 40° after adjusting to pH 3.9 with acetic acid and a final volume of 0.50 ml. Aliquots (50 μl) were removed at the indicated times, neutralized, and assayed by paper chromatography as described in Fig. 3.

with 0.1 M alkali for 60 min at room temperature. The activity was completely lost by treating HPr with the following proteinases: chymotrypsin, trypsin, papain, pronase, and pepsin. In each case, suitable controls showed that the loss in activity was a result of proteinase action on HPr, and was not the result of possible residual proteinase action on I and II.

Formation and Properties of P^{32}-HPr.—As shown in Figures 1 and 2, P^{32} was transferred from P^{32}-PEP to protein. While the *rate* of the reaction depended on the concentration of enzyme I, the *extent* of incorporation was directly proportional to the concentration of HPr. Thus, we concluded that P^{32} was transferred to HPr. In these experiments the required components were: I, HPr, P^{32}-PEP, and Mg^{++}. Omission of any of these or substitution for I and/or HPr by II gave no detectable incorporation (i.e., less than 200 cpm compared with 120,000 cpm in the complete system). Further, the addition of II to the complete system did not affect the rate of the reaction. Preliminary experiments indicated that the P-transfer from PEP to HPr was reversible.

To demonstrate reaction B (see above), the following components (in μmoles) were incubated in a final volume of 0.4 ml for 30 min at 37°: *N*-acetylmannosamine, 4.0; MgCl$_2$, 2.5; Tris-HCl, pH 7.4, 25; enzyme II, 0.25 mg; P^{32}-HPr, 22,400 cpm (specific activity, 20 $\times$ 10^6 cpm/μmole). The product, *N*-acetylmannosamine-6-P^{32}, isolated by ion-exchange chromatography, contained 17,500 cpm and gave glycolaldehyde-P^{32} on periodate oxidation. The remaining P^{32} was shown to be P_i^{32}. Omission of enzyme II, or substitution of I for II, gave less than 100 cpm in the product; further, addition of I to the complete system did not affect the transfer of P^{32} to the sugar. In a similar experiment, the incubation was conducted with unlabeled P-HPr and C^{14}-*N*-acetylmannosamine; approximately the same quantity of C^{14}-*N*-acetylmannosamine-6-P was formed as from P^{32}-HPr. In the C^{14}-experiment, the addition of unlabeled PEP to the system did not affect the

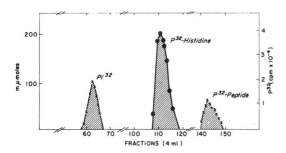

Fig. 5.—Ion-exchange chromatography of alkali-hydrolyzed P^{32}-HPr. The labeled protein was prepared by incubating the following components in 16 ml: 28 mg of purified HPr, 0.2 mg enzyme I, 100 μmoles of P^{32}-PEP (2×10^5 cpm/μmole), 1.0 mmole Tris buffer, pH 7.4, and 0.1 mmole MgCl$_2$. After 30 min at 37°, the mixture was fractionated on a column (2.5 $\times$ 80 cm) of Sephadex G-25, previously equilibrated with 0.1 M NaCl adjusted to pH 8.5 with Na$_2$-CO$_3$. The P^{32}-protein fraction (eluted with NaCl solution) contained 1.3 $\times$ 10^5 cpm, and was concentrated by pressure dialysis to 11 ml. It was adjusted with 10 M Na OH to 3 M, and was hydrolyzed in a sealed tube at 100° for 7 hr. After diluting 10-fold with water, the sample was placed on a column of Dowex-1, hydroxyl form (200–400 mesh; 2.5 $\times$ 30 cm), washed extensively with H$_2$O (no P^{32} was eluted), and eluted with a linear gradient of 0.0–1.5 M NaHCO$_3$–Na$_2$CO$_3$ buffer,[16] pH 8.5 (total volume, 1 liter). Each fraction (4 ml) was analyzed for P^{32}; the total radioactivity in the 3 peaks (indicated by crosshatching) was over 95% of that placed on the column. All peaks were analyzed for inorganic phosphate[17] ((-X-X) using 2.5-cm cells in the Cary model 14 recording spectrophotometer. Only peak 1 contained inorganic phosphate. However, when 0.5-ml samples of peaks 2 and 3 were pretreated with 0.25 ml of 4 N S$_2$SO$_4$ at 40° for 30 min, all of the P^{32} was measurable as inorganic phosphate. Histidine (-O-O-) measured fluorometrically,[18] was not detectable in any of the peaks, but was found in peak 2 after the acid treatment described above. Histidine, and 3 or 4 other amino acids were found in peak 3, using paper chromatographic methods, after 6 N HCl hydrolysis (12 hr, 100°), of the pooled fractions. Peak 3 was not observed when the 3 M NaOH hydrolysis was conducted for 12 rather than 7 hr, and there was a concomitant increase in peaks 1 and 2.

results unless it was added along with enzymes I and II, HPr, and the sugar; in this case the results were similar to those given in Table 1.

The properties of P^{32}-HPr suggested that the P^{32} was attached to a histidine residue; for example, P_i^{32} was immediately formed in acid solution, while none was detected when P^{32}-HPr was treated with 0.1 M NaOH for 1 hr at 25°. The results of two studies on the hydrolysis of P^{32}-HPr are given in Figures 3 and 4. The rates of hydrolysis were the same as those observed with P-histidine. These data, along with results indicating that P^{32}-HPr was remarkably susceptible to hydrolysis at pH 6.5 in the presence of pyridine or other organic amines, were typical of the properties reported for P-histidine[6] and for protein bound P-histidine.[15]

To characterize more fully the P^{32}-HPr, it was subjected to alkaline hydrolysis as described by Boyer et al.,[16] and the hydrolyzate fractionated as shown in Figure 5. Three radioactive peaks were observed. Peak 1 was identified as inorganic P_i^{32} by colorimetric, paper electrophoretic, and chromatographic methods.

Peak 2 was identified as phosphohistidine by the following criteria: (a) It cochromatographed with authentic P-histidine on ion-exchange columns,[15, 16] paper chromatography,[6, 16] and in two electrophoretic systems. (b) Inorganic phosphate and histidine were not detected unless the sample was first hydrolyzed with acid. (c) Analysis of peak 2 for phosphate by modification of a colorimetric method,[17] and for histidine by a fluorometric method[18] showed a ratio of 1.00 $\pm$ 0.06 in each fraction. The specific activity of the P^{32} in peak 2 was the same as the P^{32}-PEP originally used to label the P^{32}-HPr.

Peak 3 appeared to be a peptide containing P^{32}-histidine since (a) it liberated P_i^{32} under the same conditions as P-histidine, and (b) acid hydrolysis revealed the presence of several amino acids including histidine.

Comments.—Although phospho-proteins have long been known (including such proteins from *E. coli*[19]) and enzymatic transfer of P to protein has been achieved,[20] these have generally involved a serine moiety in the protein. Protein-bound phosphohistidine was first described by Boyer and his co-workers.[16, 21, 22]

The transferase system described here is unique[23] in several respects: (*a*) the initial P-donor is PEP; (*b*) the protein to which the P is transferred serves as a donor for transfer to seven sugars of the gluco- and manno-series when supplemented with the required enzyme (II). Preliminary experiments have shown that more than one enzyme II can be formed by the cells. Thus, when glucose-grown cells were washed and incubated for 6 hr in fresh medium containing galactose, a new enzyme II was isolated that transferred P from P-HPr to galactose and *N*-acetylgalactosamine. No difference in enzyme I or HPr could be detected between glucose and galactose-grown cells. Thus, it seems possible that reaction A (above) serves as a source of phosphate in a variety of phospho-transferase reactions.

While the distribution of the transferase system in various organisms is now under study, it has been detected in strains of *E. coli* (including wild type), *Aerobacter cloacae* and *aerogenes*, and *Lactobacillus arabinosus*.

The biological significance of the new phospho-transferase system is not clear[24] and fruitful speculation must await further experimental results.

Summary.—A novel phospho-transferase system was isolated from *E. coli* K235, and was detected in other bacteria. The system involved a sequential transfer of phosphate from phosphoenolpyruvate to a heat-stable protein to hexoses; the two reactions were catalyzed by two distinct enzyme fractions. Isolation and characterization of the phosphorylated protein showed that the phosphate group was linked to a histidine residue.

The expert technical assistance of Mr. Alan Jacobs in these studies is gratefully acknowledged

* The Rackham Arthritis Research Unit is supported by a grant from the Horace H. Rackham School of Graduate Studies of The University of Michigan. This investigation was supported in part by a grant from the National Institutes of Health (AM 00512-10).

† Postdoctoral fellow, Arthritis and Rheumatism Foundation.

‡ Present address: Department of Biochemistry, University College of Science and Technology, 92 Upper Circular Road, Calcutta 9, India. This work was performed under the tenure of a postdoctoral fellowship from the Helen Hay Whitney Foundation.

[1] Ghosh, S., and S. Roseman, these PROCEEDINGS, **47**, 955 (1961).

[2] Warren, L., and H. Felsenfeld, *J. Biol. Chem.*, **237**, 1421 (1962).

[3] Kundig, W., and S. Roseman, unpublished work.

[4] Barry, G. W., and W. F. Goebel, *Nature*, **179**, 206 (1957).

[5] Jourdian, G. W., and S. Roseman, *J. Biol. Chem.*, **237**, 2442 (1962); Distler, J. J., J. M. Merrick, and S. Roseman, *J. Biol. Chem.*, **230**, 497 (1958).

[6] Ratlev, C. T., and T. Rosenberg, *Arch. Biochem. Biophys.*, **65**, 319 (1956).

[7] Zervas, L., and P. G. Katsoynannis, *J. Am. Chem. Soc.*, **77**, 5351 (1955).

[8] Stokes, H. N., *Am. Chem. J.*, **15**, 198 (1893).

[9] Clark, V. M., and A. J. Kirby, *Biochim. Biophys. Acta*, **78**, 732 (1963).

[10] Mendicino, J., and M. F. Utter, *J. Biol. Chem.*, **237**, 1716 (1962).

[11] We are also very grateful to Drs. Sidney M. Colowick and Don E. Hultquist for their critical comments, and to Dr. Santiago Grisolia for a sample of yeast phosphoglyceromutase and suggestions for the preparation of P^{32}-glycerate. Dr. Fredericka Dodyk kindly helped with some of the critical experiments.

[12] Lowry, O. M., N. J. Rosebrough, A. L. Farr, and R. J. Randall, *J. Biol. Chem.*, **193**, 265 (1951).

[13] In a single experiment, disc electrophoresis in polyacrylamide gel showed the presence of two major and three minor protein bands.

[14] Roseman, S., G. W. Jourdian, D. Watson, and R. Rood, these Proceedings, **47**, 958 (1961).

[15] de Luca, M. K. E. Ebner, D. E. Hultquist, G. Kreil, J. B. Peter, R. W. Moyer, and P. D. Boyer, *Biochem. Z.*, **338**, 512 (1963).

[16] Boyer, P. D., M. de Luca, K. E. Ebner, D. E. Hultquist, and J. B. Peter, *J. Biol. Chem.*, **237**, PC3306 (1962).

[17] Martin, J. B., and D. M. Doty, *Anal. Chem.*, **21**, 965 (1947).

[18] Pisano, J. J., J. D. Wilson, L. Cohen, D. Abraham, and S. Udenfriend, *J. Biol. Chem.*, **236**, 499 (1961); Shore, P. A., A. Burkhalter, and V. H. Cohen, *J. Pharmacol. Exptl. Therap.*, **127**, 182 (1959).

[19] Rafter, G. W., *J. Biol. Chem.*, **239**, 1044 (1964).

[20] Rabinowitz, M., and F. Lipmann, *J. Biol. Chem.*, **235**, 1043 (1960); Rabinowitz, M., in *Methods in Enzymology*, ed. S. P. Colowick and N. O. Kaplan (New York: Academic Press, 1963), vol. 6, p. 218; Burnett, J., and E. N. Kennedy, *J. Biol. Chem.*, **211**, 969 (1954); Sundararajan, T. A., K. S. V. S. Kumar, and P. S. Sarma, *Biokhimiya*, **22**, 135 (1957).

[21] Peter, J. B., D. E. Hultquist, M. de Luca, G. Kreil, and P. D. Boyer, *J. Biol. Chem.*, **238**, PC 1182 (1963).

[22] Dr. Boyer has recently informed us that the protein containing phosphohistidine isolated from bovine mitochondria is related to or is, in fact, succinate thiokinase (Mitchell, R. A., L. G. Butler, and P. D. Boyer, *Biochem. Biophys. Res. Commun.*, in press). Similar results were obtained with highly purified succinate thiokinase from *E. coli* (Kreil, G., and P. D. Boyer, *Biochem. Biophys. Res. Commun.*, in press). In the experiments described by these investigators, P_i^{32} was incorporated into protein in the presence of succinyl CoA, or CoA. In the present experiments, enzymes I, II, and HPr exhibited no succinate thiokinase activity, and did not incorporate P_i^{32} into protein in the presence or absence of succinate and CoA. Experiments are in progress to determine whether or not there is any relationship between the succinate thiokinase and the system described in the present paper.

[23] Hexoses can be phosphorylated by an enzyme obtained from *E. coli* where the phosphoryl donor is phosphoramidate, phosphohistidine, or *N*-phosphoglycine [Fujimoto, A., and R. A. Smith, *Biochim. Biophys Acta*, **56**, 501 (1962)]. This enzyme was present in the crude extracts used in the present studies, but was removed during the purification procedure.

[24] As shown above, the heat-stable protein HPr serves as a phosphate "carrier" in the complete system. Another type of heat-stable protein has recently been described [Majerus, P. W., A. W. Alberts, and P. R. Vagelos, these Proceedings, **51**, 1231 (1964)] and serves to "carry" acyl residues during the biosynthesis of fatty acids. A preparation of this protein was kindly provided to us by Drs. Paul Stumpf and Robert Simoni, and when tested in the phospho-transferase system, could not replace HPr. Similarly, our preparation of HPr was tested by Drs. Stumpf and Simoni, but exhibited no activity in their system.

Copyright © 1966 by the American Society of Biological Chemists, Inc.
Reprinted from *J. Biol. Chem.*, **241**, 3243–3245 (1966)

17

Restoration of Active Transport of Glycosides in *Escherichia coli* by a Component of a Phosphotransferase System*

(Received for publication, May 4, 1966)

WERNER KUNDIG, F. DODYK KUNDIG, BYRON ANDERSON,‡ AND SAUL ROSEMAN

From the McCollum-Pratt Institute and the Department of Biology, The Johns Hopkins University, Baltimore, Maryland 21218

SUMMARY

The ability of *Escherichia coli* W2244 to concentrate methyl-β-thiogalactoside and methyl-α-glucoside was substantially reduced when the cells were subjected to osmotic shock. The shocked cells contained only a fraction of their normal content of a heat-stable protein, previously characterized as a component of a phosphotransferase system. When the shocked cells were incubated with the purified heat-stable protein, the ability to accumulate the glycosides was restored. The results suggest that the phosphotransferase system may be an important component of the glycoside permease systems.

A phosphotransferase system, recently isolated from bacteria (1), catalyzed the transfer of phosphate from phosphoenolpyruvate to certain monosaccharides. The results presented below suggest that this system may be involved in the active transport of sugars by bacteria, *i.e.* a component of the "permease system" of Rickenberg, Cohen, Buttin, Monod, Kepes, and their collaborators (for review, see Reference 2).

The phosphotransferase system has been separated into three protein fractions, Enzymes I and II, and a heat-stable protein; phospho-HPr[1] contains phosphate linked to the imidazole nitrogen of protein-bound histidine. In the presence of Mg^{++}, Enzymes I and II catalyze the following reactions.

$$\text{Phosphoenolpyruvate} + \text{HPr} \rightleftarrows \text{pyruvate} + \text{phospho-HPr} \quad (1)$$
$$\underline{\text{Phospho-HPr} + \text{sugar} \rightarrow \text{HPr} + \text{sugar-P}} \quad (2)$$

Sum: Phosphoenolpyruvate + sugar → pyruvate + sugar-P

The results of recent studies (3) pertinent to the present report may be summarized as follows. (*a*) Lysis of *Escherichia coli* spheroplasts and *Bacillus subtilis* protoplasts showed that Enzyme II was located in the "membrane fraction" while most of Enzyme I and HPr were in the soluble fraction. Enzyme I was purified approximately 300-fold, and HPr, about 10,000-fold. Polyacrylamide gel electrophoresis and ultracentrifugation studies indicated, however, that HPr was not homogeneous;

two major protein bonds were visible after electrophoresis. (*b*) The fraction designated Enzyme II has not yet been solubilized, and hereafter it will be called Fraction II; conceivably, more than one protein is involved in the transfer of phosphate from phospho-HPr to each monosaccharide. (*c*) Fraction II was responsible for the specificity exhibited by the system toward the carbohydrates, and this specificity varied with the composition of the medium used for growth of the cells. With this technique, three varieties of Fraction II were distinguished; these catalyzed the phosphorylation of methyl-α-glucoside, galactose, and methyl-β-thiogalactoside, respectively.[2] The same substrates are transported by specific "permease systems" (2, 4, 5).

A mutant strain derived from *E. coli* K-12, *E. coli* W2244 ($i^+y^+z^-gal^+$),[3] was used for the present studies; it can be induced for thiogalactoside "permease," but not for β-galactosidase, and it utilizes galactose. These cells can also accumulate methyl-α-glucoside after growth in a synthetic medium without inducer; the methyl-α-glucoside "permease system" may be related to or identical with the system involved in the transport of glucose (2).

If the phosphotransferase and permease systems are related, then cells depleted of HPr should be unable to accumulate sugars that are substrates for Fraction II. This hypothesis was tested by subjecting *E. coli* W2244 to an osmotic shock at 0° (hereafter called cold shock), which results in the loss of certain proteins from the cell, although they maintain their viability

---

* This work was supported by Grant AM-9851 from the National Institute of Arthritis and Metabolic Diseases of the National Institutes of Health. Contribution 471 from the McCollum-Pratt Institute.
‡ Supported by Predoctoral Training Grant T1 GM-57 to the Department of Biology, The Johns Hopkins University.
[1] The abbreviations used are: HPr, heat-stable protein; TMG, methyl-β-thiogalactoside.

[2] All sugars are of the D configuration, and glycosides are pyranosides. "Enzyme II" represents a family of inducible enzymes, three of which have thus far been distinguished by their specificities toward different sugars. For example, the phosphotransferase system isolated from a number of *E. coli* strains grown in the following media phosphorylated the indicated sugars: (*a*) glucose-supplemented Todd-Hewitt broth; glucose, mannose, the corresponding hexosamines, *N*-acylhexosamines, hexitols, and methyl-α-glucoside; (*b*) glycerol-salts medium; at least 16 sugars and their derivatives (monosaccharides, disaccharides, glycitols, glycosides, etc.), including galactose and TMG; (*c*) glycerol-salts medium containing 0.001 M fucose; D-fucose induces a specific galactose "permease" (4), and a concomitant increase (4- to 10-fold) in Enzyme II activity toward galactose in three strains of *E. coli* with no detectable change in activity toward the other 15 carbohydrates. The products obtained from the following compounds were isolated and characterized as the 6-phosphate esters: (*a*) glucose, mannose, the corresponding hexosamines and *N*-acylhexosamines; galactose; and methyl-β-thiogalactoside.
[3] Kindly provided by Dr. Boris Rotman, who also contributed numerous valuable suggestions during the course of this work. Dr. Rotman has also conducted experiments in which the TMG "carrier" protein was specifically labeled with sulfhydryl reagents (personal communication).

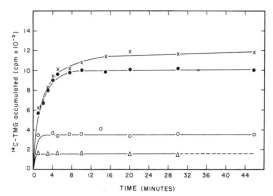

Fig. 1. Accumulation of TMG by *E. coli* W2244 induced with iso-propyl-β-thiogalactoside. ●, untreated cells; ○, cold-shocked cells incubated in Medium 63-S (± amino acids) as described in Table I; ×, cold-shocked cells incubated with HPr; △, cold-shocked cells incubated in Medium 63-S for 25 min, then for 5 min with sodium azide, before addition of TMG. The suspensions contained 1.1×10^{10} cells in 5.6 ml of the medium, and 1.6 mg of HPr or 0.04 M sodium azide where indicated. After incubation, 0.20 ml of 0.10 M TMG-^{14}C (1.5×10^5 cpm per μmole) was added; 0.5-ml aliquots were withdrawn at the indicated times, filtered, and washed twice with 1.0-ml portions of cold medium. After drying, the cells were counted by liquid scintillation techniques. Total counts per min are presented on the *ordinate*, and have not been corrected for controls (formalin, azide, etc.). As in Table I, HPr inactivated with trypsin or by photooxidation was inactive in increasing TMG accumulation in cold-shocked cells.

(6, 7).[4] As shown in Fig. 1 and Table I, the shocked cells did in fact lose from 50 to 85% of their ability to accumulate TMG and methyl-α-glucoside. Concomitant with this loss of physiological function, the cells liberated from 50 to 80% of their HPr into the surrounding medium. However, if the shocked cells were incubated with purified HPr, the active transport systems for the two glycosides were restored.

Table I, Fig. 1, and other experiments also showed the following. (*a*) Uninduced shocked cells did not accumulate TMG, and incubation with HPr showed no effect. On the other hand, the cells were responsive to HPr treatment when methyl-α-glucoside was tested. The cell membrane therefore retained its specificity through the cold shock procedure, and this specificity was not affected by incubation with HPr. (*b*) The transport systems were relatively stable to EDTA under two sets of conditions (7, 10), and the complete cold shock procedure was required to reduce sugar accumulation substantially. (*c*) HPr stimulated accumulation of TMG by induced cells only after the cells were shocked. (*d*) In one experiment, the shocked cells were washed after incubation with HPr, and retained full transport ability. (*e*) The active HPr preparations were obtained from three different lots of *E. coli* K235, and the effect of HPr was specific in that other proteins, such as serum albumin or crude *E. coli* extract (treated with charcoal to remove HPr) were inactive. (*f*) HPr became inactive for both the phosphotransferase and transport systems when treated with trypsin

or alkali or by photooxidation (9). Suitable controls, such as the addition of active HPr to the inactive preparations, showed that the latter did not contain inhibitors of the transport systems. (*g*) As observed with phenylalanine,[5] neither L-valine nor L-proline accumulation was significantly affected by the cold shock procedure. Amino acid accumulation was measured as described (11) and was sensitive to azide; *i.e.* amino acid accumulation was an energy-requiring process.

Cold-shocked cells therefore showed a reduced ability to accumulate sugars, and this ability was specifically restored by a highly purified protein, HPr. Is this a direct effect, thereby implicating the phosphotransferase in the "permease" systems, or is it indirect in the sense that HPr aids in restoring the normal energetic systems of the cell, or in the repair of a defective membrane, etc.? A conclusive answer to this question cannot yet be offered, but the following points merit consideration. (*a*) Energy-dependent amino acid accumulation was essentially unaffected by the cold shock procedure when sugar accumulation was depressed. (*b*) The effect of HPr was selective, requiring shocked cells of the proper type for the sugar under study.

The phosphotransferase system can be adapted to many of the models offered to explain the processes of active transport or facilitated diffusion (2, 12, 13). In general, these models explain facilitated diffusion by the action of specific carrier ("permease") molecules; in some cases, nonspecific transporting molecules are also invoked. Active transport results when facilitated diffusion is coupled with an energy-yielding process. In the present case, Fraction II, known to be associated with the cell membrane fraction and to be specific for certain sugars, may be regarded as the "carrier" or "permease" proteins which transport the sugars across the membrane. By coupling Fraction II with Enzyme I, HPr, and phosphoenolpyruvate, the sugars would be removed from Fraction II and active transport would result. However, this model suggests that the end product should be the sugar phosphate. Since bacterial cells generally accumulate the free sugars (2, 5, 12), it is necessary to invoke the action of a phosphatase, coupled to the phosphotransferase system, to give the final result. In this connection, Rogers and Yu reported (14) that sugar phosphates were accumulated in active transport in bacteria; in fact, the first detectable product observed with methyl-α-glucoside was its phosphate ester.[6] Furthermore, bacterial acid phosphatase is much more active with sugar phosphate than with other phosphate esters (7). We again note that Fraction II has not yet been resolved into soluble proteins of known function. The binding of the sugars and subsequent phosphorylation by this fraction may involve more than one protein; perhaps phospho-proteins are intermediates in this process, a concept that would fit current suggestions concerning nonspecific transporter molecules (13).

Several approaches have been used to characterize "transporting" or "carrier" proteins. For example, cells grown in isotopic growth media were induced to synthesize proteins controlled by the lac operon, and the labeled proteins were then fractionated (15). Another approach involved protection of the glycerol "carrier" of erythrocytes with propanediol against

---

[4] The cell preparations obtained by subjecting *E. coli* W2244 to the cold shock procedure varied substantially in terms of viability, ability to accumulate the glycosides, and response to HPr. For example, despite extensive efforts to standardize the procedure, viabilities ranged from 0 to 90%. Experiments of the type reported in Table I were conducted with 26 different cell preparations; 11 of these responded to treatment with HPr, and Table I and Fig. 1 give some typical results from four such experiments.

[5] L. Heppel, personal communication. We are most grateful to Dr. Heppel for many valuable discussions concerning the cold shock procedure. The strain used in our experiments, *E. coli* W2244, has not yet been tested in Dr. Heppel's laboratory.

[6] D. P. Kessler and H. V. Rickenberg, personal communication. Within the first minute of accumulation of methyl-α-glucoside, over 90% of the accumulated sugar is the phosphate ester.

the action of phenyl isothiocyanate, followed by exposure of the "carrier" to the reagent (16). Similarly, Kepes (17) reported that TMG "permease" was inhibited by sulfhydryl reagents and was protected by its substrate.[3] In a recent important contribution, Fox and Kennedy (18) labeled TMG "carrier" protein with isotopic N-ethylmaleimide.[7]

Whether or not the "carrier" protein and Fraction II are related remains to be determined, but preliminary experiments have shown that Fraction II (and Enzyme I, but not HPr) was inhibited by N-ethylmaleimide, and was protected against this inhibition by its substrate. Other experiments are now in progress to test further the hypothesis that the permease and phosphotransferase systems are related. For example, fluoride has been found to be a powerful and reversible inhibitor of active transport (by inhibiting enolase?); also, the data presently available suggest that metal ions that inhibit the phosphotransferase system also inhibit active transport of glycosides by *E. coli.*

TABLE I
Transport studies with E. coli W2244

All values are corrected to counts per min of ^{14}C-glycoside accumulated per 1×10^9 cells. The glycosides were ^{14}C-methyl-labeled: methyl-β-thiogalactoside, specific activity, 1.5×10^5 cpm per μmole; and methyl-α-glucoside, specific activity, 1.8×10^5 cpm per μmole. Uninduced cells were grown in a mineral medium similar to Medium 63 (8), supplemented with 0.1% Casamino acids (Difco) and 0.4% glycerol; 1 μg of thiamine per ml was also used in Experiments 1 and 2. Induced cells were grown in the same medium containing 0.001 M isopropyl-β-thiogalactoside. At approximately half their maximum growth, the cells were harvested and washed twice with 0.01 M Tris-HCl buffer, pH 7.1, containing 0.033 M NaCl, and an aliquot was suspended in the growth medium containing 50 μg of chloramphenicol succinate per ml (Medium 63-S); this aliquot, stored at 0°, represented "untreated" cells. In this solution, the cells did not divide at 25°. The remainder of the cells were subjected to the cold shock procedure used for actively growing cells (7). Stage I consisted of shaking the cells for 10 min at room temperature in 10^{-4} M EDTA containing 0.033 M Tris-HCl buffer, pH 7.1, and 0.40 M NaCl; in Stage II, the cells were shaken for 10 min at 0° with 5×10^{-4} M MgCl$_2$. After cold shock, the cells were harvested and suspended in the same medium as "untreated" cells. As indicated in Fig. 1, several experiments showed that amino acids could be omitted from Medium 63-S during the incubation and transport studies. Incubations (30 min) and transport studies were conducted at room temperature on a reciprocal shaker; after incubation, the labeled glycoside was added and the cells were shaken for 10 min (Experiments 1 and 2) or 20 min (Experiment 3). Aliquots were removed, filtered, washed with cold medium, and counted by liquid scintillation techniques. Incubations were performed only where indicated; this step was required for maximum effect with HPr. Inactive preparations of HPr were obtained as follows: (*a*) 0.6 mg of HPr was incubated at 37° for 20 min with 0.5 mg of trypsin at pH 7.3 (Tris-HCl), followed by the addition of soybean trypsin inhibitor; (*b*) 1.5 mg of HPr per ml were photooxidized (9) in the presence of 0.01 mg of rose bengal per ml, followed by passage through Dowex 1-bicarbonate resin to remove the dye; (*c*) 1.5 mg of HPr per ml in 0.1 N NaOH was maintained at 25° for 60 min, neutralized, and dialyzed. The three preparations were inactive in the phosphotransferase system while the following controls were fully active and also restored the transport ability of the shocked cells: (*a*) the trypsin and its inhibitor were mixed before addition of HPr; (*b*) the dye and HPr were mixed but not subjected to photooxidation; (*c*) the NaOH was neutralized immediately after addition of HPr. Under the conditions used here, the quantities of glycosides accumulated were proportional to cell number, had reached a maximum value after about 7 min, and were not increased when the glycoside concentrations were increased. The values given below are corrected for controls, which were as follows: Experiment 1, formalin-treated cells of all types, 1,325 to 1,500 cpm for TMG and 1,700 to 2,150 for the glucoside; Experiment 2, formalin-treated cells of all types, 281 to 389 for TMG and 403 to 471 for the glucoside; Experiment 3, sodium azide (0.043 M) used with untreated cells gave values of 306, and with shocked cells gave values of 1,200.

---

[7] A protein that shows a high affinity for inorganic sulfate (19) and is presumed to be the sulfate "carrier" has recently been isolated from the supernatant fluid obtained by the cold shock procedure, and has been purified to homogeneity (A. Pardee, personal communication).

TABLE I—*Continued*

Cell type	Incubated with	Transport studies	
		TMG	Methyl-α-glucoside
		cpm	*cpm*
Experiment 1			
Uninduced, untreated		100	8,100
Uninduced, cold-shocked		0	1,200
Uninduced, cold-shocked	HPr, 116 μg	0	8,400
Induced, untreated		7,500	
Induced, untreated	Medium 63-S	7,800	
Induced, untreated	HPr, 116 μg	7,700	
Induced (TMG) or uninduced (methyl-α-glucoside), all cold-shocked	Medium 63-S	1,800[a]	1,300[b]
	HPr, 23 μg	3,350[a]	3,400[b]
	HPr, 58 μg	5,100[a]	5,400[b]
	HPr, 116 μg	8,300[a]	8,400[b]
	HPr, 116 μg, then washed	8,100[a]	8,300[b]
Experiment 2			
Uninduced, untreated		50	4,700
Uninduced, cold-shocked		0	670
Induced, untreated		5,800	
Induced (TMG) or uninduced (methyl-α-glucoside), all cold-shocked	Medium 63-S	1,100[a]	850[b]
	Serum albumin, 113 μg	1,200[a]	960[b]
	HPr, 113 μg	11,500[a]	10,600[b]
	HPr, 113 μg, trypsin-digested	1,240[a]	970[b]
Experiment 3			
Induced, untreated		6,700	
Induced, cold-shocked		1,570	
	Medium 63-S	1,630	
	HPr, 74 μg	5,530	
	HPr, 148 μg	7,700	
	HPr, 74 μg (25 min); azide (5 min)	200	
	HPr, 74 or 148 μg, photooxidized	1,600	
	HPr, 148 μg, trypsin-digested	1,880	
	HPr, 148 μg, alkali-treated	1,450	

[a] Induced.
[b] Uninduced.

REFERENCES

1. Kundig, W., Ghosh, S., and Roseman, S., *Proc. Natl. Acad. Sci. U. S.,* **52,** 1067 (1964).
2. Kepes, A., and Cohen, G. N., in I. C. Gunsalus and R. Y. Stanier (Editors), *The bacteria,* Vol. *IV,* Academic Press, Inc., New York, 1962, p. 179.
3. Kundig, W., Dodyk Kundig, F., Anderson, B. E., and Roseman, S., *Federation Proc.,* **24,** 658 (1965).
4. Ganesan, A. K., and Rotman, B., *J. Mol. Biol.,* **16,** 42 (1966).
5. Horecker, B. L., Thomas, J., and Monod, J., *J. Biol. Chem.,* **235,** 1580 (1960).
6. Neu, H. C., and Heppel, L. A., *J. Biol. Chem.,* **240,** 3685 (1965).
7. Nossal. N. G., and Heppel, L. A., *J. Biol. Chem.,* **241,** 3055 (1966).
8. Rickenberg, H. V., Cohen, G. N., and Buttin, G., *Ann. Inst. Pasteur,* **91,** 829 (1956).
9. Westhead, E. W., *Biochemistry,* **4,** 2139 (1965).
10. Leive, L., *Proc. Natl. Acad. Sci. U. S.,* **53,** 745 (1965).
11. Britten, R. J., Roberts, R. B., and Fresch, E. F., *Proc. Natl. Acad. Sci. U. S.,* **41,** 863 (1955).
12. Christensen, H. N., *Biological transport,* W. A. Benjamin, Inc., New York, 1962.
13. Koch, A. L., *Biochim. Biophys. Acta,* **79,** 177 (1964).
14. Rogers, D., and Yu, S., *J. Bacteriol.,* **84,** 877 (1962).
15. Naono, S., Rouviere, J., and Gros, F., *Biochem. Biophys. Res. Commun.,* **18,** 664 (1965).
16. Stein, W. D., *Nature,* **181,** 1662 (1958).
17. Kepes, A., *Biochim. Biophys. Acta,* **40,** 70 (1960).
18. Fox, C. F., and Kennedy, E. P., *Proc. Natl. Acad. Sci. U. S.,* **54,** 891 (1965).
19. Pardee, A. B., and Prestidge, L. S., *Proc. Natl. Acad. Sci. U. S.,* **55,** 189 (1965).

Reprinted from *Proc. Natl. Acad. Sci. U.S.A.*, **57**, 913–919 (1967)

TWO CLASSES OF PLEIOTROPIC MUTANTS OF AEROBACTER AEROGENES LACKING COMPONENTS OF A PHOSPHOENOLPYRUVATE-DEPENDENT PHOSPHOTRANSFERASE SYSTEM*

By Shuji Tanaka and E. C. C. Lin†

DEPARTMENT OF BIOLOGICAL CHEMISTRY, HARVARD MEDICAL SCHOOL, BOSTON, MASSACHUSETTS

18

Communicated by Herman M. Kalckar, February 16, 1967

During a study of the phosphorylation of N-acyl-D-mannosamine by extracts of *Escherichia coli* and other bacteria, Kundig, Ghosh, and Roseman[1] discovered a type of phosphotransferase system consisting of three protein components acting according to the following scheme:

$$\text{PEP} + \text{HPr} \xrightleftharpoons{\text{enzyme I}} \text{P-HPr} + \text{pyruvate} \qquad (A)$$

$$\text{P-HPr} + \text{sugar} \xrightarrow{\text{enzyme II}} \text{sugar-P} + \text{HPr}, \qquad (B)$$

where PEP denotes phosphoenolpyruvate; HPr, a small heat-stable protein containing histidine; and P-HPr, the phosphorylated protein.

Their data indicated that enzyme I and HPr have broad functions serving a family of enzymes II, each specific for one or a few sugars.[1,2]

We have recently demonstrated that the dissimilation of D-mannitol by *Aerobacter aerogenes* depends upon a phosphotransferase system similar to the one described above. In the absence of an inducible enzyme II specific for D-mannitol (referred to as "enzyme II" in this work), *A. aerogenes* fails to grow on this compound.[3]

In the present report some of the phenotypic consequences of mutations affecting enzyme I or HPr are described. In contradistinction to the loss of enzyme II, which renders the cell incapable of utilizing a particular compound, the lack of enzyme I or HPr pleiotropically affects the metabolism of a number of polyhydric alcohols and sugars.

Materials and Methods.—Bacteria: Strain 5P14, a double auxotroph of *A. aerogenes* 1033 (ref. 4), provided by B. Magasanik, was used as the starting cell line. Its strict requirement for arginine and guanine permitted distinction of mutant progeny from contaminating organisms. For our convenience this strain has been called strain 2002 (ref. 3) and will be referred to as such in this communication. Strain 2006, a mutant lacking enzyme II, hence unable to grow on D-mannitol, was obtained from strain 2002 (ref. 3). Strain 2050, lacking enzyme I, and strain 2070, lacking HPr, were both isolated as D-mannitol-negative clones from populations of strain 2002 treated with ethyl methanesulfonate.[5] MacConkey agar in which D-mannitol was substituted for lactose[6] was used for the detection of these mutants. Strain 2051, a spontaneous revertant exhibiting normal growth rate on agar with D-mannose as carbon source, was derived from a population of cells of strain 2050 recycled in liquid medium with the same carbon source. Strain 2071 was isolated as a spontaneous revertant by directly plating 10^9 cells of strain 2070 on agar with D-sorbitol as carbon source.

Chemicals: 1-C14-D-mannitol (25 c/mole) was obtained from Nuclear-Chicago Corp.; D-mannitol, L-arabinose, D-mannose, and D-fructose from Eastman Kodak Co.; glycerol and maltose from Fisher Scientific Co.; D-arabitol, ribitol, D-sorbitol, D-ribose, and D-galactose from Pfanstiehl Chemical Co.; i-inositol, lactose, and acid-hydrolyzed casein from Nutritional Biochemicals Corp.; D-glucose from Merck and Co.; and PEP (sodium salt) from Calbiochem.

913

Growth of cells and preparation of components for enzyme assays: Growth rates were determined at 37° in 50 ml of mineral medium with 0.2% carbon source and 4×10^{-4} M each of arginine and guanine. The cultures were incubated in 300-ml nephelometric flasks on a rotary shaker at about 240 cycles per min. Growth was monitored by turbidity readings as before.[3] The composition of the mineral medium has also been described.[3] For assays of enzyme I, enzyme II, and HPr or for the preparation of these components for use as reagents in the assay, cells were grown to stationary phase in 250 ml of media in 2-liter Erlenmeyer flasks.

The assay of each component of the PEP-dependent phosphotransferase system was based upon the phosphorylation of labeled D-mannitol, as described below, in a reaction mixture containing the remaining two components in excess.

To obtain "enzyme II + HPr," cells from a 250-ml culture of strain 2002 fully grown on 0.2% D-mannitol were suspended in 2.5 ml of 0.1 M Tris-HCl buffer, pH 7.6, and disrupted by ultrasonic treatment for 2 min in a model 60 W MSE apparatus (Measuring and Scientific Equipment Ltd., London, England) while being chilled in a $-10°$ bath. After centrifugation at $10,000 \times g$ for 20 min at 0° to remove unbroken cells and fragments, the supernatant fraction was centrifuged at $100,000 \times g$ for 2 hr at 0°. The pellet, containing about 70% of the enzyme II activity in the original extract, was resuspended in 2.5 ml of Tris-HCl buffer. The $100,000 \times g$ supernatant fraction was heated to 100° for 10 min and clarified by centrifugation at $35,000 \times g$ for 10 min. This solution retained almost full activity of HPr and was combined with the suspension of enzyme II. The mixture contained no detectable enzyme I activity.

The procedure for obtaining "enzyme I + HPr" has already been described.[3] Cells of strain 2006 grown on 2% antibiotic medium 3 (Difco) supplemented with 1.5% D-glucose were used as the starting material. The extract contained no enzyme II activity.

To prepare "enzyme I + enzyme II," cells of strain 2002 fully grown on 0.2% D-mannitol were suspended in 2.5 ml of Tris-HCl buffer, and disrupted by sonication as above. The extract was centrifuged at $10,000 \times g$ for 20 min at 0°. The supernatant fraction was treated with 250 mg of Norit A for 30 min at 0°. The supernatant fraction after centrifugation contained less than 0.5% of the HPr activity originally present in the crude extract.

All three preparations could be stored for several weeks at $-20°$ without significant loss of activity.

Enzyme assays: All enzyme activities are expressed as mμmoles of D-mannitol phosphorylated per min per mg protein at 25°. All the reactions were carried out in the presence of 5×10^{-5} M C^{14}-D-mannitol (2.5 c/mole), 5×10^{-3} M PEP, 5×10^{-5} M MgCl$_2$, and 0.04 M Tris-HCl (pH 7.6). For measuring enzyme I activity, 0.1 ml of "enzyme II + HPr" was added; for measuring enzyme II activity, 0.1 ml of "enzyme I + HPr" was added; and for measuring HPr activity, 0.1 ml of "enzyme I + enzyme II" was added. An appropriate amount of the crude extract to be assayed was added to initiate the reaction. In each case the final volume of the mixture was 0.4 ml. The reaction was allowed to proceed for 10 min after which further phosphorylation of the labeled substrate was arrested by the addition of 0.6 ml of 0.2 M unlabeled D-mannitol. The labeled D-mannitol phosphate was separated from the free hexitol by passing the mixture through a Dowex-1 formate column. D-mannitol phosphate was eluted and counted in a planchet as previously described.[3]

Protein concentrations were measured with the biuret reaction.[7]

Assay of C^{14}-D-mannitol uptake: Cells grown overnight in 50 ml of medium with 1% casein hydrolysate as carbon source were diluted 100-fold with fresh medium containing 1% casein hydrolysate in the presence or absence of 2% D-mannitol. After reaching 100 Klett units, the cells were washed twice and resuspended at the same density in mineral medium at 0°. Two-tenths ml of this suspension was added to a test tube, equilibrated at 25°, containing 0.1 ml of 5×10^{-5} M C^{14}-D-mannitol (25 c/mole), 0.1 ml of chloramphenicol (400 μg/ml), and 0.6 ml of mineral medium. After 30 seconds the cells were quantitatively collected on a Millipore filter, washed, and counted.[8] Under the conditions used, the rate of D-mannitol uptake was linear for at least 1 min. Cells metabolically poisoned with 0.03 M azide and 0.001 M iodoacetate[9] for 1 hr at 25° immediately prior to the assay served as controls.

Results.—Strains 2050 and 2070, originally isolated as D-mannitol nonfermenting mutants, exhibited multiple impairments of growth when tested on agar plates

TABLE 1

GROWTH PATTERNS OF PARENTAL, MUTANT, AND REVERTANT
STRAINS ON VARIOUS CARBOHYDRATES

Carbon Source	Strains				
	2002 (Parental)	2050 (Mutant)	2051 (Revertant)	2070 (Mutant)	2071 (Revertant)
Glycerol	+ +	+	+ +	+ +	+ +
D-arabitol	+ +	+ +	+ +	+ +	+ +
Ribitol	+ +	+ +	+ +	+ +	+ +
D-mannitol	+ +	0	+ +	0	+ +
D-sorbitol	+ +	0	+ +	0	+ +
i-Inositol	+ +	+ +	+ +	+ +	+ +
L-arabinose	+ +	+ +	+ +	+ +	+ +
D-ribose	+ +	+ +	+ +	+ +	+ +
D-glucose	+ +	+	+ +	+	+ +
D-mannose	+ +	+	+ +	+	+ +
D-galactose	+ +	+ +	+ +	+ +	+ +
D-fructose	+ +	0	+ +	+	+ +
Lactose	+ +	+ +	+ +	+ +	+ +
Maltose	+ +	+	+ +	+ +	+ +

Growth was tested on agar plates. Normal growth rate of colonies is denoted by + +, slow growth by +, and no growth by 0.

(Table 1). Strain 2050 not only failed to grow on D-mannitol, but also on D-sorbitol and D-fructose. Its rates of growth on glycerol, D-glucose, D-mannose, and maltose were considerably slower than those of the parental strain 2002. A similar pattern was exhibited by strain 2070, except that it grew normally on glycerol and slightly on D-fructose.

Table 2 gives the doubling times of the parental and the two mutant strains with D-mannitol, D-glucose, and D-arabitol as sole carbon source in liquid medium.

To test whether the multiple impairments in the two mutants were phenotypic expressions of single mutations, a *spontaneous* revertant was obtained from each mutant by selection on a single carbon source. As shown in Tables 1 and 2, strain 2051, selected from strain 2050 on D-mannose, concomitantly regained all the other growth abilities. Similarly, full restoration of the growth properties was observed in strain 2071, selected from strain 2070 for growth on D-sorbitol.

The wild-type, mutant, and revertant strains were grown on casein hydrolysate in the presence or absence of D-mannitol and their cell-free extracts were analyzed for enzyme I, enzyme II, and HPr under conditions in which their activities were proportional to concentration. (See Fig. 1A for assay of enzyme I, Fig. 1B for assay of HPr, and ref. 3 for assay of enzyme II.) (The slight deviation of enzyme I activity from linearity at very low concentrations suggests the requirement of another factor in the phosphotransferase system or dissociation of enzyme I at high dilution. Table 3 shows that the activity of enzyme I is absent in strain 2050 and the activity of HPr is absent in strain 2070. The missing activity in each case is restored in the revertant.

TABLE 2

DOUBLING TIMES OF PARENTAL, MUTANT, AND REVERTANT STRAINS ON THREE
DIFFERENT CARBOHYDRATES

Carbon source	Strains				
	2002 (Parental)	2050 (Mutant)	2051 (Revertant)	2070 (Mutant)	2071 (Revertant)
D-mannitol	52	n.g.*	54	n.g.*	49
D-glucose	45	132†	54	160†	54
D-arabitol	47	54	50	65	55

All cells were pregrown on casein hydrolysate and transferred to fresh medium containing the new carbon source at a level of 0.2%. Doubling time is expressed in minutes.
* No growth.
† Slow growth commenced after a lag period of 3–4 hr instead of the usual lag time of about 20 min.

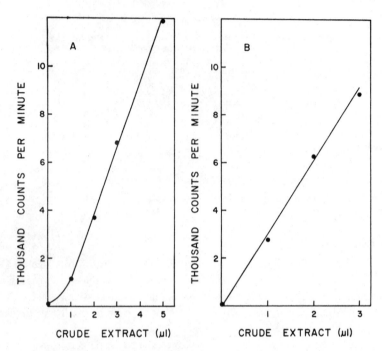

Fig. 1.—(*A*) Activity versus concentration of enzyme I. (*B*) Activity versus concentration of HPr (in this case a significant blank value of 1500 cpm has been subtracted).

Table 3 indicates that the presence of D-mannitol stimulated the production of enzymes I and II.   The effect on enzyme I is not specific; addition of glycerol, D-arabitol, D-sorbitol, D-glucose, D-fructose, or D-mannose to the casein hydroly-sate medium also resulted in similar elevations (data not shown).   The induction of enzyme II by D-mannitol is highly specific.   Glycerol, D-arabitol, D-sorbitol, D-glucose, D-fructose, or D-mannose in the growth medium caused no appreciable increase in the specific activity of this enzyme (data not shown).   There was also apparently a slight stimulation of HPr formation by D-mannitol.   Slight stimula-tion of HPr formation by other polyhydric alcohols and sugars has likewise been observed.

Table 3 also shows that D-mannitol was able to induce enzyme II in strain 2050 (with no measurable enzyme I activity) and in strain 2070 (with no measurable HPr activity).   Moreover, the lack of HPr activity did not prevent the hexitol from

TABLE 3

SPECIFIC ACTIVITIES OF ENZYME I, ENZYME II, AND HPR IN PARENTAL,
MUTANT, AND REVERTANT STRAINS

Strain	Growth on Casein Hydrolysate			Grown on Casein Hydrolysate Plus D-Mannitol		
	Enzyme I	Enzyme II	HPr	Enzyme I	Enzyme II	HPr
2002 (Parental)	3.4	0.2	8.2	14	7.4	13
2050 (Mutant)	0.0	0.3	6.6	0.0	4.8	8.8
2051 (Revertant)	2.5	0.1	7.5	12	4.7	9.7
2070 (Mutant)	4.8	0.2	0.0	14	13	0.0
2071 (Revertant)	2.7	0.2	8.2	12	6.5	14

Casein hydrolysate was added at a level of 1% and D-mannitol, when used, at 2%.   All enzyme activities are expressed as mμmoles of D-mannitol phosphorylated per min per mg protein at 25°.

TABLE 4

ACCUMULATION OF RADIOACTIVE MATERIAL BY CELLS
INCUBATED WITH C^{14}-D-MANNITOL

Strain	Grown on casein hydrolysate (cpm)	Grown on casein hydrolysate plus D-mannitol (cpm)
2002 (Parental)	822	5534
2050 (Mutant)	183	391
2051 (Revertant)	930	5304
2070 (Mutant)	14	80
2071 (Revertant)	889	6214

The incubation conditions were as in *Materials and Methods.* The counts retained by poisoned cells (21 cpm) have been subtracted from each value.

influencing the level of enzyme I in strain 2070. Hence it must be concluded that the mutant cells were sufficiently permeable to D-mannitol for induction to occur.

The question arises as to whether there is a specific transport mechanism which permits the cell to bring D-mannitol into its cytoplasm against a concentration gradient. If so, one would expect cells of strain 2050 and 2070 to retain their potential for accumulating free D-mannitol, although they can no longer phosphorylate it. Cells of parental, mutant, and revertant strains, grown in the presence and absence of the inducer, were tested for their ability to take up radioactive material when incubated with C^{14}-D-mannitol. The results show that substantial retention of labeled material (presumably as D-mannitol 6-phosphate and its metabolites) occurred only in the parental and revertant cells, and was most marked when they were induced (Table 4). The accumulation process was severely curtailed, although not entirely abolished, in the mutant cells. The fact that the uptake in the mutant cells could also be increased several-fold by induction suggests that the mutations affecting enzyme I and HPr were actually slightly leaky and that the residual function could be more readily detected by the highly sensitive *in vivo* experiment. In any event, the large disparities between the wild-type and mutant cells in the ability to accumulate radioactive substances, *with or without induction,* strongly imply that phosphorylation of D-mannitol is indispensable for the retention and that there is no mechanism for concentrating the *free hexitol* against a gradient.

Discussion.—The pleiotropic growth defects associated with the loss of enzyme I and HPr activities establish the importance of this phosphotransferase system for the utilization of not only D-mannitol, but also of D-sorbitol, D-glucose, D-mannose, and D-fructose. The residual capacity of these mutants for slow growth on D-glucose and D-mannose is probably due to the function of minor alternative pathways in this organism.[10-13] Inasmuch as the enzyme I and HPr mutants reported here were both isolated on the basis of D-mannitol fermentation, the possibility cannot be excluded that there exists more than one kind of enzyme I-HPr system in the same organism and that each system governs the metabolism of a different family of carbohydrates. Pleiotropic mutations affecting the utilization of carbohydrates in other species of bacteria, e.g., *E. coli*[14-16] and *Staphylococcus aureus*[17-19] have been reported before. It would be interesting to examine whether any of them lack a component of an enzyme I-HPr system. (In collaboration with D. G. Fraenkel we have just shown that the *E. coli* mutant MM6 (ref. 16) lacks enzyme I. A more detailed account of this finding will be given elsewhere.)

Kundig and co-workers deduced, from the retention of sugar analogues in osmotically shocked cells, that the PEP-dependent phosphotransferase system plays a direct role in carbohydrate transport.[2]   The present study provides genetic evidence of the operation of the phosphotransferase system in the accumulation of certain sugars.   The transport of D-mannitol was amenable to analysis because its first phosphorylated product is directly utilized by a specific dehydrogenase.[20, 3] Hence retention of molecules, as distinguished from their passage into the cells,[21–25] is associated with the first step of dissimilation.   In such a situation a separate active transport system for the free hexitol would not be required and is in fact not found.

The entry of the free hexitol may be facilitated by a protein, possibly by enzyme II itself.   The process of D-mannitol uptake is thus analogous to the capture of glycerol by *E. coli*, except that glycerol is believed to diffuse freely[26] and that glycerol kinase utilizes ATP as the phosphoryl donor.[27]   The presence of a specific protein mediating the equilibration of D-mannitol could be tested by the counterflow method successfully used in the study of the lactose permeation system of *E. coli* by Winkler and Wilson.[9]   If such a protein exists, its relationship to enzyme II could be examined by the use of temperature-sensitive mutants.

The basic mechanism for D-mannitol uptake has the virtue that a single high-energy phosphate bond is expended for both capture and activation of a substrate molecule.   The model proposed by Kundig and collaborators as a process of transport requires phosphorylation and immediate dephosphorylation.[2]   Such a model would therefore demand the demonstration of a highly specific phosphatase inside the cell.

Finally it should be pointed out that some sugars may be accumulated in cells by mechanisms different from those discussed here, i.e., without phosphorylation.   For example, the transport system for D-galactose also acts on D-fucose,[28] which cannot be phosphorylated at carbon six.

Summary.—Two classes of pleiotropic mutants of *Aerobacter aerogenes* with multiple defects in sugar and hexitol utilization have been isolated.   One mutant lacks the enzyme I and another lacks the histidine protein (HPr) of the phosphoenol-pyruvate-dependent phosphotransferase system.   Both mutants fail to grow on D-mannitol and D-sorbitol and grow slowly or very poorly on D-glucose, D-mannose, D-fructose, or maltose.   Neither mutant is able to accumulate significant amounts of radioactive material when incubated with C^{14}-D-mannitol, even after growth in the presence of 2 per cent D-mannitol.   A spontaneous revertant selected on a single carbon source was isolated from each mutant.   All the other pleiotropic defects were corrected in the revertant.   A model for the role of this phosphotransferase in active transport is proposed.

We are grateful to T. H. Wilson, H. M. Kalckar, and J. B. Alpers for helpful comments during the preparation of this manuscript.

* This work was aided by the National Science Foundation (GB-5854) and the U.S. Public Health Service (GM-11983).
† Supported by a Research Career Development Award from the U.S. Public Health Service.

1 Kundig, W., S. Ghosh, and S. Roseman, these Proceedings, 52, 1067 (1964).
2 Kundig, W., F. D. Kundig, B. Anderson, and S. Roseman, *J. Biol. Chem.*, 241, 3243 (1966).
3 Tanaka, S., S. A. Lerner, and E. C. C. Lin, *J. Bacteriol.*, 93, 642 (1967).
4 Magasanik, B., M. S. Brooke, and D. Karibian, *J. Bacteriol.*, 66, 611 (1953).
5 Lin, E. C. C., S. A. Lerner, and S. E. Jorgensen, *Biochim. Biophys. Acta*, 60, 422 (1962).

[6] Cozzarelli, N. R., and E. C. C. Lin, *J. Bacteriol.*, **91**, 1763 (1966).

[7] Gornall, A. G., C. J. Bardawill, and M. M. David, *J. Biol. Chem.*, **171**, 751 (1949).

[8] Hayashi, S., J. P. Koch, and E. C. C. Lin, *J. Biol. Chem.*, **239**, 3098 (1964).

[9] Winkler, H. H., and T. H. Wilson, *J. Biol. Chem.*, **241**, 2200 (1966).

[10] Magasanik, A. K., and A. Bojarska, *Biochem. Biophys. Res. Commun.*, **2**, 77 (1960).

[11] Kamel, M. Y., and R. L. Anderson, *J. Biol. Chem.*, **239**, PC3607 (1964).

[12] Kamel, M. Y., and R. L. Anderson, *J. Bacteriol.*, **92**, 1689 (1966).

[13] Kamel, M. Y., D. P. Allison, and R. L. Anderson, *J. Biol. Chem.*, **241**, 690 (1966).

[14] Doudoroff, M., W. Z. Hassid, E. W. Putman, A. L. Potter, and J. Lederberg, *J. Biol. Chem.* **179**, 921 (1949).

[15] Asensio, C., G. Avigad, and B. L. Horecker, *Arch. Biochem. Biophys.*, **103**, 299 (1963).

[16] Fraenkel, D. G., F. Falcoz-Kelly, and B. L. Horecker, these PROCEEDINGS, **52**, 1207 (1964).

[17] Egan, J. B., and M. L. Morse, *Biochim. Biophys. Acta*, **97**, 310 (1965).

[18] *Ibid.*, **109**, 172 (1965).

[19] *Ibid.*, **112**, 63 (1966).

[20] Liss, M., S. B. Horwitz, and N. O. Kaplan, *J. Biol. Chem.*, **237**, 1342 (1962).

[21] Cohen, G. N., and J. Monod, *Bacteriol. Rev.*, **21**, 169 (1957).

[22] Davis, B. D., *Arch. Biochem. Biophys.*, **78**, 497 (1958).

[23] Horecker, B. L., M. J. Osborn, W. L. McLellan, G. Avigad, and C. Asensio, in *Membrane Transport and Metabolism*, ed. A. Kleinzeller and A. Kotyk (New York: Academic Press, Inc., 1960), p. 378.

[24] Koch, A. L., *Biochim. Biophys. Acta*, **79**, 177 (1964).

[25] Fox, C. F., and E. P. Kennedy, these PROCEEDINGS, **54**, 891 (1965).

[26] Hayashi, S., and E. C. C. Lin, *Biochim. Biophys. Acta*, **94**, 479 (1965).

[27] Hayashi, S., and E. C. C. Lin, *J. Biol. Chem.*, **242**, 1030 (1967).

[28] Ganesan, A. K., and B. Rotman, *J. Mol. Biol.*, **16**, 42 (1966).

Reprinted from *Proc. Natl. Acad. Sci. U.S.A.*, **58**, 1963–1970 (1967)

GENETIC EVIDENCE FOR THE ROLE OF A BACTERIAL
PHOSPHOTRANSFERASE SYSTEM IN SUGAR TRANSPORT*

By Robert D. Simoni, Mark Levinthal, F. Dodyk Kundig,
Werner Kundig, Byron Anderson, Philip E. Hartman,
and Saul Roseman

McCOLLUM-PRATT INSTITUTE AND DEPARTMENT OF BIOLOGY, THE JOHNS HOPKINS UNIVERSITY,
BALTIMORE, MARYLAND

Communicated by W. D. McElroy, September 8, 1967

We previously reported the isolation[1] and some properties[2,3] of a bacterial phosphotransferase system, comprised of two enzymes, I and II, and a heat-stable low-molecular-weight protein, called HPr:

$$\text{I.} \quad \text{Phosphoenolpyruvate} + \text{HPr} \underset{}{\overset{Mg^{++}}{\rightleftharpoons}} \text{Pyruvate} + \text{Phospho-HPr}$$

$$\text{II.} \quad \text{Phospho-HPr} + \text{Sugar} \xrightarrow{Mg^{++}} \text{Sugar-6-P} + \text{HPr}$$

$$\text{I} + \text{II.} \quad \text{Phosphoenolpyruvate} + \text{Sugar} \xrightarrow{Mg^{++},HPr} \text{Sugar-6-P} + \text{Pyruvate}$$

Enzyme II, the membrane-bound component of the system, determines its carbohydrate specificity; five such enzymes have thus far been identified.

A close relationship between the phosphotransferase and carbohydrate permease systems was reported earlier[4] based on results obtained with osmotically shocked *Escherichia coli* W2244;[5] the cells lost most of their HPr along with their ability to transport two glycosides. Other transport systems were subsequently studied by this method.[6-8]

In the course of the present experiments, results obtained[9,10] with mutants of *E. coli* and *Aerobacter aerogenes* (lacking either enzyme I or HPr) showed that the phosphotransferase system was involved in either the transport and/or metabolism of five sugars. The present report is concerned with a single mutation in *Salmonella typhimurium* that results in an inability to grow on nine carbohydrates concomitant with a loss of enzyme I. The mutant was also seriously defective in its ability to transport sugars into the cell. We therefore conclude that the phosphotransferase system is intimately involved in sugar transport.

Materials and Methods.—Unless otherwise specified, all materials were obtained from commercial sources. The parent strain, *Salmonella typhimurium* SB497 (*rfb-816, his-1367*), was grown in liquid culture in a mineral-salts medium, as described in Table 1, or in nutrient broth (Difco). To obtain the desired mutant, the parent strain was treated with nitrosoguanidine,[11] grown in the mineral medium in the presence of melibiose and penicillin,[12] and the resulting culture spread on eosin-methylene blue plates (EMB) containing melibiose in order to obtain melibiose-negative colonies. Several such colonies were only melibiose-negative, while one mutant, designated SB703, lacked the ability to ferment eight other carbohydrates, as described below. This carbohydrate lesion was termed car⁻, by analogy with the car⁻ mutants of *Staphylococcus aureus*.[13] Fermentation tests were conducted on peptone EMB plates,[14] or on the more sensitive bromthymol blue indicator plates,[15] in the presence of 1 per cent carbohydrate.

1963

To avoid the ambiguities of the fermentation tests as an index of carbohydrate utilization, growth curves were obtained in liquid culture for SB497 and SB703, as described in Table 1.

Utilization of Carbohydrates by Strains SB497 and SB703.—Growth experiments with various carbon sources gave the results shown in Table 1. In contrast to the parent strain, SB497, the mutant SB703 was pleiotropic carbohydrate-negative since it was unable to grow on nine carbohydrates utilized by the parent strain (at high concentrations, glucose was slowly utilized by the mutant). The same results were obtained in fermentation tests on agar plates, with both the EMB and bromthymol blue media, i.e., the nine carbohydrates were not fermented.

The defect in SB703 was not likely to be in one of the enzymes of the glycolytic or tricarboxylic acid pathways, since the mutant could grow on and ferment the following sugars: galactose, glucose-6-P, gluconic and glucuronic acids, and three pentoses. (The generation times shown in Table 1 suggest that the mutant and parent strains were equally capable of growth on the pentoses, but this is not strictly true since the mutant showed a prolonged lag phase on the pentoses before measurable growth commenced.)

Enzymatic Defect in Mutant SB703.—Assay for the three protein components of the phosphotransferase system showed that the defect in SB703 was the loss of enzyme I activity (Table 2). The mutant contained enzyme II and HPr, and, in

TABLE 1
GROWTH OF *S. typhimurium* ON VARIOUS CARBON SOURCES

Carbon source*	Generation Time in Minutes	
	SB497	SB703
Glucose	73	N.d.
" , 0.5%	72	"
" , 1.0%	75	300
Glycerol, 0.2–1.0%	107	N.d.
Fructose	95	"
Mannose	120	"
N-Acetylglucosamine	80	"
Mannitol	80	"
Sorbitol	75	"
Maltose	120	"
Melibiose	90	"
L-Arabinose	95	100
Ribose	90	210
Xylose	210	210
Galactose, 0.05–1.0%	70–80	85–95
Glucose-6-P	70	73
Gluconic acid	100	80
Glucuronic Acid	85	85
L-Lactic Acid	95	95
Pyruvate	100	99

* All sugars are of the D-configuration unless otherwise indicated. The symbol n.d. = not detectable, i.e., division times exceeded 600 min. Inocula of the two strains were obtained by growth in 1% nutrient broth (Difco) to the middle of the exponential phase, the cells harvested aseptically by centrifugation at room temperature, washed twice with the mineral medium described below, and transferred to 15 ml of the mineral medium containing the indicated substances at 0.2% concentration, unless indicated otherwise. Growth was conducted in Erlenmeyer flasks (125 ml) containing Klett tube side-arms, at 37° and 250 rpm on a New Brunswick rotary shaker, and was followed for at least 10 hr by measuring turbidity in the Klett colorimeter at 500 mμ. Generation times were calculated from semilogarithm plots. The growth medium contained the following components (per liter): 0.05 mole K phosphate buffer, pH 7.3; (NH₄)₂SO₄, 2 gm; MgSO₄·7H₂O, 0.2 gm; FeSO₄·7H₂O, 0.5 mg; L-histidine, 30 mg; the specified carbon source. The latter two components were separately sterilized, the sugars by filtration. The following substances did not support growth of either strain: arabitol, ribitol, erythritol, galactitol, L-rhamnose, L- and D-fucose, trehalose, sucrose, myo-inositol, L-sorbose. raffinose, melezitose, lactose, glucoheptonic lactone, cellobiose, and D-lactate.

TABLE 2

ASSAY FOR PHOSPHOTRANSFERASE SYSTEM IN *S. typhimurium*

Strain*	Specific Activity in Crude Extracts		
	Enzyme I	Enzyme II	HPr
SB497	0.36	0.12	0.16
SB703	0.00	0.39	0.20
Transductants 1 through 9†	0.32–0.40	0.13–0.17	

* Enzyme isolations and assays were conducted by minor modifications of the described procedures,[2] using exponentially growing cells. The three protein fractions from *S. typhimurium* showed essentially the same properties as those from *E. coli*, and substituted for the latter in the assay. Purified protein fractions from SB497 were used for quantitation of each component in crude extracts of all cell types. Specific activity is defined as μmoles of methyl α-glucoside phosphorylated per mg protein in 30 min at 37°. Enzymes II from both strains catalyzed the phosphorylation of the following: glucose, mannose, galactose, fructose, and methyl β-galactoside.
† Each of the nine transductants was obtained by selection on one of the nine sugars not utilized by SB703, as described in the text. Crude extracts of these cells were not assayed for HPr.

fact, enzyme II was consistently found to be present at higher levels of activity (using methyl α-glucoside as substrate) in crude extracts of the mutant when compared to similar extracts of the parent strain. Various control experiments showed that the inability to detect enzyme I activity in extracts of the mutant resulted from the absence of the enzyme, rather than from the presence of inhibitors, or absence of cofactors, or other possible artifacts. If enzyme I was present in extracts of the mutant, then the level was less than 0.5 per cent of that observed in extracts of the parent strain, SB497.

Genetic Analysis of SB703.—To support the conclusion of this paper, that the pleiotropic car⁻ defect observed in SB703 resulted solely from an inability to synthesize enzyme I with an accompanying defect in carbohydrate permease systems, it was of critical importance to prove that the genetic defect was the consequence of a single event. These experiments were conducted as follows. SB703 was first converted from a rough his⁻ to a smooth his⁺ strain by conjugation.[16] The recombinant, which could be used for transduction experiments, showed all of the expected properties, including the car⁻ phenotype of SB703, and no detectable enzyme I in crude extracts. This recombinant was used for reversion and transduction experiments.

Reversion of smooth SB703 was induced with nitrosoguanidine, and revertants were selected by growth on minimal media agar plates containing one of the nine carbohydrates not utilized by SB703. Under these conditions, cells would grow that had either regained the car⁺ phenotype (complete revertants), or that could utilize only the sugar employed for selection (partial revertants). The latter type of revertant was observed only when glucose and maltose were used for selection, and even here, most of the revertants were car⁺. The partial revertants were not characterized other than to note that those obtained by selection on glucose medium fermented only glucose, while those obtained by selection on maltose fermented both maltose and glucose; the partial revertants did not contain detectable levels of enzyme I.

The reversion tests consisted of selecting 24 colonies from each of the 9 growth plates, and testing these colonies for their ability to ferment the sugar used for growth, as well as the remaining 8 compounds. As indicated above, most of the colonies obtained on glucose and maltose growth plates were car⁺, while all 168 colonies selected by growth on the following sugars were car⁺: mannose, fructose, glycerol, mannitol, sorbitol, *N*-acetylglucosamine, and melibiose. Further, one revertant selected on each sugar was grown on a large scale, the cells ruptured, and

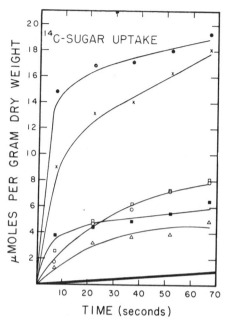

FIG. 1.—Uptake of C14-sugars by SB497 and SB703. Uninduced cells were used, i.e., both strains were grown in minimal medium containing lactate as described in Table 1. Transport experiments were conducted by minor modifications of the described procedure[3] using optimum conditions for filtration and washing.[18] C14-melibiose was unavailable for study, while C14-glycerol and L-arabinose were not taken up to a significant extent by either the parent or mutant uninduced cells. C14-mannose, unlike the six sugars shown above, was taken up to a slight but significant extent by the mutant, the rate being about one fourth that of the parent strain. The following sugars were taken up by the parent strain, SB497: glucose (●), mannitol (X), sorbitol (○), fructose (□), N-acetylglucosamine (■), and maltose (△). The lowest, heavy curve gives the average results obtained when uptake of these six sugars was studied with the mutant, SB703; at 7 sec there was no detectable uptake of any of the C14-compounds.

the crude extracts assayed; each of these extracts contained a normal level of enzyme I.

Transduction studies, which were even more convincing, were conducted with phage P-22,[17] using the selection method described above. One hundred colonies were selected from each plate (containing one of the nine sugars not utilized by SB703), and the colonies were then tested for their ability to ferment all of the nine carbohydrates. By these fermentation tests, using both EMB and bromthymol blue indicator plates, all 900 transductants, each of which had been selected by growth on a single sugar, fermented all nine sugars, and were therefore car+. Again, assay of one transductant of each type showed the presence of enzyme I (Table 2).

The results of the reversion and transduction studies clearly showed that the mutation from the car+ to car- phenotype resulted from a single genetic event, and that this event gave rise to cells lacking enzyme I of the phosphotransferase system.

Transport Studies and "Galactose Effect."—An inability to synthesize enzyme I would result in a car- phenotype if the phosphotransferase system was involved in the metabolism of the nine carbohydrates, or in their transport into the cell, or in both processes. As shown in Figure 1, incubation of the cells with various radioactive sugars resulted in a rapid accumulation of these substances by the parent strain, while uptake by the mutant cells was negligible at any time, and was not detectable at the earliest point, seven seconds. Since it will be shown that the car- mutant contained the *internal* enzyme systems necessary for the utilization of glucose, and since the mutation was a single event involving the synthesis of enzyme I, we presume that the enzyme systems for the utilization of the remaining eight substances were also intact in the mutant. Therefore, in spite of the obvious objections to the use of metabolizable substances for transport studies, the results in Figure 1 strongly suggest that the mutant was unable to transport these substances from the medium into the cell.

This important conclusion was substantiated by extensive studies with two non-metabolizable derivatives, methyl α-glucopyranoside (MeGlu) which is thought to be transported by a constitutive glucose permease,[19] and methyl β-galactopyranoside (MeGal) which is transported by an inducible methyl galactoside (MG) permease.[20]   In transport experiments under "steady-state" conditions (10 min), cells of the parent strain accumulated the glycosides against a concentration gradient, while the mutant cells did not.   These experiments were conducted at various concentrations of the glycosides, in the presence and absence of lactate,[21] and with cells containing the glucose permease (lactate-grown), or both the glucose and MG permeases (grown on galactose, or lactate plus 10^{-3} M fucose).

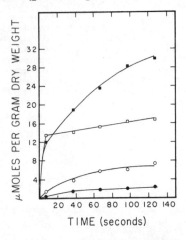

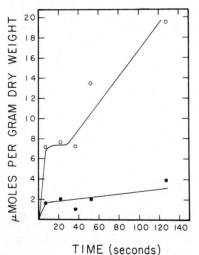

Fig. 2.—Uptake of methyl α-glucoside by SB497 and SB703. Transport studies were conducted as described in Fig. 1. Cells were grown as described in Table 1: (■), SB497 grown on lactate; (□), SB497 grown on galactose; (●), SB703 grown on lactate; (○), SB703 grown on galactose. Transport experiments were conducted with the glycoside at 6.9 × 10^{-3} M concentration. The uptake curves showed essentially the same patterns at 7 concentrations of the glycoside, ranging from 0.2 × 10^{-3} to 6.9 × 10^{-3} M.

Fig. 3.—Uptake of methyl β-galactoside by SB497 and SB703. Transport studies were conducted as described in Figs. 1 and 2. Cells were grown on galactose: (○), SB497; (●), SB703. The curves were typical of those obtained in six different experiments using the glycoside at 6.9 × 10^{-3} M concentration. Similar results were obtained at 7 different concentrations over the range from 0.2 × 10^{-3} to 6.9 × 10^{-3} M, including the unexpectedly low value at 37 sec in the case of the mutant.

A critical parameter of the physiological effect of transport is the maximum rate of entry of the sugars into the cell ($V_{\max}^{\text{in}}$).   In E. coli, for example, the rate-limiting step in lactose utilization is its rate of uptake.[19]   The $V_{\max}^{\text{in}}$ values for the glycosides were therefore determined by kinetic studies of the type illustrated in Figures 2 and 3, and Table 3.   The mutant clearly displayed a major defect in its ability to take up MeGlu via the glucose permease.   (The $V_{\max}^{\text{in}}$ values for the parent and mutant cells grown on lactate differed by 50-fold.)   This result agrees with that shown in Figure 1 for glucose uptake, and supports the interpretation given above.

The mutant was also defective in MG permease, although the quantitative effect was much less, the $V_{\max}^{\text{in}}$ values for MeGal differing only fourfold for the two

TABLE 3

KINETIC VALUES FOR TRANSPORT OF GLYCOSIDES BY *S. typhimurium**

Cell type	Carbon source used for growth	C14-Glycoside transported	Estimated Entry $K_m \times 10^3$	V_{max}
SB497	Lactate	Methyl α-glucoside	1.1	1.9
	Galactose	"	1.1	1.9
SB703	Lactate	"	?	0.04 (?)
	Galactose	"	0.2–1.0	0.2–0.3
SB497	Galactose	Methyl β-galactoside	0.87	1.5
SB703	Galactose	"	1.0	0.38

* Cells were grown as described in Table 1, and transport studies were conducted as described in Figs. 1–3, except that 7 concentrations of glycosides were used ranging from 0.20×10^{-3} to 6.9×10^{-3} M Internal concentrations of the glycosides in the mutant, SB703, were frequently less than and did not, at any time exceed the external concentrations. In contrast to kinetic data on the y permease system in *E. coli*,[22] the data in the present experiments were not readily interpretable. That is, the uptake of MeGlu was biphasic in the first 2 min and then became triphasic (Fig. 2), while the uptake of MeGal was triphasic in the first 2 min (Fig. 3). The rates of uptake were related to substrate concentrations only during the first period, i.e., at 7 sec, and these points were therefore used to determine rates of entry. In one case, MeGlu uptake by the mutant SB703 grown on lactate, the values were too low for kinetic analysis, and the V_{max} shown in the table (0.04) was estimated from the original data. The remaining values were determined by Lineweaver-Burk plots (reciprocals of the glycosides accumulated at 7 sec. vs. reciprocals of the external glycoside concentrations). Despite the experimental errors, the plots were reasonably linear with the exception of the low values obtained when MeGlu was transported by SB703 grown on galactose. In this case, the range of calculated values is given. The K_m values are presented as $M \times 10^3$ concentrations, while the V_{max} represents the maximum rate of entry in μmoles per gram dry weight per second. The V_{max} values observed for glycoside uptake by the parent strain, SB497, in the present studies, approximate those found for lactose uptake by *E. coli*.[19, 23]

cell types. It is important to note that while only negligible quantities of MeGlu were taken up by the mutant *via* the glucose permease, significant quantities were taken up by the mutant cells containing the MG permease (Fig. 2, Table 3), and that the rate of entry was approximately the same as that of MeGal. This observation agrees with the recent report that glucose can enter the cell via a galactose permease system.[24]

An apparent anomaly is suggested by the studies on the MG permease. We concluded that the mutant was defective in this system, yet galactose (which presumably enters the cell via the MG permease[20]) was utilized almost as efficiently by the mutant as by the parent strain (Table 1) (see *Note added in proof*). One explanation for these results is that the defective MG permease system in the mutant, which catalyzed only facilitated diffusion, nevertheless allowed entry of sufficient galactose to support growth. By the same token, since it was shown that the mutant MG permease system transported MeGlu at approximately the same rate as

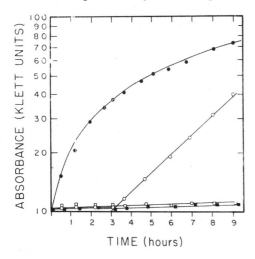

FIG. 4.—Growth of the car⁻ mutant, SB703, on glucose. The cells were first grown either in minimal medium containing 0.2% galactose (Table 1), or in nutrient broth. After harvesting and washing twice with mineral medium, the galactose-grown cells were transferred to mineral medium, containing 0.2% glucose (●), and the nutrient broth-grown cells were transferred to mineral medium containing 0.2% glucose (□), or 0.001 M fucose (■), or both sugars (O). Growth was followed as described in Table 1.

MeGal, then glucose should support limited growth of mutant cells containing the MG permease. This prediction was correct. It can be seen from Figure 4 that SB703, pregrown on galactose to induce the MG permease, was able to grow for about three generations in a glucose minimal medium. The growth curve was that expected where the permease could no longer be synthesized; maximum growth occurred at zero time, followed by a gradual decrease as the permease system became diluted. Similarly, mutant cells grown in nutrient broth, and transferred to glucose-minimal medium containing fucose as an inducer of MG permease, showed exponential growth. These experiments also prove that the mutant contained the necessary enzyme systems for glucose utilization provided that the sugar was transported into the cell at an adequate rate.

Discussion.—We conclude from the results presented above that at least nine carbohydrates are transported into bacterial cells by the action of the phosphotransferase system, and that the latter is either identical to or a key component of the carbohydrate permease systems. The specificities of these systems are determined by different enzymes II. Further, we conclude that the phosphotransferase system is crucial to the transport process, and *not* in the utilization of the sugars.

These conclusions are based on the following observations. A single genetic event resulted in a phenotypic change from car$^+$ to car$^-$, where car$^-$ is defined as an inability to grow on or ferment the following: glycerol, glucose, fructose, mannose, mannitol, sorbitol, *N*-acetylglucosamine, maltose, and melibiose. The car$^-$ mutant lacked enzyme I of the phosphotransferase system, and was seriously defective in its ability to take up sugars from the medium. The mutant was capable of growth on glucose when this sugar was transported into the cell at an adequate rate via the MG permease system, showing that the internal glucose-metabolizing enzymes were fuctioning, and that phosphorylation of glucose by the phosphotransferase system was *not* required for metabolism of this sugar.

Mechanisms for sugar transport[19, 25−27] involving the phosphotransferase system will be discussed elsewhere. Briefly, some of the concepts are as follows: (*a*) Enzymes II catalyze facilitated diffusion of some of the sugars utilized by mutants lacking enzyme I or HPr. (*b*) Phosphorylation by the transferase system is required for passage of most sugars[29] across the membranes of gram-negative bacteria, and of probably all sugars utilized by *Staphylococcus aureus.*[13, 30] (*c*) Where sugars are utilized *per se* within the cell, passage across the membrane involves two steps, phosphorylation and dephosphorylation, the latter catalyzed by specific phosphatases.[28] Current experiments are designed to test these concepts.

Summary.—A mutant of *Salmonella typhimurium* was isolated that was unable to utilize nine carbohydrates readily utilized by the parent strain. Genetic analysis showed that the mutation was a single event, and resulted in the loss of one of the enzymes of a previously described bacterial phosphotransferase system. Transport experiments showed that the physiological defect in the mutant was its inability to take up sugars from the medium. These results inextricably link the phosphotransferase and permease systems.

Note added in proof: The uptake of galactose, at a single concentration of the sugar, was compared in galactose-kinaseless strains of *S. typhimurium;* the rate of galactose accumulation in the parent (car$^+$) strain exceeded that in the car$^-$, Enzyme I-deficient mutant by a factor of 2.5 to 3.

We are greatly indebted to Miss Marie Smith for her expert help with some of the experiments, and to Dr. Harry Schachter for critical and illuminating discussions concerning the kinetic data.

* Contribution No. 514 from the McCollum-Pratt Institute. This work was supported by grants AM-9851 and AI-01650 from the National Institutes of Health, USPHS. R. D. S. was supported by NSF postdoctoral fellowship 46117; M. L. was supported by USPHS postdoctoral fellowship 5-F2-AI-24,825; B. A. was supported by USPHS predoctoral training grant TI GM-57 to the Department of Biology.

[1] Kundig, W., S. Ghosh, and S. Roseman, these Proceedings, **52**, 1067 (1964).

[2] Kundig, W., and S. Roseman, in *Methods in Enzymology*, ed. S. P. Colowick, N. O. Kaplan, and W. A. Wood (New York: Academic Press, 1966), vol. 9, p. 396.

[3] Kundig, W., F. Dodyk Kundig, B. Anderson, and S. Roseman, *Federation Proc.*, **24**, 658 (1965).

[4] Kundig, W., F. Dodyk Kundig, B. Anderson, and S. Roseman, *J. Biol. Chem.*, **241**, 3243 (1966).

[5] Nossal, N. G., and L. A. Heppel, *J. Biol. Chem.*, **241**, 3055 (1966).

[6] Piperno, J. R., and D. L. Oxender, *J. Biol. Chem.*, **241**, 5732 (1966).

[7] Pardee, A. B., *J. Biol. Chem.*, **241**, 5886 (1966).

[8] Anraku, A. B., *J. Biol. Chem.*, **242**, 793 (1967).

[9] Tanaka, S., D. G. Fraenkel, and E. C. C. Lin, *Biochem. Biophys. Res. Commun.*, **27**, 63 (1967).

[10] Tanaka, S., and E. C. C. Lin, these Proceedings, **52**, 913 (1967).

[11] Adelberg, E. A., M. Mandel, and G. Chein Ching Chen, *Biochem. Biophys. Res. Commun.*, **18**, 788 (1965).

[12] Lederberg, J., and N. Zinder, *J. Am. Chem. Soc.*, **70**, 4267 (1948).

[13] Egan, J. B., and M. L. Morse, *Biochim. Biophys. Acta*, **97**, 310 (1965); *ibid.*, **109**, 172 (1965); *ibid.*, **112**, 63 (1966).

[14] Meynell, G. G., and E. Meynell, *Theory and Practice in Experimental Bacteriology* (Cambridge: Cambridge University Press, 1965).

[15] Korman, R., and D. T. Berman, *J. Bacteriol.*, **76**, 454 (1958).

[16] Anton, D. N., personal communication (to be submitted to *J. Mol. Biol.*).

[17] Roth, J. R., D. N. Anton, and P. E. Hartman, *J. Mol. Biol.*, **22**, 305 (1966).

[18] Leder, I. G., and J. W. Perry, *Federation Proc.*, **26**, 394 (1967).

[19] Kepes, A., and G. N. Cohen, in *The Bacteria*, ed. I. C. Gunsalus, and R. Y. Stanier (New York: Academic Press, 1962), vol. 4, pp. 179–222.

[20] Galactose is transported by four different permease systems. The two systems pertinent to the present studies are a specific galactose permease, and one, called MG permease, which is induced by galactose or fucose, and which transports methyl β-galactoside, galactose, and fucose (Ganesan, A. K., and B. Rotman, *J. Mol. Biol.*, **16**, 42 (1966)). Ganesan and Rotman found the galactose permease in one strain of *E. coli*, and the MG permease in several strains; the latter results were also reported by Wu, H. C. P., *J. Mol. Biol.*, **24**, 213 (1967). In both the parent and mutant strains of *S. typhimurium*, the permease shows the properties of MG permease, being inducible by galactose or fucose, and transporting methyl β-galactoside. We therefore assume that galactose is transported by MG permease in these strains.

[21] Hoffee, P., and E. Englesberg, these Proceedings, **48**, 1759 (1962).

[22] Kepes, A., *Biochim. Biophys. Acta*, **40**, 70 (1960).

[23] Winkler, H. H., and T. H. Wilson, *J. Biol. Chem.*, **241**, 2200 (1966).

[24] Kamogawa, A., and K. Kurahashi, *J. Biochem. (Japan)*, **61**, 220 (1967).

[25] Koch, A. L., *Biochim. Biophys. Acta*, **79**, 177 (1964).

[26] Wilbrandt, W., and T. Rosenberg, *Pharmacol. Rev.*, **13**, 109 (1961).

[27] Christensen, H. N., *Biological Transport* (New York: W. A. Benjamin, Inc., 1962).

[28] Lee, Y., J. R. Sowokinos, and M. J. Erwin, *J. Biol. Chem.*, **242**, 2264 (1967).

[29] The concept that sugars are phosphorylated during transport has been suggested by many workers, and has been reviewed.[26, 27] MeGlu was shown to be accumulated by *E. coli* partly as its phosphate ester (Rogers, D., and S. Yu, *J. Bacteriol.*, **84**, 877 (1962)), and the latter was essentially the only substance detected during the first minute of transport (Winkler, H. H., *Biochim. Biophys. Acta*, **117**, 231 (1966)). In *S. typhimurium*, both MeGlu and 3-O-methylglucose were accumulated in large part as anionic substances, presumably the phosphate esters.

[30] Hengstenberg, W., J. B. Egan, and M. L. Morse, these Proceedings, **58**, 274 (1967).

Copyright © 1968 by the American Society of Biological Chemists, Inc.

Reprinted from *J. Biol. Chem.*, **243**, 3711–3724 (1968)

20

The Role of the Phosphoenolpyruvate-phosphotransferase System in the Transport of Sugars by Isolated Membrane Preparations of *Escherichia coli**

(Received for publication, January 31, 1968),

H. R. KABACK

From the Laboratories of Biochemistry, National Heart Institute, National Institutes of Health, Bethesda Maryland 20014

SUMMARY

The phosphoenolpyruvate-phosphotransferase system provides a mechanism for the passage of certain sugars through the bacterial cell membrane and their accumulation as phosphorylated derivatives. These conclusions are based on the following observations.

1. Isolated bacterial membrane preparations specifically require P-enolpyruvate for uptake of certain sugars which accumulate almost completely as phosphorylated derivatives.

2. Membranes prepared from *Escherichia coli* GN-2, a mutant lacking Enzyme I of the phosphotransferase system, are unable to take up significant quantities of α-methylglucoside.

3. ^{3}H-Glucose added to the incubation medium is phosphorylated more rapidly than free ^{14}C-glucose in the intramembranal pool, suggesting that the P-transferase system is a mechanism by which sugars penetrate the membrane.

4. There is a stoichiometric relation between ^{32}P loss from P-enolpyruvate and appearance of ^{32}P in α-methylglucoside-P.

5. The uptake and phosphorylation of α-methylglucoside exhibit saturation kinetics with an apparent K_m of about 4 × 10^{-5} M, whereas the appearance of free α-methylglucoside in the intramembranal pool shows no saturation. The initial rate of uptake of *free* α-methylglucoside is independent of the presence of P-enolpyruvate.

6. Under steady state conditions, intramembranal free α-methylglucoside and the external pool do not equilibrate regardless of the presence of P-enolpyruvate. Furthermore, α-methylglucoside in the external pool does not exchange with the α-methylglucoside previously taken up by the membranes.

Membranes prepared from glucose-grown cells take up and phosphorylate α- and β-methylglucoside, glucose, 2-deoxyglucose, fructose, galactose, and 3-O-methylglucose in the

presence of P-enolpyruvate. Ribose, arabinose, mannitol and sorbitol are not taken up or phosphorylated.

That phosphorylation is involved in sugar transport has been proposed by many workers, and the subject has been reviewed by Wilbrandt and Rosenberg (1) and Christensen (2). More recently, Rogers and Yu showed that methyl-α-D-glucopyranoside[1] is accumulated partly as the phosphate ester in *Escherichia coli* (3), and Winkler has shown that only α-methylglucoside phosphate is detected during the first minute of transport (4).

In 1964, Kundig, Ghosh, and Roseman (5) reported the isolation of a bacterial phosphotransferase system which catalyzes the transfer of phosphate from phosphoenolpyruvate to various carbohydrates according to the following reactions.

$$\text{P-enolpyruvate} + \text{HPr}^1 \xrightleftharpoons{\text{Enzyme I, Mg}^{++}} \quad (1)$$
$$\text{pyruvate} + \text{P-HRr}$$

$$\text{P-HPr} + \text{sugar} \xrightarrow{\text{Enzyme II, Mg}^{++}} \text{sugar-P HPr} \quad (2)$$

$$\text{P-enolpyruvate} + \text{sugar} \xrightarrow[\text{Enzyme, II, Mg}^{++}]{\text{HPr, Enzyme I}} \quad (1 + 2)$$
$$\text{sugar-P} + \text{pyruvate}$$

The system is composed of two enzymes, I and II, and a heat-stable, low molecular weight protein, HPr, which functions as a phosphate carrier in the over-all reaction (6). Enzyme II, the membrane-bound component of the system, is responsible for specificity with respect to the various sugars studied. In all cases thus far studied, the sugars were phosphorylated in position 6 (7).

* Preliminary reports of this work have been presented (KABACK, H. R., in *Biological interfaces, flows and exchanges* (*Proceedings of the New York Heart Association Symposium, New York, 1967*), in press; KABACK, H. R., *Fed. Proc.*, **27**, 644 (1968)).

[1] The abbreviations used are: HPr, heat-stable protein; α-methylglucoside, methyl-α-D-glucopyranoside; U, uniformly labeled.

3711

Subsequently, a report appeared from the same laboratory (8) showing that whole cells of *E. coli* W2244 subjected to the osmotic shock procedure of Neu and Heppel (9) lost the ability to take up α-methylglucoside and methyl-β-thio-D-galactopyranoside. Furthermore, Kundig *et al.* demonstrated that the ability to take up these compounds was completely restored by the addition of HPr to the reaction mixture. This study represents the first evidence that the P-enolpyruvate-P-transferase system is involved in carbohydrate transport. However, the radioactive compounds accumulated by these cold-shocked cells were not identified, leaving some question as to the nature of the HPr effect. Subsequently, genetic evidence implicated the P-transferase system in the transport, metabolism, or both of a number of carbohydrates (10, 11). Pleiotropic mutants of *Aerobacter aerogenes* and *E. coli* which failed to grow on these carbohydrates also lacked either HPr or Enzyme I activities when tested *in vitro*. Most recently, Simoni *et al.* (12) described a mutant of *Salmonella typhimurium* that was unable to utilize nine carbohydrates for growth. The mutation was a single genetic event resulting in the loss of Enzyme I activity in the phosphotransferase system, and the physiological defect was shown to be an inability to transport carbohydrates, including α-methylglucoside.

Virtually all of the carbohydrates taken up by *Staphylococcus aureus* appear in the cell as phosphorylated derivatives (13, 14), and furthermore these cells contain an inducible galactosidase which specifically splits β-galactoside phosphates (13–15).

Previous work from this laboratory reported the isolation of a membrane preparation from *E. coli* which, in the absence of soluble proteins, catalyzes the concentrative uptake of proline (16) as well as the facilitated diffusion of glycine and its ultimate conversion to phosphatidylethanolamine (17, 18). Subsequent studies with sonically treated membrane preparations showed that the proline uptake system is tightly bound to the membrane (19). These findings, in the light of the highly interesting reports regarding the P-enolpyruvate-P-transferase system, lead to the studies reported in this communication.

EXPERIMENTAL PROCEDURE

Methods

Whole Cells—*E. coli* ML 308-225 ($i^-z^-y^+$) and *E. coli* K_2lt (lacking P-enolpyruvate synthetase activity) were kindly provided by Drs. H. Winkler and H. Kornberg, respectively. *E. coli* GN-2 was reisolated from a culture provided by Dr. L. Heppel by selecting a glucose-negative colony from a MacConkey plate[2] containing 1% glucose. ML 308-225 was grown on a rotary shaker at 37° in Medium A (16) with 0.5% glucose as a carbon source. K_2lt was grown on Difco penassay broth[3] under the same conditions. GN-2 was grown on a medium containing, per liter, 17 g of Bacto-peptone (Difco), 3 g of proteose-peptone (Difco), and 5 g of NaCl, also under the conditions stated above. Stock cultures of the strains were maintained in the medium given containing 10% glycerol at −20° and on nutrient agar slants stored at 4° under sterile mineral oil.

Preparation of Spheroplasts—Large scale preparations of membranes were obtained by the lysozyme procedure described below, a more convenient method than the penicillin technique (16). In this procedure, 8- to 9-hour cultures of the appropriate

[2] *Difco Manual*, Difco Laboratories, Detroit, 1953, p. 131.
[3] *Difco Manual*, Difco Laboratories, Detroit, 1953, p. 203.

strain, grown under the conditions indicated above, were diluted 1:8 with fresh medium (at 37°), and the incubation was continued until the cells had reached approximately midlog phase (approximately 140 to 150 Klett units with a No. 60 filter or ∼0.5 mg per ml, dry weight). After centrifugation, the cells were washed twice with 0.01 M Tris-HCl, pH 8.0, at 4°, resuspended (1 g, wet weight, per 80 ml) at room temperature in 0.03 M Tris-HCl, pH 8.0, containing 20% sucrose, and placed on a magnetic stirrer. EDTA and lysozyme (Worthington, crystalline) were then added to final concentrations of 10 mM and 0.5 mg per ml, respectively. The suspensions were then incubated for 30 min at room temperature. This method is essentially that of Neu and Heppel (20), with the exception that higher concentrations of EDTA and lysozyme are used. After 30 min, the preparation was examined with a phase contrast microscope and centrifuged; unlike penicillin-induced spheroplasts, the lysozyme method did not yield extensive spherical forms with these strains or with the W strain of *E. coli*. However, this observation had no apparent relationship to their osmotic fragility. The membranes ultimately isolated by the lysozyme procedure were essentially identical with those described previously (16). Specifically, they catalyzed the concentrative uptake of proline and the uptake of α-methylglucoside as effectively as membranes prepared from penicillin-induced spheroplasts; further, the preparations contained essentially no ribosome-like structures when examined with the electron microscope (these studies were graciously carried out by Dr. V. Marchesi of the National Cancer Institute).

Preparation of Membranes—A Teflon-glass homogenizer was used to resuspend the pellet described above in the smallest possible volume of 0.1 M potassium phosphate buffer, pH 6.6, containing 20% sucrose and 20 mM $MgSO_4$. The homogenization was facilitated by the addition of DNase (Worthington, pancreatic, crystallized once) to a final concentration of 10 μg per ml (after lysis). From this point, the membranes were prepared as described previously (16) with the following exceptions: (a) the homogenized spheroplast pellet was diluted at least 200-fold by adding it to 0.05 M potassium phosphate buffer, pH 6.6, which had been equilibrated at 37°, and (b) after the membranes were prepared, they were centrifuged and washed at least five times by homogenization in cold 0.1 M potassium phosphate buffer, pH 6.6, containing 10 mM EDTA. Since ML 308-225 was relatively insensitive to lysozyme-EDTA, lysates prepared from this strain usually contained a significant number of whole cells. For this reason ML 308-225 membrane preparations were routinely purified by differential centrifugation, by centrifugation after layering over 60% sucrose as described previously (16), or by both treatments. Furthermore, freshly prepared ML 308-225 membranes were found to exhibit high endogenous α-methylglucoside uptake and phosphorylation activities so that no P-enolpyruvate effect could be shown; however, after aging for 5 to 8 days at 4° the membranes exhibited the desired effect. Subsequently, it was found that 10^{-2} M NaF inhibited the endogenous activity without inhibiting the effect of P-enolpyruvate on the system (see "Results").

E. coli K_2lt was quite sensitive to lysozyme-EDTA treatment, and osmotic lysis of this strain produced essentially complete conversion to membranes. Centrifugation of these membranes over 60% sucrose (16) resulted in the appearance of little or no pellet at the bottom of the tube, and no loss in activity from the interface fraction. Furthermore, membranes prepared from

K_2lt showed low endogenous activity with regard to uptake and phosphorylation of sugars.

All membrane preparations were finally resuspended in 0.1 M potassium phosphate buffer, pH 6.6, at about 10 to 15 mg per ml; dry weight (corresponding to 8 to 9 mg of protein per ml), and stored in small aliquots in liquid N_2. In terms of α-methylglucoside and proline uptake, the preparations were stable to at least one sequence of freezing and thawing under these conditions, and could be maintained in the frozen state for an unlimited period of time.

Uptake Studies—For all experimental procedures, unless otherwise noted, 50-μl aliquots of the membrane preparations were diluted to 100-μl final volumes containing, in final concentrations, 0.25 M potassium phosphate buffer, pH 6.6, 0.01 M $MgSO_4$, and other additions as indicated. The samples were incubated at the indicated temperatures for 15 min. At this time, radioactive substrate was added, and the incubation was continued for an appropriate period of time. To terminate the reaction, each sample was rapidly diluted 50-fold with 0.25 M potassium phosphate buffer, pH 6.6, at the same temperature as the incubation mixture, immediately filtered with Millipore HA filters (Millipore Filter Corporation, Bedford, Massachusetts), and washed once with an equal volume of buffer. The dilution, filtration, and washing procedures were conducted in less than 30 sec; the filters were immediately removed from the suction apparatus, mounted on planchets with tight fitting stainless steel rings that covered the periphery of the filters, dried, and counted in a Nuclear-Chicago gas flow counter at about 22% efficiency. No corrections were made for self absorption, and all samples were counted under the same conditions. Each set of experimental samples was corrected for a control obtained by diluting the samples before adding the radioactive substrate and omitting P-enolpyruvate. With the use of the assay conditions described above and 3.64×10^{-5} M ^{14}C-α-methylglucoside (73.4 mC per mmole), control values were 200 cpm or less. Where indicated, the filtrates from the incubations were collected and lyophilized.

Extraction, Isolation, and Assay of Conversion Products—After counting, the filters were removed from the planchets, placed in test tubes, and eluted four to six times with 1.5 ml of distilled water each at room temperature. This procedure removed more than 90% of the radioactivity from the filters. The pooled eluates were lyophilized, the residue was resuspended in 20 μl of distilled water and centrifuged, and 10 μl of supernatant (or an aliquot of the lyophilized filtrate which had been dissolved in water) were applied to the origin of a Silica Gel G thin layer plate (Mann Biochemicals, New York). The plates were then developed with either chloroform-methanol-water (60:70:26, v/v/v) or chloroform-methanol-0.1 M boric acid (60:70:26, v/v/v). The solvent was routinely allowed to migrate 14 cm (approximately 90 min). The thin layer plates were dried in air and placed in contact with Eastman Kodak blue sensitive x-ray film for 12 to 14 hours. As an example of this technique, Fig. 1 shows the radioautogram from the experiment shown in Fig. 5A (+ P-enolpyruvate). The radioactivity remaining at the origin, labeled αMGP, was quantitatively converted to a compound with the same R_F as authentic α-methylglucoside by treatment with alkaline phosphatase (described below). As shown, α-methylglucoside migrates about two-thirds of the way up the plate. With all of the carbohydrates studied in these experiments the phosphorylated derivatives were readily separated from the free sugars in a similar manner. The radioactive spots

were then scraped into scintillation vials and counted in a Nuclear-Chicago Mark I liquid scintillation spectrometer, with the use of the external standard channel ratio to detect and correct for quenching and counting efficiency. The total radioactivity of the filtered membranes (gas flow counting) represented free sugar plus P-sugar, and the absolute quantities of these substances were calculated from their relative concentrations determined by the chromatographic method described above. It is important, however, that these methods resulted in at least 90% recovery of the radioactivity in all cases.

A more rapid but less quantitative technique for the isolation of phosphorylated derivatives was also devised. Radioactive filters were placed, face down, on strips of Whatman No. 3MM paper, and the paper at the periphery of the filter was moistened with acetone. Millipore filters are extremely soluble in acetone, and this treatment caused the periphery of the filter to adhere to the paper, leaving the center intact. The paper was then moistened with the appropriate buffer and placed in a Gilson high voltage electrophoresis apparatus. Electrophoresis was carried out at approximately 40 volts per cm for 45 to 60 min in 1.88 M pyridine-acetic acid at pH 3.5. Subsequently, the paper was dried, cut into strips, and either scanned by means of a Vanguard strip counter or radioautographed. It is important to note that solubilization of the whole filter with acetone while applying it to the paper resulted in trapping and incomplete migration of phosphorylated materials from the origin.

Identification of Phosphorylated Derivatives—α-Methylglucoside-P, glucose-P, fructose-P, and galactose-P were isolated by the methods described above. The samples were then eluted, treated with alkaline phosphatase (Worthington, bacterial alkaline phosphatase) in Tris-HCl buffer, pH 8.0, at room temperature for 30 min, passed over a mixed ion exchange resin, and lyophilized. After addition of the appropriate unlabeled carrier,

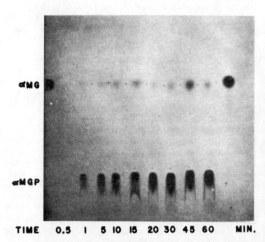

TIME 0.5 1 5 10 15 20 30 45 60 MIN.

FIG. 1. Radioautogram of experiment presented in Fig. 5A (+ P-enolpyruvate). Samples, treated as described, were chromatographed on Silica Gel G thin layer plates with chloroform-methanol-water (60:70:26, v/v/v) as described in "Methods." The tracks to the extreme *left* and *right* of the radioautogram were obtained with known ^{14}C-α-methylglucoside. αMG, α-methylglucoside; αMGP, α-methylglucoside phosphate.

the samples were chromatographed. α-Methylglucoside, glucose, and fructose were found to have sufficiently different R_F values (approximately 0.63, 0.47, and 0.35, respectively) for identification by thin layer chromatography in the borate system described above. For the identification of galactose, paper chromatography was used (21). The chromatograms were radioautographed and then stained either with alkaline AgNO₃ or by the periodate oxidation method of Weiss and Smith (22). The location of the radioactive spots was then compared directly with the location of the stain. In all cases, the radioactivity coincided with the stain.

Glucose-6-P was also identified, after isolation by one of the techniques described above, by treatment with glucose-6-P dehydrogenase (Boehringer-Mannheim) according to the method of Horecker and Smyrniotis (23). The procedure was controlled by treating known U-¹⁴C-glucose-6-P in parallel with the unknown. After a 60-min incubation at room temperature, the reaction mixtures were treated with charcoal and passed over

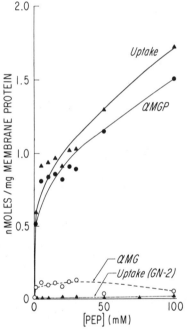

Fig. 2. α-Methylglucoside uptake by ML 308-225 and GN-2 membrane preparations as a function of P-enolpyruvate concentration. ML 308-225 or GN-2 membrane samples (100 μl), prepared as described in "Methods," were incubated at 27° for 15 min in the presence of P-enolpyruvate in the concentrations given. α-Methylglucoside-¹⁴C (73.4 mC per mmole) was then added at 3.64×10^{-5} M. The incubations were continued for 30 min, then terminated, and the samples were assayed as described in "Methods." Each experimental point was corrected for a control sample obtained as described in "Methods." *Uptake* (▲——▲), total nanomoles taken up per mg of protein; αMGP (●——●), α-methylglucoside-P isolated from the samples; αMG (○‐‐‐○), α-methylglucoside isolated from the samples; *Uptake (GN-2)* (▲····▲), total nanomoles taken up per mg of protein by membranes prepared from the GN-2 mutant.

Dowex 50 (H⁺) columns (0.5 × 4 cm), and the water eluates were lyophilized. The samples were then subjected to high voltage electrophoresis at 40 volts per cm for 30 to 60 min in 0.453 M pyridine-acetic acid at pH 5.5 after the addition of carrier 6-P-gluconate and glucose-6-P. The electrophoretograms were radioautographed and stained with AgNO₃ as described above. The radioactive spots were cut out and counted in a liquid scintillation spectrometer. The unknown was found to behave in a manner identical with that of known glucose-6-P (*i.e.* approximately 80% was converted to compounds which migrated with 6-P-gluconate and 6-P-gluconolactone).

Protein—Protein concentrations were determined by the method of Lowry *et al.* (24).

Materials

P-enolpyruvate, 2-P-glycerate, 3-P-glycerate, and 2,3-P₂-glycerate were obtained from Boehringer-Mannheim as the sodium salts. The P-enolpyruvate was found to be of stated purity by enzymatic assay.

Uniformly labeled α- and β-methyl-D-glucoside-glucose-¹⁴C were obtained from Calbiochem. D-Glucose-U-¹⁴C was obtained from New England Nuclear and Nuclear-Chicago. D-Glucose-6-³H and D-galactose-1-¹⁴C were obtained from Nuclear-Chicago. D-Fructose-U-¹⁴C, D-ribose-1-¹⁴C, D-arabinose-1-¹⁴C, D-mannitol-1-¹⁴C, and D-sorbitol-U-¹⁴C were obtained from New England Nuclear. 3-O-Methyl-D-glucose-U-¹⁴C and 2-deoxy-D-glucose-U-¹⁴C were obtained from International Chemical and Nuclear Corporation, City of Industry, California. The sugars were all found to have about the purity stated by the commercial source as judged by chromatography in at least the two thin layer solvent systems described above.

³²P-Enolpyruvate was a gift of Drs. Werner Kundig and Saul Roseman.

All other materials used in these studies were obtained from the usual commercial sources and were of reagent grade.

RESULTS

Fig. 2 shows the relationship between P-enolpyruvate concentration and α-methylglucoside uptake and phosphorylation by isolated membrane preparations. Uptake by ML 308-225 membranes increased rapidly with increasing concentrations of P-enolpyruvate from 0 to 10 mM and continued to increase, but less rapidly, up to 100 mM. More than 90% of the α-methylglucoside taken up appeared as α-methylglucoside-P. Significantly, little or no α-methylglucoside-P was found in the external medium under these conditions. Furthermore, with membranes prepared from mutant GN-2 (Enzyme I-defective) (10) uptake was negligible over the concentration range studied.

As seen in Table I, the capacity to stimulate α-methylglucoside uptake and phosphorylation was restricted to P-enolpyruvate and 2-P-glycerate. Furthermore, the stimulatory effect of P-enolpyruvate showed an absolute requirement for Mg⁺⁺ (Line 7). The lack of stimulation by the nucleoside triphosphates in this experiment probably did not result from their inability to penetrate the membrane, since they did stimulate the conversion of glycine to phosphatidylethanolamine and phosphatidylserine by similar membrane preparations from *E. coli* W (18). Neither 3-P-glycerate nor 2,3-diphosphoglycerate stimulated the system significantly when added alone or in combination. Finally, 2-P-glycerate was only two-thirds as effective as P-enolpyruvate (Lines 26 and 28).

TABLE I
Stimulation of α-methylglucoside uptake and phosphorylation by various energy sources

ML 308-225 membrane samples (100 μl), prepared as described in "Methods," were incubated at 27° for 15 min in the presence of one of the additions listed below at 0.1 M (unless otherwise indicated). ^{14}C-α-Methylglucoside (73.4 mC per mmole) was then added at 3.64×10^{-5} M and incubation was continued for 30 min. The reactions were terminated and products were assayed as described in "Methods." For K_2lt membrane samples, the incubations were carried out at 46° for 3 min.

Sample No. and additions	Total α-methyl-glucoside uptake	α-Methyl-glucoside	α-Methyl-glucoside-P
	nmoles/mg protein		
ML 308-225 membranes			
1. None	0.016	0.0143	0.0017
2. MgSO₄ absent	0.024	0.0164	0.0074
3. Glucose-6-P	0.012	0.0094	0.0026
4. Glycerol	0.024	0.0167	0.0071
5. Pyruvate	0.023	0.0152	0.0078
6. P-enolpyruvate	1.465	0.26	1.20
7. 6 minus MgSO₄	0.032	0.02	0.012
8. Lactate	0.033	0.0193	0.0137
9. Acetate	0.028	0.0154	0.0126
10. 9 + CoASH	0.026	0.0164	0.0096
11. CoASH	0.03	0.0184	0.0116
12. Acetyl-CoA	0.03	0.0189	0.0111
13. Citrate	0.021	0.0118	0.0092
14. Isocitrate	0.031	0.0169	0.0191
15. Aconitate	0.033	0.012	0.021
16. α-Ketoglutarate	0.036	0.0137	0.0223
17. Succinate	0.033	0.0139	0.0191
18. Fumarate	0.036	0.013	0.013
19. Oxalacetate	0.022	0.0111	0.0109
20. ATP	0.054	0.0351	0.0189
21. GTP	0.027	0.0157	0.0095
22. CTP	0.019	0.0108	0.0082
23. UTP	0.027	0.0139	0.0131
24. 20 to 23	0.036	0.0216	0.0144
K_2lt membranes			
25. None	0.269	0.124	0.145
26. P-enolpyruvate	1.283	0.154	1.129
27. 3-P-glycerate	0.352	0.119	0.233
28. 2-P-glycerate	0.837	0.137	0.700
29. 2,3-Diphospho-glycerate, 10^{-2} M	0.239	0.0935	0.145
30. 27 + 29	0.277	0.097	0.180

Evidence that the effect of 2-P-glycerate was exerted through its conversion to P-enolpyruvate is presented in Table II, which shows that the effect of P-enolpyruvate was not inhibited by NaF, whereas the effect of 2-P-glycerate was completely abolished. This indicates that the membranes contain sufficient enolase activity to convert 2-P-glycerate to P-enolpyruvate. This interpretation is consistent with the data in Fig. 3, which show that the uptake of α-methylglucoside began immediately after the addition of P-enolpyruvate and continued at a constant rate for at least 5 min. However, with 2-P-glycerate, there was a lag for the first 30 sec. Subsequently, uptake began and continued at a linear rate about two-thirds of the rate observed with P-enolpyruvate. Presumably, the delay observed with 2-

P-glycerate represented the time required for its conversion to P-enolpyruvate.

Effect of Temperature—Fig. 4 shows the effect of temperature on the initial rate of accumulation and steady state level of α-methylglucoside and α-methylglucoside-P by the membranes in the presence of P-enolpyruvate. With increasing temperature, the steady state level (uptake in 30 min) reached a sharp maximum at 27° and then declined rapidly. On the other hand, the initial rate (uptake in 3 min) had an equally sharp optimum at 46°. Furthermore, as much α-methylglucoside was taken up in 3 min at 46° as in 30 min at 27°. At each time and temperature, α-methylglucoside-P comprised 90% or more of the radioactivity taken up. Thus, the effects of temperature on uptake are manifestations of the phosphorylated α-methylglucoside accumulated by the membranes. The free α-methylglucoside pool in the membranes is relatively insensitive to temperautre.

Based on these data, α-methylglucoside uptake at 27° in the presence of P-enolpyruvate should reach a stable plateau at or before 30 min, while at 46°, uptake should reach a maximum at approximately 3 min and then decrease to about 50% of the peak level by 30 min. These predictions were verified by the data shown in Fig. 5, A and B. Fig. 5A shows the time course of uptake and phosphorylation at 27° with and without P-enolpyruvate. With P-enolpyruvate, the rate of uptake was linear for about the first 20 min, diminished, and reached a plateau at about 30 min. The plateau was then maintained for at least 30 min more. The total uptake was paralleled by the α-methylglucoside-P found inside the membranes, whereas free α-methylglucoside increased slowly and almost linearly, and by the end of the incubation represented less than 10% of the total uptake. Under these conditions, essentially no α-methylglucoside-P was found in the external medium. Without P-enolpyruvate, the total uptake was about 100-fold less than that of the samples incubated in its presence. Fig. 5B shows the time course of uptake and phosphorylation at 46° in the presence of P-enolpyruvate. The data are expressed as a percentage of the concentration of α-methylglucoside added to the reaction mixtures, so that the amounts of α-methylglucoside and α-methylglucoside-P recovered from the

TABLE II
Effect of NaF on P-enolpyruvate- or 2-P-glycerate-stimulated uptake and phosphorylation of α-methylglucoside

K_2lt membrane samples (100 μl), prepared as described in "Methods," were incubated at 46° for 15 min under the conditions listed below. ^{14}C-α-Methylglucoside (73.4 mC per mmole) was then added at 3.64×10^{-5} M and incubation was continued for 3 min. The reactions were terminated and products were assayed as described in "Methods."

Sample No. and additions (10^{-1} M)	NaF (10^{-2} M)	Total α-methyl-glucoside uptake	α-Methyl-glucoside	α-Methyl-glucoside-P
		nmoles/mg protein		
1. None	−	0.269	0.103	0.166
2. None	+	0.083	0.0481	0.0349
3. P-enolpyruvate	−	1.283	0.215	1.068
4. P-enolpyruvate	+	1.458	0.243	1.215
5. 2-P-glycerate	−	0.825	0.13	0.695
6. 2-P-glycerate	+	0.087	0.04	0.047

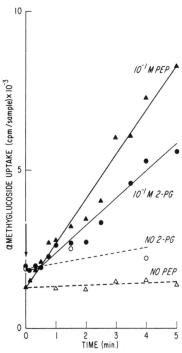

Fig. 3. Time course of P-enolpyruvate- or 2-P-glycerate-stimulated up take of α-methylglucoside by K₂lt membranes. K₂lt membrane samples (90 µl), prepared as described in "Methods," were incubated at 46° for 10 min. α-Methylglucoside-¹⁴C (73.4 mC per mmole) was then added at 3.64×10^{-5} M and incubation was continued for 5 min. At this time (0 time on the figure), 10 µl of either P-enolpyruvate (*PEP*) or 2-P-glycerate (*2-PG*) were added to give a final concentration of 0.1 M as shown. Buffer (10 µl) was added to the control samples. In addition, the samples labeled *PEP* and *NO PEP* contained 10^{-2} M NaF. Incubation was continued for the indicated times, then terminated, and products were assayed as described in "Methods." Each sample contained 0.425 mg of membrane protein. Each experimental point was corrected for a control sample obtained as described in "Methods." *PEP* and *NO PEP* (▲——▲, and △ - - -△), P-enolpyruvate and no P-enolpyruvate, respectively; *2-PG* and *NO 2-PG* (●——● and ○- - -○), 2-P-glycerate and no 2-P-glycerate, respectively.

membranes can be directly compared with that recovered from the medium (see Fig. 7A for data expressed as nanomoles per mg of membrane protein and per 0.1 ml of reaction mixture). As shown, uptake increased extremely rapidly until about 4 min, when it stopped abruptly and began to decrease. By 60 min, the membranes had lost over 60% of their radioactivity. Once again, total uptake was paralleled by the α-methylglucoside-P recovered from the membranes whereas free α-methylglucoside increased much more slowly, reaching a plateau at about 5 min.

The loss of radioactivity from the membranes at 46° appears to result from increased membrane permeability at temperatures above 27°, yielding α-methylglucoside-P in the medium. As shown by the curve labeled α*MGP* (*Medium*) in Fig. 5B, α-methylglucoside-P in the medium increased rapidly for the first

10 min of the incubation, slowed down, and finally achieved an inverse relationship to α-methylglucoside-P in the membranes. In other words, α-methylglucoside-P in the membranes had a precursor-product relationship to α-methylglucoside-P in the medium. It can be seen from the curve labeled α*MG* (*Medium*) that, by 5 min, about 95% of the α-methylglucoside added to the reaction mixtures had been phosphorylated. Thus, phosphorylation stopped at about 5 min because substrate (α-methylglucoside) became limiting; simultaneously, and for about 60 min more, α-methylglucoside-P in the membranes was redistributed from a higher (inside the membranes) to a lower (outside the membranes) concentration. Despite the large quantity of α-methylglucoside-P that appeared in the medium at 60 min, α-methylglucoside-P per unit volume was still about 25 times higher inside the membranes than in the medium (measured by the techniques described in the legend to Table III). These two compartments should equilibrate eventually if the incubation is continued.

The data presented in Fig. 6 show that induction and reversal of the leak occur extremely rapidly. When membranes were incubated with α-methylglucoside at 27° and subsequently transferred to 46°, there was an immediate decrease in radioactivity in the membranes. The rate of loss was essentially linear for the first 30 min, and decreased between 30 and 60 min.

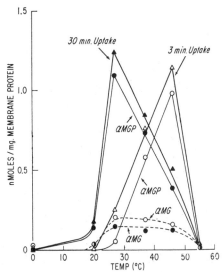

Fig. 4. Effect of temperature on initial rate and steady state uptake of α-methylglucoside by ML 308-225 membranes. ML 308-225 membrane samples (100 µl), prepared as described in "Methods," were incubated at the temperatures given for 15 min after the addition of 0.1 M P-enolpyruvate and 10^{-2} M NaF. α-Methylglucoside-¹⁴C (73.4 mC per mmole) was then added at 3.64×10^{-5} M and incubation was continued for either 3 or 30 min as shown. The assays were carried out as described in "Methods." Each experimental point was corrected for a control value obtained as described in "Methods." *30* or *3 min. Uptake*, total nanomoles taken up per mg of protein in 30-min (▲——▲) or 3-min (△——△) incubations; α*MGP*, α-methylglucoside-P recovered from the samples after 30-min (●——●) or 3-min (○——○) incubations; α*MG*, α-methylglucoside recovered from the samples after 30-min (●- - -●) or 3-min (○- - -○) incubations.

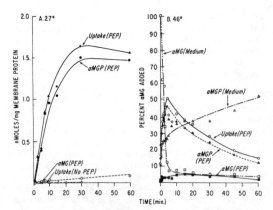

FIG. 5. The time course of α-methylglucoside uptake at 27° (A) or 46° (B) by ML 308-225 membranes. ML 308-225 membrane samples (100 μl), prepared as described in "Methods," were incubated in the presence or absence of 0.1 M P-enolpyruvate (where indicated) and 10⁻² M NaF for 15 min at either 27° (A) or 46° (B). α-Methylglucoside-¹⁴C (73.4 mC per mmole) was then added at 3.64 × 10⁻⁵ M and incubation was continued for the indicated times. The reactions were terminated and each sample was assayed as described in "Methods." Each experimental point has been corrected for a control sample treated as described in "Methods." The data are plotted as nanomoles per mg of membrane protein in A, and as a percentage of the α-methylglucoside initially added to the reaction mixtures in B. A: Uptake (PEP) (▲——▲), total uptake of α-methylglucoside by the membranes incubated in the presence of P-enolpyruvate; αMGP (PEP) (●——●), α-methylglucoside-P recovered from samples incubated in the presence of P-enolpyruvate; αMG (PEP) (○- - -○), α-methylglucoside recovered from samples incubated in the presence of P-enolpyruvate; Uptake (NO PEP) (△- - -△), total uptake of α-methylglucoside by the membranes incubated in the absence of P-enolpyruvate. B: Uptake (PEP) (○——○), total uptake of α-methylglucoside (αMG) by membranes incubated in the presence of P-enolpyruvate; αMGP (PEP) (▲- - -▲), α-methylglucoside-P recovered from samples incubated in the presence of P-enolpyruvate; αMG (PEP) (■——■), α-methylglucoside recovered from samples incubated in the presence of P-enolpyruvate; αMGP (Medium) (△-·—·△), α-methylglucoside-P recovered from the filtrates of the reaction mixtures; αMG (Medium) (□·——·□), α-methylglucoside recovered from the filtrates of the reaction mixtures.

On the other hand, when the membranes were incubated with α-methylglucoside at 46° for 5 min and then shifted to 27°, there was essentially no loss of radioactivity.

Role of P-enolpyruvate in Phosphorylation of α-Methylglucoside —The data presented in Fig. 7A show that P-enolpyruvate acts as a phosphate donor for the phosphorylation of α-methylglucoside. In this experiment, equivalent quantities of ³²P-enolpyruvate and α-methylgluucoside-¹⁴C were added to reaction mixtures at 46°. At the times indicated, the entire reaction mixture was subjected to high voltage electrophoresis to separate the various labeled components. The concentration of ³²P-enolpyruvate decreased with time, whereas there was a stoichiometric increase in ³²P found in α-methylglucoside-P. The relationship between ³²P loss from P-enolpyruvate and its appearance in methylglucoside-P (expressed as (³²P-enolpyruvate + α-methylglucoside-³²P)/(³²P-enolpyruvate_{i itial}) approached unity at each time point studied (Fig. 7B). As expected, the increase in ¹⁴C activity found in α-methylglucoside-P

was inversely related to the disappearance of α-methylglucoside-¹⁴C from the reaction mixture (Fig. 7A). It should be noted that the rate of phosphorylation was much slower here than in Fig. 5B, where higher concentrations of P-enolpyruvate were used under the same conditions. There was essentially no change in inorganic ³²P in the reaction mixtures, indicating little or no nonspecific hydrolysis of P-enolpyruvate by the membranes. The apparent incomplete disappearance of ³²P-enolpyruvate was due to contamination of the ³²P-enolpyruvate sample with 2-³²P-glycerate which was not converted to P-enolpyruvate in the presence of NaF (see Table II) and was not separated from P-enolpyruvate under the conditions of the electrophoresis.

Role of P-enolpyruvate-P-transferase System in Sugar Transport—Although previous investigations provided evidence that the P-enolpyruvate-P-transferase system is involved in sugar transport (8, 10–14), the exact nature of its role is somewhat unclear. There are essentially two alternatives: (a) the P-enolpyruvate-P-transferase system is responsible for entry of sugars (by phosphorylation at the membrane prior to entry into the internal pool) as well as their accumulation as phosphorylated derivatives (under physiological conditions), or (b) the P-enolpyruvate-P-transferase system is responsible for accumulation of sugars after entry into an internal pool by passive or facilitated diffusion.

If the first alternative is correct, the appearance of sugar-P should be independent of the appearance of free sugar, and it should be possible to demonstrate, with the use of double isotope techniques, that sugar in the internal pool does not act as a precursor for sugar-P. Because tritium-labeled α-methylglucoside of sufficiently high specific activity was not available,

TABLE III

α-Methylglucoside uptake at 27°: distribution ratios

ML 308-225 membrane samples (1.0 ml), prepared as described in "Methods," were incubated at 27° for 15 min in the presence or absence of 0.1 M P-enolpyruvate as indicated. After 15 min, α-methylglucoside-¹⁴C (73.4 mC per mmole) was added at 3.64 × 10⁻⁵ M and incubation was continued for 30 min at 27°. At this time, the samples were immediately centrifuged in the cold at about 37,000 × g for 10 min. The supernatants were carefully aspirated, the pellets were drained, and the insides of the tubes were wiped dry. The pellets were then resuspended to 1.0 ml with water and an aliquot was chromatographed as described in "Methods." An aliquot of the supernatant was subjected to the same procedure. α-Methylglucoside and α-methylglucoside-P were located and assayed as described in "Methods." Identical samples which had been incubated with ¹⁴C-inulin or ¹⁴C-dextran were used to determine intra- and extramembranal volumes as described previously (16).

Conditions	[α-Methylglucoside]_i^a	[α-Methylglucoside-PO_4]_i	[α-Methylglucoside]_o^b	Ratio, [α-methylglucoside]_i^a to [α-methylglucoside]_o^b
	cpm	*cpm*	*cpm*	
27°	7,192	7,504	25,400	0.283
27° + P-enolpyruvate	6,588	235,504	20,900	0.315

[a] Corrected for α-methylglucoside-¹⁴C trapped in pellet waters.
[b] Counts per min contained in a volume of supernatant equivalent to the intramembranal space (*i.e.* total fluid volume of the pellet minus the inulin space).

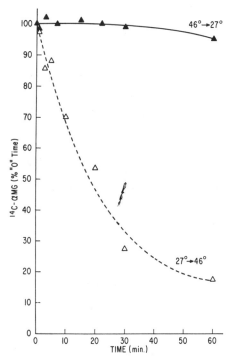

Fig. 6. Effect of temperature shift of α-methylglucoside-^{14}C (^{14}C-αMG) taken up by ML 308-225 membranes at 46° or 27°. $46° \rightarrow 27°$ (▲——▲): 100-μl ML 308-225 membrane samples containing 0.1 M P-enolpyruvate and 10^{-2} M NaF, prepared as described in "Methods," were incubated at 46° for 15 min. α-Methylglucoside-^{14}C (73.4 mC per mmole) was then added at 3.64×10^{-5} M and incubation was continued at 46° for 5 min (zero time). This sample contained approximately 1.57 nmoles per mg of membrane protein. The samples were then immediately placed in a 27° bath. The incubations were continued for the times indicated, then terminated, and samples were assayed as described in "Methods." A control sample (not shown) left at 46° for 60 min contained approximately 30% of the zero time activity. $27° \rightarrow 46°$ ($\triangle$---$\triangle$): A 1.0-ml ML 308-225 membrane sample containing 0.1 M P-enolpyruvate and 10^{-2} M NaF, prepared as described in "Methods," was incubated at 27° for 15 min. α-Methylglucoside-^{14}C (73.4 mC per mmole) was then added at 3.64×10^{-5} M and incubation was continued for 30 min. The sample was then centrifuged in the cold at about $37,000 \times g$ for 20 min. After aspiration of the supernatant, the pellet was resuspended in 0.9 ml of 0.25 M potassium phosphate buffer, pH 6.6, containing 10^{-2} M MgSO$_4$, 10^{-2} M NaF, and 0.1 M P-enolpyruvate; 0.1-ml samples were transferred to individual reaction vessels. One sample was assayed immediately (zero time); the remaining samples were incubated at 46° for the times indicated, incubations were terminated, and samples were assayed as described in "Methods." The zero time sample contained approximately 0.8 nmole per mg of membrane protein. A control sample (not shown) incubated at 27° for 60 min lost less than 5% of the activity of the zero time sample.

glucose was used. Preliminary studies on the uptake of ^{14}C-glucose by the membranes showed that its uptake at 30 min as a function of temperature was almost identical with that of α-methylglucoside (Fig. 8, *A* and *B*), with and without P-enolpyruvate. Furthermore, as seen in Fig. 9, *A* and *B*, the time

course of glucose uptake at 46° in the presence of P-enolpyruvate and the appearance of glucose-P in the medium were identical with α-methylglucoside. Under these conditions (*i.e.* 46° in the presence of 10^{-2} M NaF), glucose-6-P represented at least 90% of the metabolites that accumulated within the membranes or appeared in the medium. Glucose-6-P was identified as described in "Methods."

Based on the above observations with ^{14}C-glucose, a double isotope experiment with ^{14}C- and ^{3}H-glucose was conducted to determine whether the internal or external pools provide substrate for the P-enolpyruvate-P-transferase system. In this experiment, membrane samples were first incubated with high concentrations of ^{14}C-glucose under nonphosphorylating conditions (*i.e.* at 0° in the absence of P-enolpyruvate) in order to load the intramembranal pool with free glucose. Subsequently, external ^{14}C-glucose was removed by centrifugation or filtration, and the membranes were incubated with low concentrations of ^{3}H-glucose under phosphorylating conditions (*i.e.* at 40° in the presence of P-enolpyruvate). If glucose passes through an

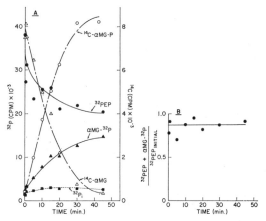

Fig. 7. Phosphorylation of ^{14}C-α-methylglucoside with ^{32}P-enolpyruvate. ML 308-225 membrane samples (20 μl), containing about 0.094 mg of membrane protein, prepared as described in "Methods," were incubated at 46° for 15 min after the addition of ^{32}P-enolpyruvate (approximately 225 mC per mmole) and NaF to give final concentrations of 2×10^{-5} M (approximately) and 10^{-2} M, respectively. α-Methylglucoside-U-^{14}C (73.4 mC per mmole) was then added at 1.84×10^{-5} M. At the times indicated, 10-μl samples were withdrawn, applied to Whatman No. 3 paper, and dried with a hair drier. Approximately 1 μmole of carrier P-enolpyruvate was added to each sample. The paper was subsequently wet with pyridine-acetic acid at pH 5.5 and subjected to electrophoresis at 5000 volts for 45 min. After the paper was dried and fluorescent spots were located, it was placed in contact with x-ray film for 2 or 4 hours in order to locate radioactive spots. The radioactive spots were cut out, and the paper was placed in scintillation vials and counted for ^{14}C and ^{32}P in a Nuclear-Chicago Mark I liquid scintillation spectrometer with the use of the appropriate settings for ^{32}P and ^{14}C counting. ^{32}P counting efficiency in the ^{14}C channel was approximately 3.8% and was subtracted from the experimental data shown. The data are plotted as counts per min of ^{14}C and ^{32}P in the compounds indicated. $\alpha MG^{32}P$ (▲——▲), α-methylglucoside-^{32}P; ^{14}C-αMG-P (○·—·○), ^{14}C-α-methylglucoside-P; ^{32}PEP (●——●), ^{32}P-phosphoenolpyruvate; ^{14}C-αMG ($\triangle$·—·$\triangle$), ^{14}C-α-methylglucoside; ^{32}P$_i$ (■---■), inorganic ^{32}P.

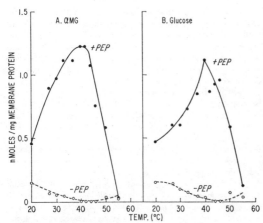

FIG. 8. The effect of temperature on the steady state uptake of α-methylglucoside (A) and glucose (B) by K₂lt membranes in the presence and in the absence of P-enolpyruvate. See legend to Fig. 4 for experimental methods. Glucose-U-¹⁴C (161 mC per mmole) or α-methylglucoside-U-¹⁴C (73.4 mC per mmole) was added at 3.64 × 10⁻⁵ M. +PEP (●——●) and −PEP (O---O), presence or absence, respectively, 0.1 M P-enolpyruvate.

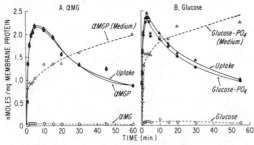

FIG. 9. The time course of α-methylglucoside (A) and glucose (B) uptake at 46° by K₂lt membranes. See legend to Fig. 5 for experimental method. Glucose-U-¹⁴C (161 mC per mmole) or α-methylglucoside-U-¹⁴C (73.4 mC per mmole) was added at 3.64 × 10⁻⁵ M. The data are plotted as nanomoles per mg of membrane protein, with the exception of αMGP (Medium) and Glucose-PO₄ (Medium) (△---△), which represent the quantity of α-methylglucoside-P or glucose-P recovered from the filtrates of the reaction mixtures, expressed as nanomoles per 0.1 ml of reaction mixture.

internal pool before phosphorylation, the ¹⁴C-glucose already present in the internal pool should be phosphorylated before external ³H-glucose. Also, ¹⁴C-glucose in the internal pool should decrease as it is phosphorylated and should be replaced by ³H-glucose. Conversely, if glucose does not pass through an internal pool prior to phosphorylation, the rate of phosphorylation of ³H-glucose might exceed that of ¹⁴C-glucose already present in the membranes. The data presented in Fig. 10, A and B, show that the latter alternative holds. In Fig. 10A, ¹⁴C-glucose was phosphorylated about as rapidly as ³H-glucose only during the first 30 sec. Subsequently, there was no increase in the amount of ¹⁴C-glucose-P recovered from the membranes. On the other hand, the amount of ³H-glucose-P in-

creased markedly throughout the incubation. Furthermore, intramembranal ¹⁴C-glucose remained constant, indicating that ¹⁴C-glucose-P formed during the first 30 sec did not arise from the intramembranal pool. It is noteworthy that the appearance of ³H-glucose in the intramembranal pool was small under these assay conditions.

Since the small but rapidly phosphorylated ¹⁴C-glucose fraction described above could represent phosphorylation from a very small, rapidly turning over pool, an alternative method of assay was devised by which essentially all of the external ¹⁴C-glucose was removed after preliminary loading. In the experiment presented in Fig. 10B, membrane samples, after preliminary loading with ¹⁴C-glucose, were rapidly filtered and washed, and a solution containing ³H-glucose and P-enolpyruvate was placed in contact with the filters. Under these conditions, there was

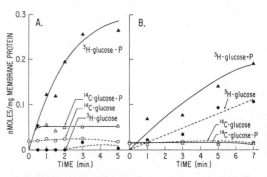

FIG. 10. Uptake and phosphorylation of glucose-6-³H by K₂lt membranes previously loaded with U-¹⁴C-glucose. For A, 100 μl of K₂lt membrane samples, prepared as described in "Methods," were incubated for 2 to 3 hours at 0° in the presence of 10⁻² M NaF and 36.4 × 10⁻⁵ M U-¹⁴C-glucose (161 mC per mmole). Each sample was then diluted 10-fold with 0.25 M potassium phosphate buffer, pH 6.6, containing 10⁻² M NaF at 0° and immediately centrifuged for 10 to 15 min at approximately 37,000 × g in the cold. Without resuspending the pellets, each tube was washed once by filling it with buffer and recentrifuging. The supernatants were discarded. The inside of each tube was then wiped dry. The pellets were rapidly resuspended in a solution of 0.25 M potassium phosphate buffer, pH 6.6, containing 10⁻² M NaF, 0.1 M P-enolpyruvate, and glucose-6-³H (550 mC per mmole) at 3.64 × 10⁻⁵ M which had been equilibrated to 40°. The samples were then incubated at 40° for the times indicated and treated as described previously. The isolated glucose and glucose-P spots from the thin layer chromatogram (see "Methods") were counted for ³H and ¹⁴C in a Nuclear-Chicago Mark I liquid scintillation spectrometer with the use of the appropriate settings for double isotope counting. ¹⁴C counting efficiency in the ³H channel was approximately 11.5%, and was the same for each sample. ¹⁴C-glucose-P (△——△) has been corrected for the zero time value. For B, samples previously loaded as described above were filtered on Millipore HA filters and washed twice with 2.0 ml of 0.25 M potassium phosphate buffer, pH 6.6, containing 10⁻² M MgSO₄ and 10⁻² M NaF. After the vacuum was released, 1.0 ml of a solution containing (final concentrations) 0.25 M potassium phosphate buffer, pH 6.6, 10⁻² M MgSO₄, 10⁻² M NaF, 0.1 M P-enolpyruvate, and glucose-6-³H (550 mC per mmole) which had been equilibrated to 40° was layered on top of the filters and incubation was continued. At the times given, the samples were filtered and washed once with 5.0 ml of 0.25 M potassium phosphate buffer, pH 6.6, containing 10⁻² M MgSO₄ and 10⁻² M NaF. The filters were immediately removed from the filtration apparatus and treated as described above. ▲——▲, ³H-glucose-P; ●---●, ³H-glucose; △——△, ¹⁴C-glucose-P; O---O, ¹⁴C-glucose.

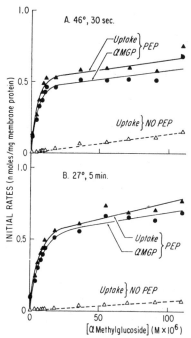

FIG. 11. Kinetics of α-methylglucoside uptake by ML 308-225 membranes at 27° (*A*) and 46° (*B*) in the presence and in the absence of P-enolpyruvate. ML 308-225 membrane samples (100 μl), prepared as described in "Methods," were incubated at 27° (*A*) or 46° (*B*) in the presence or absence of 0.1 M P-enolpyruvate, as indicated, for 15 min. α-Methylglucoside-¹⁴C (73.4 mC per mmole) was then added at the concentrations given. Incubation was continued for 5 min at 27° (*A*) or for 30 sec at 46° (*B*), and samples were assayed as described in "Methods." Each experimental point has been corrected for a control sample, incubated at the same α-methylglucoside-¹⁴C concentration, as described in "Methods." *Uptake PEP* (▲——▲), total uptake in the presence of P-enolpyruvate; *αMGP PEP* (●——●), α-methylglucoside-P recovered from the membranes in the presence of P-enolpyruvate; *Uptake NO PEP* (△---△), total uptake in the absence of P-enolpyruvate.

essentially no increase in ¹⁴C-glucose-P over the zero time value. Furthermore, in agreement with the experiment in Fig. 10*A*, there was no decrease in ¹⁴C-glucose. Although the rate of phosphorylation of ³H-glucose by filtered membranes was only about one-half of that of membranes in suspension, it was much greater than that of the ¹⁴C-glucose already present in the intramembranal pool. The increase in ³H-glucose observed in this experiment resulted from diffusion of ³H-glucose into peripheral areas of the filter which were inadequately washed on termination of the reaction.

When preliminary loading was omitted and both ³H- and ¹⁴C-glucose were presented with P-enolpyruvate during the 40° incubations, the appearance of ³H and ¹⁴C in glucose and glucose-P was essentially the same, ruling out an isotope effect. This experiment, together with those presented in Fig. 5, provides strong evidence that the P-enolpyruvate-P-transferase

system is responsible for the entry of sugars and their concentration within the membrane as phosphorylated derivatives.

The possibility that sugars are phosphorylated after penetrating the membrane by passive or facilitated diffusion is also inconsistent with the data presented in Figs. 11 to 13 and Table III. The data in Fig. 11 show the initial rates (30 sec at 46° and 5 min at 27°) of uptake and phosphorylation as a function of α-methylglucoside concentration with and without P-enolpyruvate. At low concentrations of the glycoside, its initial rate of uptake by membranes incubated at either 27° or 46° was approximately 50 times greater in the presence than in the absence of P-enolpyruvate. Uptake in the presence of P-enolpyruvate showed saturation kinetics, whereas the initial rate of uptake by membranes incubated without P-enolpyruvate increased slowly and linearly with increasing glycoside concentration. The α-methylglucoside-P isolated from each of these samples is also shown in these figures. The initial rate of phosphorylation in the presence of P-enolpyruvate as a function of increasing glycoside concentration was essentially identical with that of the uptake mechanism at the corresponding temperatures. It is obvious from these data that the rate-limiting step for phosphorylation is not a direct function of sugar concentration. Thus, it is unlikely that passive diffusion coupled to phosphorylation is the mechanism for glycoside transport.

Since the quantity of free α-methylglucoside in the membranes was too small to measure reliably, and because it is possible that the initial rates of uptake in the absence of P-enolpyruvate might saturate at higher concentrations, another kinetic study was carried out over a 1,000,000-fold concentration range. The results are presented in Fig. 12, *A* to *C*, as log-log plots so that all of the data can be presented in one figure. The initial rate of uptake at 27° or 46° with P-enolpyruvate (Fig. 12, *A* and *B*) increased rapidly with increasing glycoside concentrations from 10⁻⁷ to 10⁻⁵ M. Above 10⁻⁵ M, the initial rate of uptake with increasing concentration was independent of P-enolpyruvate (Fig. 12*C*). When each of these samples was assayed for α-methylglucoside-P and α-methylglucoside, the uptake curves in Fig. 12, *A* and *B* were resolved into two components: one, the initial rate of phosphorylation which showed saturation kinetics, and two, the initial rate of α-methylglucoside uptake which showed linear kinetics. It is important to note that the kinetics of appearance of α-methylglucoside in the internal pool was identical for samples incubated with or without P-enolpyruvate (Fig. 12*C*).

Lineweaver-Burk treatment of some of the data is presented in Fig. 13, *A* and *B*, and in each case gave a linear plot. Furthermore, in the presence of P-enolpyruvate, an apparent K_m value of about 4×10^{-6} M was calculated for both uptake and phosphorylation. Free α-methylglucoside uptake in the absence of P-enolpyruvate was also a linear function (Fig. 13*B*), but it intersected the *x* and *y* axes at the origin, implying a nonsaturable process without a K_m. It appears, therefore, that the uptake of α-methylglucoside into the internal pool occurs by passive diffusion (as opposed to facilitated diffusion), but the phosphorylation of α-methylglucoside has no relationship to this free pool.

Data presented in Table III show that, under steady state conditions, there was no equilibration between the intramembranal α-methylglucoside pool and the external medium. This observation is also inconsistent with a facilitated diffusion-

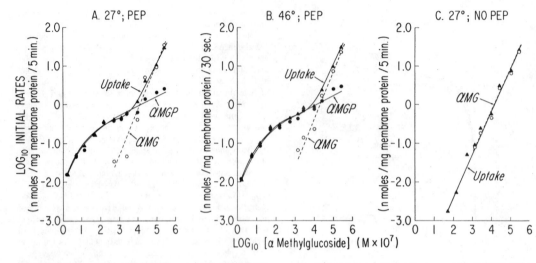

FIG. 12. Kinetics of α-methylglucoside uptake by ML 308-225 membranes at 27° (A and C) and 46° (B) in the presence (A and B) and in the absence (C) of P-enolpyruvate over a 1,000,000-fold concentration range. Experimental points were obtained as in Fig. 11, with the exception that above 7.28×10^{-5} M, the specific activity of the α-methylglucoside-^{14}C was diluted as follows: 1.182×10^{-4} M (11.2 mC per mmole), 4.182×10^{-4} M (3.18 mC per mmole), 1.0182×10^{-3} M (1.31 mC per mmole), 4.0182×10^{-3} M (0.331 mC per mmole), 1.00182×10^{-2} M (0.133 mC per mmole), and 4.00182×10^{-2} M (0.0333 mC per mmole). $Uptake$ ($\blacktriangle$——$\blacktriangle$), total uptake; αMGP ($\bullet$——$\bullet$) and αMG ($\circ$- - -$\circ$), α-methylglucoside-P and α-methylglucoside, respectively, recovered from the membranes.

phosphorylation mechanism. It can be seen that the distribution ratio for α-methylglucoside ([α-methylglucoside]$_i$/[α-methylglucoside]$_0$) was only about 0.3, regardless of the presence of P-enolpyruvate. The distribution ratio for α-methylglucoside-P was extremely high since almost none was found in the external pool.

Exchange Studies—Fig. 14A shows the results of an experiment in which membranes were incubated with α-methylglucoside-^{14}C and P-enolpyruvate for 30 min, at which time a 1000-fold excess of unlabeled α-methylglucoside was added (zero time). There was no demonstrable loss of radioactivity from the membranes for the first 5 min, followed by a slow loss so that only 20 to 30% of the label was lost by 30 min. Thus, no rapidly exchangeable pool was present under these conditions. The same results were obtained when unlabeled glucose-6-P was added.

Since it has already been shown that free α-methylglucoside is only a small percentage of the total radioactivity taken up, it is possible that exchange of only a small percentage of this radioactivity would not be discernible. However, if an exchangeable pool was built up at some time early in the course of uptake, before α-methylglucoside-P accumulated to such high levels, it might be more easily demonstrated. To investigate this possibility, membranes were incubated with α-methylglucoside and P-enolpyruvate and exchanged at various times during uptake. The results are shown in Fig. 14B. The radioactivity lost from the membranes 5 min after the addition of unlabeled α-methylglucoside was negligible at each point studied. Thus no rapidly exchangeable pool is detectable at any time during uptake.

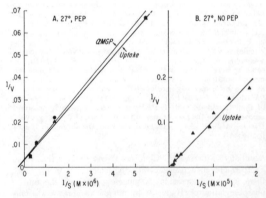

FIG. 13. Lineweaver-Burk treatment of data presented in Figs. 11 and 12. Plots were carried out in A with the use of the experimental points obtained at the lowest concentration of α-methylglucoside in Figs. 7 and 8. $Uptake$ ($\blacktriangle$) total uptake; αMGP ($\bullet$), α-methylglucoside-P recovered from the membranes; PEP, phosphoenolpyruvate.

Specificity of System—The results in Table IV show the specificity of membranes prepared from glucose-grown cells for uptake and phosphorylation of various sugars. The accumulation of β-methyl-D-glucoside, 2-deoxy-D-glucose, and D-fructose was similar to that of α-methyl-D-glucose and D-glucose, *i.e.* the uptake and phosphorylation of these sugars in response to P-enolpyruvate were marked. The uptake and phosphorylation of 3-O-methyl-D-glucose and D-galactose were stimulated by P-enolpyruvate, but they were taken up and phosphorylated less than the first group of sugars. The uptake of D-arabinose, D-ribose, D-sorbitol, and D-mannitol was minimal, with or without P-enolpyruvate. The phosphorylated derivatives of fructose and galactose were isolated, treated with alkaline phosphatase, and identified as described in "Methods." Galactose-P yielded only galactose. However, approximately 50% of the fructose taken up was converted to fructose diphosphate (presumably fructose 1,6-diphosphate) as judged by high voltage electrophoresis, treatment with alkaline phosphatase, and thin layer chromatography. The phosphate present in these derivatives is assumed to be in position 6, although this was shown only for glucose-6-P.

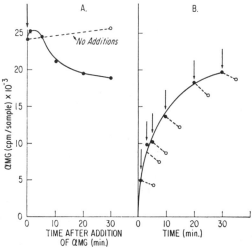

FIG. 14. Exchange of α-methylglucoside-^{14}C taken up by ML 308-225 membranes. For *A*, 100 μl of ML 308-225 membranes, containing 0.47 mg of membrane protein, prepared as described in "Methods," were incubated at 27° for 15 min after the addition of 0.1 M P-enolpyruvate and 10^{-2} M NaF. α-Methylglucoside-^{14}C (73.4 mC per mmole) was then added at 3.64 × 10^{-5} M. Incubation was continued for 30 min, at which point 10 μl of 1 M α-methylglucoside (unlabeled) were added to each reaction mixture (indicated by the *arrow*). To the sample labeled *No Additions*, 10 μl of buffer were added. Incubations was continued for the times indicated, then terminated, and samples were assayed as described in "Methods." For *B*, 2.0-ml ML 308-225 membrane sample containing P-enolpyruvate, prepared as described in "Methods," was incubated at 27° for 15 min. α-Methylglucoside-^{14}C (73.4 mC per mmole) was then added at 3.64 × 10^{-5} M. At the times indicated by *arrows*, 100-μl portions of the reaction mixture, containing 0.47 mg of membrane protein, were assayed as described in "Methods" (●——●). Another 100-μl sample, removed at the same time, was added to another reaction vessel to which had been added 10 μl of a 1 M solution of unlabeled α-methylglucoside and was incubated for another 5 min at 27°, at which time the sample was assayed as described in "Methods" (●---○). Each experimental point has been corrected for a control sample treated as described in "Methods."

TABLE IV

Uptake and phosphorylation of various sugars in presence and in absence of P-enolpyruvate

K₂lt membrane samples (100 μl), containing 10^{-2} M NaF, prepared as described in "Methods," were incubated at 40° with and without P-enolpyruvate, as indicated. One of the ^{14}C sugars listed was then added at 3.64 × 10^{-5} M and incubation was continued for 30 min. Incubations were terminated and samples were assayed as described in "Methods." The specific activities were, in millicuries per mmole, U-^{14}C-α-methylglucoside, 73.4; U-^{14}C-β-methylglucoside, 56; U-^{14}C-2-deoxyglucose, 4.0; U-^{14}C-glucose, 161; U-^{14}C-fructose, 177; U-^{14}C-3-O-methylglucose, 4.36; galactose-1-^{14}C, 35.4; ribose-1-^{14}C, 8; arabinose-1-^{14}C, 2.05; sorbitol-1-^{14}C, 4; and mannitol-U-^{14}C, 2.68.

Sugar	P-enol-pyruvate (10^{-1} M)	Total uptake	Free sugar	Sugar-P
		nmoles/mg protein		
α-Methylglucoside[a]	−	0.728	0.39	0.338
α-Methylglucoside[a]	+	2.0	0.51	1.49
β-Methylglucoside[a]	−	0.74	0.215	0.525
β-Methylglucoside[a]	+	2.34	0.52	1.82
Glucose	−	0.156	0.0242	0.1318
Glucose	+	1.67	0.067	1.603
2-Deoxyglucose	−	0.65	0.198	0.452
2-Deoxyglucose	+	3.14	0.867	2.273
Fructose	−	0.252	0.0113	0.2407
Fructose	+	1.49	0.0595	1.3305
3-O-Methylglucose[a]	−	0.198	0.0625	0.1355
3-O-Methylglucose[a]	+	0.7	0.199	0.501
Galactose	−	0.0228	0.00535	0.01745
Galactose	+	0.393	0.0706	0.3224
Ribose	−	0.0815		
Ribose	+	0.12		
Arabinose	−	0.0795		
Arabinose	+	0.009		
Sorbitol	−	0.023		
Sorbitol	+	0.036		
Mannitol	−	0.067		
Mannitol	+	0.062		

[a] These experiments were carried out with ML 308-225 membranes. The samples were incubated at 46° without the addition of NaF, and the reactions were terminated at 3 min.

DISCUSSION

The data presented in this paper provide evidence for involvement of the P-enolpyruvate-P-transferase system in transport of sugars sharing affinity for the so-called "glucose permease" system. α-Methylglucoside uptake and phosphorylation by isolated membrane preparations from two strains of *E. coli* was almost totally dependent on the presence of P-enolpyruvate and only P-enolpyruvate. Although 2-P-glycerate was partially able to support uptake and phosphorylation, its effect was mediated by a mechanism sensitive to NaF, indicating conversion to P-enolpyruvate prior to stimulation of transport. This hypothesis is consistent with the time lag observed when 2-P-glycerate was used as an energy source. The requirement for high concentrations of P-enolpyruvate, a highly charged compound, may result from its poor ability to penetrate the membrane. Kinetic studies on P-enolpyruvate uptake by the membranes may provide insight into this problem. The requirement for high concentrations of P-enolpyruvate suggests that Enzyme I and HPr are bound to the interior surface of the membrane. If they were exposed to the outside, much lower concentrations of P-enolpyruvate should produce maximal effects on the uptake and phosphorylation of α-methylglucoside.

In further support of the argument that the P-enolpyruvate-P-transferase system is involved, membranes prepared from E. coli GN-2, a mutant which cannot grow on glucose (25) and which lacks Enzyme I activity in the P-transferase system (10), did not take up or phosphorylate α-methylglucoside with or without P-enolpyruvate.

Experiments are presented which show a stoichiometric relation between the disappearance of ^{32}P-enolpyruvate and the appearance of ^{32}P in α-methylglucoside-^{14}C-P, and which suggest that P-enolpyruvate provides energy for the simultaneous uptake and phosphorylation of α-methylglucoside, a mechanism by which sugars are transported and accumulated within the cell. This hypothesis is supported by the double isotope experiment presented in Fig. 10, A and B. In membranes which had been loaded with ^{14}C-glucose, the P-enolpyruvate-P-transferase system exhibited a marked preference for ^{3}H-glucose added to the outside simultaneously with P-enolpyruvate. If glucose first went into an intramembranal pool and then was phosphorylated, ^{14}C-glucose already present in this pool should have appeared as ^{14}C-glucose-P more rapidly than ^{3}H-glucose added to the external medium. Furthermore, ^{14}C-glucose-P should have been formed at the expense of the intramembranal ^{14}C-glucose pool. Neither result was observed. Instead, only ^{3}H-glucose was phosphorylated, the intramembranal ^{14}C-glucose pool did not change, and the amount of ^{14}C-glucose-P remained constant after 30 sec. Although (Fig. 10A) a small amount of ^{14}C-glucose was rapidly phosphorylated, it was not intramembranal ^{14}C-glucose. This suggests that a small amount of ^{14}C-glucose was bound to the exterior surface of the membrane, and was the source of ^{14}C-glucose-P. This ^{14}C-glucose fraction was not observed when the membranes were filtered and washed prior to the addition of ^{3}H-glucose and P-enolpyruvate (Fig. 10B).

The possibility that the P-enolpyruvate-P-transferase system phosphorylates sugars which have entered an internal pool by facilitated or passive diffusion is inconsistent with two more observations. (a) The initial rates of α-methylglucoside uptake and phosphorylation exhibited saturation kinetics with an apparent K_m of about 4×10^{-6} M, whereas the uptake of α-methylglucoside into the intramembranal pool exhibited linear kinetics. Thus, neither passive nor facilitated diffusion can be rate-limiting steps for phosphorylation. The K_m for α-methylglucoside uptake and phosphorylation by these membrane preparations is less than 0.1 that of whole cells.[4] In their ability to concentrate

sugars, membrane preparations are more than 5 times as effective as the whole cells studied by Winkler (4). (b) The intramembranal pool contained only one-third of the α-methylglucoside concentration of the external pool at the steady state, and the appearance of α-methylglucoside-P preceded that of free α-methylglucoside (Figs. 5 and 9). Both observations provide evidence against the facilitated diffusion hypothesis (26), which assumes a rapid equilibration between internal and external pools. Although α-methylglucoside did enter the intramembranal pool by passive diffusion, it is concluded that this pool is unrelated to the phosphorylation mechanism.

The absence of rapid exchange in the membranes is interesting, since Winkler (4) has demonstrated this phenomenon in whole cells. Possibly, the membranes did not carry out exchange because they lack a phosphatase which may be involved in the exit process. The recent report of specific sugar phosphatases (27) may be important in this respect. Since glucose-6-P did not exchange with sugar-P in the membranes, it is unlikely that the α-methylglucoside-P formed during transport was adsorbed to the exterior of the membranes.

The temperature effects displayed by the uptake system are unusual. In addition to sharp temperature optima, different optima were found for the steady state levels and the initial rates of α-methylglucoside and glucose uptake and phosphorylation. At 27° in the presence of P-enolpyruvate, α-methylglucoside uptake and phosphorylation reached a plateau at 30 min, whereas at 46° uptake and phosphorylation reached the same level in 3 min. However, this level decreased to about 30% of the maximum by the end of the incubation. The loss of radioactivity at 46° appears to result from leakage of α-methylglucoside-P from the membranes, and this temperature-induced leakage is reversible. It is noteworthy that proline uptake by a membrane preparation of E. coli W (16) manifested similar temperature effects with the exception that the steady sta'e level and the initial rate showed the same optima at 27°.[5] The leakage may be explained by phase changes in phospholipids such as those described by Luzzati and Husson (28), who showed by x-ray diffraction that aqueous dispersions of phospholipids undergo a series of abrupt phase changes with increasing temperature. They postulated that such changes should result in marked permeability changes in phospholipid-containing structures (e.g. membranes). Interestingly, the temperature optimum for α-methylglucoside uptake at 30 min by membranes prepared from E. coli $K_2$1t was 40° (see Fig. 8) whereas for membranes prepared from E. coli ML 308-225 it was 27° (see Fig. 4). Although not shown, the initial rate optimum for glycoside uptake by $K_2$1t membranes was at 46°. Thus $K_2$1t membranes do not begin to leak significantly until the temperature is above 40°, as opposed to 27° with ML 308-225 membranes. The reasons for this difference are under investigation.

Regarding the specificity of the system, two points are noteworthy. One is the finding that fructose was converted to fructose-P and fructose diphosphate. If phosphate in C-1 of what was presumably fructose 1,6-diphosphate was derived from P-enolpyruvate via the P-enolpyruvate-P-transferase system, it would be the first instance in which the system has been found to phosphorylate a sugar at position 1. Furthermore, such a finding would imply that the P-enolpyruvate-P-transferase

---

[4] S. Roseman and W. Kundig, personal communication.

[5] H. R. Kaback, unpublished information.

system may be involved in subsequent metabolism of sugars as well as their transport. Alternatively, this may be related to sugar-P uptake mechanisms such as those described by Winkler (4) and Fraenkel, Falcoz-Kelly, and Horecker (25). The other noteworthy point is the phosphorylation of galactose during transport. This finding is especially interesting because the whole cells from which these membranes were prepared were not induced with galactose. Although in whole cells galactose is not reported to have an affinity for the glucose uptake system (29), the activity toward galactose of the membranes used in these experiments is presumably a property of the glucose uptake system. In view of these observations, perhaps a role for sugar phosphates in enzyme induction should be re-examined.

Finally, although this system was stimulated by P-enolpyruvate, there was no requirement for exogenous HPr or Enzyme I. This observation is somewhat surprising in that Kundig *et al.* (8) demonstrated that a defect in α-methylglucoside and methyl-β-thio-D-galactopyranoside uptake in cold-shocked whole cells of *E. coli* W2244 could be alleviated by addition of HPr. Since the membranes used here were subjected to procedures much harsher than the cold shock procedure, not only P-enolpyruvate, but also HPr and possibly Enzyme I, might have been expected to be rate limiting. These membrane preparations have been assayed for HPr and Enzyme I by Dr. Werner Kundig, and small but significant quantities were found. This anomaly may have several explanations: (*a*) despite a much lower K_m value and greater capacity to concentrate α-methylglucoside, the membranes have a V_{max} for entry which is about three orders of magnitude lower than that of whole cells,[4] implying that HPr, Enzyme I, or both may be rate limiting, (*b*) the P-enolpyruvate-P-transferase system (or at least HPr and Enzyme I) may be involved in other metabolic pathways which are present in cold-shocked cells (thus, in whole, cold-shocked cells, P-HPr may be used in other reactions making it rate-limiting for transport), and (*c*) there may be differences among strains of *E. coli* in the binding of HPr and Enzyme I to Enzyme II in the membrane such that these soluble proteins are easily released from some strains but not from others. This problem is currently under investigation, in collaboration with Drs. Saul Roseman and Werner Kundig.

Note Added in Proof—Membranes prepared from *Salmonella typhimurium* and *Bacillus subtilis* behaved like those prepared from *E. coli*. Uptake and phosphorylation of α-methylglucoside by these membrane preparations were also markedly stimulated by P-enolpyruvate and showed similar phenomena with respect to temperature. Furthermore, membranes prepared from *S. typhimurium* mutants SB 703 (Enzyme I-defective) (12), Car 5 (Enzyme I-defective), and Car 6 (HPr-defective) did not take up or phosphorylate α-methylglucoside over a P-enolpyruvate concentration range of 0 to 0.1 M. These mutants were the generous gift of Drs. Robert Simoni and Saul Roseman.

Acknowledgments—The author would like to thank Drs. H. Winkler, H. Kornberg, and L. Heppel for the strains of *E. coli* used in these experiments. He would also like to express his appreciation to Drs. Earl Stadtman, Saul Roseman, Werner Kundig, Robert Simoni, and Herbert Winkler for their helpful advice and discussion during the course of these investigations. Finally, the author would like to thank Dr. Bernard Babior for his help in the initial conception of the ^{3}H-, ^{14}C-glucose experiment, and Drs. Drue Denton and Saul Roseman for their valuable editorial advice.

REFERENCES

1. WILBRANDT, W., AND ROSENBERG, T., *Pharmacol. Rev.*, **13**, 109 (1961).
2. CHRISTENSEN, H. N., *Biological transport*, W. A. Benjamin, Inc., New York, 1962.
3. ROGERS, D., AND YU, S.-H., *J. Bacteriol.*, **84**, 877 (1962).
4. WINKLER, H. H., *Biochim. Biophys. Acta*, **117**, 231 (1966).
5. KUNDIG, W., GHOSH, S., AND ROSEMAN, S., *Proc. Nat. Acad. Sci. U. S. A.*, **52**, 1067 (1964).
6. KUNDIG, W., AND ROSENMAN, S., in S. P. COLOWICK, N. O. KAPLAN, AND W. A. WOOD (Editors), *Methods in enzymology*, Vol. IX, Academic Press, New York, 1966, p. 396.
7. KUNDIG, W., KUNDIG, F. D., ANDERSON, B. E., AND ROSEMAN, S., *Fed. Proc.*, **24**, 658 (1956).
8. KUNDIG, W., KUNDIG, F. D., ANDERSON, B., AND ROSEMAN, S., *J. Biol. Chem.*, **241**, 3243 (1966).
9. NEU, H. C., AND HEPPEL, L. A., *J. Biol. Chem.*, **240**, 3685 (1965).
10. TANAKA, S., FRAENKEL, D. G., AND LIN, E. C. C., *Biochem. Biophys. Res. Commun.*, **27**, 63 (1967).
11. TANAKA, S., AND LIN, E. C. C., *Proc. Nat. Acad. Sci. U. S. A.*, **52**, 913 (1967).
12. SIMONI, R. D., LEVENTHAL, M., KUNDIG, W., ANDERSON, B., HARTMAN, P. E., AND ROSEMAN, S., *Proc. Nat. Acad. Sci. U. S. A.*, **58**, 1963 (1967).
13. EGAN, J. B., AND MORSE, M. L., *Biochim. Biophys. Acta*, **97**, 310 (1965); **109**, 172 (1965); **112**, 63 (1966).
14. HENGSTENBERG, W., EGAN, J. B., AND MORSE, M. L., *Proc. Nat. Acad. Sci. U. S. A.*, **58**, 274 (1967).
15. KENNEDY, E. P., AND SCARBOROUGH, G. A., *Proc. Nat. Acad. Sci. U. S. A.*, **58**, 225 (1967).
16. KABACK, H. R., AND STADTMAN, E. R., *Proc. Nat. Acad. Sci. U. S. A.*, **55**, 920 (1966).
17. KABACK, H. R., AND KOSTELLOW, A. B., *J. Biol. Chem.*, **243**, 1384 (1968).
18. KABACK, H. R., AND STADTMAN, E. R., *J. Biol. Chem.*, **243**, 1390 (1968).
19. KABACK, H. R., AND DEUEL, T. F., *Fed. Proc.*, **26**, 393 (1967).
20. NEU, H. C., AND HEPPEL, L. A., *J. Biol. Chem.*, **239**, 3893 (1964).
21. COLOMBO, P., CORBETTA, D., PIROTTA, A., RUFFINI, G., AND SARTORI, A., *J. Chromatogr.*, **3**, 343 (1960).
22. WEISS, J. B., AND SMITH, I., *Nature*, **215**, 638 (1967).
23. HORECKER, B. L., AND SMYRNIOTIS, P. Z., in S. P. COLOWICK AND N. O. KAPLAN (Editors), *Methods in enzymology*, Vol. I, Academic Press, New York, 1955, p. 323.
24. LOWRY, O. H., ROSEBROUGH, N. J., FARR, A. J., AND RANDALL, R. J., *J. Biol. Chem.*, **193**, 265 (1951).
25. FRAENKEL, D. G., FALCOZ-KELLY, F., AND HORECKER, B. L., *Proc. Nat. Acad. Sci. U. S. A.*, **52**, 1207 (1964).
26. PARK, C. R., in A. KLEINZELLER AND A. KOTYK (Editors), *Membrane transport and metabolism, proceedings of a symposium held in Prague*, Academic Press, New York, 1961, p. 19.
27. LEE, Y.-P., SOWOKINOS, J. R., AND ERWIN, M. J., *J. Biol. Chem.*, **242**, 2264 (1967).
28. LUZZATI, V., AND F. HUSSON, F., *J. Cell Biol.*, **12**, 207 (1962).
29. GANESAN, A. K., AND ROTMAN, B., *J. Mol. Biol.*, **16**, 42 (1966).

Binding Proteins

V

Editor's Comments on Papers 21 Through 28

21 **Pardee, Prestige, Whipple, and Dreyfuss:** *A Binding Site for Sulfate and Its Relation to Sulfate Transport into* Salmonella typhimurium

22 **Neu and Heppel:** *The Release of Enzymes from* Escherichia coli *by Osmotic Shock and During the Formation of Spheroplasts*

23 **Piperno and Oxender:** *Amino Acid Binding Protein Released from* Escherichia coli *by Osmotic Shock*

24 **Anraku:** *The Reduction and Restoration of Galactose Transport in Osmotically Shocked Cells of* Escherichia coli

25 **Boos and Sarvas:** *Close Linkage Between a Galactose Binding Protein and the β-Methylgalactoside Permease of* Escherichia coli

26 **Ames and Lever:** *Components of Histidine Transport: Histidine-Binding Proteins and* hisP *Protein*

27 **Ames and Lever:** *The Histidine-Binding Protein* J *Is a Component of Histidine Transport*

28 **Hazelbauer and Adler:** *Role of the Galactose Binding Protein in Chemotaxis of* Escherichia coli *Toward Galactose*

In 1965, Dreyfuss and Pardee observed that *Salmonella typhimurium* mutants incapable of transporting sulfate bound small quantities of this ion at the cell surface. Binding activity was repressed by growth on cysteine, as was transport activity in wild-type cells. The conversion of wild-type cells into spheroplasts resulted in the loss of sulfate transport and the appearance of sulfate binding activity in the supernatant (Pardee and Prestige, 1965). Pardee, Prestige, Whipple, and Dreyfuss (Paper 21) verified and extended these results and showed, in addition, that binding activity was released when transport-negative cell were subjected to the osmotic shock procedure of Neu and Heppel (Paper 22). The decrement in binding activity in the shocked cells was similar in magnitude to the observed decrease in sulfate transport when wild-type cells were given the same treatment. Neu and Heppel's (Paper 22) procedure, which involves plasmolyzing the cells in 20 percent sucrose in the presence of EDTA and then quickly resuspending them in cold water or 0.5 mM MgSO$_4$, causes the release of a variety of enzymes thought to be localized within the "pericytoplasmic space," the region between the cytoplasmic membrane and the cell wall in gram-negative bacteria (Heppel, 1967, 1971). Pardee et al. (Paper 21) suggested that binding of sulfate to the surface component released by osmotic shock might represent the first step in the active transport of sulfate.

Thus began the study of "binding proteins" and their relationship to transport. In rapid succession, proteins specifically binding leucine-isoleucine-valine, galactose, arabinose, phosphate, arginine, histidine, glutamine, and leucine, among others, were isolated from the supernatant of osmotically shocked cells (see reviews by Lin, 1971;

Oxender, 1972b). Nearly all the binding proteins so far isolated have molecular weights in the range 30,000–45,000 and dissociation constants of the protein–substrate complex of 10^{-6} M or less.

Until recently, the evidence that binding proteins are involved in transport has been indirect. Piperno and Oxender (Paper 23) found that the dissociation constants for leucine, isoleucine, and valine binding were similar to the reported K_m's for transport. Boos (1969) found a good correlation between the affinities of the galactose binding protein and the β-methylgalactoside transport system in E. $coli$. Similar results were reported for many other systems as well (Oxender, 1972b). Moreover, the effectiveness of competitive inhibitors of transport correlates with their abilities to inhibit binding (Pardee et al., Paper 21; Boos, 1969; Ames and Lever, Paper 27).

In nearly every instance where binding proteins are released by osmotic shock, a simultaneous reduction in transport activity toward the binding substrate has been reported (see, for example, Pardee et al., Paper 21; Piperno and Oxender, Paper 23; Anraku, Paper 24). In one case, the discovery of a protein specifically binding leucine preceded and in fact led to the discovery of a leucine-specific transport system in E. $coli$ (Furlong and Weiner, 1970). As stressed in a review by Kaback (1970), however, these results do not necessarily mean that the observed decrement in transport is due to the release of binding proteins; osmotic shock releases many other substances, the acid-soluble nucleotide and amino acid pools, for instance, that might be involved, directly or indirectly, in the transport process.

Anraku (Paper 24) reported that galactose transport activity in shocked cells could be restored by incubating the cells with shock fluid containing, among other things, the galactose binding protein. Purified galactose binding protein also stimulated transport but was most effective when added with another component of the shock fluid that showed no galactose binding activity (Anraku, 1968). Purified phosphate binding protein stimulated phosphate transport in shocked cells of E. $coli$ and in transport-negative mutants lacking the phosphate binding protein (Medveczky and Rosenberg, 1969, 1970). Successful reconstitution of transport with purified binding proteins appears to be the exception rather than the rule, however; Pardee et al. (Paper 21), Piperno and Oxender (Paper 23), and many others (Oxender, 1972b) have reported negative results.

The most direct evidence supporting a role for binding proteins in transport comes from genetic studies. These, however, have been hampered by the inability to select directly for mutants lacking binding proteins; transport-negative mutants have generally been selected instead in an attempt to correlate binding activity with transport phenotypes. In transport-negative strains, binding activity may or may not be present, indicating the participation of other components in the transport process. In cases where binding activity is absent, it is not always possible to demonstrate unequivocally that the associated loss of transport activity derives from the absence of binding protein. The minimal requirement for such a demonstration is that binding and transport activities be lost simultaneously as the result of a single mutation in the structural gene for the binding protein.

Pardee et al. (Paper 21) reported that about one-third of the isolated sulfate transport-negative mutants of S. $typhimurium$ showed lowered sulfate binding activity. Since the residual sulfate binding proteins in these mutants were later found to have the same properties as the wild-type proteins, the mutations probably affected regulatory genes (Ohta, Galsworthy, and Pardee, 1971). No mutants have yet been isolated with defects in the structural gene for the sulfate binding protein (Ohta et al., 1971).

Boos and Sarvas (Paper 25) isolated 28 different mutants with defects in the "methylgalactoside permease," the high-affinity transport system for galactose in *E. coli*. All but one of the mutants showed either reduced levels of galactose binding activity or altered reactivity toward antibodies to the binding protein. In addition, a mating experiment suggested that certain mutations affecting the permease and the binding protein were in closely linked, if not identical, genes; deletions, polar mutations, or regulatory defects could not be excluded as possible explanations for the data, however. Subsequent studies indicated that the biosyntheses of the binding protein and the methylgalactoside permease are co-regulated (Lengeler, Hermann, Unsold, and Boos, 1971). Recently, Boos (1972) has reported that one of the permease-negative mutants (EH3039) isolated by Boos and Sarvas (Paper 25) is defective in the structural gene for the galactose binding protein, as shown by the mutant protein's low affinity for galactose and altered fingerprint pattern. A revertant showed simultaneous restoration of transport, high-affinity binding, and the wild-type fingerprint pattern.

Ames and Lever (Paper 26) found that at least three proteins are involved in the high-affinity transport system for histidine in *S. typhimurium*. Mutations in the *hisP* gene resulted in complete loss of high-affinity histidine transport; *hisJ* mutants retained some residual histidine uptake activity and remained sensitive to a histidine analogue transported by the high-affinity system. The third component (*K*) was thought to be responsible for the residual uptake in *hisJ* mutants and to operate in parallel with the *J* protein, the activity of both components depending upon the presence of the *P* pretein. A histidine binding protein could be isolated from wild-type cells of *hisP* mutants but was absent in *hisJ* mutants. A third mutation, *dhuA*, resulted in increased levels of the histidine binding protein and correspondingly increased transport rates. The three genes, *hisP*, *hisJ*, and *dhuA*, were found by transduction analysis to be closely linked.

Recently a revertant of the *hisJ* mutant has been isolated that exhibits an increased temperature sensitivity for histidine uptake and growth on D-histidine (Ames and Lever, Paper 27). Histidine binding activity was present at wild-type levels in revertant cells grown at 30°C but was severely depressed by growth at 37°C. Furthermore, the histidine binding protein isolated from revertant cells lost activity at 100°C more rapidly than the wild-type protein, differed in its chromatographic behavior, and showed a lower affinity for histidine than the wild-type protein. Genetic mapping suggested that the "reversion" was due to a second mutation in the *hisJ* gene. These results appear to demonstrate conclusively that the *hisJ* gene is the structural gene for the histidine binding protein and that the binding protein is a component of the histidine transport system in *S. typhimurium*.

The function of binding proteins in transport is completely unknown. Recent reports indicate that the galactose and glutamine binding proteins undergo a conformational change on combining with their respective substrates, a property which might be expected of carrier proteins (Boos and Gordon, 1971; Boos, Gordon, Hall, and Price, 1972; Weiner and Heppel, 1971). Substrate-induced conformational changes were not observed, however, in the isoleucine-leucine-valine binding protein (Penrose, Zand, and Oxender, 1970). Furthermore, although intuitive arguments are notoriously unreliable, it seems unlikely that membrane-bound carriers would be as highly soluble in water or as easily detached from the cell as are the binding proteins.

A second possibility is that binding proteins facilitate the diffusion of their substrates through the outer membrane or the pericytoplasmic space. Robbie and Wilson (1969) have presented kinetic evidence consistent with the existence of a diffusion barrier external to the cytoplasmic membrane in *E. coli*. Lin (1971) has proposed that the passage of substrates through the pericytoplasmic space might be assisted by binding proteins in much the same way that concentrated hemoglobin solutions facilitate the diffusion of oxygen (Wittenberg, 1959, 1970; Scholander, 1960). Oxender (1972a) has pointed out, however, that many transport systems function perfectly well without binding proteins.

The galactose-binding protein, whatever its function in transport, appears to participate in chemotaxis as well (Hazelbauer and Adler, Paper 28; Kalckar, 1971). Chemotaxis refers to the ability of motile bacteria to be attracted by certain metabolites (Adler, 1966, 1969). In *E. coli*, chemotaxis toward galactose does not require its utilization, since nonmetabolizable galactose analogues also serve as effective attractants, and mutants unable to utilize galactose show the same response as wild-type cells; nor does it require transport, since some mutants that cannot take up galactose are nevertheless attracted by it (Adler, 1966; Hazelbauer, Mesibov, and Adler, 1969). On the other hand, mutants that are defective in galactose chemotaxis also show defects in galactose uptake and revertants for chemotaxis also regain transport activity, suggesting a common factor for both functions (Hazelbauer et al., 1969).

Hazelbauer and Adler (Paper 28) found that chemotaxis mutants lacked the galactose binding protein. One mutant (AW550), which showed a chemotactic response only at elevated galactose concentrations, produced a binding protein with an affinity for galactose much lower than that of the wild-type protein. Moreover, osmotically shocked cells, having lost their attraction for galactose, regained the chemotactic response when supplemented with purified binding protein. Binding protein from the low-affinity mutant AW550 also restored chemotaxis to osmotically shocked cells, but in this case the response was displaced toward higher galactose concentrations. Aksamit and Koshhland (1972) recently reported that chemotaxis toward ribose was restored to osmotically shocked cells by the addition of a purified ribose binding protein. It seems certain, however, that not all binding proteins are involved in chemotaxis; *E. coli* is not attracted by leucine, isoleucine, or valine, for example (Hazelbauer and Adler, Paper 28), although a protein binding these amino acids has been isolated (Piperno and Oxender, Paper 23).

References

Adler, J. (1966). Chemotaxis in bacteria. *Science,* **153,** 708–716.
Adler, J. (1969). Chemoreceptors in bacteria. *Science,* **166,** 1588–1597.

Aksamit, R. R., and D. E. Koshland, Jr. (1972). Ribose chemotaxis and a ribose binding protein. *Fed. Proc.* **31**, 458 abs.

Anraku Y. (1968). Transport of sugars and amino acids in bacteria. III. Studies on the restoration of active transport. *J. Biol. Chem.,* **243**, 3128–3135.

Boos, W. (1969). The galactose binding protein and its relationship to the β-methylgalactoside permease from *Escherichia coli. Europ. J. Biochem.,* **10**, 66–73.

Boos, W. (1972). Structurally defective galactose-binding protein isolated from a mutant negative in the β-methylgalactoside transport system of *Escherichia coli. J. Biol. Chem.,* **247**, 5414–5424.

Boos, W., and A. S. Gordon (1971). Transport properties of the galactose-binding protein of *Escherichia coli. J. Biol. Chem.,* **246**, 621–628.

Boos, W., A. S. Gordon, R. E. Hall, and H. D. Price (1972). Transport properties of the galactose binding protein of *Escherichia coli.* Substrate induced conformational changes. *J. Biol. Chem.,* **247**, 917–924.

Dreyfuss, J., and A. B. Pardee (1965). Evidence for a sulfate-binding site external to the cell membrane of *Salmonella typhimurium. Biochim. Biophys. Acta,* **104**, 308–310.

Furlong, C. E., and J. H. Weiner (1970). Purification of a leucine-specific binding protein from *Escherichia coli. Biochem. Biophys. Res. Commun.,* **38**, 1076–1083.

Hazelbauer, G. L., R. E. Mesibov, and J. Adler (1969). *Escherichia coli* mutants defective in chemotaxis toward specific chemicals. *Proc. Natl. Acad. Sci. USA,* **64**, 1300–1307.

Heppel L. A. (1967). Selective release of enzymes from bacteria. *Science,* **156**, 1451–1455.

Heppel L. A. (1971). The concept of periplasmic enzymes. In L. I. Rothfield (ed.), *Structure and Function of Biological Membranes.* Academic Press, New York, pp. 223–247.

Kaback, H. R. (1970). Transport. *Am. Rev. Biochem.,* **39**, 561–598.

Kalckar, H. M. (1971). The periplasmic galactose binding protein of *Escherichia coli. Science,* **174**, 557–565.

Lengeler, J., K. O. Hermann, H. J. Unsold, and W. Boos (1971). The regulation of the β-methylgalactoside transport system and of the galactose binding protein of *Escherichia coli* K12. *Europ. J. Biochem.,* **19**, 457–470.

Lin, E. C. C. (1971). The molecular basis of membrane transport systems. In L. I. Rothfield (ed.), *Structure and Function of Biological Membranes.* Academic Press, New York, pp. 285–341.

Medveczky, N., and H. Rosenberg (1969). The binding and release of phosphate by a protein isolated from *Escherichia coli. Biochim. Biophys. Acta,* **192**, 369–371.

Medveczky, N., and H. Rosenberg (1970). The phosphate binding protein of *Escherichia coli. Biochim. Biophys. Acta,* **211**, 158–168.

Ohta, N., P. R. Galsworthy, and A. B. Pardee (1971). Genetics of sulfate transport by *Salmonella typhimurium. J. Bacteriol.,* **105**, 1053–1062.

Oxender, D. L. (1972a). Discussion in J. F. Woessner and F. Huijing (eds.), *The Molecular Basis of Biological Transport.* Academic Press, New York, p. 287.

Oxender, D. L. (1972b). Membrane transport. *Ann. Rev. Biochem.,* **41**, 777–814.

Pardee, A. B. and L. S. Prestidge (1965). Cell-free activity of a sulfate binding site involved in active transport. *Proc. Natl. Acad. Sci. USA,* **55**, 189–191.

Penrose, W. R., R. Zand, and D. L. Oxender (1970). Reversible conformational changes in a leucine-binding protein from *Escherichia coli. J. Biol. Chem.,* **245**, 1432–1434.

Robbie, J. P., and T. H. Wilson (1969). Transmembrane effects of β-galactosides on thio-methyl-β-galactoside transport in *Escherichia coli. Biochim. Biophys. Acta,* **173**, 234–244.

Scholander, P. F. (1960). Oxygen transport through hemoglobin solutions. *Science,* **131**, 585–590.

Weiner, J. H., and L. A. Heppel (1971). A binding protein for glutamine and its relation to active transport in *Escherichia coli. J. Biol. Chem.,* **246**, 6933–6941.

Wittenberg, J. B. (1959). Oxygen transport—a new function proposed for hemoglobin. *Biol. Bull.,* **117**, 402.

Wittenberg, J. B. (1970). Myoglobin facilitated oxygen diffusion, role of myoglobin in oxygen entry into muscle. *Physiol. Rev.,* **50**, 559–636.

Copyright © 1966 by the American Society of Biological Chemists, Inc.
Reprinted from *J. Biol. Chem.*, **241**, 3962–3969 (1966)

A Binding Site for Sulfate and Its Relation to Sulfate Transport into *Salmonella typhimurium**

21

(Received for publication, January 10, 1966)

Arthur B. Pardee, Louise S. Prestidge,‡ Mettie B. Whipple,§ and Jacques Dreyfuss¶

From the Biology Department, Princeton University, Princeton, New Jersey 08540

SUMMARY

Mutants of *Salmonella typhimurium* classified by nutritional and enzymatic tests as transport-negative for sulfate are capable of combining with small amounts of this anion (up to 10^4 molecules per bacterium). The following results support the suggestion that the bacteria possess highly specific sulfate binding sites near the cell surface.

1. Under growth conditions sulfate combines with the bacteria according to a typical adsorption isotherm, which reaches half-saturation at 0.004 mM.

2. Various anions structurally similar to sulfate are specific inhibitors of binding and transport.

3. Inability of the mutants to grow on sulfate or thiosulfate suggests that the anions cannot enter the cells, but are bound near the cell surface.

4. The cells lose their ability to bind sulfate after they are converted to spheroplasts or are osmotically shocked. With the use of a new assay for the measurement of binding to soluble entities, the binding ability lost from the cells is found in solution.

5. The properties of the binding system suggest that it might be a part of the active transport system for sulfate. The surface location of the binding material as well as the following results are consistent with this hypothesis. Both binding activity and active transport are repressed by the growth of bacteria on cysteine as a sulfur source. Both functions are similarly (but not identically) inhibited by various anions. Both functions are lost, in proportion, upon osmotic shock of transport-positive cells. Both functions are simultaneously lost in a class of chromate-resistant mutants, and are regained upon transduction with PLT-22 phage or reversion.

A system for the transport of sulfate and thiosulfate into *Salmonella typhimurium* has been described in a previous paper

(1). Transport occurs against a concentration gradient and energy is required. It is highly temperature-dependent. The initial rate follows Michaelis-Menten kinetics; thiosulfate and sulfite are strong inhibitors. Sulfate transport is repressed in cysteine-grown bacteria.

Both growth on sulfate and transport require the integrity of at least the *cys A* region of *S. typhimurium*. This region contains three linked cistrons which are not involved in other known cellular functions (2, 3). *Cys A* mutants do not transport measurable amounts of sulfate, compared to transport-positive strains, at the high sulfate concentrations used in the initial experiments (1). Their combination with small amounts of sulfate was readily shown at lower sulfate concentrations (4). Observations already reported (4), which led to the hypothesis that this combination represents binding to the cell surface and is related to active transport, are the following. (*a*) The quantity of sulfate bound by the mutants was independent of the external sulfate concentration; it reached saturation at less than 0.05 mM sulfate. (*b*) The rate of binding was rapid, reaction being complete in less than 20 sec. (*c*) In cysteine-grown cells binding activity was absent, as is active transport in transport-positive cells (1). (*d*) Spheroplasts lost transport activity, suggesting loss of some component of the surface of the cell.

Binding activity has recently been shown with cell-free extracts (5). The binding substance has now been isolated and identified as a protein of molecular weight 32,000.[1] In the present communication, "binding" refers to combination of sulfate with cell fractions or with transport-negative cells.

This paper presents evidence for the existence of specific binding sites for sulfate, their localization at the cell surface, and their role in active transport.

EXPERIMENTAL PROCEDURE

Materials—"Carrier-free" $H_2{}^{35}SO_4$ was obtained from Oak Ridge National Laboratory. Its radioactivity was more than 99.99% precipitable by Ba^{++}. L-Djenkolic acid and L-cysteine hydrochloride were obtained from Calbiochem, sodium selenate and *N*-methyl-*N*-nitroso-*N*¹-nitroguanidine from K and K Laboratories, Inc., and 2,5-diphenyloxazole and 1,4-bis[2-(5-phenyloxazoyl)]benzene from Packard. Chloramphenicol was a gift from Parke, Davis.

Methods—Most mutant strains used in these studies as well as their culture conditions have been previously described (1). *Cys CD-519* is a transport-positive deletion mutant, which lacks

* This work was aided by Grants AI-04409 and 5 T1 GM 962 from the United States Public Health Service.

‡ Present address, Scripps Clinic and Research Foundation, La Jolla, California.

§ Present address, Department of Agronomy, Kansas State University, Manhattan, Kansas 66504.

¶ Present address, Squibb Institute for Medical Research, New Brunswick, New Jersey.

[1] A. B. Pardee, in preparation.

3962

the first two enzymes required for utilization of sulfate. *Cys A* mutants, such as *cys Aabc-20*, are all transport-negative.

The indirect measurement of sulfate binding was performed by incubating at 37° a dense cell suspension of *S. typhimurium* (about 20 mg of protein per ml of Medium E (6)) containing 2 mg per ml of glucose, 0.03 mg per ml of chloramphenicol, and desired concentrations of sodium sulfate and $^{35}SO_4^-$. The incubation period was terminated by filtering the cell suspension through a 0.45-μ Millipore filter, on top of which was placed a layer of Celite (Johns-Manville) prepared 1 day earlier and left at room temperature, as previously described (1). With this technique, a dense cell suspension can be filtered so that approximately 0.2 ml of clear filtrate is recovered in 5 to 10 sec. Radioactivity in this filtrate was then counted. When kinetics were not desired, centrifugation for 5 min at 10,000 × *g* replaced filtration. In some experiments, the supernatant fluid and also the sedimented cells were counted.

For the measurement of the amount of $^{35}SO_4^-$ present in a sample, 0.04 to 0.2 ml of sample was placed in a polyethylene vial (Packard) containing 10 or 20 ml of the scintillation fluid previously described (1). The vials were shaken and chilled before being counted in a Packard Tri-Carb liquid scintillation spectrometer, model 314-EX. A background of about 20 cpm was subtracted.

The $^{35}SO_4^-$ transported was calculated from the equation

$$\% \text{ transported} = \frac{C - I}{C} \times 100$$

where C represents the counts per min per ml of medium before the addition of cells, and I represents the counts per min per ml of the reaction mixture filtrate.

Binding activity of cell-free extracts was measured by the ability of the preparations to release sulfate into solution from a purified Dowex 1 resin (Bio-Rad Laboratories analytical grade anion exchange resin AG 1-X8, Cl⁻ form, 100 to 200 mesh). Dowex 1-X8, used previously (5), was less satisfactory, since it inhibited binding by the cell preparations by about 85%. The cell preparation in 0.4 ml of the salts of Medium E was mixed with 0.05 ml of 0.04 mM sulfate (1.5 × 10⁵ cpm $^{35}SO_4^-$) in E salts. Then the resin (0.3 g, dry weight) in 1 ml of H_2O was added and mixed at intervals for several minutes. The resin was allowed to settle for 5 min, and 0.08 ml of the supernatant fluid was sampled and counted (resin method).

Protein was determined according to the method of Lowry *et al.* (7).

Transport-negative mutants were obtained as follows. The bacteria were grown on nutrient broth plus 0.2 mM L-cysteine to about 1.5 × 10⁹ cells per ml, and were centrifuged and resuspended at a concentration of approximately 3 × 10¹⁰ per ml in 0.2 M acetate buffer at pH 5.0. A portion, 0.5 ml, of this suspension was incubated with 0.12 ml of *N*-methyl-*N*-nitroso-*N*¹-nitroguanidine (4 mg per ml) in acetate buffer for 3 hours. Then 0.01 ml was diluted to 10 ml with nutrient broth, and was aerated at 37° for 20 hours. The cells were removed by centrifugation, resuspended in 50 ml of Medium E containing djenkolic acid and glucose, and incubated for 5 hours (final cell concentration, 5 × 10⁸ per ml). Then sterile 0.5 mM Na₂CrO₄ (final concentration) was added and the culture was incubated for 24 hours at 37°. The cells were plated on nutrient agar with cysteine. Representative colonies were tested for resistance to chromate on agar plates containing Medium E and djenkolate,

to which sterile glucose and 0.25 mM Na₂CrO₄ were added after autoclaving. Cells resistant to chromate were tested for their inability to transport sulfate and to grow on sulfate or thiosulfate.

RESULTS

Combination of Sulfate with Transport-negative Bacteria—The test previously used to measure binding is indirect, since it depends on measuring the quantity of sulfate which disappears from the medium. Therefore, a more direct method was devised to show that sulfate is combined with the cells. The procedure tried first involved the separation of the cells on a Millipore filter and counting them. This was unsuccessful because bound sulfate was rapidly lost from the cells on rinsing with sulfate-free medium. A method which proved successful was the separation of the bacteria by centrifugation and counting them. In one experiment 0.36 ml of Medium E containing 0.02 mM Na₂SO₄ and 70,000 cpm of $^{35}SO_4^-$ was mixed with cells of mutant *cys Aabc-20* representing 16.2 mg of protein. After centrifugation the cells plus intercellular fluid in the pellet were found to contain 41,700 cpm, of which about 4,500 were calculated to be in the intercellular space (assuming this to be 30% of the pellet (8)). Thus, 37,000 cpm were associated directly with the cells. This is in reasonable agreement with 33,000 cpm which were shown to disappear from the medium when the indirect method was used. In another experiment, 10,100 cpm were found with the cells, compared to 8,700 cpm which disappeared from the medium. Therefore, the indirect method actually measures combination of sulfate with bacteria.

The ability of several mutants to remove sulfate from the medium, following growth under different conditions, is shown in Table I. The transport-positive cells *cys CD-519* when grown under conditions of derepression (with djenkolate which permits only slow growth as a sulfur source) removed all of the sulfate from the medium. These cells grown on cysteine removed negligible sulfate. All *cys A* mutants grown on djenkolate removed approximately 30% of the total sulfate. None of the individual cistrons of the *cys A* region appear to control this activity. It also was repressed by growth of mutant *cys Aabc-20* on cysteine.

Specificity Studies—Several lines of evidence suggest that the combination of sulfate with *cys A* cells is specific, and cannot be attributed to random trapping of sulfate on, in, or between cells. The strength and quantity of sulfate combined per cell suggest high specificity. A typical saturation curve with a strong affinity constant was obtained (Fig. 1). Saturation was reached at about 10⁴ sulfate molecules per cell, calculated from a plot of these data in the form of the ratio of bound to free SO_4^- against bound SO_4^-. The affinity constant was 4 × 10⁻⁶ M.

Specificity was investigated further by comparison of the inhibitory effects of anions on sulfate transport in mutant *cys CD-519* and on binding to *cys Aabc-20* (Table II). Inhibition was quite specific in both cases. These anions probably "inhibit" transport by an exchange reaction since they drive sulfate out of the cells (Fig. 2). In general, the anions showed similar effects on the two activities. Chromate provides an interesting example. Chromate and sulfate are tetrahedral ions with similar bond distances (1.64 A and 1.54 A, respectively). Chromate also activates ATP-cleaving activity by acting as a sulfate analogue, although it is inferior to WO_4^- and MoO_4^- in this reaction (9). The main exception was seen with thiosulfate which prevented transport in *cys CD-519* but less strongly in-

TABLE I

Sulfate removal by some cysteine-requiring mutants

Cells were grown in Medium E containing either 0.2 mM cysteine or 0.15 mM djenkolate, harvested, and washed once with Medium E. Cells to be starved of their endogenous glucose reserves were incubated for 2½ hours in glucose-free Medium E containing 0.15 mM djenkolate, and glucose was omitted from the final incubation mixture. To measure the amount of sulfate bound, cells representing 50 mg of protein (0.35 ml) were incubated at 37° for 2 min with 1.65 ml of Medium E containing 0.2 mg per ml of glucose, 0.03 mg per ml of chloramphenicol, 3.8×10^6 cpm of $^{35}SO_4^-$ per ml, and 0.01 mM sulfate. The cells were removed by filtration through Celite in most samples, but by centrifugation in the experiment with KCN. Cell suspensions and cell-free solutions were counted for radioactivity as described under "Methods."

Mutant	Sulfur source during growth	Sulfate removed per mg of protein
		mμmoles
Transport-positive		
cys CD-519	Djenkolate	0.33
cys CD-519	Cysteine	0.03
Transport-negative		
cys Aabc-20	Djenkolate	0.11
cys Aa-201	Djenkolate	0.09
cys Ab-21	Djenkolate	0.10
cys Ac-22	Djenkolate	0.11
cys Aabc-20	Cysteine	0.00
cys Aabc-20	Djenkolate, glucose-starved	0.11
cys Aabc-20	Djenkolate	0.25
cys Aabc-20	Djenkolate, 0.02 M KCN present during assay	0.18

TABLE II

Inhibition of transport and binding

Transport: The experiment was preformed with *cys CD-519* as described in Fig. 2. The competing anion was added at 60 sec. The data are presented as sulfate removed at 80 or 100 sec, relative to that removed at 60 sec. Binding: Mutants *cys Aabc-20* or *cys Aa-210* (as indicated) were used. The procedure was the same as for measuring transport, except that 0.01 mM sulfate was added, followed after 30 sec by the competing anion, except as indicated. In the binding experiment, L-cysteine was 1.5 mM.

Anion (0.5 mM)	% maximum		
	Transport at		Binding at 80 sec
	80 sec	100 sec	
SO_4^-	23	13	14
SO_3^-	32	15	0[a]
CrO_4^-	32	16	4, 2[b]
$S_2O_3^-$	24	24	88,[a] 80[a, b]
SeO_4^-	62	52	41[a]
MoO_4^-	84	76	81
VO_4^-	102	79	60
WO_4^-	91	84	87
S^-	98	79	
L-Cysteine	95	83	70

[a] *Cys Aa-210* used.

[b] Anion added 20 sec prior to sulfate.

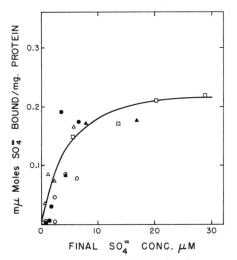

Fig. 1. Saturation curve for removal of sulfate by a transport-negative mutant. The samples from *cys Aabc-20* grown on djenkolate were assayed essentially as described in Table I, except that the sulfate concentration was varied. The reaction mixtures were sampled to determine radioactivity, incubated for 5 min at 37°, and centrifuged for 10 min at $10,000 \times g$, and the supernatants were sampled for remaining radioactivity. Each *symbol* represents results from a different experiment. Specific activity was kept constant within a single experiment only.

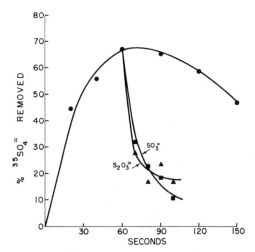

Fig. 2. Exchange of transported $^{35}SO_4^-$ with sulfite and thiosulfate in *cys CD-519*. Cells were grown overnight with 0.15 mM djenkolate, harvested, and washed once with Medium E. Cells representing 40 mg of protein (0.22 ml) were incubated, as in Table I, for the time intervals shown. Nonradioactive sulfite or thiosulfate was added to some samples after 60 sec, to give a final concentration of 0.5 mM. After removal of the bacteria by filtration through Celite on a Millipore filter, $^{35}SO_4^-$ was determined in the filtrate.

TABLE III

Effects of growth with cysteine or djenkolic acid

Mutant *cys A abc-20* (transport-negative) was grown overnight on either 0.15 mM djenkolic acid or 0.2 mM L-cysteine. The bacteria were centrifuged, washed, and resuspended in fresh media to allow further growth. They were then assayed as described in Table VII, except that centrifugation was used to remove the bacteria.

Sulfur source during growth		Amount of $SO_4^=$ removed per mg of protein
Primary	Secondary	
		mμmole
Djenkolic	Djenkolic, 20 min	0.18
Djenkolic	Cysteine, 20 min	0.13
Cysteine	Cysteine, 2 hrs; djenkolic, 5 min	0.0
Cysteine	Djenkolic, 2 hrs	0.05

hibited the binding reaction with mutant *cys Aabc-20*. Even 1.0 mM thiosulfate, added prior to sulfate, inhibited the latter by only 32%. Vanadate differed by affecting transport less strongly than binding. These results suggest that combination and transport are related, although not identical processes.

Repression of Binding—The specificity of binding is also indicated by the fact that transport-negative cells grown on cysteine did not bind sulfate in contrast to the same cells grown on djenkolate (Table I). Activity did not change greatly with the age of the culture, although cysteine-grown cells in the late resting stage acquired some ability to combine with sulfate.

Cysteine most probably represses synthesis of the binding material. However, the possibility exists that the extent of binding is modified by metabolites derived from cysteine. This is unlikely in extracts (see later) in which the metabolites would be considerably diluted. Dilution of the added $^{35}SO_4^=$ by sulfate derived from cysteine is unlikely, since the pathway of cysteine formation from sulfate is irreversible, and no other pathway connecting these compounds is known in *S. typhimurium*. But other metabolic pools may differ in cysteine- and djenkolate-grown bacteria. As a test of this possibility, the bacteria were first grown overnight on the two sulfur sources and then subcultured for a fraction of a division time on the alternative sulfur source (Table III). Growth for a fraction of a cell division should strongly modify pools of metabolites and only moderately change repressible properties. Only moderate changes in combining activity were observed when the cells were grown for lengths of time permitting up to 40% increases in turbidity of the cultures. These data favor effects of repression on the binding sites, rather than inhibition or activation of pre-existing sites by metabolites.

Growth on Different Sulfur Sources—The sulfate combined with transport-negative bacteria could either be bound to the exterior of the cells or it could be contained within them. If the latter were the case, two predictable consequences are that the cells should be able to grow with sulfate as a sulfur source, and that disrupted cells should not be able to combine with sulfate. The *cys A* mutants possess all enzymes required to reduce sulfate to sulfide, and also can grow on sulfite, supplied as cysteinesulfinic acid (1, 3). These transport-negative mutants failed to grow appreciably with either sulfate or thiosulfate as sulfur sources, either on agar plates or in liquid medium in 45 hours (Fig. 3). The possibility exists that these transport-negative mutants are able to bring sulfate into the cells so slowly that no appreciable

growth can be seen in 2 days, yet fast enough to permit full combination with the cells in less than 20 sec (4). It can be calculated that this rate of binding would support growth at least 10% as rapidly as the wild type rate if sulfate penetrates. Therefore, sulfate cannot penetrate the cells rapidly enough to account for the observed binding; it would appear to be bound extracellularly.

Binding by Bacterial Fractions—Disrupted cell components are able to bind sulfate (5). A new assay required for these studies depends on the ability of binding substances to shift the equilibrium between sulfate ions adsorbed to the anion exchange resin Dowex 1 and sulfate in the supernatant fluid. As a test of this method, Fig. 4 shows that the quantity of sulfate released

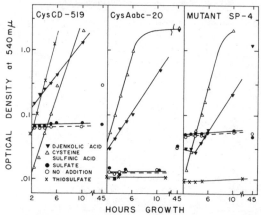

FIG. 3. Growth curves of mutants. Media containing the indicated sulfur sources (0.15 to 0.50 mM) were inoculated with the mutants. The cultures were shaken at 37°, and optical densities at 540 mμ were taken at intervals.

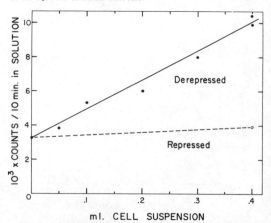

FIG. 4. Binding by a transport-negative mutant, as measured by the release of sulfate from resin. Mutant *cys A abc-20* was grown overnight on djenkolate or cysteine, washed by centrifugation, and resuspended in E salts at a concentration of 50 mg of cell protein per ml. Volumes of cell suspension indicated were diluted to 0.4 ml with E salts and assayed by the resin method, as described under "Methods." Results are given as counts obtained per 0.08 ml per 10 min, to show the type of raw data obtained.

TABLE IV

Binding by fractions of a transport-negative mutant

In these three experiments *cys Aabc-20* was grown overnight on cysteine or djenkolic acid. The cells were washed and concentrated by centrifugation. In Experiment 1 the cells (0.45 g per ml, wet weight) in E salts were disrupted in a Branson Sonifier for $\frac{1}{2}$ to 1 min at setting 7, to provide clarified broken cell preparations. These were centrifuged for 10 min at 10,000 $\times$ *g* to provide extracts. Material from 0.09 g of wet cells was used in each assay, performed as described under "Methods," except that Dowex 1-X8 was used in this experiment. All results are given as counts per min from 0.08 ml of supernatant fluid. The final column presents counts per min minus those obtained in absence of sample (without cells). In Experiment 2, broken cells were obtained as described for Experiment 1. To obtain the spheroplast preparation, 0.24 g of wet cells in 5 ml of 0.004 M Tris, pH 7.5, plus 10.3% sucrose at 37° was mixed with 0.2 mg of lysozyme, and 15 min later 0.52 ml of 0.4 M Tris, pH 8.0, and 0.052 ml of 0.2 M EDTA were added. This spheroplast preparation was assayed directly in its medium. A portion was centrifuged for 10 min at 11,000 $\times$ *g* to provide the spheroplast supernatant. The pellet was resuspended in E salts and disrupted with the Sonifier to provide the broken spheroplasts. Binding activity was measured by the resin method, as described under "Methods." Material obtained from 0.02 g of original wet cells was used in each assay. In Experiment 3, 1.05 g of wet cells were suspended in 20 ml of 20.6% sucrose plus 0.03 M Tris, pH 8.0; 0.30 ml of 0.2 M EDTA was added and the suspension was shaken at 37° for 10 min. The cells were centrifuged in the cold and resuspended in 20 ml of ice water, followed by stirring for 10 min. The cells were removed by centrifugation to provide the shocked cell supernatant fluids. These were assayed directly, and also after addition of E salts. The pellets were taken up in 20 ml of 0.004 M Tris, pH 7.5, plus 10.3% sucrose to provide the shocked cell suspensions. A similar suspension of unshocked cells was made as a control. Portions of these preparations corresponding to 0.02 g of original wet cells were assayed by the resin method, as described under "Methods."

Material	Medium	Without cells	Repressed cells	Derepressed cells	
		cpm	*cpm*	*cpm*	*net cpm*
1. Intact cells	E salts	343	325	587	244
Broken cells	E salts	343	329	541	198
Extract	E salts	343	311	519	176
2. Spheroplast preparation	Tris-sucrose-EDTA-lysozyme	25		346	321
Spheroplast supernatant	Tris-sucrose-EDTA-lysozyme	25		303	278
Broken cells	E salts	227		580	353
Broken spheroplasts	E salts	227		303	76
3. Intact cells	Tris-sucrose	3	13	228	225
Shocked cells	Tris-sucrose	3	59	99	96
Shocked cell	Water	3	29	235	232
supernatant	E salts	306	281	584	278

from the resin was proportional to the number of intact cells added. These cells combined with about 0.1 as much sulfate per g of protein as did cells assayed by the methods used in the previous experiments. The difference is due to the low concentration of free sulfate in equilibrium with the resin (about 0.2 μM), which only partly saturates the binding ability at the ionic strength used.[1]

With this assay, binding by extracts of sonically disrupted *cys CD-519* transport-positive bacteria was reported (5). Similar results were obtained with the transport-negative mutant *cys Aabc-20* (Table IV, part 1), and with a double mutant, *SP-4*, derived from *cys CD-519* (see section on mutant isolation) which failed to grow on sulfate or thiosulfate (Fig. 3). It has been suggested that these results could be explained if sulfate entered into intact cells and also could be bound, to an equal extent, to an enzyme released by disruption of the cells. The most likely enzyme is adenosine triphosphate:sulfate adenyltransferase (EC 2.7.7.4), since it is the first to react with sulfate and is needed for all subsequent reactions. But although this enzyme is missing from *cys CD-519* owing to a deletion, and also from *SP-4* which was derived from it, extracts of these cells bind. The results with disrupted cells thus favor binding to a cell component rather than transport into mutant cells.

A third line of evidence that binding is associated with the cell surface is obtained by modification of surface components. Cells of mutant *cys CD-519* lost their ability to transport sulfate when spheroplasts were prepared with the use of lysozyme (4). In similar experiments, by the indirect method of assay, spheroplasts made from the transport-negative mutant *cys Aabc-20* lost 80% of the original binding activity. Further experiments were performed with the resin assay for binding (Table IV, Experiment 2). The unfractionated spheroplast preparation retained binding activity, 87% of which was found in the supernatant fluid after centrifugation. The spheroplasts themselves retained only 22% of the activity of the broken cells. Binding to cell fractions, like binding to resin, was strongly affected by the ionic strength. It was weaker in E salts than in the lysing medium. Spheroplast formation was incomplete in these dense cell suspensions, as indicated by an optical density drop at 400 mμ from 1.71 to only 1.20 upon dilution of the preparation into water (relative to dilution into a sucrose solution). Nevertheless, the data show that loss by spheroplasts of most of the binding activity is due to release of this activity into the supernatant fluid.

The third experiment in Table IV shows that cells lost binding activity when they were osmotically shocked by the method of Neu and Heppel (10). The activity appeared in the supernatant fluid, as determined in both the original water extract and after addition of E salts. Repressed cells and their components did not possess activity under any conditions. Although osmotic shock was apparently not complete in this dense cell suspension, it released most of the binding activity. These data suggest that most of the binding activity is located on easily released material, probably near the cell surface.

The ability to transport anything into cells is of course a process which cannot occur with disrupted cells. After sonic oscillation, transport-positive cells lost about 75% of their ability to combine with sulfate (5). The residual 25% activity was equal to the amount bound by disrupted preparations of transport-negative mutants; it must represent binding to subcellular material.

The effects of osmotic shock on transport and binding were compared by measuring transport with intact cells and binding with sonically disrupted preparations of the same transport-positive cells (Table V). Osmotic shock reduced both by 80%. The binding activity lost by the shocked cells was recovered quantitatively in the supernatant fluid. This result is consistent with the hypothesis that binding is required for transport, since the two were decreased to the same extent.

287

TABLE V

Loss of transport and binding by osmotic shock of transport-positive cells

Mutant *cys CD-519* (200 ml) was grown overnight with 0.15 mM djenkolate. A portion of the cells (27 mg of protein) was shocked as described in Table IV, with final resuspension in 22.5 ml of distilled water. The cells were sedimented, resuspended in 2 ml of E salts, and portions were disrupted in the Branson Sonifier. The supernatant fluid of the shocked cells was concentrated approximately 100 times by lyophilization, and E salts were added. Portions of these preparations and controls of unshocked cells were assayed for protein and for binding activity by the resin method, as described under "Methods." Data are presented as amount of sulfate bound per mg of protein of the original cell suspension. The assay tubes for untreated and shocked cells contained the same quantity of protein (5.3 mg); that for supernatant fluid contained 0.2 mg.

Condition of cells	Amount of SO$_4$$^=$ bound per mg of protein		
	Intact cells	Broken cells	Supernatant fluid
	mµmole		
Untreated	0.094	0.035	
Shocked	0.022	0.007	0.026

Effect of Energy Supply—Active transport of sulfate requires glucose, presumably as an energy source (1). The part of transport which is energy-dependent must be different from the binding activity, since binding occurred equally well in glucose-starved cells or in the presence of 20 mM KCN (Table I).

Binding activity of intact cells can be measured in the absence of transport by adding KCN; this technique provides a simple screening test for binding by intact cells. Intact cells of the transport-negative mutant *cys Aabc-20* were little affected by KCN except for a modest inhibition (possibly due to competition of CN$^-$ with SO$_4$$^=$ for the binding site) (Table VI). Active transport by *cys CD-519* was eliminated by KCN and was enhanced by providing glucose. KCN had little effect on binding by extracts of either organism.

Mutants with Modified Binding Ability—Mutation should modify binding ability; transport should simultaneously be modified if it is dependent on binding. Since a method for selecting binding-negative mutants directly was not known, transport-negative mutants were obtained and then were examined for their inability to bind sulfate. To isolate these mutants, advantage was taken of an observation that growth of transport-positive mutants was severely inhibited by chromate, whereas transport-negative mutants grew normally, probably because they were not able to bring chromate into the cells via the sulfate-transporting system (Table II). About 80% of over 100 chromate-resistant mutants, selected as described under "Methods," failed to grow with sulfate as a sulfur source. One-third of these possessed binding ability, and are probably *cys A* mutants. The rest bound less strongly. An example of such a mutant is *cys SP-4*, which was obtained from *cys CD-519*. Its growth characteristics were those expected of transport-negative, sulfate activation-negative mutants (Fig. 3). The binding ability of this particular mutant was somewhat reduced by mutation, as well as its being transport-negative (Table VII).

Chromate-resistant mutants were also isolated from the wild type strain, LT-2. Most of these were unable to grow on sulfate. About one-sixth of these mutants bound very little or no sulfate. Representative examples of this class are *cys SP-25* and *cys*

SP-30. They did not grow on agar plates or in liquid medium with sulfate or thiosulfate as sulfur sources. They grew normally with cysteine, djenkolate, or cysteinesulfinic acid (equivalent to sulfite). Their binding ability in the presence of KCN is shown in Table VIII. It was also very low when assays were made with intact cells in the absence of KCN, with sonically disrupted cells, or with supernatant fluids from osmotically shocked cells. These are considered to be binding-negative and transport-negative mutants.

To determine whether one mutational event was responsible for both of these losses, back-mutation and transduction (11) were used to see if both properties were restored at once. Spontaneous revertants capable of growing on sulfate plates were obtained at a frequency of about 10^{-8} from *cys SP-25*, but fewer were obtained from *cys SP-30*. Transduction of either mutant with phage PLT-22 (H-1 or H-4) grown on LT-2 also resulted in bacteria capable of growth on sulfate, with a frequency of about 10^{-7}. About one-third of the revertants, like *cys SP-25R4*,

TABLE VI

Effect of KCN on transport and binding

The bacteria were grown overnight with 0.15 mM djenkolate. They were washed and resuspended at an optical density at 540 mµ of 30 in 20 mM potassium phosphate buffer, pH 6.8. Portions were disrupted in the Branson Sonifier, and then were centrifuged to provide "Extracts." Assays for binding activity were performed by the resin method, with additions to the reaction mixture listed above. An unknown quantity of cyanide was adsorbed to the resin, so the concentration given is not that actually in solution during the assay. Protein was estimated from the optical densities of the washed cell suspensions.

Conditions during assay	SO$_4$$^=$ bound per mg of protein			
	cys CD-519		*cys Aabc-20*	
	Intact	Extract	Intact	Extract
	mµmole		*mµmole*	
Glucose, 2 mg per ml, + chloramphenicol, 0.03 mg per ml	0.28		0.069	
No additions	0.13	0.078	0.054	0.11
KCN, 5 mM	0.051	0.070	0.053	0.091
KCN, 10 mM	0.048	0.050	0.049	0.056
KCN, 15 mM	0.043		0.039	

TABLE VII

Sulfate removal by transport-negative mutants

Both mutants were grown overnight with 0.15 mM djenkolic acid. The bacteria were centrifuged, washed, and assayed for binding ability by incubation for about 3 min with 0.01 mM ^{35}SO$_4$$^=$ (specific activity 1.4 × 10^8 cpm per mµmole), followed by centrifugation or filtration through Celite and determination of the counts removed from the medium. The presence of cells enhanced the count rate by 10% for which correction was not made.

Mutant	Protein	SO$_4$$^=$ removed per mg of protein	
		Centrifuged	Filtered
	mg/ml	*mµmole*	
cys Aabc-20	4.8	0.30	0.25
cys Aabc-20	12.0	0.27	0.22
cys SP-4	4.3	0.12	0.09
cys SP-4	10.8	0.12	0.11

TABLE VIII

Binding by various mutants

The bacteria were grown with 0.3 mM djenkolate into the resting state. They were prepared as described in Table VI, and assayed as whole cells with 10 mM KCN present, by the resin binding method. Protein was estimated from the optical densities of the suspensions.

Mutant	$SO_4^=$ bound per mg of protein
	mμmole
cys CD-519	0.048
cys Aabc-20	0.040
cys SP-25	0.001
cys SP-30	0.004
cys SP-25R4	0.046
cys SP-30T1	0.060
cys SP-25R1	0.005
cys SP-25R3	0.010
cys SP-30T4	0.011
cys SP-30T7	0.021

or transductants, like *cys SP-30T1*, gave large colonies on agar and reacted like LT-2 in every way (Table VIII). These are presumably true reversions or transductions at the site of the original lesion. Their existence suggests that a single lesion is involved in loss of both transport and binding ability.

The other two-thirds of the revertants and transductants formed smaller colonies, and had binding abilities that ranged from those of the original mutants to levels close to the wild type. Representative values are shown in Table VIII. They generally combine with some sulfate (measured in the absence of KCN). Their growth was slow in liquid medium (division time, about 2 hours) irrespective of the sulfur source. The properties of these revertants are like those of the suppressor mutants of *cys A* mutants studied genetically by Howarth (11), who demonstrated suppressions by mutations that map outside of the *cys A* region, and which were not specific for the original lesion. She suggested an alternative pathway as an explanation. Possibly a defective, leaky cell membrane could explain both the appearance of ability to use sulfate, and poor growth; lysogenization has often been shown to change bacterial surface properties. The study of binding ability in these strains is made difficult by the slow growth rate, not limited by djenkolate, which establishes an undetermined amount of repression. Since the "true" reversions and transductions showed a genetic connection between binding and transport in the original strains, irrespective of complex suppressor effects, the latter were not pursued further at this time.

DISCUSSION

The main objectives of this article are to establish the existence of specific sulfate-binding components in *S. typhimurium*, and to suggest that these components are parts of the sulfate-transporting system. Such a specific binding component could be shown by direct tests. Both intact cells and solubilized material combined with sulfate. The binding component has been highly purified, and is a protein of molecular weight about 32,000. Physical, chemical, and binding properties will be presented in a later communication.

Bound sulfate seems to be localized near the cell surface, because certain mutants possess the metabolic apparatus necessary

for sulfate utilization once it enters the cells, and yet sulfate does not permit appreciable growth (1, 3). Also, osmotic shock or spheroplast formation, treatments which are believed to release surface components (10), release the binding activity into solution.

The surface location of the binding site suggests that it might be involved in an initial step of transport, as was surmised for a lactose-binding site in *Escherichia coli* (12). Four similar properties of the sulfate-transporting system and the binding component support the idea that the two are related. Both are similarly though not identically inhibited by a series of anions (Table II). Both are present in derepressed cells but absent in repressed cells. Both are lost from the cells to similar extents upon osmotic shock (Table V). Both can be simultaneously lost by mutation, and regained on reversion or transduction (Table VIII).

These results do not prove that binding is related to transport. The relationships just presented might be coincidental. It should be noted that the binding material is a relatively major component of cells grown under derepressed conditions. It can amount to about 1% of the total cell protein.[1] However, it is dispensable in some mutants, and in cells grown under repressed conditions. This suggests that it should have some function not essential under all physiological conditions. An enzyme which would fulfill many of these requirements is a repressible aryl-sulfate sulfohydrolase (EC 3.1.6.1), found in *Aerobacter aerogenes* but not in *S. typhimurium* (13). (We have confirmed the absence of this enzyme in our strain, grown under conditions which produce ample binding material, and assayed under conditions used for determining binding and also those described by Rammler, Grado, and Fowler (13).)

Although binding appears to be related to transport, the two activities are not identical. The transport system has other parts. This is shown (*a*) by the ability of many transport-negative mutants to bind, (*b*) by the energy requirement for transport but not for binding (Table VI), (*c*) by transport requiring at least 10 times more sulfate for half-saturation than does binding, and (*d*) by different inhibition of the two activities by thiosulfate.

Investigations of transport have usually been limited to intact cells (14–16), because of the problems of determining transport in subcellular preparations. Exceptions are Na^+ and K^+ transport by reconstituted erythrocyte ghosts (see Reference 16), the assay of ATPase activity associated with Na^+ and K^+ transport (16), and radioactive labeling of a site involved in the transport of β-galactosides (12, 17). Isolation and chemical studies on the sulfate-binding component, coupled with physiological studies (18), should help the understanding of at least one transport system.

Acknowledgment—We are indebted to Mrs. Sondra Karipides for assistance with some of these experiments.

REFERENCES

1. DREYFUSS, J., *J. Biol. Chem.*, **239**, 2292 (1964).
2. MIZOBUCHI, K., DEMEREC, M., AND GILLESPIE, D. H., *Genetics*, **47**, 1617 (1962).
3. DREYFUSS, J., AND MONTY, K. J., *J. Biol. Chem.*, **238**, 1019 (1963).
4. DREYFUSS, J., AND PARDEE, A. B., *Biochim. Biophys. Acta*, **104**, 308 (1965).
5. PARDEE, A. B., AND PRESTIDGE, L. S., *Proc. Natl. Acad. Sci. U. S.*, **55**, 189 (1965).

6. VOGEL, H. J., AND BONNER, D. M., *J. Biol. Chem.*, **218**, 97 (1956).
7. LOWRY, O. H., ROSEBROUGH, N. J., FARR, A. L., AND RANDALL, R. J., *J. Biol. Chem.*, **193**, 265 (1951).
8. PARDEE, A. B., *J. Bacteriol.*, **73**, 376 (1957).
9. BANDURSKI, R. S., WILSON, L. G., AND SQUIRES, C. L., *J. Am. Chem. Soc.*, **78**, 6408 (1956).
10. NEU, H. C., AND HEPPEL, L. A., *J. Biol. Chem.*, **240**, 3685 (1965).
11. HOWARTH, S., *Genetics*, **43**, 404 (1958).
12. FOX, C. F., AND KENNEDY, E. P., *Proc. Natl. Acad. Sci. U. S.*, **54**, 891 (1965).
13. RAMMLER, D. H., GRADO, C., AND FOWLER, L. R., *Biochemistry*, **3**, 224 (1964).
14. KEPES, A., AND COHEN, G. N., in I. C. GUNSALUS AND R. Y. STANIER (Editors), *The bacteria, Vol. 4*, Academic Press, Inc., New York, 1962, p. 179.
15. BRITTEN, R. J., AND MCCLURE, F. T., *Bacteriol. Rev.*, **26**, 292 (1962).
16. HOKIN, L. E., AND HOKIN, M. R., *Ann. Rev. Biochem.*, **32**, 553 (1963).
17. KOLBER, A. R., AND STEIN, W. R., *Nature*, **209**, 691 (1966).
18. DREYFUSS, J., AND PARDEE, A. B., *J. Bacteriol.*, **91**, 2275 (1966).

Copyright © 1965 by the American Society of Biological Chemists, Inc.
Reprinted from *J. Biol. Chem.*, **240**, 3685–3692 (1965)

22

The Release of Enzymes from *Escherichia coli* by Osmotic Shock and during the Formation of Spheroplasts

Harold C. Neu* and Leon A. Heppel

*From the National Institute of Arthritis and Metabolic Diseases, National Institutes of Health,
United States Public Health Service, Bethesda, Maryland 20014*

(Received for publication, March 29, 1965)

When *Escherichia coli* cells are converted into spheroplasts by means of ethylenediaminetetraacetate and lysozyme (1), most of the inducible alkaline phosphatase (2, 3) is released into the surrounding sucrose-tris(hydroxymethyl)aminomethane-HCl medium (4, 5), and a large fraction of the "latent" ribonuclease (6–8) and the ribonucleic acid inhibited deoxyribonuclease (9) is also set free (10).[1] We have since found that three additional enzymes are liberated: a Co++-stimulated 5′-nucleotidase, not previously described in *E. coli*; an acid phosphatase; and a cyclic phosphodiesterase (11, 12). A total of 10 other enzymes have been examined and found to remain entirely within the spheroplasts. Various lines of evidence suggest that the enzymes which are released occur at or near the cell surface.

We have also developed a new method for releasing most of these enzymes in high yield without greatly impairing the viability of the cells. The procedure involves osmotic shock. *E. coli* are first suspended in a concentrated solution of sucrose, which does not penetrate the cells, and ethylenediaminetetraacetate is added. Then they are suddenly shifted to a medium of low osmotic strength. This causes the release of the phosphatases and of the cyclic phosphodiesterase in a yield of 70% or more.

A preliminary report of this work has appeared (13).

EXPERIMENTAL PROCEDURE

Materials

E. coli soluble ribonucleic acid was purchased from General Biochemicals. Bis(*p*-nitrophenyl)phosphate, *p*-nitrophenyl phosphate, isopropyl-β-D-thiogalactopyranoside, and *o*-nitrophenyl galactoside were obtained from commercial sources. DEAE-cellulose (Selectacel, type 20, 1 meq per g) was obtained from Brown. Chelex 100 (analytical grade chelating resin) was purchased from Calbiochem.

Methods

Growth of Cells—Stock cultures of *E. coli*, including strains B, K12s, U7, K12λ, C90, and C₄F₁, were maintained on nutrient agar slants. Strains C₄F₁ and C90 are constitutive for alkaline phosphatase and inducible for β-galactosidase, whereas U7 carries

---

* Present address, Department of Medicine, Columbia-Presbyterian Medical Center, New York, New York 10032.

[1] In addition to being set free on spheroplast formation, a large fraction of the enzyme is also released simply by suspending cells in 20% sucrose-1 × 10⁻³ M EDTA, in the absence of lysozyme (R. J. Hilmoe, unpublished observations). The cells must be in log phase.

a deletion for alkaline phosphatase. These strains were kindly provided by Dr. A. Garen.

The low phosphate medium (Medium A) contained 0.12 M Tris, 0.08 M NaCl, 0.02 M KCl, 0.02 M NH₄Cl, 0.003 M Na₂SO₄, 0.001 M MgCl₂, 2 × 10⁻⁴ M CaCl₂, 2 × 10⁻⁶ M ZnCl₂, and 0.5% Difco Bacto-peptone adjusted to pH 7.5 (14). Medium B consisted of 0.04 M K₂HPO₄, 0.022 M KH₂PO₄, 0.08 M NaCl, 0.02 M NH₄Cl, 0.003 M Na₂SO₄, 1 × 10⁻⁴ M MgCl₂, 1 × 10⁻⁵ M CaCl₂, 2 × 10⁻⁶ M ZnCl₂, and 0.5% Difco Bacto-peptone adjusted to pH 7.1. The media were supplemented with 0.6% glycerol. Isopropyl-β-D-thiogalactopyranoside was used to induce β-galactosidase in the inducible strains. Organisms were incubated at 37° on a rapid rotary shaker, and cells were harvested during the exponential or stationary phase by centrifugation in the cold. They were washed three times with about 40 parts (w/w) of cold 0.01 M Tris-HCl, pH 8.1.

Cells were subjected to osmotic shock by the following standard procedure: Cells (1 g, wet weight) were suspended in 80 ml of 20% sucrose-0.03 M Tris-HCl, pH 8, at 24°, yielding about 10¹⁰ cells per ml. The suspension was treated with disodium EDTA to give a concentration of 1 × 10⁻³ M and mixed in a 1-liter flask on a rotary shaker (about 180 rpm). After 10 min[2] the mixture was centrifuged for 10 min at 13,000 × g in a cold room. The supernatant fluid was removed, and the well drained pellet was rapidly mixed with a volume of cold water equal to that of the original volume of the suspension. The suspension was mixed in an ice bath on a rotary shaker for 10 min and centrifuged, and the supernatant fluid, hereafter termed the cold water wash, was removed. For comparison, a separate fraction of the cells was usually converted into spheroplasts by treatment with EDTA and lysozyme (10). Sonic extracts were made by treatment for 10 min in a Raytheon 10-Kc sonic oscillator.

Viability was determined on serial dilutions of cells in a minimal salts medium, plated on tryptone agar.

Enzyme Assays—Previously published methods were used for measuring RNase (10, 15), DNase (16), β-galactosidase (4), and alkaline phosphatase (17). Pancreatic RNase (50 μg) was present in DNase assays. For assay of cyclic phosphodiesterase, the reaction mixture (0.05 ml) contained 0.06 μmole of adenosine 2′,3′-cyclic phosphate, 0.25 μmole of MgCl₂, 0.05 μmole of CoCl₂, 2.5 μmoles of sodium acetate buffer (pH 6), and excess purified alkaline phosphatase from *E. coli* (usually 1 unit (17)). Anraku (11) has pointed out that, over a 900-fold range of puri-

---

[2] Actually, a period of 5 min was found to be long enough for contact with the sucrose solution.

fication of cyclic phosphodiesterase, the ratio of rate of hydrolysis of 2'3'-cyclic AMP to that of bis(p-nitrophenyl)phosphate was constant. Accordingly, the following, more convenient assay was used when specified. The incubation mixture (0.1 ml) contained 0.5 μmole of $MgCl_2$, 0.1 μmole of $CoCl_2$, 5 μmoles of sodium acetate buffer (pH 6), 0.1 mg of bis(p-nitrophenyl)phosphate, and enzyme. After 18 min at 37° the reaction was stopped with 1.0 ml of 0.1 N NaOH, and the absorbance at 410 mμ was measured.

For assay of 5'-nucleotidase the mixture (0.1 ml) contained 0.4 μmole of 5'-AMP, 1.0 μmole of $CaCl_2$, 0.1 μmole of $CoCl_2$, and 10 μmoles of sodium acetate, pH 5.8. After addition of enzyme the mixture was incubated for 20 min at 37° and P_i was measured (18). For determination of acid phosphatase the conditions were similar except that the reaction mixture (0.1 ml) contained 0.5 μmole of glucose 6-phosphate and 9 μmoles of sodium acetate, pH 5.65. These two enzymes could be assayed only under conditions that caused the formation of alkaline phosphatase to be suppressed.

Activities were expressed in terms of micromoles hydrolyzed per hour except when bis(p-nitrophenyl)phosphate was the substrate. In the latter case, 1 unit represents a change in absorbance at 410 mμ of 2.0 per 20 min; this unit is numerically almost equivalent to the others. The data are presented as units of enzyme released per g of cells, wet weight.

Protein was determined according to Lowry *et al.* (19).

RESULTS

Release of Enzymes by Osmotic Shock

General Observations

The enzymes being considered here were not released from *E. coli* during ordinary handling. Thus, only traces of cyclic

TABLE I

Effect of time of stirring with cold water on release of enzymes and other ultraviolet-absorbing substances

E. coli K12s were grown to stationary phase in Medium B. Method A: Standard procedure for osmotic shock (see "Methods"). Method B: Cells were suspended in 20% sucrose-0.2 M Tris (pH 7.2)-0.005 M EDTA and then shocked. The shocked cells were stirred with cold water, and at various times a portion was centrifuged with rapid acceleration. Assays of the cold water wash are shown. Leakage of β-galactosidase was less than 1% of that contained in the cells. During the treatment with the sucrose solutions very little 5'-nucleotidase, cyclic phosphodiesterase, or acid phosphatase was released (less than 3% of that subsequently set free on stirring with cold water).

Method	Time of stirring	A_{260}	Total protein	5'-Nucleo-tidase[a]	Cyclic phospho-diesterase[b]	Acid phospha-tase[c]
	min		*mg/ml*		*units/g*	
A	5	0.292	0.023	1300	480	110
A	20	0.305	0.023	1370	450	120
A	40	0.319	0.025	1470	450	120
B	5	0.330	0.033	2060	535	135
B	20	0.360	0.033	1800	550	140
B	40	0.370	0.037	1800	550	140

[a] Intact cells contained 1850 units per g.

[b] Sonic extract contained 500 units per g. Substrate was bis(p-nitrophenyl)phosphate.

[c] Sonic extract contained 275 units per g.

TABLE II

Effect of EDTA, sucrose, and Tris buffer on release of enzymes by osmotic shock

E. coli C_4F_1 were grown to stationary phase in Medium A. The standard procedure for osmotic shock was followed except for the use of 2×10^{-3} M EDTA, and as noted below. Data are also presented on the release of enzymes into the medium surrounding spheroplasts made with lysozyme and EDTA (10). Cells subjected to osmotic shock showed 90% survival in Experiments 1 and 2. Absorbance at 260 mμ of various fractions is also presented.

Fraction	Alkaline phos-phatase	Cyclic phospho-diesterase	β-Ga-lacto-sidase	RNase	A_{260}
		units/g			
Experiment 1					
Sucrose-Tris-EDTA superna-tant	65	24	9	370	0.184
Cold water wash[a]	3600	600	16	40	0.600
Experiment 2					
Sucrose-EDTA supernatant[b]	16	0	9	0	0.118
Cold water wash	480	95	0	0	0.327
Experiment 3					
Tris-EDTA supernatant[c]	16	0	0	400	0.250
Cold water wash	8	0	0	0	0.210
Experiment 4					
Medium, spheroplasts	3600	600	12	3600	0.360
Spheroplast lysate	500	100	1650	2600	
Experiment 5					
Intact cells, sonic extract	4000	600	1520	5100	

[a] When the concentration of EDTA was reduced to 5×10^{-4} M, almost identical results were obtained. In other, comparable experiments, only 30% of alkaline phosphatase was released with 1×10^{-4} M EDTA and 5% in the absence of EDTA.

[b] Tris buffer was omitted from the medium in which cells were initially suspended.

[c] Sucrose was omitted from the medium in which cells were initially suspended.

phosphodiesterase and of the phosphatases were excreted into the growth medium,[3] and no significant release took place during the washing procedure. The cells were then suspended in 20% sucrose-0.03 M Tris (pH 8)-0.001 M EDTA, causing them to shrink, but again there was little or no loss of enzymes or other ultraviolet-absorbing material into the medium (Tables I to III). The mixture was centrifuged; the excess sucrose solution was decanted, and the pellet was rapidly dispersed in cold water. As observed by phase contrast microscopy, the cells quickly increased in size, and at the same time there occurred a release of protein and of other ultraviolet-absorbing material into the medium (Tables I to III).[4] Protein set free into the cold water

[3] Cultures were examined 12 hours after the onset of stationary phase to see if selective secretion of enzymes into the medium might occur at this late stage. The results were negative, and therefore the enzymes here considered do not appear to be extracellular enzymes.

[4] It was important to use well washed cells. *E. coli*, washed once, released only 60% as much alkaline phosphatase during osmotic shock as did the cells washed three times with 0.01 M Tris, pH 8. The pellet of cells could be stored at 3° for 8 hours before the third wash without reducing the yield of enzymes liberated by osmotic shock. Dr. David Schlessinger has successfully stored cells for 2 days (personal communication). In this way large quantities of cells can be processed in small batches.

TABLE III

Release of enzymes by osmotic shock from cells harvested in exponential phase of growth

E. coli, C₄F₁,[a] were grown to midexponential phase in Medium A. The standard procedure for osmotic shock was followed except as noted. Because the original samples were lost, the assays for alkaline phosphatase represent a separate experiment run under identical conditions. Cells subjected to osmotic shock showed 50% survival in Experiments 1 and 2. Data on release of enzymes during formation of spheroplasts with lysozyme and EDTA (10) are also presented.

Fraction	Alkaline phosphatase	Cyclic phosphodiesterase	β-Galactosidase	RNase	A_{260}
		units/g			
Experiment 1					
Sucrose-Tris-EDTA[b] supernatant	20	55		500	0.410
Cold water wash	1120	500	56	500	0.710
Experiment 2					
Sucrose-Tris-EDTA supernatant[c]	30	44		200	0.360
Cold water wash	1100	520	40	400	0.735
Experiment 3					
Sucrose-2 × 10⁻³ M EDTA supernatant[d]		25		0	0.235
Cold water wash		235	16	0	0.360
Experiment 4					
Medium, spheroplasts	1140	550	60	3250	0.440
Spheroplast lysate	136	25		2620	
Experiment 5					
Intact cells, sonic extract	1280	500	1360	5260	

[a] Comparable experiments with *E. coli* B gave similar results.
[b] EDTA, 2 × 10⁻³ M.
[c] EDTA, 5 × 10⁻⁴ M.
[d] The 0.033 M Tris buffer was omitted for this experiment. In a separate trial, in which sucrose was left out, there was essentially no release of enzymes.

amounted to 4 to 8% of that found in an equivalent, uncentrifuged sonic extract. Most of the ultraviolet-absorbing material was derived from the acid-soluble fraction of the cell, which was reduced by one-half to two-thirds as a result of osmotic shock.

The process of release was nearly complete within 5 min after contact with cold water; prolonged stirring thereafter had little effect (Table I). For cells of a given strain and phase of growth the amount of protein released varied within narrow limits, even when the concentration of sucrose in the original suspending medium was increased from 12% to 37% and that of EDTA from 5 × 10⁻⁴ M to 5 × 10⁻³ M. Furthermore, the process was selective in character; starch gel electrophoresis revealed a pattern of 15 discrete bands in the cold water wash[5] compared with more than 50 bands seen with the entire sonic extract. Also, certain enzymes were released in high yield into the cold water wash, whereas others appeared only in trace amounts.

Most of the work was done with cells harvested in the stationary phase, but similar results were obtained with *E. coli* in the exponential phase (compare Tables II and III). Viability was measured on 15 suspensions, representing four strains of *E. coli*, that had been subjected to osmotic shock. Survival

[5] We are indebted to Dr. Elliot Vesell for doing this experiment.

TABLE IV

Effect of EDTA and sucrose concentrations on release of enzymes by osmotic shock

E. coli C₄F₁ were grown to stationary phase in Medium A. The standard procedure for osmotic shock (see "Methods") was varied by changing the composition of the sucrose-Tris-EDTA solution, as indicated below. Assays of the cold water wash are presented.

Sucrose	Tris	EDTA	Alkaline phosphatase	Cyclic phosphodiesterase[a]	β-Galactosidase	Total protein
%	M	mM		*units/g*		*mg/g*
20	0.03	1	1370	820	8	6.5
12	0.03	1	1230	675	4	5.3
11[b]	0.11	0	95	130	8	2.2
23[b]	0.11	0	500	410	4	5.7
37[b]	0.11	0	400	360	8	6.0
Sonic extract of untreated cells			1500	1350	2200	79

[a] Assayed with bis(*p*-nitrophenyl)phosphate.
[b] Stirring with sucrose-Tris solution for 40 min instead of the usual 10-min period gave similar results.

compared with the original suspensions varied from 50 to 95%, with most of the values above 70%.

Comparison with Results Obtained on Spheroplast Formation

Alkaline phosphatase, cyclic phosphodiesterase, acid phosphatase, and 5′-nucleotidase were also released into the medium when *E. coli* were converted into spheroplasts by means of lysozyme and EDTA (Tables II, III, and VI) (13). The yields varied from 70 to 100% in a large number of experiments, and it is of interest that greater release on spheroplast formation was generally associated with an increased yield by the osmotic shock procedure and vice versa. In contrast to the results with osmotic shock, at least 50% of the RNase was set free when spheroplasts were made (Tables II, III, and VI) (10, 13).

Release of Alkaline Phosphatase

Effect of EDTA and Other Chelating Agents—From 70 to 100% of the alkaline phosphatase was usually found in the cold water wash[6] when the standard procedure was applied, but this was reduced to 5 to 15% when EDTA was omitted from the sucrose mixture (Tables II and IV). With strain C90 the release of this enzyme decreased from 90% to 2% in the absence of EDTA.

Other chelating agents were less effective. Thus, a solution of 20% sucrose-0.03 M Tris, pH 8, received various supplements, and the release of alkaline phosphatase in the subsequent cold water wash was measured. Yields were as follows: 30% with 0.01 M citrate, 40% with 8 × 10⁻³ M ATP, 90% with 1.8 × 10⁻² M ATP, and 25% with 2 × 10⁻³ M hydroxyquinoline. A quantitative release of alkaline phosphatase into the cold water wash was obtained after the following treatment. A 1.25% suspension of cells in 20% sucrose-0.1 M Tris was shaken for 3 min with approximately 0.1 volume of Chelex resin that has been

[6] When the standard procedure was altered by substituting cold tryptone broth for water as the shock medium, the yield of alkaline phosphatase was reduced from 80% to 12%. Similarly, treatment of cells with 5 × 10⁻³ M magnesium acetate instead of water reduced the yield of alkaline phosphatase from 80% to 30%.

TABLE V

Effect of repeated washes with cold water on release of enzymes from E. coli

Experimental conditions were as in Table VI except that the wash with cold water, to produce osmotic shock, was followed by two further treatments with cold water.

Fraction	Alkaline phosphatase	Cyclic phosphodiesterase	β-Galactosidase	RNase
		units/g		
Sucrose-Tris-EDTA supernatant	100	Trace	10	Trace
First cold water wash	2,200	840	4	350
Second cold water wash	320	30	3	500
Third cold water wash	40	16	2	Trace
Sonic extract, untreated cells	3,100	1,540	2,300	10,000

of Tris was helpful (Table II), and at a concentration of 0.1 M it partially replaced the need for EDTA (Table IV). The results were similar at pH 8 and pH 7.1.

Release of 5'-Nucleotidase

Effect of EDTA—The 5'-nucleotidase was liberated more easily by osmotic shock than was alkaline phosphatase. Thus 65% of the activity was released in the cold water wash when cells were initially suspended in 20% sucrose-0.03 M NaCl without EDTA (Table VII). Nearly quantitative release was obtained with 0.2 M Tris in sucrose solution or by the suspension of cells in 20% sucrose-0.01 M Tris, pH 8, followed by treatment with Chelex resin as described above. The ease of release without the use of EDTA may have been due to the fact that the

TABLE VI

Effect of repeated cycles of osmotic shock on release of enzymes

E. coli C_4F_1 were grown to stationary phase in Medium A. The standard procedure for osmotic shock (see "Methods") was varied in that 1 g of cells was suspended in 20 ml of 20% sucrose-2 × 10⁻³ M EDTA-0.03 M Tris, pH 8. The complete cycle of treatment with the sucrose solution and then with cold water was repeated.

Data are presented on the release of enzymes in the cold water wash for each cycle. For comparison, release of enzymes into the medium on spheroplast formation is shown in the lower section of the table. In this experiment there was 80% conversion of cells into spheroplasts.

Fraction	Alkaline phosphatase	Cyclic phosphodiesterase[a]	β-Galactosidase	RNase	Total protein
			units/g		
First cycle					
Sucrose-Tris-EDTA supernatant	16	15	3.2	320	1.2
Cold water wash	1660[b]	820[b]	2	60	7.4
Second cycle					
Sucrose-Tris-EDTA supernatant	190	60	3	1000	2.4
Cold water wash	1100	180	2	690	5.8
Sonic extract, treated cells	100	170	1400	5100	63
Sonic extract, untreated cells	3200	1400	1400	9000	90
Sucrose medium, spheroplasts	2000	930	20	2900	11
Spheroplast lysate	300	360	1800	4000	80

[a] Tested with bis(p-nitrophenyl)phosphate.
[b] A corresponding wash with water at 24° released only 400 units of alkaline phosphatase and 200 units of cyclic phosphodiesterase.

equilibrated with the same buffer. After the resin was removed by brief centrifugation at 500 × *g* the cells were sedimented, and the pellet was stirred with cold water as usual.

Effect of Sucrose—The concentration of sucrose could be safely reduced to 12% (Table IV), but in the complete absence of sucrose the release of enzyme was barely detectable (Table II).[7]

Effect of Temperature—Treatment with sucrose-EDTA solutions could be carried out at 2° or at room temperature, but the subsequent shock required cold water. With distilled water at 24° the release of alkaline phosphatase was only 25%.

Effect of Other Conditions—When the release of alkaline phosphatase was less than quantitative on first contact with distilled water, very little was gained by prolonged stirring or repeated washing (Table V). However, additional enzyme could be released by subsequent repetition of the entire cycle of treatment with sucrose-Tris-EDTA and water (Table VI). The presence

[7] We also tried incubating whole cells at 37° with 0.01 M Tris (pH 7.2)-0.005 M EDTA without the use of osmotic shock. This resulted in very slow release of enzymes into the medium and the process was not selective.

high phosphate Medium B contained less $MgCl_2$ and $CaCl_2$ than did the medium used for work on alkaline phosphatase.

Effect of Sucrose—Sucrose was essential and could not be omitted (Table VII), nor could it be replaced by glycerol.[8]

Effect of Other Conditions—Repeated washing of cells harvested in the stationary phase improved the yield of 5'-nucleotidase from 1500 units per g to 2400 units per g for osmotic shock carried out on cells that had been washed once and three times, respectively. Replacement of Medium B with Bacto-Pennassay Broth gave a similar release of this enzyme by osmotic shock. Media low in P_i could be explored only with strains such as U7 (alkaline phosphatase deletion) because no specific inhibitor for alkaline phosphatase is known. With strain U7 grown on Medium B (high P_i) release of 5'-nucleotidase was 2000 units per g, and growth on Medium A (low P_i) gave 1800 units per g. Re-

[8] Dr. Nancy Nossal has recently observed in this laboratory that sucrose may be replaced by 0.4 M NaCl with results that are nearly as good. This treatment causes the cells to shrink, and the membrane is observed to separate from the more rigid cell wall (20).

placement of glycerol by glucose as a carbon source did not affect the release of 5′-nucleotidase or cyclic phosphodiesterase during osmotic shock.

Release of Cyclic Phosphodiesterase and of Acid Phosphatase

In general, these enzymes behaved like 5′-nucleotidase. Most of the activity was removed by a cold water wash following 20% sucrose-0.03 M Tris (pH 8)-0.001 M EDTA. Yields were reduced by 30 to 50% when EDTA was omitted, and this was partially corrected by increasing the concentration of Tris buffer, or by supplements of 1×10^{-3} M *o*-phenanthroline or hydroxyquinoline.

The term "osmotic shock" emphasizes that the osmotic transition should be abrupt. Thus, when the sucrose treatment was followed by dropwise addition of cold water over a period of 30 min, release of cyclic phosphodiesterase was 55% of that obtained by rapid mixing and, in the case of acid phosphatase, only 40%. Under conditions involving a smaller osmotic transition the extent of release of cyclic phosphodiesterase was decreased (Table IV).

Release of RNase and DNase—During the first cycle of osmotic shock the release of RNase varied from a trace to 8% of that found in sonic extracts. It was difficult to obtain reproducible values for the RNA-inhibited DNase of sonic extracts (9). However, in eight experimental runs with cells in the *stationary* phase, the release of this enzyme during osmotic shock appeared to vary from 2 to 7%. By contrast, the yield of DNase in the fluid surrounding EDTA-lysozyme spheroplasts varied from 50 to 100%.

Examination of Other Enzymes

Ten other enzymes were examined and found not to be liberated in significant amounts either by osmotic shock or by conversion of *E. coli* to spheroplasts with lysozyme and EDTA. Thus strain K12s contained 500 units per g of histidine-activating enzyme (21), of which only 5 units per g were set free during osmotic shock or on spheroplast formation. Also, the cells contained 9000 units per g of inorganic pyrophosphatase, of which only 120 units per g were released.[9] Leakage of β-galactosidase in most cases was 1% or less. Other enzymes found to remain within the cell include polynucleotide phosphorylase, RNA phosphodiesterase (22, 23), glucose 6-phosphate dehydrogenase (4), and glutamic dehydrogenase (4). Less than 5% of leucine aminopeptidase (24) and thiogalactoside transacetylase (25)[10] were released. In two experiments the sonic extract contained 11,280 and 12,000 units of DNA polymerase per g (26) while the cold water wash contained only 60 and 50 units per g, respectively.

Evidence for Surface Localization of Certain Enzymes

Several lines of indirect evidence suggest that the enzymes being considered here are located at or near the cell surface. First, they are released on spheroplast formation, whereas 10 other enzymes were found to remain within the spheroplasts.

[9] This low activity may represent slow hydrolysis by a nonspecific phosphatase. Units of activity are expressed as micromoles of P_i formed per hour at 37°.

[10] We are indebted to Dr. David E. Alpers for assistance with these assays.

TABLE VII

Effect of EDTA and sucrose on release of enzymes by osmotic shock

E. coli K12s were grown to the stationary phase in Medium B. The standard procedure (see "Methods") for osmotic shock was carried out, except for changes in the composition of the 20% sucrose-Tris-EDTA solution. Data on release of enzymes and protein in the cold water wash are presented. In a separate but comparable experiment, release of enzymes into the medium surrounding EDTA-lysozyme spheroplasts (10) was not significantly different from that shown in the first line.

Tris	EDTA	5′-Nucleo-tidase	Cyclic phospho-diesterase	Acid phosphatase	β-Galacto-sidase	Total protein
M	mM		*units/g*			*mg/g*
0.03	1	1950	408	106	4	1.7
0.03	None	1190	240	90	5	1.7
None[a]	None	1300	180	80	6	1.2
0.20	None	1700	270	110	6	2.4
0.03[b]	1	30	0	4	4	Trace
Sonic extract, cells		980[c]	390	140	510	70

[a] Replaced with 0.03 M NaCl.

[b] Sucrose omitted.

[c] Suspension of intact cells assayed 2000 units per g. Extracts made in various ways show low 5′-nucleotidase activity. The assay for this sonic extract was increased to 1500 units per g by raising the concentrations of Co++ and Ca++ to 0.005 and 0.02 M, respectively.

In addition, the unique property of being released by osmotic shock may reflect a relatively superficial location. The evidence now to be presented was obtained with intact cells and does not involve osmotic shock.

Comparison of Activity of Intact Cells and Cell Extracts

Suspensions of intact *E. coli* K12S and C_4F_1 were used as a source of enzyme and compared with equivalent amounts of cell extract in assays for 5′-nucleotidase, cyclic phosphodiesterase, acid phosphatase, and alkaline phosphatase. A high fraction of these activities was measurable with intact cells (13), whereas enzymes such as β-galactosidase could be detected only in broken cell preparations or by treatment with agents such as chloroform. These observations have been confirmed with other strains, and release of enzymes into the medium by cell lysis has been ruled out. It must be admitted that, in general, no more than 50 to 70% of the total activity can be measured with intact cells. The reason for this is under investigation.

Stimulation by metals was observed with intact cells. Thus, a suspension of U7 assayed 23.7 units per ml with the usual test for 5′-nucleotidase, exhibited 6.8 units per ml in the absence of Co++, and was without activity in citrate buffer.

Especially significant are results on the inhibition of cyclic phosphodiesterase by RNA. Anraku (12) first demonstrated that the purified enzyme is inhibited by RNA from a variety of sources but not by mononucleotides. With strain C_4F_1 we observed 50% inhibition by *E. coli* soluble ribonucleic acid in a concentration of 0.3 mg per ml, both in assays with purified enzyme and with intact cells.

As reported previously (13) quite similar values for K_m were obtained, whether the measurements were made with purified

TABLE VIII

Activity of 5'-nucleotidase against different substrates

The assay is described in "Methods," and a fraction obtained by DEAE-cellulose chromatography was the source of enzyme. Compounds were tested at concentrations of 4×10^{-3} to 6×10^{-3} M.

Compound	Relative activity (5'-AMP = 100)	Compound	Relative activity (5'-AMP = 100)
5'-AMP	100	dTMP	59
5'-GMP	155	dGMP	106
5'-UMP	45	dAMP	89
5'-CMP	78	dCMP	52
3'-AMP	3	β-Glycerophos-	
2'-AMP	0	phate	3
2'(3')-UMP	0	Galactose 6-phos-	
Adenosine 3',5'-di-		phate	1.5
phosphate	0	Ethanolamine	
Uridine 2'(3'),5'-		phosphate	1.5
diphosphate	6		

TABLE IX

Hydrolysis of hexose esters by intact cell suspensions of E. coli

E. coli U7 (alkaline phosphatase-negative) was grown to the stationary phase on Medium A, washed with 0.01 M Tris, pH 8, and suspended in 0.01 M MgCl₂-0.01 M Tris, pH 8. The suspension was incubated at 37° with shaking, with 50 mM acetate, pH 6.0, 5 mM MgCl₂, and 5 mM hexose ester. The reaction was terminated with perchloric acid and P_i was measured (18).

Substrate	Activity of cell suspension[a]
	units/g, wet wt
Glucose 6-phosphate	143
α-D-Galactose 1-phosphate	127
β-D-Galactose 1-phosphate	0
β-D-Glucose 1-phosphate	0

[a] A comparable sonic extract showed 140 units per g against glucose 6-phosphate. When intact cells were centrifuged, no activity was detected in the supernatant solution; apparently no significant lysis of cells occurred. At pH 7 and pH 8 the activity was 60 and 45%, respectively, of that at pH 6.

extracts or intact cells. Similar pH optima were also observed in the two cases.

Growth of Cells with Supplements of Nucleotides and Nucleosides

Preliminary experiments were carried out in which cells of strain U7 were grown in a basal medium deficient in a carbon source, and this was supplemented in various ways. Stimulation of growth was observed with adenosine, uridine, and certain nucleotides presumed to be hydrolyzed by surface enzymes. With 0.002 M substrate, the relative rates of growth were as follows: glycerol, 100; 5'-AMP, 60; uridine 2',3'-cyclic phosphate, 65; adenosine 2',3'-cyclic phosphate, 60; adenosine, 40; 3'-AMP, 40; and 2'-AMP, 0. This result is consistent with the presence of enzymes able to hydrolyze all of the esters except 2'-AMP. Further experiments are contemplated in which the basal medium is deficient only in phosphate.

Partial Purification of Cyclic Phosphodiesterase, 5'-Nucleotidase and Acid Phosphatase by Use of Osmotic Shock

E. coli, strain K12s, were grown to the stationary phase in Medium B, washed three times with cold 0.01 M Tris-Cl, pH 8 (20 ml per g), and suspended in 20% sucrose-0.03 M Tris-Cl, pH 8 (1 g per 80 ml). The standard procedure (see "Methods") was used for osmotic shock except that the concentration of EDTA was 2×10^{-3} M. Portions of the cold water wash were fractionated on DEAE-cellulose under somewhat different conditions on four occasions. The best results were achieved as follows. A column of DEAE-cellulose (2.5 × 12 cm), free of fine particles, was prepared in a cold room and washed with 500 ml of 5×10^{-3} M Tris (pH 7.4)-0.001 M MgCl₂. The cold water wash (100 ml) was applied at a rate of about 2 ml per min, followed by 100 ml of the same Tris buffer-MgCl₂ mixture. A linear gradient was applied in which the mixing vessel contained 150 ml of the starting buffer, and the reservoir contained 150 ml of 5×10^{-3} M Tris (pH 7.4)-0.001 M MgCl₂-0.2 M NaCl. Fractions of 4.7 ml each were collected at a flow rate of 2 ml per min, and three well separated peaks of enzyme activity were eluted.

The yield of the three enzymes in the cold water wash was quantitative in this experiment, and since the wash fluid contained only 5.4 mg of protein per g of cells, or 7.1% of that in an equivalent sonic extract, we obtained a 14-fold purification for this step. Fractions 18 to 23 from the column were pooled to give 70% of the cyclic phosphodiesterase with an over-all purification of 80-fold. Similarly, Fractions 27 to 33 yielded 70% of the 5'-nucleotidase with a 70-fold over-all purification; and the acid phosphatase in Fractions 44 to 48 was pooled to give an over-all purification of 120-fold with a 70% recovery.

The final preparations of acid phosphatase and 5'-nucleotidase were stable for at least several weeks when stored at −15°. However, the cyclic phosphodiesterase lost one-third of its activity in 2 days. Some improvement resulted from the use of mercaptoalbumin, but more work is needed to ensure stable fractions when this scheme of purification is used.

Properties of 5'-Nucleotidase

The fraction obtained by DEAE-cellulose chromatography appeared to be relatively specific for 5'-ribonucleotides and 5'-deoxyribonucleotides (Table VIII). It is likely that the low activity observed with other compounds will disappear on further purification, and this procedure is planned for future work. The enzyme was almost inactive in the absence of divalent metal cations. Various combinations were tried; the most effective stimulation was noted with 0.01 M Co⁺⁺ and 0.01 M Ca⁺⁺, present together. Relative rates were as follows: Co⁺⁺ plus Ca⁺⁺, 100; Co⁺⁺ alone, 46; Ca⁺⁺ alone, 23; Mg⁺⁺ alone, 7; and no addition, 1.

The pH optimum was pH 5.8 with about 50% as much activity at pH 7 and at pH 4.8. There was no inactivation after exposure to 60° for 10 min at pH 6, 7, or 8, but at pH 5 three-fourths of the activity was destroyed.

At pH 8 and with no metal additions other than Mg⁺⁺, activity of 5'-nucleotidase could not be detected in the cold water wash. This is of interest because it means that magnesium-requiring systems can be tested at this pH without interference from this enzyme. Surprisingly, 5'-nucleotidase was inhibited by only about 50% in the presence of 0.05 M potassium phosphate, and substantial activity remained in 0.1 M phosphate. This contrasts with alkaline phosphatase, which is completely inhibited by 0.01 M phosphate.

Properties of Acid Phosphatase

This fraction may well consist of several separable enzymes. After a 120-fold purification the following activities, in units per ml, were observed with a fraction containing 0.02 mg of protein per ml (under conditions for acid phosphatase assay in "Methods"): D-glucose 6-phosphate, 11.1; α-D-galactose 1-phosphate, 12.7; β-D-galactose 1-phosphate, 0.0; D-glucosamine 6-phosphate, 11.2; α-D-glucose 1-phosphate, 10.1; β-D-glucose 1-phosphate, 0.0; α-D-xylose 1-phosphate, 8.0; β-D-xylose 1-phosphate, 0.0; α-D-mannose 1-phosphate, 9.4; D-galactose 6-phosphate, 10.0; D-mannose 6-phosphate, 2.4; lactose 1-phosphate, 0.0; α-N-acetylglucosamine 1-phosphate, 10.0; β-N-acetylglucosamine 1-phosphate, 0.0; β-glycerophosphate, 2.0; 5'-AMP, 0.8; 3'-AMP, 0; and ATP, 0. The original cold water wash showed a similar specificity with respect to these sugar esters; no detectable hydrolysis of β-linked esters was observed. In fact, even with suspensions of intact cells hydrolysis of α-D-galactose 1-phosphate was noted, but no cleaving of β-D-galactose 1-phosphate could be detected (Table IX).

There was no metal requirement; full activity was observed in the presence of 0.01 M EDTA. Considerable variation was encountered in the amount of activity among different strains of *E. coli*, but the level of enzyme appeared to be unaffected by the concentration of P_i in the growth medium.

DISCUSSION

This work has indicated that a number of degradative enzymes can be released from *E. coli* by osmotic shock, and this occurs with most of the cells remaining viable. These are not extracellular enzymes, for they are not excreted into the medium during any phase of the growth cycle. It is quite likely that other enzymes are also released during osmotic shock; we have preliminary evidence that an ATPase is set free. It should be possible to exploit the method for various kinds of biological experiments, for the cells have been rendered unusually permeable, they have lost most of their phosphatase activity, and the bulk of the acid-soluble pool has been removed. In addition, the procedure affords a substantial purification of the enzymes that are set free; in recent work with strain K12s a 30-fold increase in purity resulted from this simple treatment. Furthermore, the enzymes are obtained in a very dilute aqueous solution of sucrose, and subsequent steps are easier than in the case of the spheroplast procedure (4), where the release takes place into 0.5 M sucrose solution. Finally, it should be mentioned that the shocked cells can be extracted in conventional ways to give enzyme preparations that are comparatively free of contaminating phosphatase activities. Osmotic shock of a milder degree has been used to remove the amino acid pool from *E. coli* (27, 28).

On the basis of the present results and earlier work (4, 10, 13), we draw the tentative conclusion that a family of degradative enzymes are located at or near the cell surface of *E. coli*. These include alkaline phosphatase, an acid phosphatase, 5'-nucleotidase, a DNase, an RNase, and cyclic phosphodiesterase. Evidence for their location is admittedly indirect. The literature on surface enzymes in microorganisms has been reviewed (29), and it includes rather extensive studies on the invertase and phosphatase of yeast (30–34), the invertase of *Neurospora crassa* (35), and the membrane-bound enzymes of *Streptococcus faecalis* (36). Various types of experimental evidence have been published, such as selective release by an external agent (32); susceptibility of the activity of intact cells to inhibition by agents believed not to penetrate freely, *e.g.* as antibodies (37); activity of intact cells against very large molecules (38); and the isolation of membrane fragments containing enzymic activity.

The acid phosphatase that is released by osmotic shock is probably similar to one of the fractions described by Rogers and Reithel (39), although specificities do not match exactly. As for 5'-nucleotidase, it has been described in a number of species of bacteria (40) but not in *E. coli*. Presumably the enzyme has been missed because of the unusual nature of its metal requirements and because of inhibitory material present in crude extracts. As mentioned earlier, treatment with EDTA in the absence of sucrose does not result in significant release of these enzymes. However, this treatment does allow the uptake of actinomycin D by whole cells (41) and also permits entry of deoxyribonucleoside triphosphates (42). Finally, in evaluating experiments with glucose 6-phosphate and intact cells, one must bear in mind that this ester has recently been shown to penetrate *E. coli* (43).

SUMMARY

A group of enzymes can be released from *Escherichia coli* without destroying their viability. A procedure is described in which the bacterial cells are treated with a solution of tris(hydroxymethyl)aminomethane and ethylenediaminetetraacetate in hypertonic sucrose and then are rapidly dispersed in cold water. The sudden osmotic transition releases the following enzymes in high yield: an alkaline phosphatase, a cyclic phosphodiesterase, an acid phosphatase fraction that is most active with hexose phosphates, and a 5'-nucleotidase. The same group of enzymes are set free into the sucrose medium when spheroplasts are made by means of lysozyme and ethylenediaminetetraacetate. Ribonuclease activity is released in large amount only by the spheroplast procedure. It is suggested that this family of degradative enzymes may be located at or near the surface of cells.

A substantial purification results when the bacterial phosphatases are released into cold water. Some properties of the partially purified enzymes are described. The 5'-nucleotidase has not been reported previously in *E. coli*. It is greatly stimulated by various metals, especially the combination of divalent cobalt and calcium. The acid phosphatase fraction is able to hydrolyze hexose esters in α linkage but not in β linkage.

Acknowledgments—We are indebted to Dr. Bruce Ames for assistance with the assays for histidine-activating enzyme and for many helpful discussions, and to Dr. Herbert Tabor for a number of useful suggestions and criticisms.

REFERENCES

1. REPASKE, R., *Biochim. et Biophys. Acta*, **30**, 225 (1958).
2. HORIUCHI, T., HORIUCHI, S., AND MIZUNO, D., *Nature*, **183**, 1529 (1959).
3. TORRIANI, A., *Biochim. et Biophys. Acta*, **38**, 460 (1960).
4. MALAMY, M., AND HORECKER, B. L., *Biochem. and Biophys. Research Commons.*, **5**, 104 (1961).

5. MALAMY, M. H., AND HORECKER, B. L., *Biochemistry*, **3,** 1889 (1964).
6. ELSON, D., *Biochim. et Biophys. Acta*, **36,** 372 (1959).
7. BOLTON, E. T., BRITTEN, R. J., COWIE, D. B., McCARTHY, B. J., McQUILLEN, K., AND ROBERTS, R. B., *Carnegie Institution of Washington Yearbook, Vol. 58*, Carnegie Institution, Washington, D. C., 1959, p. 259.
8. SPAHR, P. F., AND HOLLINGWORTH, B. R., *J. Biol. Chem.*, **236,** 823 (1961).
9. LEHMAN, I. R., ROUSSOS, G. G., AND PRATT, E. A., *J. Biol. Chem.*, **237,** 819 (1962).
10. NEU, H. C., AND HEPPEL, L. A., *J. Biol. Chem.*, **239,** 3893 (1964).
11. ANRAKU, Y., *J. Biol. Chem.*, **239,** 3412 (1964).
12. ANRAKU, Y., *J. Biol. Chem.*, **239,** 3420 (1964).
13. NEU, H. C., AND HEPPEL, L. A., *Biochem. and Biophys. Research Commun.*, **17,** 215 (1964).
14. LEVINTHAL, C., SIGNER, E. R., AND FETHEROLF, K., *Proc. Natl. Acad. Sci. U. S.*, **48,** 1230 (1962).
15. NEU, H. C., AND HEPPEL, L. A., *Biochem. and Biophys. Research Commun.*, **14,** 109 (1964).
16. WEISSBACH, A., AND KORN, D., *J. Biol. Chem.*, **238,** 3383 (1963).
17. HEPPEL, L. A., HARKNESS, D. R., AND HILMOE, R. J., *J. Biol. Chem.*, **237,** 841 (1962).
18. AMES, B. N., AND DUBIN, D. T., *J. Biol. Chem.*, **235,** 769 (1960).
19. LOWRY, O. H., ROSEBROUGH, N. J., FARR, A. L., AND RANDALL, R. J., *J. Biol. Chem.*, **193,** 265 (1951).
20. MITCHELL, P., in T. W. GOODWIN AND O. LINDBERG (Editors), *Biological structure and function, Vol. II*, Academic Press, Inc., New York, 1961.
21. BERG, P., BERGMANN, F. H., OFENGAND, E. J., AND DIECKMANN, M., *J. Biol. Chem.*, **236,** 1726 (1961).
22. SPAHR, P. F., *J. Biol. Chem.*, **239,** 3716 (1964).
23. SINGER, M. F., AND TOLBERT, G., *Science*, **145,** 593 (1964).
24. GOLDBARG, J. A., AND RUTENBURG, A. M., *Cancer*, **11,** 283 (1958).
25. ALPERS, D. H., APPEL, S. H., AND TOMKINS, G. M., *J. Biol. Chem.*, **240,** 10 (1965).
26. PRICER, W. E., JR., AND WEISSBACH, A., *J. Biol. Chem.*, **239,** 2607 (1964).
27. COWIE, D. B., AND McCLURE, F. T., *Biochim. et Biophys. Acta*, **31,** 236 (1959).
28. HALVORSON, H. O., AND COWIE, D. B., in A. KLEINZELLER AND A. KOTYK (Editors), *Membrane transport and metabolism*, Academic Press, Inc., New York, 1961, p. 479.
29. POLLOCK, M. R., in I. C. GUNSALUS AND R. Y. STANIER (Editors), *The bacteria, Vol. IV*, Academic Press, Inc., 1962, p. 121.
30. DEMIS, D. J., ROTHSTEIN, A., AND MEIER, R., *Arch. Biochem. Biophys.*, **48,** 55 (1954).
31. SUTTON, D. D., AND LAMPEN, J. O., *Biochim. et Biophys. Acta*, **56,** 303 (1962).
32. McLELLAN, W. R., JR., AND LAMPEN, J. O., *Biochim. et Biophys. Acta*, **67,** 324 (1963).
33. SCHMIDT, G., BARTSCH, G., LAUMONT, M., HERMAN, T., AND LISS, M., *Biochemistry*, **2,** 126 (1963).
34. WEIMBERG, R., AND ORTON, W. L., *J. Bacteriol.*, **88,** 1743 (1964).
35. METZENBERG, R. L., *Biochim. et Biophys. Acta*, **77,** 455 (1963).
36. ABRAMS, A., AND McNAMARA, P., *J. Biol. Chem.*, **237,** 170 (1962).
37. POLLOCK, M. R., *J. Gen. Microbiol.*, **15,** 154 (1956).
38. STANIER, R. Y., *Bacteriol. Revs.*, **6,** 143 (1942).
39. ROGERS, D., AND REITHEL, F. J., *Arch. Biochem. Biophys.*, **89,** 97 (1960).
40. KOHN, J., AND REIS, J. L., *J. Bacteriol.*, **86,** 713 (1963).
41. LEIVE, L., *Biochem. and Biophys. Research Commun.*, **18,** 13 (1965).
42. BUTTIN, G., BERTSCH, L., AND KORNBERG, A., *Federation Proc.*, **24,** 349 (1965).
43. FRAENKEL, D. G., FALCOZ-KELLY, F., AND HORECKER, B. L., *Proc. Natl. Acad. Sci. U. S.*, **52,** 1207 (1964).

Copyright © 1966 by the American Society of Biological Chemists, Inc.
Reprinted from J. Biol. Chem., **241**, 5732–5734 (1966)

Amino Acid-binding Protein Released from *Escherichia coli* by Osmotic Shock*

23

(Received for publication, August 15, 1966)

Jeanette R. Piperno‡ and Dale L. Oxender§

From the Department of Biological Chemistry, The University of Michigan, Ann Arbor, Michigan 48104

SUMMARY

The ability of *Escherichia coli* K-12 to take up leucine, isoleucine, or valine against apparent concentration gradients is considerably reduced when the cells are subjected to osmotic shock in the cold. When the lyophilized supernatant fluid derived from that treatment was dialyzed against labeled amino acids, binding of leucine, isoleucine, or valine was observed. With leucine or isoleucine binding as an assay, a protein was isolated and highly purified. The dissociation constants for the leucine and isoleucine complexes were found to be indistinguishable from their respective K_m values for cellular uptake. The results suggest that the isolated protein may well be part of an amino acid transport system of *E. coli*.

Kinetic studies of solute transport through cell membranes have demonstrated that the transfer is membrane mediated. Several approaches have been attempted to characterize the effective molecular structures. Recently Neu and Heppel (1) and Nossal and Heppel (2) showed that cold osmotic shock treatment of bacteria causes the loss of certain enzymes and proteins usually associated with the cell envelope.

Following these reports we subjected *Escherichia coli* K-12 cells to osmotic shock treatment at 0° to determine if any of the neutral amino acid transport systems (3–5) would be impaired or lost. Table I shows that the transport activity of *E. coli* as measured by initial rates of accumulation is considerably reduced for leucine and valine, while that of alanine and proline is relatively unaffected. We next looked for amino acid-binding activity in the fluid centrifuged from the cells after the osmotic shock treatment. That fluid was lyophilized and the solid residue remaining was dissolved in 0.1 M phosphate buffer, at pH 7. The solution was dialyzed overnight to remove small molecules, such as amino acids, which might interfere with measurement of the binding activity. To measure binding activity, aliquots of the dissolved and dialyzed residue were further dialyzed overnight against labeled amino acids. The results are presented in Table II. Of the amino acids tested L-leucine, L-isoleucine, and L-valine were the only ones bound measurably by the preparation. These three amino acids are known to share a common transport system in this organism (3–5). Somewhat weaker binding activity was observed for the D isomers of valine and leucine. This observation compares favorably with the weak transport activities of the cells for these isomers (4, 5).

* This work was supported in part by United States Public Health Service Grants GM-11024 and HD-01233 from the National Institutes of Health.

‡ Predoctoral Trainee of the United States Public Health Service. These studies are taken from a thesis submitted by J. R. Piperno in partial fulfillment of the requirement for the degree of Doctor of Philosophy in Biological Chemistry, The University of Michigan, Ann Arbor. Present address, Department of Biochemistry, Michigan State University, East Lansing, Michigan.

§ To whom requests for reprints should be addressed.

TABLE I

Transport studies with E. coli K-12 in stationary phase[a]

All values are expressed as micromoles of amino acid taken up per 30 sec per g, wet weight, of cells. One gram, wet weight, of cells corresponds to about 1×10^{12} cells. The ^{14}C-labeled amino acids were: L-leucine-1-^{14}C, specific activity 8.3 mC per mmole; uniformly labeled L-alanine-^{14}C, specific activity 100 mC per mmole; uniformly labeled L-proline-^{14}C, specific activity 146 mC per mmole. Cells were grown on minimal media. Ten milliliters of a stationary culture were diluted with 100 ml of fresh sterile medium and the suspension was incubated for 4 hours. At this time 300 ml of additional medium were added, and after 2 additional hours of incubation the cells were harvested and washed twice with 0.03 M Tris, pH 8.2. The washed cells (250 mg, wet weight) were resuspended in 20 ml of 20% sucrose-Tris buffer, pH 8.0, containing 10^{-3} M EDTA. An aliquot of these cells was withdrawn and mixed with an equal volume of 0.05 M Tris buffer, pH 8.0, containing 0.001 M MgCl$_2$. These cells were centrifuged and represent the control cells for the uptake studies. The remaining cells were centrifuged in the cold and then subjected to the osmotic shock treatment (1) by resuspension in 20 ml of ice-cold distilled water. After 2 min a second portion was withdrawn and again added to an equal volume of Tris-magnesium buffer. Cells obtained from this treatment represent the "shocked with Mg^{++}" cells in the table. A third portion of cells from the ice-cold water treatment represent the "shocked cells" used in the studies. Transport studies were conducted at 37° with 30 sec-incubations of cell suspensions with labeled amino acids. Incubations were terminated by filtration on a Millipore filter. The filters were glued on aluminum planchets and counted in a gas flow counter. Under the conditions used each filter contained about 0.8 mg of cells, wet weight, corresponding to 1×10^9 cells. Measurement of cell viability following osmotic shock under various conditions indicated that if magnesium was added within a few minutes after suspending the stationary cells in the distilled water good cell viability could be maintained. Less than 5% of exponentially grown cells were viable following the same treatment and proved to be unsatisfactory for showing loss of transport activity while still maintaining viability.

Treatment	L-Leucine (10^{-6} M)	L-Valine (10^{-6} M)	L-Alanine (10^{-6} M)	L-Proline (0.2 × 10^{-6} M)	Cell viability
	μmoles/30 sec/g cells				%
Control cells	0.64	0.27	0.113	2.57	100
Shocked cells with Mg^{++}	0.016	0.02	0.11	2.14	100
Shocked cells without Mg^{++}	0.012	[b]	0.08	1.38	36

[a] The viability studies and some of the transport studies were done by William Penrose of this laboratory.

[b] No value.

The binding activity was not sedimented by centrifuging for 1 hour at 40,000 × *g*. Binding activity was destroyed during heating for 5 min at 70°. The binding activity in the shock fluid was precipitated by ammonium sulfate between 70 and 100% of saturation. These properties taken along with nondializability suggest that the activity is associated with a macromolecular entity presumably protein in nature. With the use of conventional protein fractionation procedures the binding protein has been extensively purified. The results are shown in Table III. After purification the binding protein moved as one major band during electrophoresis on cellulose acetate strips in 0.05 M potassium phosphate buffer at pH 7.4. An approximate molecular weight of 45,000 was obtained for it by comparing the R_F values of the binding activity during passage through Sephadex G-100 and Bio-Gel P-60 columns with the R_F values of hemoglobin and cytochrome *c* on these same columns.

Dissociation constants were determined for the amino acid-protein complex by equilibrium dialysis, varying the concentration of the amino acid and thereby obtaining typical saturation curves. Fig. 1 shows the millimicromoles of leucine bound as a function of its concentration. Table IV shows the constants for the dissociation and the inhibition of formation of the leucine,

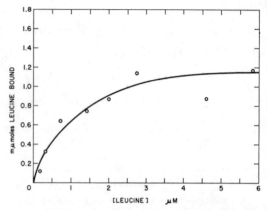

FIG. 1. Leucine binding to isolated protein. The purified protein was dissolved in 0.05 M potassium phosphate buffer, pH 7, to give a protein concentration of 0.17 mg per ml. The methods used to assay binding activity have been described in the legend to Table II. The millimicromoles of leucine bound are shown on the *ordinate* and the concentrations of leucine used are listed on the *abscissa*.

TABLE II

Amino acid-binding activity of substance removed during osmotic shock

The fluid obtained from shock treatment of *E. coli* was lyophilized and the dry residue was redissolved in water and dialyzed overnight. Portions of the solution, 0.4 ml, containing about 1.5 mg of protein per ml, were dialyzed overnight at 4° against 6 ml of the labeled amino acid. The radioactivity inside and outside of the tubing was measured by liquid scintillation counting. Under the above conditions dialysis equilibrium was usually reached in about 6 hours.

Amino acid	Final external concentration	Distribution ratio
	$M \times 10^6$	*cpm inside/cpm outside*
L-Isoleucine	0.78	2.2
L-Leucine	0.40	1.7
L-Valine	0.90	1.2
L-Phenylalanine	0.58	0.9
L-Methionine	0.35	1.0
L-Tryptophan	0.52	1.0
L-Alanine	1.16	1.0
Glycine	0.63	0.9

TABLE III

Purification of binding protein

For purification steps 316 g, wet weight, of stationary phase *E. coli* K-12 cells were subjected to cold osmotic shock treatment. Binding activity was measured by dissolving the protein in 0.05 M potassium phosphate buffer, pH 7.0, and dialyzing overnight against 0.8×10^{-6} M labeled L-isoleucine. The total millimicromoles of isoleucine bound were calculated by extrapolating the saturation curves to infinite concentration.

Purification step	Total protein	Total ¹⁴C-L-isoleucine bound at saturating concentration	Specific activity[a]
	mg	*mµmoles*	
100% ammonium sulfate	6240		
70% ammonium sulfate	2565	1460	0.570
DEAE-cellulose column	126	839	6.66
Sephadex G-100 column	71	31	4.36[b]

[a] Specific activity is defined as millimicromoles of amino acid bound at a saturating concentration per mg of protein.
[b] Low value reflects instability of highly purified binding protein.

TABLE IV

A comparison of kinetic constants for amino acid-binding activity of binding protein with those for cellular transport by E. coli

Harvested and washed stationary phase *E. coli* K-12 bacteria were suspended in minimal media, pH 7.0, at 37°. The kinetic constants were determined from reciprocal plots of initial velocities of transport as measured by 30 sec-incubations. The details are given in the legend to Table I. The constants for the dissociation and the inhibition of formation of the amino acid complexes were measured with protein isolated after the DEAE-cellulose step in the fractionation procedure. The protein solutions used in these studies contained about 0.17 mg per ml of the binding protein. Binding was determined as described in the legend to Table II.

Amino acid	Transport studies		Binding studies	
	K_m	K_i	Dissociation constant	Half-maximal inhibition of binding
	µM		*µM*	
L-Leucine	1.1	0.5 (Ile)	1.1	2.1 (Ile)
L-Isoleucine	1.2	1.2 (Leu)	2.2	1.7 (Leu)
L-Valine	8.0	4.5 (Leu)	12.0	7.3 (Leu)
		3.4 (Ile)		8.9 (Ile)

valine, and isoleucine complexes with the protein, as obtained by the usual linear plots. These constants are compared with the K_m and K_i values for the cellular uptake of these amino acids.

Conversion of the labeled amino acids into other labeled substances following binding could not be detected. In trying to discover an enzymatic function for the binding material we assayed it for leucyl ribonucleic acid synthetase activity. Whereas an extract from *E. coli* containing leucyl-RNA synthetase activity prepared according to the procedure described by Berg (6) activated 0.23 µmole of leucine in 1 min, 0.2 mg of the binding protein gave a value similar to the water blank used in the assay. The following three facts seem to rule out the possibility that the binding protein and the leucyl-RNA synthetase are identical. (*a*) No enzymatic activity was observed. (*b*) The molecular weight found for the aminoacyl-RNA synthetases all have been near 100,000 (7). (*c*) Isoleucine and leucine are both bound to the isolated protein with similar affinities.

Studies carried out with other systems have attempted to isolate transport-reactive structures. For example, a reactive site serving in glycerol entry into human red blood cells is protected from phenyl isothiocyanate by the presence of an excess of a glycerol analogue (8). The structural proteins of the lac operon have been separated, and one of them, presumably the y-gene product, was shown to have no recognizable enzymatic activity (9, 10). A membrane fraction labeled with N-ethylmaleimide and presumably carrying the binding site (no longer active) for β-galactoside transport has been separated (11). A protein with a high affinity for inorganic sulfate and presumed to be the transport carrier for sulfate in *Salmonella typhimurium* has been isolated and highly purified from the supernatant fluid obtained after osmotic shock (12, 13). Recently the β-galactoside transport activity of *E. coli* was shown to be diminished by osmotic shock, and then restored by incubating the cells with a purified component of a phosphotransferase system, a result suggesting a relationship between this enzymatic system and β-glycoside transport (14).

The results of the present studies suggest but do not prove that the binding protein is a component of the transport system responsible for concentrative uptake of leucine, isoleucine, and valine by *E. coli* K-12 (5). The location of the protein in the cell envelope and the close similarity of the kinetic constants for cellular uptake and for binding by the purified protein supports that interpretation. All attempts to restore transport activity to shocked cells by adding back the purified binding protein have so far been unsuccessful. Additional confirmation of the role of the binding protein is being sought.

REFERENCES

1. Neu, H. C., and Heppel, L. A., *J. Biol. Chem.*, **240**, 3685 (1965).
2. Nossal, N. G., and Heppel, L. A., *J. Biol. Chem.*, **241**, 3055 (1966).
3. Cohen, G. N., and Rickenberg, H. V., *Ann. Inst. Pasteur*, **91**, 693 (1956).
4. Piperno, J., *Federation Proc.*, **25**, 801 (1966).
5. Piperno, J., Ph.D. thesis, The University of Michigan, 1966.
6. Bergmann, F. H., Berg, P., and Dieckmann, M., *J. Biol. Chem.*, **236**, 1735 (1961).
7. Baldwin, A. N., and Berg, P., *J. Biol. Chem.*, **241**, 831 (1966).
8. Stein, W. D., *Nature*, **181**, 1662 (1958).
9. Naono, S., Rouviere, J., and Gros, F., *Biochem. Biophys. Res. Commun.*, **18**, 664 (1965).
10. Kolber, A. R., and Stein, W. D., *Nature*, **209**, 691 (1966).
11. Fox, C. F., and Kennedy, E. P., *Proc. Natl. Acad. Sci. U. S.*, **54**, 891 (1965).
12. Pardee, A. B., and Prestidge, L. S., *Proc. Natl. Acad. Sci. U. S.*, **55**, 189 (1966).
13. Pardee, A. B., Prestidge, L. S., Whipple, M. B., and Dreyfuss, J., *J. Biol. Chem.*, **241**, 3962 (1966).
14. Kundig, W., Dodyk Kundig, F., Anderson, B., and Roseman, S., *J. Biol. Chem.*, **241**, 3243 (1966).

Copyright © 1967 by the American Society of Biological Chemists, Inc.

Reprinted from *J. Biol. Chem.*, **242**, 793–800 (1967)

The Reduction and Restoration of Galactose Transport in Osmotically Shocked Cells of *Escherichia coli*

24

(Received for publication, August 12, 1966)

Yasuhiro Anraku

From the Laboratory of Biochemistry and Metabolism, National Institute of Arthritis and Metabolic Diseases, National Institutes of Health, Bethesda, Maryland 20014

SUMMARY

D-Galactose transport in two mutant strains of *Escherichia coli* K12 lacking galactokinase has been described. These cells showed a reduction of about 50% in the capacity for galactose uptake and in the rate of exit as a consequence of osmotic shock. In the shock procedure the cells were first treated with 0.5 M sucrose containing 1×10^{-4} M ethylenediaminetetraacetate, after which the pelleted cells were rapidly dispersed in cold 5×10^{-4} M $MgCl_2$. A factor was released into the $MgCl_2$ solution (shock fluid) which formed a complex with galactose *in vitro*, detected by Sephadex chromatography. The material was heat labile and nondialyzable and behaved in several respects like a protein. More of this material appeared when cells high in galactose transport activity were used.

The reduced uptake of galactose in shocked cells could be restored by first incubating them with dialyzed shock fluid. The shock fluid could be removed before measuring uptake, suggesting that some factor in the shock fluid needed for uptake could reassociate with the shocked cells. Restoration of activity was always limited to an amount less than the original level.

These observations suggest that a factor necessary for galactose transport has been partially released by osmotic shock into the shock fluid. Further, the results suggest that the factor can reassociate with the cell, thereby restoring the reduced uptake of galactose nearly to the original level, but not exceeding it.

Recent work has shown that a number of metabolic functions and synthetic processes occur closely associated with or near the cell membrane (1–5). One unique function of the membrane is concerned with the selective movement of substances into and out of the cell, even against a concentration gradient. The mechanism governing entrance and exit of materials is termed active transport. According to one current hypothesis, the whole system consists of several components including highly stereospecific carrier protein (or proteins) and an energy-coupling process leading to a mediated movement of substrate against a concentration gradient (6–12). Although no direct evidence on the location of the active transport system exists, several kinds of experiments suggest an association with the cell membrane (13–15).

Current studies in our laboratory have revealed that when *Escherichia coli* are subjected to a form of osmotic shock a number of enzymes and other uncharacterized proteins are selectively released into the surrounding medium (shock fluid), together with acid-soluble constituents of the cell (16–18).[1] Decisive proof for localization of these proteins is lacking, but histochemical (19, 20) and other kinds of evidence (21, 22) suggest that they are near the cell surface. The osmotically shocked cells resist changes in osmotic pressure and maintain their viability under a variety of nutritional conditions (17).[2]

Recovery of shocked cells is not immediate; there is a lag period before growth is resumed, and this depends on the nature of the culture medium.[2] Accordingly, we decided to investigate whether there might be some impairment of a transport system. The present paper reports a reduction in galactose transport as a result of osmotic shock and the restoration of galactose uptake by shocked cells after they are treated with dialyzed shock fluid. These results may be interpreted in terms of the partial release of a protein component active in galactose transport and its subsequent restoration. It has also been observed that a protein released into the shock fluid binds ^{14}C-galactose and it is suggested that this represents the formation *in vitro* of a protein-galactose complex that may possibly play a role in the transport phenomena.

An effect of osmotic shock in causing reduced uptake of certain sugars (23) and in causing releasing of a protein that binds sulfate (24) has been reported.

EXPERIMENTAL PROCEDURE

Materials

Bacteria—Two strains of *E. coli* K12 were used. W3092, which was isolated by Dr. E. Lederberg and obtained through Dr. I. G. Leder, lacks galactokinase and is lysogenic by phage lambda. W3350, which lacks galactokinase and galactose 1-phosphate phosphotransferase, was obtained from Dr. M. B.

---

[1] In this procedure the bacterial cells are suspended in sucrose-Tris-EDTA medium and centrifuged. The supernatant fluid is referred to as the sucrose fluid. The pellet of cells is then rapidly dispersed in cold 5×10^{-4} M $MgCl_2$ solution and the mixture is again centrifuged. This supernatant fluid is called the shock fluid.

[2] Y. Anraku, and L. A. Heppel, unpublished observations.

Yarmolinsky. Both strains are constitutive for a galactose permease, but the level of activity is much lower in W3350.

Medium—The synthetic medium of Cohen and Rickenberg (25) was routinely supplemented with trace metals. The medium (Medium CR, pH 7.3) thus consists of (per liter): 13.6 g of KH_2PO_4, 2.0 g of $(NH_4)_2SO_4$, 0.2 g of $MgSO_4 \cdot 7H_2O$, 0.0005 g of $FeSO_4 \cdot 7H_2O$, and 1 ml of the trace metal solution containing 480 mg of $FeCl_3 \cdot 6H_2O$, 280 mg of $MnCl_2 \cdot 4H_2O$, 270 mg of $CaCl_2$, 2000 mg of $ZnCl_2$, 290 mg of H_3BO_3, and 130 mg of $CoSO_4$ per liter. Glycerol or sodium succinate (pH 7.3), each at 0.3%, was added to the medium before use.

Chemicals—^{14}C-D-Galactose (1.61×10^7 cpm per µmole), ^{14}C-thiomethylβ-D-galactoside (1.93×10^6 cpm per µmole), and leucine (1.67×10^8 cpm per µmole) were obtained from New England Nuclear; ^{14}C-phenylalanine (2.16×10^8 cpm per µmole) was purchased from Nuclear-Chicago. D-Fucose was from Mann, and D-galactose (Fisher) was reagent grade and was used without further purification. Tris was Sigma's Trisma Base and sucrose was from Fisher. Pancreatic ribonuclease and deoxyribonuclease were obtained from Worthington. Pronase was from Calbiochem.

Methods

Growth of Cells—Bacteria were grown in Medium CR on a rotary shaker at 37°. Cells in the middle of logarithmic phase were collected and washed twice with 0.01 M Tris-0.03 M NaCl, pH 7.3. A large scale cultivation was carried out in a carboy (15-liter portion), with vigorous aeration with a stream of cotton-filtered air at 37°.

Osmotic Shock Procedure—The method described recently (17) was used in this study. In Stage I cells were exposed, at 22°, to 0.033 M Tris-1 × 10⁻⁴ M EDTA-20% sucrose (pH 7.3) at a concentration of 1:80 (1 g, wet weight, of cells per 80 ml). The medium was removed by centrifugation (sucrose fluid), and in Stage II the pellet of cells was rapidly dispersed in cold 5×10^{-4} M $MgCl_2$. Selective release of proteins occurred at this stage and once more the medium was removed by centrifugation (shock fluid). The shocked cells, as well as unshocked control cells, were finally made to a 1:80 suspension in 0.033 M Tris, pH 7.3 ($8 \pm 1 \times 10^9$ cells per ml). Viability of the shocked cells is routinely $100 \pm 5\%$ compared with unshocked normal cells.

Assay for Galactose Uptake—The reaction mixture contained (per ml): 0.2 ml of the cell suspension described above, 0.6 ml of Medium CR (without a carbon source), 40 µg of chloramphenicol, 0.023 µmole of ^{14}C-D-galactose, and water. The mixture without substrate was shaken gently in a 25-ml Erlenmeyer flask at 22° for 5 min, and then substrate was added to start the assay. Incubation at 22° was continued and, at various time intervals, samples of 0.2 ml were filtered on Millipore HA membranes (0.45 µ pore size), followed by washing with 5 ml of an ice-cold solution consisting of 0.01 M Tris, 0.15 M NaCl, and 0.5 mM $MgCl_2$, pH 7.5. Sampling and washing were completed within 20 sec. Filters were mounted on counting planchets and dried under an infrared lamp, and the radioactivity was measured on a low background end window counter (Nuclear-Chicago, 8703 series). Nonspecific adsorption of radioactivity during this treatment was corrected by a suitable control run. Results were expressed as micromoles of the substrate accumulated by 1 g of wet cells.

Preparation of Cell-free Extract Which Was Active in Restoring Galactose Uptake—Osmotic shock was carried out with 20 g of W3092 in a 1:40 suspension with 2×10^{-4} M EDTA (see above).

The fluids obtained at Stage I and Stage II of the procedure (sucrose fluid and shock fluid) were dialyzed against 5×10^{-4} M Tris-5×10^{-4} M $MgCl_2$, pH 7.5, for 48 hours and 24 hours, respectively, with several changes of buffer. Each fraction was then centrifuged, lyophilized, and dissolved in 20 ml of 0.01 M Tris, pH 7.3.

No formation of bacterial colonies was detected when 0.01 ml of the above solutions was spread on nutrient agar plates.

Paper Chromatography—Two solvent systems were used: A, *n*-butyl alcohol-acetic acid-H_2O (40:10:50) (26); B, pyridine-ethyl acetate-H_2O (10:36:11.5) (27). Descending chromatography at room temperature with Whatman No. 1 filter paper was employed.

Column Chromatography—Sephadex G-100 (medium) or G-25 (coarse) was allowed to swell in glass-distilled water with several decantations to remove small particles. The gel was packed into a column and washed with 0.01 M Tris-5 mM $MgCl_2$, pH 7.5.

Protein determination (28) and assays for cyclic phosphodiesterase (16), 5'-nucleotidase (17), hexose phosphatase (16), and uridine diphosphoglucose pyrophosphatase (18) have been described.

RESULTS

General Patterns of Galactose Transport—Fig. 1 shows the transport activities for galactose in the two mutant strains. In good agreement with the results of Horecker, Thomas, and Monod (9), the rate of uptake was found to level off after 7 to 10 min. Maintenance of the steady state level for over 20 min and rapid displacement of the intracellular radioactivity by an excess of nonlabeled galactose, as in Fig. 1, are characteristic of the transport system. In mutant W3092 the intracellular concentration of galactose under steady state conditions was estimated as 2×10^{-3} M, which exceeds the external concentration by a factor of 100. Galactose uptake in W3350 was limited to one-eighth of that of W3092; this strain is considered to be partially constitutive for galactose permease because D-fucose (1×10^{-3} M) does effect an elevation of transport activity in W3350 (Table I). Properties of the two mutants are summarized in Table I.

The effect of variation in concentration of external galactose on its entrance was studied. A Lineweaver-Burk plot shows that the K_t values of entrance (6.6×10^{-5} M) were similar in both strains.

Accumulated radioactivity was extracted by boiling the washed cells in buffer. After removal of cells by centrifugation, the supernatant fluid was examined by paper chromatography. The radioactive material migrated as galactose in two solvent systems. The same results were obtained with samples derived from shocked cells.

Effect of Osmotic Shock on Galactose Uptake—The rate of galactose uptake in osmotically shocked cells was compared with that in unshocked control cells (Fig. 2). The plateau level of galactose attained by shocked cells, as well as the initial rate of uptake, decreased by about 50% compared with unshocked cells. The effect was reproducible; thus, in separate experiments decreases of 70, 50, 50, 40, and 40% (W3092) and 50, 40, 40, and 30% (W3350) were observed.

The effect of various concentrations of $MgCl_2$ in the shock fluid was tested. Cold solutions of $MgCl_2$ that were either 5×10^{-4} M or 1×10^{-2} M gave a similar reduction in galactose uptake.

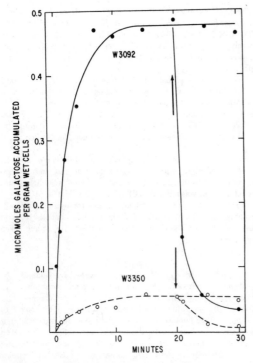

FIG. 1. Uptake of ^{14}C-galactose and displacement by unlabeled galactose. Bacteria were grown to midlogarithmic phase in Medium CR with 0.3% sodium succinate. The cells were collected and washed twice with 0.01 M Tris (pH 7.3)-0.03 M NaCl. For assay, see "Experimental Procedure." *Arrows* indicate the time when a portion of the reaction mixture received unlabeled galactose at a concentration of 1.7×10^{-2} M.

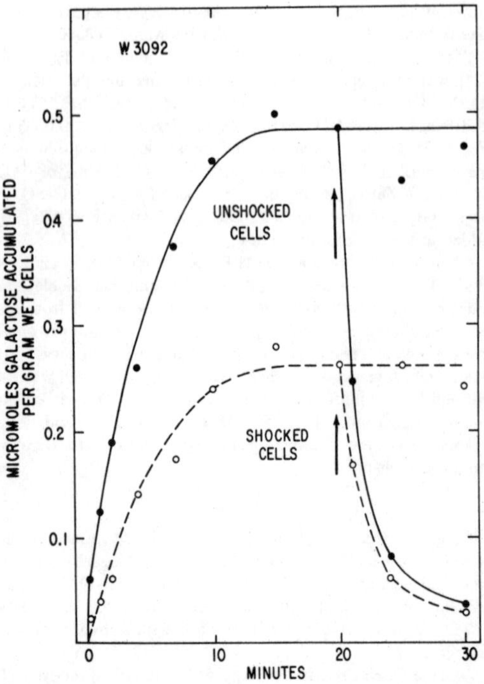

FIG. 2. Effect of osmotic shock on the uptake of galactose by W3092. Cells were collected at midlogarithmic phase and washed as described. A portion of the washed cells was treated with 0.033 M Tris (pH 7.3)-20% sucrose-1×10^{-4} M EDTA, followed by osmotic shock with cold 5×10^{-4} M MgCl$_2$. The shocked cells were centrifuged and resuspended in 0.033 M Tris, pH 7.3. Standard assay conditions for uptake of galactose were employed. *Arrows* indicate the time when a portion of the reaction mixture received unlabeled galactose at a concentration of 1.7×10^{-2} M.

TABLE I

Properties of E. coli K12 derivatives W3092 and W3350 with respect to uptake of galactose and thiomethyl-β-D-galactoside

The bacteria were grown in Medium CR with 0.3% of a carbon source and with addition as indicated, harvested in the middle of a logarithmic phase, and washed twice as described. Uptake of galactose and thiomethyl-β-D-galactoside by bacteria was measured under standard conditions for 15 min at 22°. Concentration of substrates used was: galactose, 2.3×10^{-5} M; thiomethyl-β-D-galactoside, 1×10^{-4} M.

Experiment	Strain	Conditions of growth		Substrate	Substrate accumulated
		Carbon source	Addition		
					μmole/g wet cells
I	W3092	Succinate	None	Galactose	0.47
		Glycerol	None	Galactose	0.51
II	W3092	Glycerol	None	Galactose	0.37
		Glycerol	D-Fucose, 1×10^{-3} M	Galactose	0.37
		Glycerol	None	Thiomethyl-β-D-galactoside	0.14
III	W3350	Succinate	None	Galactose	0.086
		Succinate	D-Fucose, 1×10^{-3} M	Galactose	0.088
		Succinate	D-Galactose, 1×10^{-3} M	Galactose	0.184
		Succinate	None	Thiomethyl-β-D-galactoside	0.14
		Succinate	D-Fucose, 1×10^{-3} M	Thiomethyl-β-D-galactoside	0.14
		Succinate	D-Galactose, 1×10^{-3} M	Thiomethyl-β-D-galactose	0.40

The K_t value of shocked cells was measured in several experiments and found not to be altered. This suggested to us that osmotic shock may have caused the partial release into the shock medium of a fraction involved in active transport. These considerations prompted us to examine possible binding between galactose and a component of the shock fluid.

Evidence for Presence of Protein Component Released by Osmotic Shock Which Binds Galactose—Concentrated shock fluids were prepared (see "Experimental Procedure"). Shock fluid was incubated in the presence of 2×10^{-5} M ^{14}C-galactose and 5 mM $MgCl_2$ in 0.01 M Tris, pH 7.5, for 40 min at 22°. The reaction mixture was chilled and analyzed on a Sephadex G-100 column (Fig. 3). A minor peak, which is distinctly removed from the major peak of free galactose, was eluted in tubes 10 to 15. This peak was not observed with the sucrose fluid. The peak of activity was associated with protein (Fig. 3) and the radioactive material was identified as unchanged galactose. Judging from the fact that cyclic phosphodiesterase, also present in the shock fluid, was eluted in tubes 9 to 14, the protein fraction capable of

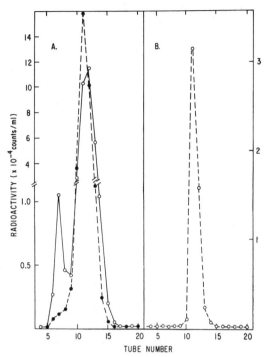

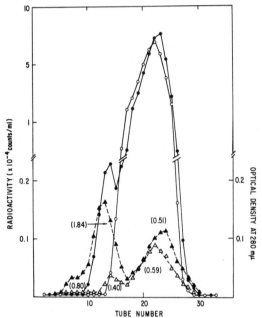

FIG. 3. Demonstration of a protein-galactose complex on Sephadex G-100. Sucrose fluid and shock fluid were prepared from W3092 as described under "Experimental Procedure." Concentrated, dialyzed fluids were used in these binding studies. The reaction mixture contained 0.01 M Tris (pH 7.5), 5 mM $MgCl_2$, 0.023 μmole of ^{14}C-galactose per ml, and 60 μg of protein per ml (sucrose fluid) or 250 μg of protein per ml (shock fluid). Incubation was at 22° for 40 min. The chilled reaction mixture was layered on a Sephadex G-100 column (1 $\times$ 40 cm), which had been washed, and was eluted with cold 0.01 M Tris-5 mM $MgCl_2$ solution, pH 7.5. Fraction volume, 2 ml; flow rate, 0.2 ml per min. Radioactivity of ^{14}C-galactose in the reaction mixtures with sucrose fluid (O) and with shock fluid (●). Measurement of material absorbing at 280 mμ with sucrose fluid (△) and with shock fluid (▲). *Figures in parentheses represent $A_{280}:A_{260}$ of the indicated peaks.*

FIG. 4. Comparison of shock fluid from strains W3092 and W3350 with respect to the ability to form a complex with galactose. Shock fluid protein, 500 μg, derived from W3092 or W3350 was incubated under the conditions described in the legend for Fig. 3. A Sephadex G-25 column (0.6 $\times$ 40 cm) was used in order to detect formation of a complex. Experimental methods were similar to those of Fig. 3. *A,* radioactivity pattern with shock fluid from W3092 (O) and W3350 (●). The substrate was 0.023 mM ^{14}C-galactose. *B,* radioactivity pattern with shock fluid from W3092 in the presence of ^{14}C-thiomethyl-β-D-galactoside (0.1 mM).

forming a complex with galactose was assumed to have a molecular weight of less than 70,000 (30).

Further studies on the binding of galactose by shock fluid were carried out with Sephadex G-25, which gave better resolution of the complex and free galactose. The shock fluid from W3092 again showed considerable binding, with 6.3% of the total radioactivity in the minor peak. By contrast, shock fluid from W3350 showed only a small skewing in this position of the elution diagram (Fig. 4), suggesting a parallelism between binding ability *in vitro* and the level of transport activity *in vivo*. Shock fluid contains a number of enzymes (16, 17, 31), two of which (hexose phosphatase (32, 33) and uridine diphosphoglucose pyrophosphatase (31)) might bind galactose. However, there was no significant difference in the levels of these two enzymes in shock fluids obtained from W3092 and W3350, supporting the hypothesis that the ability of shock fluid to bind ^{14}C-galactose is related to transport activity of the host cells. When shock fluid from W3092 was incubated with 1×10^{-4} M thiomethyl-β-D-galactoside, no complex with this sugar appeared (Fig. 4). In a few experiments we observed that osmotic shock had less effect on the transport of thiomethyl-β-D-galactoside than of galactose in both strains; preparations of shocked cells showed decreases of 10 and 20% for thiomethyl-β-D-galactoside with respect to 40 and 50% for galactose.

Binding ability could be destroyed by heating the fluid (1.36 mg of protein in 1 ml of 0.01 M Tris-5 mM MgCl₂, pH 7.5) or treating it with Pronase (100 μg per ml). Incubation for 60 min at 22° with RNase (50 μg per ml) and DNase (50 μg per ml) had no effect.

Shock fluid from W3092 was observed to form a complex with certain amino acids when the same techniques were used. Thus, with concentrations of 1 × 10⁻⁴ M, 3.7 and 1.7% of radioactive leucine and phenylalanine, respectively, were combined to form a complex. These findings have not been studied further.

Restoration of Reduced Galactose Uptake in Shocked Cells— Fig. 5 shows the pattern of restoration of galactose uptake in shocked cells of strain W3350 when treated with shock fluid from W3092. Shocked cells were preincubated with the shock fluid, after which ¹⁴C-galactose was added to start the assay. A distinct effect on shocked cells can be observed with respect to restoration of uptake of galactose, whereas the shock fluid had very little effect on unshocked cells. Subsequent work showed that when cells were preincubated with smaller concentration of shock fluid protein than indicated for Fig. 5 greater restoration was achieved, so that almost the original transport activity was obtained (Fig. 6). The reason for these differences is made clear by Fig. 6, which shows that low concentrations of shock fluid protein stimulate transport activity, whereas increased concentrations lead to a significant inhibition. In the experiments of Fig. 6 it was again noted that there was little effect when unshocked cells were preincubated with the shock fluid.

The factor in shock fluid responsible for restoration of galactose transport activity was nondialyzable and it was found to be heat labile. Thus, shock fluid from W3092 (85 μg of protein per ml of the standard reaction mixture without substrate and

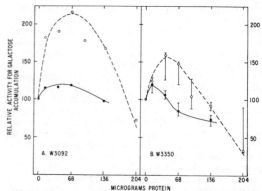

Fig. 6. Effect of different concentrations of shock fluid from W3092 on galactose uptake. Cells from strains W3092 or W3350 were preincubated for 20 min at 22° in the standard reaction mixture in the presence of the indicated amounts of dialyzed shock fluid from W3092 (expressed as concentration of protein). Assays were begun by addition of ¹⁴C-galactose (0.023 μmole). Incubation was for 20 min at 22° under standard conditions. Accumulation of radioactivity in unshocked control cells (●) and in shocked cells (○). *Vertical lines* in *B* show the deviation obtained in three experiments.

TABLE II

Effect of excess of shock fluid on restoration of uptake of galactose by shocked cells

Bacteria were preincubated in the presence of the indicated amount of protein of shock fluid from W3092 for 20 min at 22°. After preincubation, the reaction mixtures were divided into two portions. The cells in one of them were collected and washed with 2 ml of ice-cold 0.033 M Tris, pH 7.3, on a Millipore filter. Washed cells were then resuspended in the same reaction mixture. Assays for uptake of galactose were carried out under the standard conditions. Initial stage and "steady state" represent 2 and 20 min, respectively, after the onset of the uptake reaction.

Experiment	Strain	Treatment	Addition of shock fluid during preincubation	Presence of shock fluid during incubation	Galactose accumulated	
					Initial stage	Steady state
			μg protein/ml		*μmole/g wet cells*	
I	W3350	Shocked	0	−	0.020	0.036 [a]
			136	+	0.020	0.035
			204	+	0.019	0.036 [a]
			136	−	0.020	0.066
			204	−	0.026	0.061 [a]
		Unshocked	0	−	0.033	0.057
			136	+	0.029	0.057
			136	−	0.044	0.062
II	W3092	Shocked	0	−	0.025	0.078
			204	+	0.015	0.050
			204	−	0.044	0.114
		Unshocked	0	−	0.098	0.320
			136	+	0.093	0.315
			136	−	0.100	0.335

[a] Three different experiments were carried out with shocked W3350 cells and 204 μg of protein per ml of shock fluid. The uptake of galactose at the steady state was: 100% without addition of shock fluid; 70 ± 30% when shock fluid was not removed by filtration; 145 ± 20% when Millipore filtration was carried out.

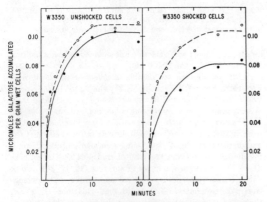

Fig. 5. Restoration of reduced uptake of galactose by shocked cells after a prior incubation with shock fluid derived from W3092. Cells of strain W3350 were preincubated in the standard reaction mixture without ¹⁴C-galactose in the presence and absence of shock fluid from W3092 (100 μg of protein per ml of reaction mixture) for 20 min at 22°, then ¹⁴C-galactose (0.023 μmole) was added, and radioactivity incorporated into the cells was measured periodically. Radioactivity accumulated in the cells in the presence (○) and in the absence (●) of shock fluid from W3092.

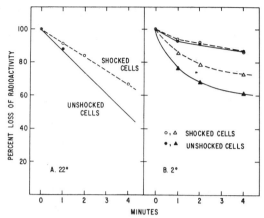

FIG. 7. Comparison of kinetics for exit of galactose in shocked and unshocked cells of strain W3092. Shocked cells and unshocked cells were preincubated in the presence of 0.23 and 0.12 μmole per ml of ^{14}C-galactose, respectively, for 20 min at 22° in the standard reaction mixture in order to have cells preloaded at the same internal concentration of galactose. After being centrifuged in the cold to remove an excess of free galactose in the reaction mixture, shocked and unshocked cells were shown to have been preloaded with 1.42×10^4 cpm and 1.79×10^4 cpm, respectively, per 1.6×10^9 cells. The preloaded cells were then rapidly resuspended in the standard reaction mixture without ^{14}C-galactose. Rate of exit of galactose was determined in several different ways. Techniques used are described under "Experimental Procedure." *A*, rate of exit at 22°. *B*, rate of exit at 2° of shocked cells (○) and of unshocked cells (●). Rate of exit at 2° in the presence of 4×10^{-3} M unlabeled galactose in the reaction mixture: △, shocked cells; ▲, unshocked cells.

cells) was heated; at 98° for 5 min, activity was destroyed, whereas a factor remained active at 80° and 60° for 5 min.

The restorative effect of shock fluid does not depend upon its continued presence during the uptake measurements. This was shown by preincubation of shocked cells with a large excess of shock fluid. The cells were then collected and washed on Millipore filters and resuspended in the standard reaction mixture with ^{14}C-galactose. Table II shows that such cells show recovery of galactose uptake. Other cells from which the shock fluid had not been removed showed no stimulation because inhibition levels had been reached (see Fig. 6). These data suggest that a factor required for uptake is removed from the shock fluid by the shocked cells. Several attempts were made to test whether a factor had been removed from shock fluid after preincubation with shocked cells. As an assay, we looked for a reduction in ^{14}C-galactose binding in Sephadex G-25 chromatography (see Fig. 4). The results were negative, perhaps because of lack of sensitivity of the assay.

Effect of Osmotic Shock on Exit Process—The evidence presented thus far suggests that a component of the galactose transport system can be released by osmotic shock. A decrease in the plateau level of galactose in shocked cells, however, does not necessarily indicate a decrease in entrance activity. The same result would be achieved if the exit process were to be enhanced in shocked cells. In order to test this possibility, shocked and unshocked cells were preloaded with the same internal concentration of ^{14}C-galactose. Exit activities of both types of cells

were then examined under different conditions. Fig. 7 indicates that the shocked cells show a reduction of about 50% in the rate of exit of galactose.

DISCUSSION

In the present investigation we have studied the galactose transport system in two mutant strains of *E. coli* lacking galactokinase. Accordingly, galactose is accumulated in the cell as a gratuitous substrate because of lack of its metabolic enzyme. This makes it possible to study the transport of galactose uncomplicated by its further metabolism. The mutants are considered to be galactose permease constitutive because D-fucose, a good inducer for galactose permease (29), does not change the level of transport activities in these cells. It would be desirable to study a galactokinase-less mutant with an inducible galactose transport system, and we hope to obtain such an organism in the future. Both strains showed the same K_t values for galactose uptake, although the internal concentrations at the plateau level differed by 8-fold. We assume that the galactose transport systems in the two strains are similar in nature.

Osmotic shock reduces both the extent of galactose uptake and the rate of exit. Concentrated, dialyzed shock fluid restores uptake almost to the control level and also shows the ability to bind ^{14}C-galactose. The question arises whether these separate effects are due to the same component. At this point, all that we can say is that much more of the ^{14}C-galactose-binding factor appeared in the shock fluid derived from cells with a high level of transport activity (W3092) than was true for shock fluid from cells poor in transport activity (W3350). We considered the possibility that either hexose phosphatase or uridine diphosphoglucose pyrophosphatase might bind galactose, but the levels of these enzymes were quite similar in shock fluids whose galactose binding ability varied greatly.

The observations recorded here lead us to assume that a protein component responsible for galactose transport can be released by osmotic shock. It is attractive to speculate that this component may function as a *protein carrier*, in line with a current hypothesis for active transport (see Reference 34), but more evidence is required to support this idea.

The osmotic shock procedure appears to be a useful technique for transport studies since the shocked cells are viable and osmotically stable in spite of certain drastic changes, such as loss of 3% of the cellular protein. Preliminary studies with the electron microscope have not as yet revealed changes in the surface structure of shocked cells.[3] It should be noted that the extent of release of total protein and of some of the enzymes is sharply reduced with increased concentrations of $MgCl_2$ in the shock fluid (17). However, in the present study, the reduction in galactose uptake of shocked cells was quite similar with shock fluids containing 5×10^{-4} M and 1×10^{-2} M $MgCl_2$. In this connection, the findings made by Rogers (35) are interesting; he has shown that a protein component can be solubilized when *E. coli* cells are incubated in a growth medium at pH 7.3, but not at pH 5.3, in the presence of chloramphenicol, and that this causes loss of glucose permeability of the cells.

It is difficult to be certain of the location of the proteins that are released by osmotic shock or on spheroplast formation with EDTA and lysozyme. All of the released enzymes, with the exception of a fraction of the ribonuclease (36), appear in the

---

[3] B. K. Wetzel, personal communications.

soluble fraction when broken cell preparations are made. Histochemical studies indicate a surface localization, although there are differences in the distribution for different enzymes (19, 20). Alkaline phosphatase has been assigned to the periplasmic space between wall and membrane (21). Operationally, a protein component of the galactose transport system behaves as if it had a surface localization, but again no direct evidence is at hand.

In this study, the released protein which is able to bind ^{14}C-galactose appeared to reassociate with shocked cells. It should be noted that, even with an optimal concentration of the protein, the maximal restoration in both strains approaches a value no greater than 2-fold, thereby restoring transport activity nearly to the original levels (Fig. 6). Thus, the repair process leads to a similar percentage increase in transport in two strains which differ by 8-fold in the uptake of galactose. This causes us to speculate that the mechanism involved in restoration may be rather specific, perhaps involving a reassociation of the protein fraction to sites from which it had been released.

Preliminary experiments with 5'-nucleotidase also indicated that the enzyme could be adsorbed by shocked cells. Shock fluid from W3092 containing 900 units of 5'-nucleotidase was incubated with about 8×10^{10} shocked cells (W3092) for 30 min at $22°$. About 20% of the activity was found to be associated with the shocked cells after being washed twice, compared with 5% for control cells. It could be calculated that the amount of 5'-nucleotidase adsorbed by the shocked cells restored the bound enzyme to the original level. It remains to be established whether this represents a significant reassociation or only a nonspecific adsorption, but at least the possibility exists that released proteins may re-enter shocked cells under certain conditions.

We have shown that what is apparently a protein component released into shock fluid can bind ^{14}C-galactose, and the amount of this material is greater in the mutant which shows a greater uptake of galactose by whole cells. We speculate that the binding may represent the formation of a protein-galactose complex *in vitro*. Binding of ^{14}C-galactose with shock fluid from W3092 on Sephadex chromatography was always observed under the conditions employed. Incubation of the shock fluid with ^{14}C-galactose at $22°$ or $37°$ seemed to be important for the formation of this complex. It is surprising that the complex did not dissociate during the separation of protein and galactose on the column. Perhaps this is due to a slow rate of dissociation at $2°$, the temperature used in the chromatography. Further studies on the nature of the binding are under way. On the assumption that a complex was indeed being measured, certain rough calculations can be made. Thus, about 1.4 mμmoles of galactose were bound by 500 μg of protein in the dialyzed shock fluid (Fig. 4). When this same fluid was tested for its ability to restore transport activity to shocked cells, it was found that 13.6 μg of protein could carry 0.16 mμmole of galactose into the shocked cells (Fig. 6). The binding ability under the conditions *in vitro* was 4-fold less than in the experiment *in vivo*. We have been unable to recover the complex after precipitation with cold trichloracetic acid, after disc gel electrophoresis, or following treatment with an anion exchange resin. It remains to be established that the factor (factors) in shock fluid responsible for restoration of galactose uptake *in vivo* is the same as the factor (factors) that bind galactose *in vitro* as detected by Sephadex chromatography. The main evidence so far is that the levels of the two activities in W3092 and W3350 can be correlated.

Attempts to purify the factor (factors) are under way. The sulfate-binding factor, which is also released by osmotic shock, has already been fractionated by conventional methods (24).

Kundig *et al.* (37) have described a novel phosphotransferase system in *E. coli* and ascribed to it a role in the active transport of sugars (23). The elements involved are a phosphoprotein kinase, a heat-stable and acid-stable phosphoprotein, and a phosphotransferase. In their scheme a sugar phosphate would be formed inside the cell membrane which presumably would be hydrolyzed. Kundig *et al.* (23) have also shown that *E. coli* subjected to osmotic shock show reduced uptake of thiomethyl-β-D-galactoside and α-glucoside, and the system could be restored by treating the shocked cells with the purified heat-stable protein. In their work, cells were exposed to 0.4 M NaCl-1 × 10^{-4} M EDTA in Stage I of the shock procedure, rather than a sucrose medium. The factor in shock fluid that restored galactose uptake for our mutant cells and formed a complex with galactose was found to be unstable to heating. This strongly suggests that our factor is different from the heat-stable protein, but evidence is hardly conclusive. It will be worthwhile to further investigate whether the model proposed by these workers is operative for galactose uptake in our mutants.

Acknowledgments—I am most grateful to Dr. L. A. Heppel for his continual interest and valuable discussions throughout this work and in the preparation of the paper. I wish also to thank Drs. W. B. Jakoby, B. N. Ames, and G. FerroLuzzi Ames for their critical reading of the manuscript and helpful discussions.

REFERENCES

1. COHEN, G. N., AND MONOD, J., *Bacteriol. Rev.*, **21**, 169 (1957).
2. SALTON, M. R. J., *Bacteriol. Rev.*, **25**, 77 (1961).
3. LAMPEN, J. O., in M. R. POLLOCK AND M. H. RICHMOND (Editors), *Function and structure in microorganisms*, Cambridge University Press, London, 1965, p. 115.
4. JACOB, F., BRENNER, S., AND CUZIN, F., *Cold Spring Harbor Symp. Quant. Biol.*, **28**, 329 (1963).
5. LARK, K. G., *Bacteriol. Rev.*, **30**, 3 (1966).
6. RICKENBERG, H. W., COHEN, G. N., BUTTIN, G., AND MONOD, J., *Ann. Inst. Pasteur*, **91**, 829 (1956).
7. ROTMAN, B., *Biochim. Biophys. Acta*, **32**, 599 (1959).
8. KEPES, A., *Biochim. Biophys. Acta*, **4C**, 70 (1960).
9. HORECKER, B. L., THOMAS, J., AND MONOD, J., *J. Biol. Chem.*, **235**, 1580 (1960).
10. KOCH, A. L., *Biochim. Biophys. Acta*, **79**, 177 (1964).
11. PRESTIDGE, L. S., AND PARDEE, A. B., *Biochim. Biophys. Acta*, **100**, 591 (1965).
12. WINKLER, H. H., AND WILSON, T. H., *J. Biol. Chem.*, **241**, 2200 (1966).
13. SISTROM, W. R., *Biochim. Biophys. Acta*, **29**, 579 (1958).
14. FOX, C. F., AND KENNEDY, E. P., *Proc. Natl. Acad. Sci. U. S.*, **54**, 891 (1965).
15. KABACK, H. R., AND STADTMAN, E. R., *Proc. Natl. Acad. Sci. U. S.*, **55**, 920 (1966).
16. NEU, H. C., AND HEPPEL, L. A., *J. Biol. Chem.*, **240**, 3685 (1965).
17. NOSSAL, N., AND HEPPEL, L. A., *J. Biol. Chem.*, **241**, 3055 (1966).
18. DVORAK, H. F., ANRAKU, Y., AND HEPPEL, L. A., *Biochem. Biophys. Res. Commun.*, **24**, 628 (1966).
19. SPICER, S. S., WETZEL, B. K., AND HEPPEL, L. A., *Federation Proc.*, **25**, 539 (1966).
20. DONE, J., SHOREY, C. D., LOKE, J. P., AND POLLAK, J. K., *Biochem. J.*, **96**, 27c (1965).
21. MALAMY, M., AND HORECKER, B. L., *Biochem. Biophys. Res. Commun.*, **5**, 104 (1961).

22. BROCKMAN, R. W., AND HEPPEL, L. A., *Federation Proc.*, **25**, 591 (1966).

23. KUNDIG, W., KUNDIG, D., ANDERSON, B., AND ROSEMAN, S., *J. Biol. Chem.*, **241**, PC3243 (1966).

24. PARDEE, A. B., AND PRESTIDGE, L. S., *Proc. Natl. Acad. Sci. U. S.*, **55**, 189 (1966).

25. COHEN, G. N., AND RICKENBERG, H. W., *Ann. Inst. Pasteur*, **91**, 693 (1956).

26. PARTRIDGE, S. M., *Biochem. J.*, **42**, 238 (1948).

27. HICKMAN, J., AND ASHWELL, G., *J. Biol. Chem.*, **241**, 1424 (1966).

28. LOWRY, O. H., ROSEBROUGH, N. J., FARR, A. L., AND RANDALL, R. J., *J. Biol. Chem.*, **193**, 265 (1951).

29. BUTTIN, G., *J. Mol. Biol.*, **7**, 164 (1963).

30. ANRAKU, Y., in G. L. CANTONI AND D. R. DAVIES (Editors), *Procedures in nucleic acid research*, Harper and Row, New York, 1966, p. 130.

31. MELO, A., AND GLASER, L., *Biochem. Biophys. Res. Commun.*, **22**, 524 (1966).

32. ROGERS, D., AND REITHEL, F. J., *Arch. Biochem. Biophys.*, **89**, 97 (1960).

33. VON HOFSTEN, B., AND PORATH, J., *Biochim. Biophys. Acta*, **64**, 1 (1962).

34. KEPES, A., AND COHEN, G. N., in I. C. GUNSALUS AND R. Y. STANIER (Editors), *The bacteria*, Vol. *4*, Academic Press, New York, 1962, p. 179.

35. ROGERS, D., *J. Bacteriol.*, **88**, 279 (1964).

36. ANRAKU, Y., AND MIZUNO, D., *Biochem. Biophys. Res. Commun.*, **18**, 462 (1965).

37. KUNDIG, W., GHOSH, S., AND ROSEMAN, S., *Proc. Natl. Acad. Sci. U. S.*, **52**, 1067 (1964).

309

Copyright © 1970 by the Federation of European Biochemical Societies

Reprinted from *Eur. J. Biochem.*, **13**, 526–533 (1970)

Close Linkage between a Galactose Binding Protein and the
β-Methylgalactoside Permease in *Escherichia coli*

25

Winfried Boos and Matti O. Sarvas

Department of Biological Chemistry, Harvard Medical School, and Biochemical Research Laboratory,
Massachusetts General Hospital, Boston

(Received November 26, 1969)

It has been found that a gene necessary for the production of the galactose binding protein is closely linked, if not identical with the gene(s) determining the β-methylgalactoside permease. This was determined by crossing a galactose binding protein negative, β-methylgalactoside permease negative strain of *E. coli* K-12 with wild type donor strains.

In addition, after mutagenation of a permease positive, binding protein positive strain of *E. coli* K-12 with *N*-methyl, *N*-nitroso, *N'*-nitroguanidine, some of the β-methylgalactoside permease defective mutants showed concomitant changes in the galactose binding protein. Out of 28 independent permease defective mutants two had lost their ability to produce cross reacting material against antibodies specific for the galactose binding protein; four mutants showed temperature sensitive synthesis of cross reacting material; five mutants had immunochemically somewhat changed binding protein. The immunochemical assay showed no changes in the binding protein of the rest of the 17 mutants. However, the galactose binding activity of the crude extract of these mutants varied to a great extent (10—90%) from the activity found in the crude extract of the wild type.

These results suggest a close relationship between the galactose binding protein and the β-methylgalactoside permease.

Cold osmotic shock, following treatment with Tris-EDTA liberates from the cell envelope of bacteria, among other components, binding proteins with a specific binding activity for a certain number of low molecular weight substances such as sugars, amino acids, and ions [1—4].

Besides their binding activity no other function of binding proteins has been demonstrated so far. However, an increasing amount of evidence has led to the assumption that binding proteins may play an essential role in the transport of low molecular weight substances through the cell membrane [5]. Several findings which point to the function of binding proteins in transport are: (a) The apparent K_m values of uptake of the different transport systems for their substrates are comparable to the dissociation constants of the different binding proteins towards the same substrates [2,6,7]. (b) Addition of the purified binding protein to the osmotically shocked cells restores reduced uptake [1]. (c) There has been demonstrated a direct correlation between the status of derepression of the transport systems and the amount of binding protein found [6].

In a previous publication [7] we demonstrated the close relationship of the galactose binding protein and a galactose transporting system, the β-methylgalactoside permease (MeGal-permease). It was found that the K_m values for sugars transported by this

permease were comparable to the dissociation constants of the binding protein towards the same sugars. The uptake of galactose *in vivo* was inhibited by other sugars to the same extent as was the binding of galactose to the binding protein *in vitro*. By testing several mutants for the permease and the binding protein, it was found that all *mgl+* mutants had the binding protein, whereas in the *mgl−* mutants the binding protein was either absent or present. (Genetic symbols are listed in Table 1.)

In the present publication we demonstrate genetic evidence suggesting that the galactose binding protein is a necessary component of the MeGal-permease.

MATERIALS AND METHODS

Bacterial strains used are listed in Table 1.

Media and Growth of Bacteria

The bacteria were grown in nutrient broth (8 g nutrient broth, Difco, 4 g sodium chloride per liter) at 25° or 40° in large bottles which were shaken vigorously. Selective medium was either Vogel-Bonner minimal medium [28] or modified Hutner's mineral base [29] supplemented with 0.1% glucose. The indicator media for sugar fermentation was eosin-methylene blue supplemented with different

Table 1. E. coli K12 strains used

Genetic symbols: ara = arabinose; gal = galactose; his = histidine; lac = lactose; mal = maltose; xyl = xylose; met = methionine; mgl = MeGal-permease; pur = purine; pyr = pyrimidine; str = streptomycin; thi = thiamine; tyr = tyrosine; —for absence or inability, + for presence or ability to utilize or ferment; — s for sensitivity; — r for resistance. Hfr indicates donor strains. Cistron symbols are according to Demerec et al. [34].

Strain number	Genetic markers	Origin
W3092cy⁻	F⁻ lac⁻ (y⁻) galK⁻	H. C. P. Wu [17]
W4345	F⁻ lac_del (i⁻ z⁻ y⁻) galK⁻ mal⁻ xyl⁻ met⁻ ara₂⁻ T₆—r mgl⁻	Rotman [18]
EH3001	his⁻ derivative of W4345	This study
EH3004	his⁻ derivative of W4345	This study
EC6	Hfr, pro⁻, thi⁻, str — s,	J. R. Beckwith [9]
EC26	Hfr	J. R. Beckwith [9]
EC35	Hfr	J. R. Beckwith [9]

sugars [30]. For the permease test the cells were resuspended in synthetic medium A [31].

MeGal-Permease Assay

Cells grown at either temperature were washed once with medium A and resuspended with medium A to an absorbance of 0.5 at 650 nm. 0.6 ml of this suspension were incubated at either 25° or 40° for 10 min. 10 µl of [³H]galactose (1 C/mmole) was added to a final concentration of 0.5 µM. After 2 min 0.5 ml were filtered through Millipore HA 0.65 µ pore size filters without washing. The filters were then counted in a Packard scintillation counter (Model 314 EX) with 10 ml of Fluoralloy (Beckman) and 0.5 ml H₂O.

Genetic Methods

Mutagenation was performed with N-methyl-N-nitroso-N'-nitroguanidine according to Adelberg et al. [10]. Histidine requiring mutants were obtained by screening for small colonies on minimal medium plates containing limiting amounts of histidine (0.4 µg/ml). Crosses were made by the method of sexual conjugation according to Mäkelä [32]. Screening for mgl⁻ mutants was done according to Zwaig and Lin [11]; cells were grown at 40° to single cell colonies on nutrient broth agar plates containing 0.1 µM [¹⁴C]galactose (35 mC/mmole). The colonies were replicated on sterile filter paper. The dry filter paper was exposed to Medical X-ray film (Kodak) for three days. mgl⁺ strains result in black spots, mgl⁻ strains in pale ones.

Preparation of Bacterial Extracts

The different strains were grown overnight in 150 ml nutrient broth. They were harvested by centrifugation at 0° and resuspended in 1 ml of 0.01 M Tris, pH 7.3. The suspension was sonicated for 3 min (Raytheon sonic oscillator) at 0°. The sonicate was centrifuged at 13 000 × g for 4 min. The supernatant was used directly for the immunochemical test and for the galactose binding activity test. Protein determination was done according to Lowry et al. [33].

Immunological Assay of the Galactose Binding Protein

Antibodies against the galactose binding protein were obtained by immunizing rabbits with the shock fluid of the mgl⁺ strain (W3092cy⁻). Then the serum was absorbed with shock fluid of the galactose binding protein negative strain W4345 as follows: 400 mg of lyophilized shock fluid of strain W4345 was added to 10 ml of the antiserum, the mixture was incubated at 4° for 12 hours and the precipitate removed by centrifugation for 10 min at 15000 × g.

After the procedure was repeated three times the serum yielded two small and one strong precipitation line in Ouchterlony immunodiffusion against proteins from a sonicate of a binding protein positive strain. The strong line was identical with the one obtained against purified binding protein.

Binding Activity of the Galactose Binding Protein

The galactose binding activity of the crude bacterial extract was evaluated by preparative polyacrylamide electrophoresis using gels which were polymerized in the presence of 1 µM [³H]galactose (1 C/mmole). The system contains only lower gel (1.2 ml 10% polyacrylamide in 0.4 M Tris-Cl buffer pH 8.9) and the sample of 0.2 ml crude extract (14—16 mg protein/ml) in 20% sucrose. The PD21/70 column of the Canalco system was used. Tris (3.0 g/l)-glycine (14.4 g/l) was used as anode and cathode buffer. The electrophoresis was performed at 250 V, 5 mA and 25°. 4.5 ml fractions were collected. The radioactivity of 0.5 ml aliquots was counted in a Packard scintillation counter (Model 314 EX) with 10 ml Fluoralloy (Beckman).

RESULTS

Close Linkage of the Gene Loci of the MeGal-Permease and the Galactose Binding Protein

Ganesan and Rotman [8] found that a mutation responsible for the defect in the MeGal-permease in strain W4345 of E. coli mapped in the mgl locus, which was 70% linked to the his operon in crosses

311

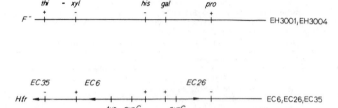

Fig. 1. *Approximate origin and orientation of Hfr strains used in crossing with* mgl⁻, *binding protein⁻ strains*. In each cross the recipient alleles of *pro* and *thi* were selected

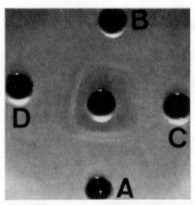

Fig. 2. *Immunodiffusion test for the galactose binding protein.* The outer wells contain the crude extracts (14—16 mg protein/ml) of the following strains: (A) W3092cy⁻ (*mgl⁺, binding protein⁺*); (B) W4345 (*mgl⁻, binding protein⁻*); (C, D) recombinants of the crosses EC6-X EH3004. The center well contains the antibodies

with Hfr-mediated conjugation. From previous studies it was known that this strain also lacked the galactose binding protein [7]. In order to find out if the mutation affecting the galactose binding protein in W4345 is linked to the permease gene we crossed *his⁻* derivatives of W4345 (EH3001, EH3004) with three *mgl⁺, binding protein⁺* Hfr strains (EC6, EC26, EC35). Origin and direction of the Hfr strains used are shown in Fig. 1. *his⁺* recombinants were selected and tested for MeGal-permease activity as well as the galactose binding protein.

The test for the galactose binding protein is based on immunoprecipitation of the protein and demonstrated in Fig. 2. An Ouchterlony double diffusion plate is shown in which the precipitation lines between the specific antibodies, the preparation of a galactose binding protein positive and a negative control strain and of two different *his⁺* recombinants can be seen. Appearance of the precipitation line present in the control positive and absent in the

control negative preparation indicated the presence of the binding protein in the recombinant. This precipitation line is identical with that obtained against purified galactose binding protein [1]. Rabbit antiserum against the shock proteins of an *mgl⁺* strain after absorption with the shock proteins of an *mgl⁻ binding protein⁻* strain was used.

The MeGal-permease test was performed as described in Methods and Materials. Under standard conditions a value of more than 25000 counts/min on the filter was considered *mgl⁺*, a value of less than 3000 as *mgl⁻*. All *his⁺* recombinants had either 25000—30000 counts/min or 1500—3000 counts/min. No intermediate value was observed.

The results of three crosses are shown in Table 2. A total of 75 *mgl⁺* and 27 *mgl⁻* recombinants were analyzed. Irrespective of the Hfr strain used, all *mgl⁺* recombinants were binding protein positive, all *mgl⁻* recombinants were binding protein negative.

These results show that the mutation responsible for the lack of the binding protein is closely linked to, if not identical with, the gene, *mgl*, determining, or else necessary for the expression of the MeGal-permease. From these crosses it cannot be learned whether the MeGal-permease defect in strain W4345 is due to a defect in the binding protein since there is the possibility that the genes for the MeGal-permease and for the galactose binding protein in W4345 fall in a common deletion or are simultaneously affected by a polar or regulator mutation.

Isolation of mgl⁻ Mutants

The function of the galactose binding protein in MeGal-permease mediated transport could possibly be demonstrated by isolating *mgl⁻* mutants carrying a defect in the galactose binding protein. A culture of the *mgl⁺* strain W3092cy⁻ was mutagenized with *N*-methyl, *N*-nitroso, *N'*-nitroguanidine and then diluted into several tubes of broth. After overnight growth each culture was plated onto a nutrient agar plate containing 0.1 μM [¹⁴C]galactose so as to obtain single colonies. Plates were incubated at 40° for 24 hours and *mgl⁻* mutants were then screened for

Table 2. *Characteristics of recombinants from genetic crosses*
In all crosses *his*+ recombinants were selected

Cross	Number of *his*+ Recombinants	Genotype of recombinants			
		MeGal-permease	galactose binding protein	*gal*	*xyl*
EC6(Hfr)-X EH3004	37	+	+	−	−
	15	−	−	−	−
EC35(Hfr)-X EH3004	11	+	+	−	−
	3	+	+	−	+
	4	−	−	−	−
	1	−	−	+	+
	2	−	−	+	−
EC26(Hfr)-X EH3001	20	+	+	−	−
	4	+	+	+	−
	5	−	−	−	−

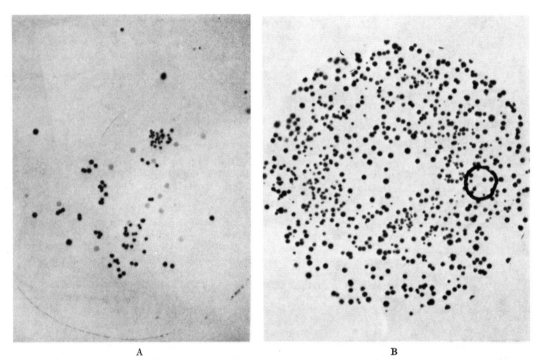

A   B

Fig. 3. *Screening for* mgl+ *and* mgl⁻ *mutants.* The bacteria are grown to single cell colonies on nutrient broth agar containing 0.1 µM [¹⁴C]galactose, 35 mC/mmole. The cells are replicated on sterile filter paper. The filters are dried and exposed to X-ray film. Time of exposure is 3 days. (A) Autoradiography of a mixture of the *mgl*+ strain, W3092cy⁻, and the *mgl*⁻ strain, W4345; (B) Autoradiography of a mutagenized culture of strain W3092cy⁻

their inability to concentrate radioactive galactose. The amount of radioactivity accumulated by each colony was detected by replicating the plates with a filter paper and subsequent autoradiography of the paper.

Fig. 3 A shows the autoradiography of a control experiment performed with a mixture of the *mgl*+ strain W3092cy⁻ and the *mgl*⁻ strain W4345. *mgl*+ colonies result in dark spots, *mgl*⁻ in pale spots. Fig. 3 B shows the autoradiography of a plate carrying a mutagenized culture of strain W3092cy⁻. As indicated by the black circle the population contains a mutant which is deficient in the accumulation of [^{14}C]galactose.

By this screening technique 28 independently derived mutants were isolated and tested for MeGal-permease activity. Table 3 shows the MeGal-permease activity of the mutants under standard conditions (0.5 µM [^{3}H]galactose, 37° and 2 min incubation time) when grown in nutrient broth at 25° or 40°. MeGal-permease activities ranging from 5—30% of the wild-type activity were observed. Some of the mutants had slightly less activity when grown at 40°. No mutant showed any difference in the activity when the test was performed either at 25° or 40° after growth at either temperature. It should be noted however, that there are differences in MeGal-permease activity in the same mutants when grown in different batches. The maximum deviation observed is about 20%. The permease test itself repeatedly performed with the same stock suspension stored over a period of 6 hours has a deviation of 3—5%.

The Galactose Binding Protein Isolated from Independently Derived mgl⁻ Mutants

The galactose binding protein of the *mgl⁻* mutants described above was tested by looking for cross reacting material with immunoprecipitation and by determination of the binding activity in the crude extract of sonicated bacteria. To determine the presence of cross reacting material the mutants were grown in nutrient broth at 25° and 40°, harvested and sonicated. The supernatants (protein concn. 14—16 mg/ml) were used for the double immunodiffusion technique. Four typical results are shown in Fig.4. The center well contains the specific antibodies previously described. The other wells contain respectively: a preparation of a permease and binding protein positive strain used as a control (1), preparation of mutants grown at 40° (2) and 25° (3). Four different types of mutants were found: (A) No change in cross reacting material. A precipitation line was found which was identical to the characteristic precipitation line of the galactose binding protein. Seventeen mutants (No. 12—28) showed this behavior. (B) Absence or decreased amount of cross reacting material. The precipitation line of the mutant grown at both temperatures was either absent (No. 1, 2) or much weaker (No. 7—11) and nearer the antigen well than the control line. It is not clear from the experiment whether this behavior indicates a change in protein synthesis or a change in the structure of the polypeptide chain resulting in a changed cross reactivity. Five mutants with this behavior were found. (C, D) Temperature sensitive synthesis of cross reacting material. The precipita-

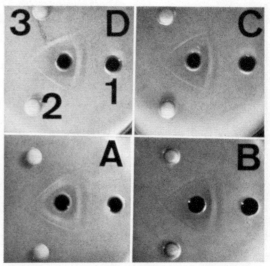

Fig.4. *Cross reactivity of the galactose binding protein from independently isolated* mgl⁻ *mutants*. The outer wells contain the crude extracts of the following strains; (A) EH3045; (B) EH3031; (C) EH3046; (D) EH3039; (2) preparation when grown at 40°, (3) preparation when grown at 25°; (1) preparation of *mgl⁺* strain W3092cy⁻. The center well always contains the antibodies

tion lines of these mutants (No. 3—6) grown at 25° were almost identical to the control line but absent when grown at 40° (D). One mutant (C) seems to produce small amounts (or a changed polypeptide chain) of cross reacting material when grown at 25° and none when grown at 40°. Four mutants with this type of temperature sensitivity were found.

As described above 17 mutants were found, the binding protein of which was not changed in its immunochemical response (Fig.4A). Since the strength and location of the precipitation lines were not changed in relation to the wild-type, we decided to evaluate the binding activity of these mutants. The interpretation of the results of the binding experiments is based on the assumption that there was no change in the amount of binding protein produced by the mutants.

The binding activity in the crude extracts is measured by preparative polyacrylamide electrophoresis. The electrophoresis was performed on a small scale with 0.2 ml protein solution (14—16 mg/ml) and 1.2 ml acrylamide gel at a concentration of 1 µM [^{3}H]galactose. Under this standardized condition, which allowed the performance of three runs per day, the background radioactivity was 450 to 550 counts/min whereas the fraction with the highest binding activity contained about 2000 counts/min. Binding activity is expressed as nmoles [^{3}H]galactose

Table 3. *Properties of independently isolated* mgl⁻ *mutants*
n. t. = not tested

Strain	Mutant No.	MeGal-permease test		Galactose binding protein immunoresponse		Binding activity of the crude extract
		Grown at 25°	Grown at 40°	Grown at 25°	Grown at 40°	Grown at 25°
		counts/min				nmoles [^{3}H]Gal/mg protein
W3092cy⁻	(wild-type)	35 812	35 613	+ + +	+ + +	0.031
EH3027	1	2 121	1 156	−	−	n. t.
EH3030	2	2 610	3 924	−	−	n., t.
EH3039	3	5 710	4 737	+ + +	−	n. t.
EH3040	4	6 734	4 172	+ + +	−	n. t.
EH3046	5	7 980	4 033	+ +	−	n. t.
EH3049	6	8 813	n. t.	+ +	−	n. t.
EH3028	7	1 870	1 897	+	+	0.031
EH3031	8	2 541	3 587	+	+	n. t.
EH3032	9	2 860	5 083	+	+	n. t.
EH3036	10	4 693	8 480	+	+	n. t.
EH3048	11	8 570	13 459	+ +	+ +	n. t.
EH3029	12	1 965	5 332	+ + +	+ + +	0.028
EH3033	13	3 165	6 521	+ + +	+ + +	0.014
EH3034	14	3 270	5 457	+ + +	+ + +	0.014
EH3035	15	4 345	4 497	+ + +	+ + +	0.015
EH3037	16	5 289	4 252	+ + +	+ + +	0.010
EH3038	17	5 020	5 394	+ + +	+ + +	0.020
EH3041	18	6 674	5 898	+ + +	+ + +	0.011
EH3042	19	6 829	4 155	+ + +	+ + +	0.010
EH3043	20	7 273	7 535	+ + +	+ + +	0.011
EH3045	21	8 009	6 254	+ + +	+ + +	0.010
EH3047	22	8 480	4 143	n. t.	+ + +	n. t.
EH3050	23	8 666	5 384	+ + +	n. t.	0.009
EH3053	24	8 450	5 473	+ + +	+ + +	0.014
EH3054	25	9 575	5 132	+ + +	+ + +	0.010
EH3055	26	9 715	3 945	+ + +	+ + +	0.014
EH3056	27	9 747	3 758	+ + +	+ + +	0.008
EH3057	28	10 450	6 710	+ + +	+ + +	0.004

bound per mg protein in the crude extract. The deviation of the repeated experiment is within 15—20%.

Table 3 shows the result of the comparison. Two mutants were not changed at all in comparison to the wild-type. Most of the mutants showed 30—50% of the wild-type activity, only three mutants had activities of 30% or less.

DISCUSSION

To demonstrate the function of an isolated protein in the transport of a low molecular weight substance through a plasma membrane requires the reconstitution of the transport phenomenon in a rather well defined *in vitro* system. The evidence for such an *in vitro* transport has not yet been accomplished. However, experiments of this kind seem not to be impossible. Kaback [12] could demonstrate that membrane preparations from *E. coli* depleted of their cytoplasmic content still showed phosphoenol pyruvate-dependent transport of α-methylglucoside through the membrane coupled with phosphorylation of the sugar. The same membrane preparations also were found to transport the amino acid proline, even without addition of any energy source [13].

Indirect but nevertheless strong evidence for the function of a protein as a necessary component of the transport of a particular substrate is derived from genetic correlations. Fox and Kennedy [14] have described the M protein as being the gene product of the y gene of the lactose operon in *E. coli*. This gene is known to affect the active transport of lactose and other β-galactoside derivatives. Another example of genetic correlation are mutants defective in Enzyme I or HPr of the phosphotransferase system in *E. coli* or *Salmonella typhimurium* which have been reported to carry a pleiotropic transport defect for a number of sugars [15,16].

In another well characterized transport system, that of β-methylgalactoside permease, one mutant (W4345) was found, among several permease defective mutants, which did not have the galactose binding protein [7]. The purpose of this work was to find a genetic correlation between the galactose binding protein and the β-methylgalactoside permease. We found that when this mutant was crossed by conjugation with *mgl⁺ binding protein⁺* Hfr strains, there was a very high linkage between the permease activity and binding protein. In fact, among 75 *mgl⁺* recombinants there were none of the type *mgl⁺, binding protein⁻*. Also, out of 27 *mgl⁻* there were none of the type mgl^-, *binding protein*$^+$. This indicates, although does not prove, that the recipient strain has a single mutation in a gene determining the Gal-binding protein. An alternative explanation could be that there are two very closely linked mutations; one in a gene determining the binding protein and another in a gene essential to the permease. The mutant might also carry a polar mutation or a large deletion covering the genes of the permease as well as of the binding protein. Also, the mutation does not need to necessarily involve the structural gene of the galactose binding protein; it could indicate a mutation of a positive regulation mechanism.

The regulation mechanism of the β-methylgalactoside permease is rather unclear. The permease appears to be inducible with D-galactose and D-fucose in gal$^+$ strains of *E. coli* and endogenously induced in galK$^-$ strains [17—19]. Thus the MeGal-permease seems to be simultaneously regulated with the galactose operon, even though the gene locus for the permease is located outside the gal operon [8,17]. A regulator gene *mglR* has been postulated by Ganesan and Rotman [8]. However, a mutation in *mglR* is always found simultaneously with a defect in galK$^-$, which is more characteristic for the endogenous induction of the gal operon than for a true constitutivity due to a mutation in a regulator gene. Yet there are mutants carrying large deletions in the gal operon which have lost their endogenous inducibility, and are still fully constitutive in MeGal-permease [20].

In an attempt to further clarify the genetic relationship between the galactose binding protein and the MeGal-permease, as well as their regulation, we isolated independently derived transport negative mutants. In the present publication we describe the galactose binding protein of these mutants in terms of immunochemical behavior as well as the galactose binding activity in the crude extracts of the sonicated bacteria.

Different types of mutants were found: (a) Mutants which had no cross reacting material against antibodies specific for the galactose binding protein; (b) Mutants whose synthesis of cross reacting material was temperature sensitive; (c) Mutants with small changes in the immunochemical behavior of the binding protein, indicating either a reduced amount or changed structure. (d) Mutants with a change in binding activity of the binding protein for galactose but no change in immunochemical response.

These results show that the mutation to defective transport in most cases simultaneously has affected the synthesis or the binding activity of the galactose binding protein.

All mutants were isolated as single step mutants after treatment with *N*-methyl, *N*-nitroso, *N*'-nitroguanidine which is not expected to cause deletions or additions [21]. Thus it seems unlikely that all these mutants have either two separate mutations or widespread genetic defects affecting both galactose binding protein and MeGal-permease simultaneously; rather the result is an additional indication that the galactose binding protein is an essential part of the MeGal-permease system. However, lacking definitive proof for this hypothesis in an *in vitro* system, alternative possibilities for the function of the galactose binding protein and other periplasmic binding proteins must be discussed.

One plausible explanation for the function of binding proteins is regulation. In this respect it would be instructive to consider the regulation of alkaline phosphatase in *E. coli*. Garen and Otsuji [22] found that the R2a protein was a gene product of the R2 gene which is known to be one of the regulator genes for alkaline phosphatase. However, even though the R2a protein is defined as the gene product of a regulator gene, it appears to be repressed and derepressed simultaneously with the alkaline phosphatase. Since the R2a protein also belongs to the group of periplasmic proteins and in addition binds phosphate [5], a possible explanation for this behavior might be that the R2a protein is identical with the phosphate binding protein [23] which is believed to function in phosphate transport. In terms of MeGal-permease regulation, one would expect the galactose binding protein to function in a positive regulation, *i.e.* necessary for the expression of the MeGal-permease. This can be concluded since no mutant has been found so far which lacks the galactose binding protein and is still fully active in transport.

Whatever the actual function of the galactose binding protein in the mechanism of the MeGal-permease is, it is most likely that the binding protein cannot be the only component of the transport system [7], since in all systems involving binding proteins, mutants have been found defective in the function of the permease but still active in respect to the corresponding binding protein [5]. Thus, what are the other components necessary for the function of the MeGal-permease?

The phosphotransferase system described by Kundig *et al.* [24] has certainly to be discussed in this respect. Roseman [25] suggested that a variety of sugars including galactose are transported by the phosphoenol-pyruvate dependent phosphotransferase system. According to this theory, the specificity of the MeGal-permease resides in a specific Enzyme II complex. Enzyme I mutants of the phosphotransferase system in *Salmonella typhymurium* do grow on galactose though, while defective in growth on the other sugars. This abnormality has been explained by the assumption that galactose enters the cell by a defective MeGal-permease which allows facilitated diffusion but not active transport. In addition, Wang *et al.* [26] reported that *ctr* mutants in *E. coli* do not grow on galactose. This finding could also be interpreted as supporting the function of the MeGal-permease in the phosphotransferase system since *ctr* mutants map close to the gene locus of Enzyme I in *E. coli* and lack Enzyme I activity [26]. However, other observations do not support this relationship between the MeGal-permease and the phosphotransferase system. Under normal conditions galactose does not get phosphorylated during transport and a highly active Gal-6-*P* phosphatase, as proposed by Roseman [25], has not yet been demonstrated. Also inhibitors of the active transport of galactose, such as dinitrophenol or sodium azide, do not inhibit the phosphotransferase system. More recently it was found that the MeGal-permease in *Salmonella typhimurium*, as well as the galactose binding protein, are unaffected by a HPr or Enzyme I mutation of the phosphotransferase system [27]. Thus it is likely that the MeGal-permease is not part of the phosphotransferase system.

A detailed study of different MeGal-permease mutants is therefore necessary to determine components other than the galactose binding protein needed for the function of the permease. As an initial step the mapping of the mutants described in this paper are under study.

We wish to thank Dr. H. M. Kalckar for his generous hospitality and his encouragement during this work. We are grateful to Dr. J. R. Beckwith for bacterial strains. Discussions with Dr. B. Rotman and Dr. T. H. Wilson were very helpful. We are indebted to Dr. A. Gordon for her help in preparing the manuscript. This work was supported by grants from the National Institutes of Health, the National Science Foundation and the Wellcome Trust. One author (W. Boos) was supported by a King Trust Fellowship.

REFERENCES

1. Anraku, Y., *J. Biol. Chem.* 243 (1968) 3116, 3123, 3128.
2. Piperno, J. R., and Oxender, D. L., *J. Biol. Chem.* 241 (1966) 5732.
3. Hogg, R. W., and Engelsberg, E., *J. Bacteriol.* 100 (1969) 423.
4. Pardee, A. B., *J. Biol. Chem.* 241 (1966) 5886.
5. Pardee, A. B., *Science*, 162 (1968) 632.
6. Pardee, A. B., Prestidge, L. S., Whipple, M. B., and Dreyfuss, J., *J. Biol. Chem.* 241 (1966) 3962.
7. Boos, W., *Eur. J. Biochem.* 10 (1969) 66.
8. Ganesan, A. K., and Rotman, B., *J. Mol. Biol.* 16 (1966) 42.
9. Beckwith, J. R., Signer, E. R., and Epstein, W., *Cold Spring Harbor Symp. Quant. Biol.* 31 (1966) 393.
10. Adelberg, E. A., Mandel, M., and Chen, G. C. C., *Biochem. Biophys. Res. Commun.* 18 (1965) 788.
11. Zwaig, N., and Lin, E. C. C., *Biochem. Biophys. Res. Commun.* 22 (1966) 414.
12. Kaback, H. R., *J. Biol. Chem.* 243 (1968) 3711.
13. Kaback, H. R., and Deuel, T. F., *Arch. Biochem. Biophys.* 132 (1969) 118.
14. Fox, C. F., and Kennedy, E. P., *Proc. Natl. Acad. Sci. U. S.* 54 (1965) 891.
15. Tanaka, S., Fraenkel, D. G., and Lin, E. C. C., *Biochem. Biophys. Res. Commun.* 27 (1967) 63.
16. Simoni, R. D., Levinthal, M., Kundig, F. D., Kundig, W., Anderson, B., Hartman, P. E., and Roseman, S., *Proc. Nat. Acad. Sci. U.S.* 58 (1967) 1963.
17. Wu, H. C. P., *J. Mol. Biol.* 24 (1967) 213.
18. Rotman, B., Ganesan, A. K., and Guzman, R., *J. Mol. Biol.* 36 (1968) 247.
19. Wu, H. C. P., Boos, W., and Kalckar, H. M., *J. Mol. Biol.* 41 (1969) 109.
20. Lengler, J., unpublished observation.
21. Whitfield, H. J., Martin, R. G., and Ames, B. N., *J. Mol. Biol.* 21 (1966) 335.
22. Garen, A., and Otsuji, N., *J. Mol. Biol.* 8 (1964) 841.
23. Medvecxky, N., and Rosenberg, H., *Biochim. Biophys. Acta*, in press.
24. Kundig, W., Ghosh, S., and Roseman, S., *Proc. Natl. Acad. Sci. U. S.* 52 (1964) 1067.
25. Roseman, S., *Gen. Physiol.* 54 (1969) 138.
26. Wang, R. J., Morse, H. G., and Morse, M. L., *J. Bacteriol.* 98 (1969) 605.
27. Boos, W., unpublished observations.
28. Vogel, H. J., and Bonner, D. M., *J. Biol. Chem.* 218 (1956) 97.
29. Cohen-Bazire, G., Sistrom, W. R., and Stanier, R. Y., *J. Cell. Comp. Physiol.* 49 (1957) 25.
30. Lederberg, J., *Methods Med. Res.* 3 (1950) 5.
31. Hartman, P. E., *Carnegie Inst. Wash. Publ.* 612 (1956) 35.
32. Mäkelä, P. H., *J. Bacteriol.* 91 (1966) 1115.
33. Lowry, O. H., Rosebrough, N. J., Farr, A. L., and Randall, R. J., *J. Biol. Chem.* 193 (1951) 265.
34. Demerec, M., Adelberg, E. A., Clark, A. J., and Hartman, P. E., *Genetics* 54 (1966) 61.

W. Boos
Biochemical Research Laboratory
Massachusetts General Hospital
Fruit Street, Boston, Massachusetts 02114, U.S.A.

M. O. Sarvas' present address:
Department of Bacteriology and Immunology
University of California
Berkeley, California 94720, U.S.A.

35*

Reprinted from *Proc. Natl. Acad. Sci. U.S.A.*, **66**, 1096–1103 (1970)

Components of Histidine Transport: Histidine-Binding Proteins and *hisP* Protein*

26

Giovanna Ferro-Luzzi Ames† and Julia Lever‡

BIOCHEMISTRY DEPARTMENT, UNIVERSITY OF CALIFORNIA (BERKELEY)

Communicated by W. Z. Hassid, May 28, 1970

Abstract. The high-affinity ($K_m = 3 \times 10^{-8}$ M) transport system for histidine in *Salmonella typhimurium* has been resolved into three components: J, K, and P. J, which is a histidine-binding protein released by osmotic shock, is specified by the *hisJ* gene: *hisJ* mutants lack the binding protein and are defective in histidine transport. Another class of mutants—*dhuA*, which is closely linked to *hisJ*—has five times the normal level of binding protein and has an increased rate of histidine transport. P, which is a protein specified by the *hisP* gene, is required for the J binding protein to be operative in transport. *hisP* mutants, though defective in transport, have normal levels of J binding protein. K, a third transport component, works in parallel to J, and also requires the P protein in order to be operative in transport. A second histidine-binding protein has been found but its relation to K is unclear. *hisJ*, *dhuA*, and *hisP* have been mapped and are in a cluster (near *purF*) on the *S. typhimurium* chromosome.

Introduction. Previous work from this laboratory[1,2] on histidine transport in *Salmonella typhimurium* had demonstrated the presence of two systems involved in histidine uptake: the histidine specific permease ($K_m \cong 10^{-8}$ M, involving the *hisP* gene) and the general aromatic permease ($K_m \cong 10^{-4}$ M; *aroP* gene). Mutants defective in each of these systems were described. Here we show that uptake through the histidine-specific permease occurs (as shown in Fig. 1)

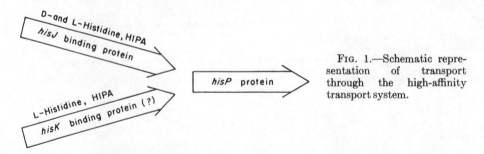

FIG. 1.—Schematic representation of transport through the high-affinity transport system.

through at least two components, J and K, functioning in parallel, and that transport through either of these components is dependent upon the presence of a protein specified by the *hisP* gene. Transport through the J component requires an intact *hisJ* gene, which codes for a histidine-binding protein.

A class of proteins located on the surface of gram-negative bacteria and easily released by mild osmotic shock[3] binds a variety of small molecules. After the pioneering work from the laboratories of Pardee, Heppel, and Oxender,[4] these "binding proteins" have been implicated in the active transport of those small molecules (recently reviewed[5]). Among these, the amino acid-binding proteins have been indirectly linked to amino acid transport. This paper demonstrates that one of these proteins which binds histidine is an indispensable part of a histidine transport system, thus confirming the relationship between transport and binding proteins, at least as far as histidine is concerned.

Materials and Methods. Strains and genetic tests: All strains used were derived from *S. typhimurium* strain LT-2 and are described in Table 1. All strains, except

TABLE 1. *Bacterial strains.*

Strain	Genotype*	D-histidine growth	HIPA-resistance	J binding protein
TA831	*hisF645*	−	Sensitive	+
TA271	*hisF645 dhuA1*	+	Supersensitive	+ (increased)
TA1014	*dhuA1*		Supersensiive	+ (increased)
TA1646	*hisF645 dhuA1 hisJ5601*	−	Sensitive	−
TA1650	*dhuA1 hisJ5601*		Sensitive	−
TA1647	*hisF645 dhuA1 hisJ5602*	−	Sensitive	−
TA1651	*dhuA1 hisJ5602*		Sensitive	−
TA1648	*hisF645 dhuA1 hisJ5603*	−	Sensitive	−
TA1652	*dhuA1 hisJ5603*		Sensitive	−
TA1649	*hisF645 dhuA1 hisJ5604*	−	Sensitive	−
TA1653	*dhuA1 hisJ5604*		Sensitive	−
TA1008	*hisF645 dhuA1 hisP5503*	−	Resistant	not assayed
TA1195	*dhuA1 hisP5503*		Resistant	+ (increased)
hisP1661	*hisP1661*		Resistant	+
TA1613	*purF145 hisHB22*	−	Sensitive	not assayed
Wild type	histidine-independent trans- ductant of TA831 or LT-2		Sensitive	+

* All *his* mutations except *hisJ* and *hisP* are defective in histidine biosynthesis. Only the histidine-requiring strains, but not their prototrophic transductants, can be assayed for D-histidine growth.

TA831 and TA271, obtained from P. E. Hartman and T. Klopotowski respectively, were constructed in this laboratory, grown, and analyzed genetically by transduction with phage P22-L4 as described earlier.[2] The selection and test for *hisP* mutations has been described previously.[2] Strains carrying *hisJ* mutations were isolated from TA271 (*hisF645 dhuA1*, a strain capable of utilizing D-histidine as the source of L-histidine) as D-histidine nonutilizers which were still sensitive to 2-hydrazino-3-(4-imidazolyl) propionic acid (HIPA)[6,7] (gift of F. A. Kuehl, Jr., of Merck, Sharp and Dohme Research Laboratories, Rahway, N.J.). TA271 was mutagenized with ICR 191,[8] then exposed to penicillin in minimal medium containing 2×10^{-5} M D-histidine: about 15% of the D-histidine nonutilizers thus obtained (assayed by radial streak[2] against 1 μmol of D-histidine) were HIPA-sensitive, i.e., *hisJ* mutants.

Column chromatography: DEAE-Sephadex (Pharmacia, Piscataway, N.J.) columns were prepared according to the Pharmacia instruction manual and equilibrated in 0.01 M Tris·HCl, pH 7.4. The shock fluid was dialyzed overnight against the equilibrating buffer and applied to the column. After washing the column with two bed volumes of equilibrating buffer, the protein was eluted with a linear NaCl gradient from 0 to 0.2 M, with a flow rate of 20 ml/hr.

Assays: The uptake of ³H-histidine (New England Nuclear Corp., Boston, Mass., 2 Ci/mmol) was determined by the "growing-cells" method[1] (about 2 μg of dry weight of cells per milliliter). The trichloroacetic acid precipitates were filtered on glass fiber filters (Gelman, Ann Arbor, Mich., type A), washed with 10% trichloroacetic acid, then with 95% ethanol, air dried, and counted in POP-toluene (Spectrafluor, Amersham/ Searle) with an efficiency of 30%. Linear initial rates of uptake are expressed as μmoles of ^{3}H-histidine incorporated per minute per gram dry weight.

Shock fluid was prepared[3] from cells harvested at a density of 5×10^8 cells/ml. The shock fluid was concentrated 10- to 60-fold by ultrafiltration (Union Carbide no. 8 dialysis tubing), and then filtered through a Millipore filter type HA, 0.45 μm (Millipore Filter Corp., New Bedford, Mass.). Histidine-binding activity was assayed by dialyzing (for 18 hr at 4°C) 0.3 ml of the shock fluid against a large volume of 0.1 M NaCl–0.01 M Tris· HCl, pH 7.4, containing 1×10^{-8} M ^{3}H-L-histidine. Small dialysis sacs were made from Union Carbide dialysis tubing (no. 8) preheated in water at 60°C for 3 hr. Aliquots of 0.2 ml from sacs and external buffer were counted in 10 ml Bray's solution[9] with an efficiency of 20%. One unit of binding activity is defined as 1 pmole of histidine bound at 1×10^{-8} M histidine. This assay allows proportionality of activity between 0.5 and 140 units of histidine-binding activity. Acid phosphatase[10] and protein concentration[11] were also measured.

Results. Mutations affecting the high-affinity histidine transport system: (1) *hisP*. This class of mutants has been described previously[2,7] as being resistant to the inhibitory histidine analogue HIPA. Their resistance is caused by a defect in the high-affinity L-histidine transport system. The *hisP* gene codes for a protein, as shown by the isolation of amber mutants in this gene.[2]

(2) *dhuA*. Histidine-requiring strains (e.g., deletion *hisF645*) are unable to utilize D-histidine as the source of L-histidine, but mutations arise at high frequency which confer the ability to utilize it. A class of these mutants allowing D-histidine growth (*dhuA*) has been isolated by Klopotowski and his collaborators.[12] They showed that the *dhuA* mutations are adjacent to the known *hisP* gene, and cause an increased D- and L-histidine uptake. The *dhuA* mutants have an increased sensitivity to HIPA (Table 1), as a consequence of the increased transport of this inhibitor.

(3) *hisJ*. These mutants have been selected (as described in *Materials and Methods*) in *dhuA1 hisF645* for the loss of the ability to grow on D-histidine, while still retaining HIPA sensitivity (Table 1).

(4) *dhuA hisP*. The *hisP* mutation has been introduced also in the *dhuA* mutant strain by selecting for resistance to HIPA. The resulting double mutants have lost the capacity to grow on D-histidine simultaneously to the loss of sensitivity to HIPA (Table 1).[12] In agreement with the properties of these double mutants, 30 *hisP* mutants (containing a histidine operon deletion) were tested and found to be unable to mutate to D-histidine utilization. Thus the *hisP* gene is required for transport of both D- and L-histidine and the analogue HIPA.

Genetic mapping: Both *dhuA*[12] and *hisP*[2] mutations were confirmed to be about 40% cotransducible with the *purF* locus, by transducing on L-histidine medium the recipient TA1613 (*purF145 hisHB22*), with phage prepared either on TA271 (*hisF645 dhuA1*) or on 30 independently isolated *hisP*-carrying strains. The *Pur+ His−* recombinants were tested for D-histidine growth and HIPA-resistance respectively.

Moreover, the *dhuA* and *hisP* loci can be shown to be very close together by exposing TA1613 (on L-histidine medium) to phage grown on TA1008 (*hisF645 dhuA1hisP5503*). Only 2% of the *Pur+ His−* recombinants inherit *dhuA1* alone from the donor: this rare type requires a recombinational event *between dhuA* and *hisP* because the *hisP* mutation eliminates D-histidine growth.

In a similar cross the *dhuA* and *hisJ* mutations can be shown to be closely linked. Strain TA1613 was exposed (on L-histidine medium) to phage grown on *dhuA hisJ* double mutants TA1646, TA1647, TA1648, and TA1649. Only 2% of the *Pur+ His−* recombinants inherit the *dhuA* mutation without *hisJ*, i.e., are D-histidine growers. The *hisP*, *hisJ*, and *dhuA* sites are therefore very closely linked to each other.

Histidine-binding proteins in the wild type and in mutant strains: By giving bacterial cells a mild osmotic shock treatment,[3] histidine-binding activity was liberated in the shock fluid. Fractionation of the wild type shock fluid on DEAE-Sephadex revealed that 95% of the binding activity (at 10^{-8} M L-histidine) was eluted as a peak (J protein) at 0.04 M NaCl. A second peak (tentatively designated as K protein), representing about 5% of the activity, was eluted at 0.15 M NaCl.

The histidine-binding activity in the shock fluid of the various mutants is shown in Table 2. It is clear that the level of J binding protein is increased in

TABLE 2.   *Levels of J binding protein.**

Strain	Genotype	No. of expts.	Units of J Binding Protein per:		
			Unit of acid phosphatase	Protein released (mg)	Dry weight of bacteria (g)
	Wild type	6	8 (1)†	42 (1)	302 (1)
TA1014	*dhuA1*	5	42 (5)	165 (4)	1,520 (5)
TA1650	*dhuA1 hisJ5601*	3	<0.45 (<0.06)	<2.4 (<0.06)	<26 (<0.08)
TA1651	*dhuA1 hisJ5602*	3	<0.45 (<0.06)	<2.4 (<0.06)	<26 (<0.08)
TA1652	*dhuA1 hisJ5603*	4	<0.45 (<0.06)	<2.4 (<0.06)	<26 (<0.08)
TA1653	*dhuA1 hisJ5604*	1	<0.45 (<0.06)	<2.4 (<0.06)	<26 (<0.08)
hisP1661	*hisP1661*	4	8 (1)	37 (0.9)	375 (1)
TA1195	*dhuA1 hisP5503*	4	39 (5)	158 (4)	3,592 (12)

* In all strains, except the *hisJ* containing strains, 95% or more of the binding activity has been shown by chromatography to be caused by the J binding protein. Therefore all quantitative assays in this table were done on the total shock fluid. Because of variability in the efficiency of shocking, the data are presented relative to the weight of protein released and the activity of acid phosphatase released, in addition to the weight of starting bacteria. The results averaged in this table were from completely independent experiments.

† The numbers in parentheses are relative to the wild type.

strains containing a *dhuA* mutation, and absent in strains containing a *hisJ* mutation. The binding protein from *dhuA1* has the same chromatographic properties and binding affinity ($K_D \sim 2 \times 10^{-7}$ M at 4°C) for histidine as the protein from the wild type. We exclude the possibility that the *hisJ* and *dhuA* mutations have caused a change in the total amount of protein released by shocking because all the strains assayed release about the same amount of either total protein or acid phosphatase per gram dry weight. Introduction of a *hisP* mutation in either the wild type or in a *dhuA*-containing strain (see *hisP1661* and TA1195 in Table 2), does not affect the level of J binding protein. The level of the K binding protein appears to be unchanged in the strains assayed in Table 2, but these experiments will be performed on a larger scale for accurate quantification.

Numerous D-histidine utilizing revertants of *hisJ* mutants have been isolated and all have recovered the J binding protein. The binding protein from one of these revertants[13] appears to be temperature sensitive, as compared to the wild type protein (from *dhuA1*).   The mutant protein loses about 61% of the histidine-binding activity after being heated at 100°C for 50 min, while the wild type protein loses about 20% of the activity (average value of several experiments). Further analysis of this mutant protein is in progress.

Uptake of L-histidine in the wild type and in transport mutants: The rate of uptake and the affinity of the transport system for L-histidine in the wild type, in *dhuA1 hisJ5601* (missing the J binding protein), and in *hisP1661* (with a normal level of J binding protein) are shown in Figures 2A and 2B.   Both *hisJ* and *hisP*

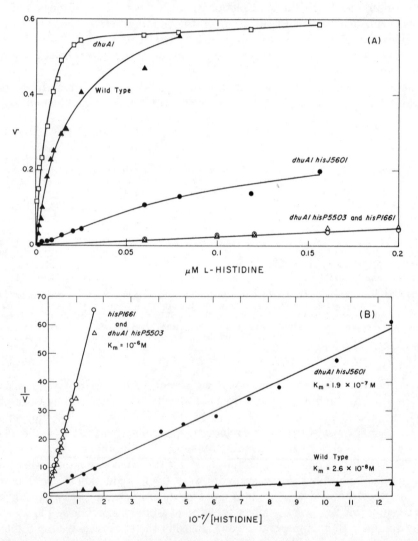

Fig. 2.—(*A*) Initial rates of uptake of L-histidine as a function of histidine concentration. (*B*) Lineweaver-Burk plots of L-histidine uptake.   The K_m for *dhuA1* is 6.6×10^{-9} M and has been calculated, together with the K_m for the wild type, from a separate plot at an appropriate scale. *dhuA1* (TAIO14, □); wild type (▲); *dhuA1 hisJ5601* (TA1650, ●); *dhuA1 hisP5503* ((TA1195, △); *hisP1661* (○).

mutant strains are defective in uptake *although to a different extent*.   The affinities of these residual transport activities for L-histidine are 2×10^{-7} M and 10^{-6} M respectively.   It is clear that these differences in residual rates and affinities are significant and characteristic of completely defective mutants in *hisJ* or *hisP*, as the same results were obtained upon analysis of three *hisJ* frameshift mutants and two *hisP* mutants, one of which is an amber mutant.   We conclude that the loss of the J binding protein in *hisJ* mutants results in the loss of a transport component, J, with a very high affinity ($K_m \sim 10^{-8}$ M, as assayed in either the wild type or in the *dhuA* mutant).

Despite the loss of this component, *hisJ* mutants can still transport histidine very efficiently ($K_m = 2 \times 10^{-7}$ M, Fig. 2), presumably through additional transport components.   Thus, the K_m of 2×10^{-7} M in *hisJ* mutants is ascribed to a component tentatively designated K.   Mutations eliminating transport through K *only* have not yet been obtained.   Mutations in *hisP* have apparently eliminated transport through both the J and the K components: this is indicated by the finding that the sum of the residual rates in *hisJ* and *hisP* mutants is less than the wild type rate (Fig. 2A).   Therefore, J and K are acting in parallel, and apparently both require the *hisP* product in order to transport histidine.

The *dhuA1* mutant, which increased binding protein fivefold, shows a marked increase in transport (Fig. 2A), and has a higher affinity ($K_m = 6.6 \times 10^{-9}$ M) than the wild type ($K_m = 2.6 \times 10^{-8}$ M).   If the wild type rate is truly the sum of the K and J activities, the preponderance of the J in the *dhuA* mutant could be responsible for this shift in the apparent affinity.

A double mutant, *dhuA1 hisP5503*, has the same rate of transport and poor affinity for histidine uptake as *hisP* mutant alone.

Additional transport systems of low affinity: As shown in Figures 2A and 2B, *hisP* mutants have a greatly decreased rate of L-histidine uptake.   The remaining activity still gives Michaelis-Menten kinetics with a K_m of about 10^{-6} M. This has been shown by kinetic and genetic analysis to be caused by at least three more transport systems of low affinity, one of which is the general aromatic permease.[1]   A complete study of these systems will be reported at a later date.

Discussion.   It has been demonstrated previously that mutants in the *hisP* gene are lacking the high-affinity ($K_m = 10^{-8}$ M) transport system for L-histidine in *S. typhimurium*.[2,7]   In this paper we show: (a) that this is a complex transport system made up of three components distinguishable kinetically and genetically; (b) that a histidine-binding protein, J (coded for by the *hisJ* gene), is one of the components of this system; (c) a number of aspects of D- and L-histidine transport through these components.   These observations led us to propose the scheme of Figure 1.

J binding protein is involved in transport: Osmotic shock of the bacteria releases from the cell surface two proteins that bind L-histidine: J (about 95% of total) and K (about 5% of total).   Mutants (*dhuA*), selected for increased utilization of D-histidine, have a fivefold increase in the J binding protein, thereby causing an increased transport for both D- and L-histidine, and an increased sensitivity to the histidine analogue HIPA.   The *dhuA* mutation, which maps adjacent to the *hisJ* gene, increases the level of the J protein, possibly by affecting the regulation of its synthesis. In agreement with this, the J protein appears to be the same in *dhuA1* and in the wild type on the basis of affinity for histidine and chromatographic properties.

The *hisJ* mutants, which were selected from strains carrying a *dhuA* mutation, have lost the ability to utilize D-histidine and completely lack the J binding protein. As a consequence they have a decreased transport for L-histidine and have lost the extra sensitivity to HIPA.

Further evidence that the *hisJ* gene is the structural gene for the J binding protein, rather than a regulatory gene, is that a D-histidine-utilizing revertant of a *hisJ* mutant appears to have an altered J binding protein.

The large difference in affinity for L-histidine between the transport *in vivo* ($K_m = 6.6 \times 10^{-9}$ M in *dhuA1*) and the J binding protein ($K_D = 2 \times 10^{-7}$M) can be rationalized as caused by the conditions of the binding assay. This is done at 4°C on a protein removed from its *in vivo* environment and under arbitrary salt and pH conditions.

hisP protein is necessary for the J binding protein to function in transport: *hisP* mutants selected by HIPA resistance (presumably unable to transport HIPA) completely lack high-affinity transport for L-histidine. These mutants have normal levels of histidine-binding proteins that, because of the defect in the *hisP* protein, are not functional in the *in vivo* transport. That the *hisP* gene codes for a protein has been deduced by the finding of *hisP* amber mutants.[2]

A functional *hisP* protein is also necessary for the transport mediated through the increased level of the J binding protein in *dhuA* strains. Thus, a *dhuA hisP* double mutant still has the expected fivefold-increased level of binding protein associated with the *dhuA* mutation, yet it lacks the high-affinity transport of L-histidine. This double mutant no longer utilizes D-histidine, nor is it HIPA sensitive, which gives additional evidence that the J binding protein is nonfunctional in transport in the absence of the *hisP* protein. In accordance with our interpretation of the nature of this double mutant, it has been found that *hisP* mutants will not mutate to D-histidine utilization.

hisP protein is necessary for the function of another transport component, K: All *hisJ* mutants have residual high-affinity L-histidine transport ($K_m = 2 \times 10^{-7}$ M), unlike the *hisP* mutants which have no high-affinity L-histidine transport. We define the residual activity in *hisJ* mutants as the K component of transport, which we deduce also to be dependent on a functional *hisP* protein, because *hisP* mutations eliminate this activity. The K component must also transport HIPA, as *hisJ* mutants are still HIPA sensitive, while *hisP* mutants are not. In support of the existence of two components J and K, both dependent on a functional *hisP* protein, is the fact that all HIPA-resistant mutants selected from the wild type are *hisP* mutants. One possibility is that the second, minor, L-histidine-binding protein, which is still present in *hisJ* and *hisP* mutants, is the K component. We have selected for HIPA resistance in *hisJ* mutants and these double mutants are being analyzed for the K binding protein to see whether they are *hisJ hisK* or *hisJ hisP* double mutants.

Other transport systems for L-histidine: The *hisP* mutations eliminate all high-affinity histidine transport and the Michaelis constant of the remaining transport is about 10^{-6} M.   This remaining transport is completely inhibited by aromatic amino acids, unlike the *hisP* system, and has been resolved into at least three components, one of which is the previously described *aroP* system.

We thank B. N. Ames for his constant interest and advice, T. Klopotowski for discovering and supplying us with the key strain, TA271, and M. A. Liggett for excellent technical assistance in the kinetic experiments.

* Supported by USPHS grant AM12121, to C. E. Ballou, whom we thank for support.

† Requests for reprints may be addressed to Dr. G. F. Ames, Biochemistry Department, University of California, Berkeley, Calif. 94720.

‡ Predoctoral trainee.

[1] Ames, G. F., *Arch. Biochem. Biophys.*, **104**, 1 (1964).

[2] Ames, G. F., and J. R. Roth, *J. Bacteriol.*, **96**, 1742 (1968).

[3] Nossal, N. G., and L. A. Heppel, *J. Biol. Chem.*, **241**, 3055 (1966); Heppel, L. A., *Science*, **156**, 1451 (1967).

[4] Pardee, A. B., and L. S. Prestidge, these PROCEEDINGS, **55**, 189 (1966); Anraku, Y., and L. A. Heppel, *J. Biol. Chem.*, **242**, 2561 (1967); Piperno, J. R., and D. L. Oxender, *J. Biol. Chem.*, **241**, 5732 (1966).

[5] Pardee, A. B., *Science*, **162**, 632 (1968); Kaback, H. R., *Ann. Rev. Biochem.*, in press.

[6] Abbreviations used: HIPA, 2-hydrazino-3-(4-imidazolyl) propionic acid; K_m, Michaelis constant; K_D, dissociation constant.

[7] Shifrin, S., B. N. Ames, and G. F. Ames, *J. Biol. Chem.*, **241**, 3424 (1966).

[8] Ames, B. N., and H. J. Whitfield, *Cold Spring Harbor Symposia on Quantitative Biology*, vol. 31, (1966), p. 211; Oeschger, N. S., and P. E. Hartman, *J. Bacteriol.*, **101**, 490 (1970).

[9] Bray, G. A., *Anal. Biochem.*, **1**, 279 (1960).

[10] Neu, H. C., *J. Biol. Chem.*, **242**, 3896 (1967).

[11] Lowry, O. H., N. J. Rosenbrough, N. C. Farn, and R. J. Randal, *J. Biol. Chem.*, **193**, 265 (1951).

[12] Krajewska-Grynkiewicz, K., W. Walczak, and T. Klopotowski, *J. Bact.*, in preparation.

[13] This revertant, induced by ICR 191 in a strain carrying a *hisJ* mutation which was also induced by ICR 191, does not grow on D-histidine as well as the *dhuA* grandparent. Presumably it is a frameshift revertant of the "+ −" type in *hisJ*.

Copyright © 1972 by the American Society of Biological Chemists, Inc.
Reprinted from *J. Biol. Chem.*, **247**, 4309–4316 (1972)

27

The Histidine-binding Protein J Is a Component of Histidine Transport

IDENTIFICATION OF ITS STRUCTURAL GENE, hisJ*

(Received for publication, January 17, 1972)

GIOVANNA FERRO-LUZZI AMES AND JULIA E. LEVER‡

From the Department of Biochemistry, University of California, Berkeley, California 94720

SUMMARY

The histidine-binding protein J, previously shown to be involved in histidine transport in *Salmonella typhimurium* (AMES, G. F., AND LEVER, J. (1970) *Proc. Nat. Acad. Sci. U. S. A.* 66, 1096), is shown unequivocally to be the product of the *hisJ* gene. A *hisJ* mutant with an altered J protein and a correspondingly altered histidine transport has been isolated and characterized. The J protein from this strain has an increased temperature sensitivity, besides having altered chromatographic and electrophoretic mobilities. The in vivo effect of the altered J protein is expressed as an increased temperature sensitivity of histidine transport. Our data indicate that the *hisJ* gene is the structural gene for the J protein and that the J protein is an obligatory component of histidine transport.

As expected, there is an excellent correlation between the specificity of transport in the wild type strain and the specificity of binding of the wild type J protein for a variety of amino acids, amino acid analogues, and inhibitors.

Salmonella typhimurium transports L-histidine through at least five permeases with different patterns of specificity and affinity (1–3). The histidine permease with the highest affinity (J-P permease, with a K_m of about 10^{-8} M) has been shown to be composed of at least two proteins, J and P (3), and has been analyzed kinetically, biochemically, and genetically. The J protein is a periplasmic histidine-binding protein, released by osmotic shock (3, 4) and coded for by the *hisJ* gene (3). The P protein is essential for histidine transport by the J protein and is coded for by the *hisP* gene (2, 3) but has not been identified biochemically. The P protein is also necessary for the functioning of another histidine permease (the K-P permease, with a K_m of about 10^{-7} M), which works in parallel to the J-P system (3).

We demonstrated earlier a direct correlation between the

activity of the histidine permease and the levels of the J protein (3). Strains with a mutation in the *hisJ* gene are defective in histidine transport and correspondingly lack the J protein. Strains with a mutation in the *dhuA* site, which is thought to be a control locus for histidine transport, have elevated histidine transport and are correspondingly elevated in the J protein. The proof that *hisJ* mutations are in the structural gene for the J protein, rather than in a control locus for the genes of histidine transport, was obtained by the characterization of a temperature-sensitive mutation which simultaneously alters both the J protein and transport through the J-P permease. This mutant is described in detail in this paper together with other evidence on the role of the J protein in transport. The J-P permease also transports other substances (among them D-histidine, L-arginine, D-2-hydrazino-3-(4-imidazolyl)propionic acid) with lower affinity; the correlation between the specificity of transport and the binding specificity of the J protein is presented.

The purification and some of the properties of the wild type J protein have been described (4).

EXPERIMENTAL PROCEDURE

Chemicals—All chemicals from commercial sources were of the highest purity available. Amino acids were obtained from Sigma, Calbiochem, and Nutritional Biochemicals Corporation. DL-Lysine was obtained from Fox Chemical Company. *O*-Diazoacetyl-L-serine was obtained from the National Cancer Institute (Cancer Chemotherapy National Service Center, Bethesda, Md.).

D-HIPA[1] was synthesized from L-histidine (5). The product was homogeneous chromatographically, using paper chromatography in 1-propanol-1 N HCl (3:1, v/v), and 1-propanol-1 N NH₃ (3:1, v/v), with detection by imidazole spray (6) and ninhydrin. L-[³H]Histidine was obtained from New England Nuclear Corporation, with a specific activity of 2.4 Ci per mmole. It was stored at −18° in 50% ethanol. Histidine concentration was verified by amino acid analysis, and radiochemical purity was monitored by paper chromatography in the two 1-propanol systems above, using radioautography for detection.

Possible histidine contamination in amino acids and analogues used in the competition experiments was monitored using a Beckman/Spinco model 121 amino acid analyzer with expanded range. Imidazole pyruvic acid (Calbiochem), L-histidinol

* This research was supported by United States Public Health Service Grants AM12121 and AM12092.

‡ Present address, Department of Medicine, BSB 2034, University of California, San Diego, La Jolla, California 92037. Part of these studies are taken from a thesis submitted in partial fulfillment of the requirements for the degree of Doctor of Philosophy in Biochemistry at the University of California, Berkeley.

[1] The abbreviations used are: D-HIPA, D-2-hydrazino-3-(4-imidazolyl)propionic acid; J^{ts} protein, temperature-sensitive J protein; azaserine, *O*-diazoacetyl-L-serine.

(Cyclo Chemical Company), and L-histidylhistidine (Mann) were purified using descending preparative paper chromatography with Whatman No. 3MM paper, in a solvent system of equal parts collidine-2,4-lutidine-H$_2$O, made 2% in diethylamine freshly before use. Compounds were eluted from the paper with water and lyophilized to dryness. Solutions of these compounds were standardized using a quantitative diazo reaction (7). L-Arginine was freed of contaminating histidine (0.01%) by column chromatography, using AG-50 W-X4 resin (Bio-Rad) in the NH$_4$+ form (8).

Assays—L-Histidine uptake was measured by the "growing cells" method (1) (10 to 20 µg, dry wt, of cells per ml), with the modification for the washing of the samples as already described (3). The J protein was purified, assayed, and characterized as described (3, 4). [³H]Histidine binding activity was assayed either by the dialysis or by the filtration method (3, 4).

Bacterial Strains and Media—All strains were derived from *S. typhimurium* strain LT-2 and are listed in Table I. Conditions for growth have been described (2). Except for TA271 which was obtained from T. Klopotowski (9), all strains were constructed in this laboratory and analyzed genetically by transduction with phage P22 *int-4* as described earlier (2). Growth on D-histidine (1 µmole on disc) and sensitivity to D-HIPA (in the presence of L-histidinol for histidine auxotrophs) were assayed on Petri plates by the radial streak method (2).

Genetic Analysis of Bacterial Strains—Strains TA271 (*dhuA1 hisF645*) and TA1014 (*dhuA1*) have been described (3). Strain TA1768 (*dhuA1 hisJ5617 hisF645*) was obtained by mutagenizing TA271 with the frameshift mutagen, ICR-191 (10), and then selecting for a mutation in *hisJ* (such mutation renders the strain unable to grow on D-histidine, but still D-HIPA sensitive), as described previously (3). Mutation *hisJ5617* so obtained was shown to be closely linked (about 98%) to *dhuA1* by the same methods used for other *hisJ* mutations (3). Strain TA1789 (*dhuA1 hisJ5617 hisJ5620 hisF645*; briefly mentioned in (3)) was isolated as a revertant of TA1768, which was capable of growing on D-histidine at room temperature, but not at 40°.

Strains TA1771 and TA1791 were obtained by transducing the

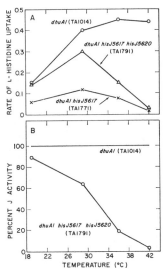

Fig. 1. Temperature sensitivity of L-histidine uptake in the *hisJ* revertant TA1791. *A*, initial rates of uptake of L-histidine (micromoles per min per g, dry wt) as a function of temperature. *B*, L-histidine uptake through the J component as a function of temperature. The initial rates of uptake at each temperature by TA1771 (*i.e.* through the K component (3)) are subtracted from the corresponding rates of uptake by TA1014 and TA1791. The rates thus corrected (uptake through the J component) are expressed as percentages, taking the corrected rate of uptake by TA1014 as 100% at each temperature. The L-[³H]histidine concentration is 4 × 10⁻⁸ M.

histidine auxotrophs TA1768 and TA1789, respectively, to prototrophy with wild type phage.

RESULTS

Mutant Strain with Qualitative Alteration of Both J Protein and Histidine Transport—Strain TA1789 (*dhuA1 hisJ5617 hisJ5620 hisF645*) is a revertant of TA1768 (*dhuA1 hisJ5617 hisF645*) and was chosen for further study because of its temperature-sensitive phenotype. It carries two mutations in *hisJ*, both induced by the frameshift mutagen ICR-191 (10): the initial mutation, *hisJ5617*, causes loss of both the J component of transport and the J protein, and *hisJ5620*, a mutation at a second site in the *hisJ* gene, restores partial J protein function.

As a consequence of the double mutation in the *hisJ* gene, strain TA1789 (and its prototrophic derivative TA1791) has a temperature-sensitive J component of transport which can be analyzed by measuring either L-histidine uptake or the ability to grow on D-histidine. Concomitantly with these altered properties of transport, TA1789 (and TA1791) produces a J protein with several altered properties including an increased temperature sensitivity.

Genetic Mapping—Strain TA1789 (*dhuA1 hisJ5617 hisJ5620 hisF645*) was shown by the following genetic tests to contain the mutation, *hisJ5620*, responsible for the second site reversion, in the *hisJ* gene. Mutation *hisJ5620* was first shown to be cotransducible (42%) with *purF145*, by growing phage on TA1789

TABLE I

Bacterial strains

Strain	Genotype[a]	D-Histidine growth	D-HIPA resistance	J protein
TA271	*dhuA1 hisF645*	+	Supersensitive	Elevated
TA1014	*dhuA1*		Supersensitive	Elevated
TA1768	*dhuA1 hisJ5617 hisF645*	−	Sensitive	Absent
TA1771	*dhuA1 hisJ5617*		Sensitive	Absent
TA1789	*dhuA1 hisJ5617 hisJ5620 hisF645*	+[b]	Sensitive	Elevated[b]
TA1791[c]	*dhuA1 hisJ5617 hisJ5620*		Sensitive	Elevated[b]

[a] Only the histidine-requiring strains, but not their prototrophic transductants, can be assayed for D-histidine growth. Only the *hisF645* causes a defect in histidine biosynthesis.

[b] Temperature-sensitive phenotype.

[c] Preliminary results with this strain have been published previously (3).

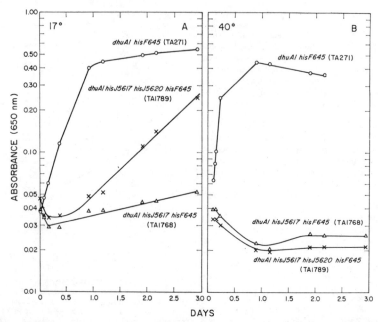

FIG. 2. Growth on D-histidine (*A*) at 17° and (*B*) at 40°. All cultures were grown overnight at 17° in minimal medium containing 7×10^{-5} M L-histidine; this histidine concentration is exhausted during growth of the cultures to stationary phase. At zero time each culture is diluted 50-fold in minimal medium containing 2×10^{-5} M D-histidine and grown at the indicated temperature. The TA271 culture stops growing at optical density of about 0.500 because the D-histidine concentration is limiting.

and transducing the double mutant *purF145 hisF645* to Pur⁺ on L-histidine medium and showing that 47 out of 111 of the Pur⁺ His⁻ recombinants had a temperature-sensitive phenotype for D-histidine growth. The double mutation causing temperature sensitivity was then placed very close to *dhuA1* (90% cotransducible) by the following cross. Phage grown on TA1789 was used to transduce the double mutant *purF145 hisF645* to growth at 23° on D-histidine (0.3 mM), in the presence of adenine and thiamine (0.2 mM and 0.02 mM, respectively; adenine plus thiamine are required by strains containing the *purF145* mutation); thus, the selection was only for *dhuA1*, and not for the temperature-sensitive mutation or for Pur⁺. The recombinants growing on these plates were then tested for D-histidine growth at 23° and at 40°: recombinants growing at both temperatures inherited only the *dhuA1* mutation, those growing only at 23° inherited also the temperature-sensitive mutation. Ninety per cent (72 out of 80) of the D-histidine-growing recombinants were temperature sensitive; the remaining 10% grew well at either temperature, confirming the presence of *dhuA1* in this strain. Thus mutation *hisJ5620* is 42% cotransducible with *purF145* and 90% cotransducible with *dhuA1*: these results firmly establish its position in the histidine permease cluster.[2]

[2] It has been recently found (Govons and Ferro-Luzzi Ames, unpublished results) that the linkage between *purF145* and the permease cluster is 2% when *purF145* is the unselected marker in the transduction experiments, as opposed to a linkage of about 40% when Pur⁺ is the selected marker. These data could be explained if *purF145* were a large deletion.

Studies on Histidine Transport in Mutant Strain

Temperature Sensitivity of L-Histidine Uptake—Fig. 1, *A* and *B*, show that increasing temperature causes a decay in the rate of uptake of labeled L-histidine through the temperature-sensitive J component in TA1791 (*dhuA1 hisJ5617 hisJ5620*).

The assays were done by the "growing cells" method (1) on cultures which had been growing at the assay temperatures for many generations and thus had the level of transport compatible with each temperature.[3] The initial linear rates of uptake were determined at each temperature and are plotted in Fig. 1*A*. The rates in TA1014 (*dhuA1*) and TA1791 have then been corrected (by subtraction) for the rates due to the K-P component of transport (3), which was simultaneously assayed in TA1771 (*dhuA1 hisJ5617*) at each temperature. The rates in TA1791 were then expressed as a percentage of the unaltered J activity, taking the corrected values of *dhuA1* as 100% (Fig. 1*B*).

[3] TA1791 cells grown at 29° for many generations and shifted to higher temperatures, have a lower rate of uptake than TA1014 cells treated in an identical way. For example, after 30 min at 43° TA1791 has only 1% of the TA1014 activity. This demonstrates that it is the activity, rather than the synthesis of the J protein, that is temperature sensitive. We have preferred to show that transport is temperature sensitive as in Fig. 1, because both strains lose some transport activity when they undergo a sudden temperature shift. This loss is probably due to the temperature shock inflicted upon other components of transport or protein synthesis when transport is assayed in growing cells under physiological conditions; however, the loss is always larger in the temperature-sensitive mutant (TA1791) than in TA1014.

TABLE II

Level of J protein in TA1791

Cells (1 to 14 g, dry wt; 58 mg, dry wt, for TA1771) were harvested in midexponential phase, and treated by the Anraku procedure (11) or by the Nossal and Heppel procedure (12), which gave comparable results on this scale. Binding activity in the shock fluid was assayed by dialysis (3).

Strain	Genotype	Growth temperature	Histidine binding activity units per		
			g, dry wt, cells	*phosphatase unit*	*mg protein*
TA1014	*dhuA1*	30°	2275	20	89
		37	2170	24	122
TA1771	*dhuA1 hisJ5617*	30	18	0.32	2.0
		37	11	0.13	1.4
TA1791	*dhuA1 hisJ5617 hisJ5620*	30	2331	16	91
		37	645	7	52

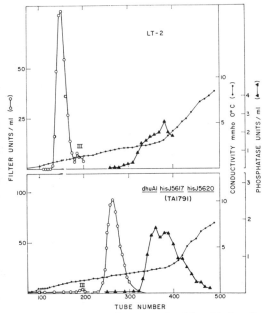

FIG. 3. DEAE-cellulose chromatography of J protein from the wild type and from the *hisJ* revertant TA1791. Shock fluids from 160 g, wet wt, of wild type (LT-2) cells and 35 g, wet wt, of TA1791 (*dhuA1 hisJ5617 hisJ5620*) cells (late exponential phase), prepared by the Anraku procedure (11) are chromatographed successively on the same DEAE-cellulose column under identical conditions (4) with a salt gradient from 0 to 0.2 M NaCl. Fractions (15 ml) are collected. The histidine binding activity (*ordinates*) is estimated by the filter assay (4). The small peaks labeled *III* represent the histidine binding Peak III.

Temperature Sensitivity of D-Histidine Growth—It has been demonstrated previously (3, 9) that growth on D-histidine (as a source of L-histidine for a histidine auxotroph) requires a mutation at the *dhuA* site and an intact J protein. This is illustrated

in Fig. 2. TA831 (*hisF645*) is the parent strain which requires histidine because of a deletion in the histidine operon; it does not grow on D-histidine. TA271 (*dhuA1 hisF645*) grows well on D-histidine at both 17° and 40°. Strain TA1768 (*dhuA1 hisJ5617 hisF645*) which has a mutation in *hisJ*, is unable to grow on D-histidine at either temperature, as expected from its lack of J protein. The significance of Fig. 2 is in the growth pattern of the temperature-sensitive strain, TA1789 (*dhuA1 hisJ5617 hisJ5620 hisF645*); this strain is able to grow at 17° (although not as well as the grandparent TA271), but it is completely unable to grow on D-histidine at 40°.

Evidence for Altered J Protein in Mutant Strain

Histidine Binding Activity Levels—Binding activity levels in the shock fluids from TA1791 (*dhuA1 hisJ5617 hisJ5620*) and from the control strains TA1014 (*dhuA1*) and TA1771 (*dhuA1 hisJ5617*) are given in Table II. Since the J protein from TA1791 is heat labile, cells were grown at both 37° and 30° for comparison. Strain TA1771 has no J protein activity, as previously demonstrated (3) for bacteria with mutations in the *hisJ* gene; the small binding activity detectable is significant and is due to other histidine-binding proteins (3, 4). The *dhuA1* mutation causes elevated levels of J protein and is present in all strains. TA1791 shows the normal activity levels expected for a *dhuA1* strain when grown at 30°, but it has greatly decreased activity (28% residual activity) when grown at 37°.

Chromatographic Properties—The chromatographic properties of J protein from TA1791 (J[ts] protein) differ markedly from those of wild type J protein isolated from either LT-2 or *dhuA1* as illustrated in Fig. 3.[4] A comparison of the elution profiles on DEAE-cellulose shows that the J[ts] protein from TA1791 elutes at a higher salt molarity than the wild type J protein. The elution pattern of this column is highly reproducible for different shock fluid preparations and for repacked columns. The position of the wild type protein was identical in four experiments, while the J[ts] protein from TA1791 was eluted at a higher salt molarity in two independent preparations. The positions of acid phenyl phosphatase activity (which is partially resolvable into two peaks) and of histidine binding activity Peak III, as well as the conductivity measurements, are highly reproducible markers. The reason for such a large difference in mobility between two proteins with the same molecular weight and isoelectric point is under investigation. The J[ts] protein presumably differs from the wild type protein in a sequence of several amino acids, because TA1791 was obtained by reversion of a frameshift mutant with ICR-191.

Hydroxylapatite chromatography of wild type and TA1791 J proteins using gradient elution from 1 to 100 mM sodium phosphate (pH 7.0) also revealed a difference in mobility. Wild type J protein was eluted at 40 mM phosphate, whereas J[ts] protein was eluted at 10 mM phosphate, using the same batch of adsorbant.

Both proteins are eluted similarly on Sephadex G-100 indicating a similar molecular weight.

Dissociation Constant for Histidine—Fig. 4 shows that the J[ts] protein from TA1791 has a higher dissociation constant for

---

[4] J protein isolated from *dhuA1* has been shown to be identical with that isolated from the wild type (4). Therefore the name of wild type protein refers to J protein from either *dhuA1* or wild type.

histidine (K_D = 2 μM) than wild type J protein (K_D = 0.11 μM). These values were determined with the filtration assay at room temperature. A difference in affinity for histidine is also observed using the dialysis assay, which is performed at 4°: under these conditions the J protein from TA1791 has a K_D = 2.5 μM and the wild type J protein has a K_D = 0.14 μM (4).

Temperature Stability—The greater heat lability of the J^{ts} protein (from TA1791) as compared with the wild type protein is shown in Fig. 5. Binding proteins are remarkably heat stable. The activity remaining after 10 min at 100° is about 70% for the wild type binding protein and 13% for the mutant protein.[5] Inactivation of a mixture of the two proteins is intermediate between that of revertant and wild type proteins alone.

The instability of the J^{ts} protein is also indicated by a faster decay during storage at 4° than the wild type protein, which is stable for several months. We have had difficulty obtaining pure J^{ts} protein because of this instability. The J protein from either strain is stable to quick-freezing in ethanol-Dry Ice and storage at −10°.

Disc Gel Electrophoresis—Disc gel electrophoresis differentiates wild type from mutant J protein. The pure wild type J protein gives a single band in the low pH system (4). A preparation of J^{ts} protein that was 80% pure shows only one major band which moves faster than the wild type protein. A mixture of the two proteins gives two bands corresponding in position to each of the single bands.

Isoelectric pH—J proteins from *dhuA1* and from TA1791 have the same isoelectric pH of 5.5.

Immunology—Shock fluid from TA1791, grown either at 30° or at 37°, cross-reacts with antiserum prepared against the wild type J protein. Shock fluid from several *hisJ* mutants lacking the binding protein contained no cross-reacting material. This demonstrates that the protein produced by TA1791 is not an altogether different protein.

Correlation between Specificity of in Vivo Histidine Transport and Isolated J Protein

The J protein has been definitively demonstrated to be a component of high affinity histidine transport by the use of mutants. Consistent with this the binding properties of the J protein *in vitro* correlate well with the properties of the transport system *in vivo* with respect to D- and L-histidine, arginine, lysine, citrulline, ornithine, HIPA, imidazole pyruvic acid, and azaserine.

In Vivo Specificity

Table III shows that L-arginine inhibits L-[³H]histidine uptake. Essentially complete inhibition is obtained at the highest arginine concentration used, which suggests that not only the J-P permease, but also the K-P permease is inhibited by arginine. Competitors of the transport system can also be identified by the inhibition of growth on D-histidine. D-Histidine is itself a much poorer substrate of the transport system than L-histidine (9). Arginine, citrulline, lysine, and ornithine completely inhibit growth of *dhuA1 hisF645* on D-histidine (Fig. 6). Inhibition of growth by these compounds only occurs when D-histidine is the substrate because the J protein is an obligatory step in D-histidine

[5] The discrepancy between these residual activities and those published in Reference 3 could be due to the higher purity of the proteins used here.

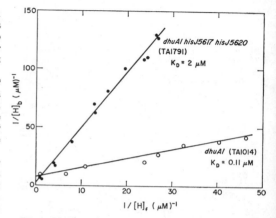

FIG. 4. Binding of histidine by the J protein from *dhuA1* and from the *hisJ* revertant TA1791. Binding is assayed by the filter assay (4). J protein from *dhuA1* had been purified by DEAE-cellulose and hydroxylapatite chromatography. J protein from TA1791 had been chromatographed on DEAE-cellulose.

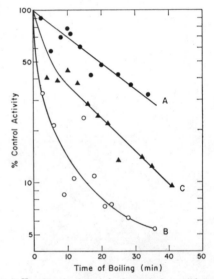

FIG. 5. Heat inactivation of the J protein from wild type and from the *hisJ* revertant TA1791. Solutions of proteins partially purified by DEAE-cellulose chromatography and of similar specific activity are heated at 100° in 0.01 M sodium phosphate, pH 7. At various times 100-μl aliquots are removed and added to 20 μl of 5 μM [³H]histidine in 0.05 M sodium phosphate, pH 7, chilled in ice water for 1 to 2 min, and assayed by filtration. *A*, J protein from the wild type (0.05 mg per ml, 51.5 filter units per ml). *B*, J protein from *hisJ* revertant TA1791 (0.12 mg per ml, 130 filter units per ml). *C*, a mixture of equal parts of *A* plus *B*.

uptake, while L-histidine is also transported by other, parallel systems (1–3).

Histidine-requiring strains are not able to transport sufficient imidazole pyruvic acid to serve as a source of L-histidine, unless

L-Arginine competition of L-histidine uptake

Bacteria (wild type) were added to a mixture of L-[³H]histidine (0.02 μM) and L-arginine (at the indicated concentrations). The arginine was shown to contain less than 0.005% contaminating histidine using an amino acid analyzer (Beckman/Spinco model 121).

L-Arginine	Rate of uptake	Control
μM	*μmole/g, dry wt/min*	*%*
0	0.36	100
10	0.097	27
100	0.013	4
500	0.0053	2

TABLE IV

Specificity of the J protein

Pure J protein from either wild type or *dhuA1* was used. Binding was assayed by filtration (4). All values are averages of duplicate assays. The L-[³H]histidine concentration was 1 μM when the additions were in 10- and 1,000-fold excess; 0.1 μM when the additions were in 10,000-fold excess. All compounds tested, except D-histidine and D-HIPA, contained less than 0.01% contaminating histidine, either after repurification or as available commercially. L-Histidine contamination in the unlabeled D-histidine was shown to be less than 0.1% by a radioisotope dilution assay of L-[³H]-histidine binding by one of the other histidine-binding proteins. D-HIPA could not be checked for small histidine contamination because of technical difficulties; it is possible that D-HIPA contains as much as 0.1% D- or L-histidine.

Addition	% Control activity		
	Molar ratio (addition/histidine)		
	10	1,000	10,000
	%		
L-Arginine	74	0	0
L-Lysine	84	34	13
DL-Lysine	*a*	*a*	24
L-Citrulline	*a*	100	55
DL-Ornithine	*a*	100	61
D-HIPA	91	67	20
D-Histidine	*a*	81	35
Azaserine	*a*	*a*	28
Imidazole pyruvic acid	*a*	*a*	83

a Not assayed.

they also contain a *dhuA* mutation (9). As is the case for D-histidine, growth on imidazole pyruvic acid is also completely dependent upon transport through the J protein, because introduction of a *hisJ* mutation eliminates simultaneously the ability to grow on imidazole pyruvic acid and on D-histidine.

We have previously demonstrated (3) that the histidine analogue HIPA is transported by the J-P permease.

Azaserine is also transported through the elevated high affinity J-P permease of the *dhuA1* mutant. This inhibitory analogue has been demonstrated to be a substrate for the aromatic permease (*aroP*) (1, 2). The inhibition of growth produced by azaserine on the wild type is completely reversed by any of the aromatic amino acids. However, introduction of the *dhuA1* mutation causes the bacteria to acquire sensitivity to azaserine, even in the presence of an aromatic amino acid or of an *aroP*

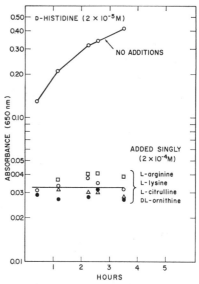

FIG. 6. Inhibition of growth on D-histidine by amino acids. TA271 (*dhuA1 hisF645*) was grown overnight and diluted as described in the legend to Fig. 2. L-Arginine, or L-lysine, or L-citrulline, or DL-ornithine were added at zero time where indicated. Growth was at 37°.

mutation. This has been interpreted to mean that azaserine is transported through the J-P histidine permease in sufficient amount to inhibit growth, when this system is elevated. In agreement with this interpretation, it was found that introduction of a *hisP* mutation in the *dhuA1* strain causes complete resistance to this new azaserine sensitivity, but only when an aromatic amino acid is present. In the absence of an aromatic amino acid, transport of azaserine still occurs through the aromatic permease in a *dhuA hisP* double mutant and causes inhibition. These data allowed us to predict that the J protein would have an affinity for azaserine, despite its lack of obvious resemblance to histidine.

Specificity of J Protein

The affinity of the pure J protein for a wide variety of amino acids and histidine analogues was measured by competition of the unlabeled compound with L-[³H]histidine binding (Table IV). The following compounds have a significant affinity for the J protein: L-histidine, arginine, lysine, D-HIPA, azaserine, D-histidine, citrulline, and ornithine. All these compounds are known to be substrates of the high affinity histidine permease. No other natural amino acid inhibited histidine binding up to 10,000-fold excess; cysteine consistently gave a slight stimulation of histidine binding; this might indicate a sulfhydryl activation of the J protein.

The relatively high affinity of the J protein for L-arginine is in agreement with the effect of L-arginine upon transport of L- and D-histidine (Table III and Fig. 6); a 1000-fold excess of L-arginine completely inhibits L-[³H]histidine-binding. A K_D of 10 μM for arginine can be calculated from the competition data. L-[³H]-Arginine was shown to bind to the J protein (data not shown);

a 1000-fold excess of unlabeled L- or D-histidine completely inhibits the binding of L-[³H]arginine. D-Arginine, however, does not inhibit L-histidine binding at 1000-fold excess.

It should be emphasized that the basic amino acids would compete more effectively with D-histidine than with L-histidine, because of the relatively poor affinity of D-histidine for the J protein. This explains the strong inhibition of D-histidine growth (Fig. 6) by citrulline and ornithine, which do not compete with L-histidine binding at 1000-fold excess.

The K_D of the D-histidine-J protein complex, measured by competition of L-[³H]histidine binding, is 500 μM. The maximal number of L-histidine binding sites on the J protein is the same in the presence and absence of D-histidine.

DISCUSSION

A necessary prerequisite for a biochemical approach to a study of active transport is the isolation of a presumed transport component and its unequivocal identification as an obligatory component of the transport system under study. Numerous laboratories have recently isolated a variety of such presumed components (reviewed in Reference 13). Among these, a class of proteins called the "binding proteins," which bind a variety of small molecules and are thought to be located on the surface of gram-negative bacteria (14), has been implicated indirectly in the active transport of these molecules. The lack of direct evidence concerning the role of binding proteins in transport has been discussed (13).

We feel that the J protein has been shown conclusively to be a component of histidine transport. We previously demonstrated (3) a direct correlation between the activity of the histidine permease and the levels of the histidine-binding protein J, by showing that (*a*) mutation in the *dhuA* site causes the simultaneous elevation of the J protein and of the rate of L-histidine transport; (*b*) mutation in the *hisJ* gene causes loss of the J protein and a decrease in the rate of L-histidine transport. The identity of the J protein as the J component of histidine transport is now firmly established by the finding that both are coded for by a single structural gene. The properties of a temperature-sensitive strain, TA1791, provide this evidence.

TA1791 was obtained as a revertant of a strain containing a *hisJ* mutation (and therefore defective in histidine transport). However, TA1791 is not a true revertant because it has not recovered the properties of a strain with an intact *hisJ* gene; the reversion is due to a second site mutation in the *hisJ* gene itself, as shown by the finding of an altered J protein and by the genetic analysis. The J protein isolated from TA1791 differs from the wild type J protein in temperature stability, affinity for histidine, chromatographic properties, and disc gel electrophoretic behavior. The production of altered J protein has been correlated with the *in vivo* temperature sensitivity of TA1791; both the L-histidine transport and the ability to grow on D-histidine are temperature sensitive. Both these activities require functional J protein.

The possibility that the mutant protein is a new, completely different protein is excluded. The two proteins have several properties in common, although we chose to stress the differences. The temperature-sensitive mutation maps in the same place as mutations causing loss of the J protein; both proteins bind histidine and function in the transport of L-histidine and D-histidine; they have exactly the same molecular weight and isoelectric pH; they run in the same position in some gel electrophoresis systems;

they are both dependent on *hisP* gene product for function; the mutant protein cross-reacts with antiserum to the wild type J protein.

Many other revertant strains have been similarly obtained from a variety of *hisJ*-containing strains. Preliminary experiments indicate that many of these also produce J proteins with altered properties.

It should be pointed out that the temperature sensitivity of the J protein in TA1791, as assayed *in vitro* by histidine binding, may not necessarily account for the *in vivo* temperature sensitivity of growth on D-histidine. If there is an interaction between the J protein and the P protein (or any other transport protein), the temperature-sensitive mutation could have affected such *in vivo* interaction, rather than the binding of the substrate to the J protein.

The evidence obtained from the genetic analysis that the *hisJ* gene is the structural gene for the J protein agrees with the excellent correlation between the properties of the wild type transport *in vivo* and of the wild type J protein *in vitro*. We presented here information concerning the correlation in specificity for substrates and analogues.

The highest affinity of both *in vivo* transport and of the J protein is for L-histidine. None of the other amino acids, added at 10,000-fold excess, competed with histidine binding to the J protein except for those shown in Table IV. In fact, only those compounds which have been shown to have a physiological effect on transport inhibit histidine binding, thus demonstrating the correlation between the specificity of the J protein and of transport. The J protein has a relatively high affinity for arginine and lysine. This is consistent with the known effects of these compounds on transport *in vivo*; both arginine and lysine are good competitors of L-histidine uptake and D-histidine growth. Even though the affinity of the J protein for arginine is very good, it should be emphasized that it still is considerably lower than its affinity for histidine (K_D for arginine = 10 μM; K_D for histidine = 0.1 μM). The affinity of the J protein for D-histidine, citrulline, and ornithine, and for the analogues, D-HIPA and azaserine, also correlates well with the affinity that these compounds have for the J-P permease *in vivo*. As a consequence of this affinity, D-histidine, D-HIPA and azaserine can enter the cell. Therefore, D-histidine can act as growth substrate, while D-HIPA and azaserine act as inhibitors. The affinity of citrulline and ornithine for the J-P permease is shown by their inhibition of growth on D-histidine.

Recently (15) a periplasmic protein which binds histidine with much poorer affinity than the J protein has been isolated and purified from *S. typhimurium*. This protein differs in several biochemical properties from the J protein (discussed in Reference 4), and in all likelihood it is a different protein altogether. Moreover, no genetic characterization of this protein has been presented, thus rendering it difficult to draw any conclusion concerning the relationship between this protein and either the J protein or histidine transport.

Acknowledgments—We would like to thank Bruce N. Ames and Sydney Govons Küstü for their advice and for innumerable stimulating discussions; Betty I. Kirk for painfully analyzing the purity of the compounds in Table IV; C. E. Ballou for his financial support through United States Public Health Service Grant AM12121.

REFERENCES

1. AMES, G. F. (1964) *Arch. Biochem. Biophys.* **104**, 1
2. AMES, G. F., AND ROTH, J. R. (1968) *J. Bacteriol.* **96**, 1742–1749
3. AMES, G. F., AND LEVER, J. (1970) *Proc. Nat. Acad. Sci. U. S. A.* **66**, 1096
4. LEVER, J. (1972) *J. Biol. Chem.* **247**, 4317–4326
5. SLETZINGER, M., FIRESTONE, R. A., REINHOLD, D. F., ROONEY, C. S., AND NICHOLSON, W. H. (1968) *J. Med. Pharmacol. Chem.* **11**, 261
6. AMES, B. N., AND MITCHELL, H. K. (1952) *J. Amer. Chem. Soc.* **74**, 252
7. RAY, W. J., JR. (1967) *Methods Enzymol.* **11**, 490
8. HIRS, C. H. W., MOORE, S., AND STEIN, W. H. (1952) *J. Biol. Chem.* **195**, 669
9. KRAJEWSKA-GRYNKIEWICZ, K., WALCZAK, W., AND KŁOPOTOWSKI, T. (1971) *J. Bacteriol.* **105**, 28–37
10. AMES, B. N., AND WHITFIELD, H. J. (1966) *Cold Spring Harbor Symp. Quant. Biol.* **31**, 211
11. ANRAKU, Y. (1968) *J. Biol. Chem.* **243**, 3116–3122
12. NOSSAL, N. G., AND HEPPEL, L. A. (1966) *J. Biol. Chem.* **241**, 3055–3062
13. KABACK, H. R. (1970) *Ann. Rev. Biochem.* **39**, 561
14. HEPPEL, A. L. (1967) *Science* **156**, 1451
15. ROSEN, B. P., AND VASINGTON, F. D. (1971) *J. Biol. Chem.* **246**, 5351–5360

Reprinted from *Nature New Biology*, **230**, 101–104 (1971)

28

Role of the Galactose Binding Protein in Chemotaxis of *Escherichia coli* toward Galactose

GERALD L. HAZELBAUER & JULIUS ADLER

Departments of Biochemistry and Genetics, University of Wisconsin, Madison, Wisconsin 53706

The galactose binding protein is the part of the galactose chemoreceptor which recognizes the attractants galactose, glucose, and a number of structurally related chemicals.

THE bacterium *Escherichia coli* is chemotactic, that is, motile cells are attracted by certain chemicals such as oxygen, some sugars, and some amino-acids[1,2]. The chemicals are detected by "chemoreceptors", sensing devices which recognize an attractant without metabolizing it[2]. There are at least eight different chemoreceptors in *E. coli*[2]. Each chemoreceptor should have a recognition component—a part that has affinity for the chemicals detected by that chemoreceptor—and possibly other components to link the recognition component to a pathway, common to all the chemoreceptors[2,3], that leads to the response.

We have isolated "specifically nonchemotactic" mutants, which are defective in a particular chemoreceptor, so that they do not respond to a group of structurally related compounds but exhibit normal responses to other attractants[4]. For example, "galactose taxis" mutants[4] have no detectable taxis toward D-galactose, 1-D-glycerol-β-D-galactoside, D-fucose, methyl-β-D-galactoside, L-arabinose, D-xylose, or L-sorbose (the sugars recognized only by the galactose chemoreceptor), and a reduced response toward D-glucose, methyl-α-D-glucoside and 2-deoxy-D-glucose (the sugars recognized by both the galactose and the glucose chemoreceptors). Taxis toward other sugar attractants (such as D-ribose) and toward amino-acids is normal in such a mutant. We measure chemotaxis by determining how many bacteria enter a capillary tube containing an attractant[2,4]. Fig. 1 shows examples of responses for a typical galactose taxis mutant, AW543. These mutants are also defective in the transport of galactose at low concentrations ($\leq 10^{-6}$ M). This suggests[4] that there is some component which is involved in both chemotaxis and transport.

To examine the composition and mechanism of action of the chemoreceptors, we have investigated the galactose taxis mutants further. We now present studies which show that the recognition component of the galactose chemoreceptor is the galactose-binding protein[5-7], a substance which binds galactose

without altering it[5] and which is released from cells by mild osmotic shock[8,9]. This protein has been identified[7,10-12] as a component of a galactose transport system, the methyl-β-galactoside permease[13,14]. Our studies also show that the galactose chemoreceptor and the galactose transport system have independent elements as well as this common component.

We wish to thank Dr Herman M. Kalckar for having led us in this direction by sending us a mutant known[10] to be defective in the galactose binding protein, W4345 (discussed later), and suggesting that we test it for chemotaxis toward galactose.

Galactose Binding Protein

Fig. 1 and ref. 4 show examples of chemotactic responses for two typical galactose taxis mutants. Responses like these are found for all the mutants listed in Table 1 except AW550. The mutants listed in Table 1 are also all defective in the transport of galactose at and below 10^{-6} M. Table 1 shows that the mutants, with the exception of AW550, have a greatly reduced level of galactose-binding activity in their osmotic shock fluid[8,9]. The cells were grown on galactose because this is required to achieve a full level of galactose-binding protein and galactose taxis.

Strains W4345 [10,14], EH3027 [11] and EH3030 [11] were first characterized as defective in galactose transport and lacking the galactose binding protein; they also fail to show taxis toward galactose. AW526, a derivative of W4345 which can metabolize galactose, still has the defective taxis (Table 1) and transport properties of its parent.

Revertants selected from galactose taxis mutants for normal galactose taxis also regain the galactose binding activity (Table 1) and galactose transport[4]. AW527 is a partial revertant for galactose taxis, binding activity, and transport.

The reduced galactose-binding activities present in shock fluid of the mutants (except AW550) have dissociation constants which are the same as those of wild type activity (Table 1). This suggests that the mutants produce wild type galactose binding protein but in reduced amounts. The mutations in these strains are probably therefore in a control gene that influences both the binding protein gene and some unknown gene(s) involved in chemotaxis. In that case the correlation observed between galactose taxis and the level of galactose binding protein could be merely coincidental. This seems unlikely, however, because of the behaviour of the mutant strain AW550.

Altered Galactose Binding Protein

AW550 responds to galactose (Fig. 1), but the threshold concentration (the lowest concentration at which a response is detected in the chemotaxis assay) is elevated about fifty-fold. The threshold for fucose is shifted at least 10^4-fold. This suggests that the galactose chemoreceptor is still functioning but with less than normal sensitivity.

Galactose binding activity in the AW550 shock fluid has a dissociation constant for galactose of 100 μM (Table 1), about thirty-fold higher than the dissociation constant listed for other strains; the activity is present at essentially wild type levels (Table 1). Binding of galactose in a shock fluid from the wild type strain W3110 was unaffected by addition of shock fluid from AW550. This makes it unlikely that the low affinity of the AW550 galactose binding protein is caused by an inhibitor. The mutant presumably has a defect in the galactose binding protein itself.

accumulate or metabolize galactose at the concentration used[2,14,15]. Attractants to which galactose taxis mutants respond normally, that is, chemicals detected by a different chemoreceptor, do not inhibit galactose taxis.

We also determined the concentration of attractant required to achieve 50% inhibition of binding of galactose by the galactose binding protein from the same strain, 20SOK⁻ (column B, Table 2). The order of effectiveness of sugars in inhibiting taxis toward galactose and the order of inhibiting binding of galactose by the galactose binding protein are the same.

For each sugar, the half-inhibition of taxis occurs at a concentration about two hundred times less than the half-inhibition of binding. This may be simply the result of the way the chemotaxis assay is designed. Because the 5×10^{-6} M galactose in the capillary diffuses out to form a gradient, the concentrations to which the bacteria actually respond are lower than 5×10^{-6} M.

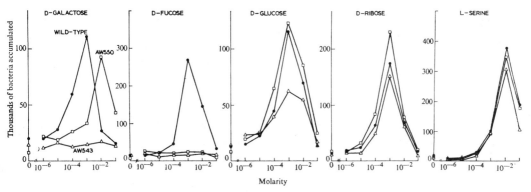

Fig. 1 Chemotaxis assays comparing responses of two galactose taxis mutants AW543 (Δ) and AW550 (□) to the parental strain B275 (●). As described previously[4], cultures were grown in galactose minimal medium (or ribose in the case of ribose taxis) containing 0.001 g/ml. threonine, leucine, methionine or histidine as required by the strains, and the cells were washed and resuspended to a concentration of 7×10^7 cells/ml. in 10^{-2} M potassium phosphate buffer, pH 7.0, 10^{-4} M EDTA, and 10^{-6} M L-methionine (chemotaxis medium). Chemotaxis was measured by the following procedure described in detail elsewhere[2,4]. A 1 μl. micropipette, sealed at one end and containing the attractant at various concentrations in chemotaxis medium, was placed open-end first into about 0.25 ml. of bacterial suspension contained in a small chamber on a slide warmer at 30° C. After 60 min the micropipette was removed, the outside was rinsed off, and the contents were plated to determine the number of bacteria that had accumulated inside the micropipette. The determination is reproducible within 10–20%. The strains used in this figure and Table 1 were: B275, a K-12 strain of *E. coli* that requires threonine, leucine, and methionine for growth, described previously[26]. B275*his⁻* is a derivative of B275 that also requires histidine for growth. AW541 and AW543 are independent isolates from B275 treated with N-methyl-N-nitro-N-nitrosoguanidine[27]. AW550 and AW551 (see text) are independent isolates from B275*his⁻* treated in the same way. Procedures for obtaining mutants will be described in later publications. AW520, a *gal⁺* revertant of the *gal⁻* strain W3109[4], was described previously[4]. AW526 is a *gal⁺* revertant of the *gal⁻* strain W4345[14]. AW521 and AW527 were selected as galactose taxis revertants from AW520 and AW526 respectively by the procedure described earlier[4].

If the galactose binding protein plays a role in galactose transport, this function should also be altered in AW550. The mutant is indeed defective in galactose transport at low concentrations of galactose ($\leq 10^{-6}$ M).

Chemicals detected by a chemoreceptor can be identified either by finding which chemicals no longer attract (or attract less well) a taxis mutant or by testing attractants for their ability to inhibit taxis toward another attractant. All the attractants detected by one chemoreceptor inhibit each other's activity, presumably by competing for the binding site on the chemoreceptor, while attractants detected by different chemoreceptors do not[2]. For measuring inhibition, a second attractant is present in both the bacterial suspension and the capillary. The effectiveness of the inhibitor is determined by varying its concentration at a constant concentration of the first attractant in the capillary and determining the concentration which inhibits the accumulation of bacteria by 50%.

Table 2 shows the inhibition of taxis toward 5×10^{-6} M galactose (column A) by 20SOK⁻, a strain that does not

Restoration of Chemotaxis

If the galactose binding protein is needed for galactose taxis, then cells which have lost the binding protein during the osmotic shock procedure[8,9] should be unable to carry out taxis toward galactose. Shocked cells are no longer motile, but motility can be restored to 30–50% of the initial level by incubation for approximately 1 h at 35° C with stage 1 supernatant (see legend to Fig. 2). Restoration of motility occurs even when chloramphenicol is present to block protein synthesis. The explanation for the loss of motility or for its restoration is unknown.

Shocked cells with motility restored in the absence of chloramphenicol show some taxis toward galactose, but shocked cells with motility restored in the presence of chloramphenicol (added to prevent synthesis of galactose binding protein) show almost no taxis (Fig. 2). Addition of the 60–95% ammonium sulphate fraction of shock fluid containing galactose binding protein from the same strain, AW521, restores

335

galactose taxis (Fig. 2). The results ranged from about half restoration to the nearly full restoration shown in Fig. 2. A control experiment using the 60–95% ammonium sulphate fraction of shock fluid from the galactose taxis mutant AW520,

Table 1 Content of Galactose Binding Protein in Galactose Taxis Mutants

Strain	Parent	Galactose taxis	Galactose binding protein		
			pmol/mg protein in shock fluid	pmol/pmol leucine binding protein in shock fluid	Dissociation constant (µmolar)
B275		Wild type	2,900		3
AW541	B275	Mutant	160		3
AW543	B275	Mutant	380		4
AW550	B275*his⁻*	Mutant	1,600		100
AW520		Mutant	70	0.05	3
AW521	AW520	Wild type	1,500	2.4	2
AW526		Mutant	< 10	< 0.01	
AW527	AW526	Wild type (partial)	280	0.25	2

Quantitative assays of galactose taxis were performed as described in the legend to Fig. 1. The strains used are also described here. Shock fluid was prepared from cells grown to late log phase (A_{590} of 2; roughly 1.5×10^9 cells/ml.) in 1 l. of minimal galactose salts medium[3] (containing 0.001 g/ml. threonine, leucine, methionine, or histidine when required) in a 6 l. flask with shaking. The cells were washed and osmotically shocked by the Nossal and Heppel[9] modification of the Neu and Heppel procedure[8]. This involves transfer from a Tris-EDTA–20% sucrose solution to ice cold 5×10^{-4} M MgCl$_2$, and removal of the cells by centrifugation. The supernatant (shock fluid) was concentrated about $\times 40$ in an Amicon Model 52 ultrafiltration cell fitted with a UM-10 membrane and then dialysed twice against 0.01 M Tris-HCl (pH 7.3), 2 mM MgCl$_2$, and 0.25 mM dithiothreitol (buffer A). A 65% and then a 95% ammonium sulphate precipitation of the dialysed, concentrated shock fluid was carried out, and the two precipitates were dissolved in and dialysed against buffer A. The binding assay mixture consisted of 10–50 µg of protein and UL ^{14}C-galactose (Mallinckrodt Chemical Works, Orlando) at a series of concentrations ranging from 10^{-6} M to 10^{-5} M in 0.1 ml. of buffer A. The mixture was left for 10 min at room temperature and then filtered using an ultrafiltration cell (Metaloglass, Boston) fitted with UM-10 membranes (Amicon, Lexington) by the method of Paulus[24]. After filtration the bottom of each membrane was washed with 5 ml. ethylene glycol and then blotted dry. The membranes were then placed in 0.3 ml. distilled water; 10 ml. of scintillation fluid (5.5 g 2,5-diphenyloxazole (PPO) in 660 ml. toluene and 330 ml. ethylene glycol monomethyl ether) was added, and the radioactivity was counted. The concentrations and dissociation constants of binding proteins were determined from the intercepts of the linear plots of l/bound galactose against l/free galactose using the relation $1/BG = K_d/B_t \ (1/G) + 1/B_t$ where $BG =$ bound galactose conc., $G =$ free galactose conc., $B_t =$ total binding protein and $K_d =$ dissociation constant of the galactose binding protein–galactose complex. As a check on the ultrafiltration method, equilibrium dialysis was used to determine the dissociation constant and level of galactose binding activity of one shock fluid from a wild type strain W3110. There was good agreement between the values determined by ultrafiltration, $K_d = 2.1 \times 10^{-6}$ M, $B_t = 7.7 \times 10^{-7}$ M, and by equilibrium dialysis, $K_d = 3.0 \times 10^{-6}$, $B_t = 6.3 \times 10^{-7}$ M. Binding was tested in the dialysed concentrated shock fluid and the ammonium sulphate fractions. Over 90% of the galactose binding activity of the shock fluid was present in the 65–95% ammonium sulphate fraction. There was usually an increase in total activity in the fractions over that in the shock fluid, perhaps indicating inhibition of binding in the shock fluid. The higher activity is reported in the table. Protein was determined by the method of Lowry *et al.*[25]. Between 3 and 8 mg protein/10^{12} cells was released by the shock procedure from both mutant and wild type strains. Leucine binding activity was similarly determined by the ultrafiltration method described above using 4,5-³H-leucine (Schwartz). For leucine auxotrophs growth had to be in the presence of leucine, which represses production of leucine binding protein[16], and thus leucine binding was not tested for the first four strains listed in Table 1. Leucine binding activity was similar in total activity and dissociation constant for each of the leucine prototrophs tested.

which contained 1/200 the binding activity present in the wild type shock fluid, did not restore galactose taxis.

Fig. 2 also shows restoration using galactose binding protein from the galactose taxis mutant AW550. This yields a response to galactose shifted to higher galactose concentrations, characteristic of AW550. This shows that the properties of the binding protein dictate the character of the chemotaxis in the restored bacteria.

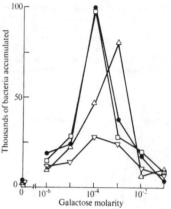

Fig. 2 Restoration of galactose taxis in osmotically shocked cells. AW521, a strain showing wild type galactose taxis (see Table 1), was grown in 10 ml. galactose minimal medium[4] to an A_{590} of 0.8, washed twice in chemotaxis medium (Fig. 1) by centrifugation, and then subjected to osmotic shock by the procedure of Nossal and Heppel[9]. The pellet of shocked cells was divided into three portions of about 3×10^8 cells each and suspended in 0.5 ml. of (1) ten-fold concentrated, dialysed stage 1 supernatant[9] in buffer A + 200 µg/ml. chloramphenicol; (2) the solution described in (1) + the concentrated dialysed 65–95% ammonium sulphate fraction of AW521 shock fluid in buffer A containing approximately 3 nmol galactose binding protein (about 1,000-fold more than is usually released from 3×10^8 cells); (3) solution (1) + the 60–95% concentrated, dialysed fraction of AW550 shock fluid in buffer A containing approximately 3 nmol galactose binding protein. After the suspensions were shaken for one hour at 35° C in a rotary shaker, about 40% of the cells were motile (before the shock procedure all the cells were swimming; directly after the shock almost none were). These cell suspensions, along with a sample of unshocked cells which had been suspended in Tris buffer (0.033 M, pH 7.3) during the shock procedure and then shaken in solution (1) along with the other suspensions, were then washed twice by centrifugation in 5 ml. of chemotaxis medium containing chloramphenicol. Finally, they were tested for galactose taxis in the continued presence of 200 µg/ml. ●, Unshocked; □, shocked + wild type galactose binding protein; △, shocked + mutant galactose binding protein; ▽, shocked, no galactose binding protein added.

Attempts to obtain normal galactose taxis in shocked cells of the galactose taxis mutant AW520 by adding wild type binding protein have so far been unsuccessful. The mutants which have a greatly reduced level of normal binding protein may also lack functional "sites" for the normal protein to attach.

Model for Chemoreception and Transport

Fig. 3 *See p. 338.* summarizes our information about chemotaxis and the galactose binding protein. The central feature is the binding protein, which is involved in both transport of and chemotaxis toward the compounds recognized by a chemoreceptor.

We have recently found osmotically shockable binding activities for ribose and maltose that seem to serve the corresponding chemoreceptors. The ribose binding activity is

*See p. 338.

induced by growth on ribose and the binding of ribose is not inhibited by galactose, glucose, or maltose as predicted from studies of ribose taxis. Similarly, the maltose binding activity is induced by growth on maltose and is not inhibited by galactose or glucose, again as predicted from studies of maltose taxis. The dissociation constant for ribose is about 2 μM and for maltose about 5 μM.

But a binding protein is utilized for chemoreception only in some instances. Some amino-acids for which binding proteins have been identified in *E. coli* are not attractants, such as leucine[5,16,17], isoleucine[16], valine[16], arginine[18], glutamine[18], and phenylalanine[19].

Table 2 Specificity of the Galactose Chemoreceptor and the Galactose Binding Protein as determined by Inhibition Studies

| | Concentration (μM) required for 50% inhibition of | | |
	(A) Taxis towards 5 μM galactose	(B) Binding of 5 μM galactose	Ratio B/A
D-Glucose	0.005	1.0	200
D-Galactose	0.036	7.0	190
1-D-Glycerol-β-D-galactoside	0.15	25	170
D-Fucose	6.2	1,100	180
Methyl-β-D-galactoside	30	3,500	120
L-Arabinose	95	17,000	180
D-Xylose	120	18,000	150

For chemotaxis studies, strain 20SOK⁻ was grown on glycerol plus 10^{-3} M D-fucose in a minimal salts medium according to the procedure described previously[4]. The cells were washed and resuspended and then chemotaxis was measured, according to the procedures of Fig. 1, except that 2 μl. micropipettes were used. Each micropipette contained 5×10^{-6} M galactose, and inhibitor at various concentrations was present in both micropipette and cell suspension. Accumulation of cells in the micropipettes was expressed as the per cent of uninhibited accumulation and was plotted against the log of the inhibitor concentration. The 50% concentration was determined from the intercept of the straight line (least squares fit) with the 50% accumulation line. Inhibition of binding was determined using the binding assay described in the legend to Table 1. The assay mixture contained UL ^{14}C-galactose at a concentration of 5×10^{-6} M, inhibitor sugar at various concentrations, and 2×10^{-6} M galactose binding protein. This protein was the 65–95% fraction of the shock fluid prepared according to the procedure of Table 1 from 20SOK⁻ bacteria grown on the medium described above. Half-inhibition concentrations were determined by plotting the data as described for the chemotaxis study. The galactose used was "substantially glucose free" (Sigma); the 1-D-glycerol-β-galactoside was a gift from W. Boos and the other sugars were all purified by procedures described previously[2].

Transport itself is not required for chemotaxis, because there are mutants which have normal galactose taxis but are blocked in the transport of galactose, as the result of a defect in the methyl-β-galactoside permease, the system that contains the galactose binding protein. These mutants are 20SOK⁻ (refs. 2, 14 and 15 for transport; ref. 2 for chemotaxis) and W3092i (ref. 13 for transport; chemotaxis data of J. Adler). These strains contain normal levels of the galactose binding protein (this article and ref. 10 for 20SOK⁻; ref. 10 for W3092i) and may therefore lack one or more additional components required for transport but not for chemotaxis *. The model thus has two separate branches both beginning with the binding protein; one branch leads to transport and the other to chemoreception and a chemotactic response.

The branch leading to chemoreception may have at least one additional component essential for chemotaxis but not for transport. A recently isolated mutant, AW551, may be defective in such a component. It fails to show taxis toward galactose, yet it has a full level of normal galactose binding protein and is normal in the transport of galactose *.

A gradient of an attractant is somehow detected by a chemoreceptor and this brings about a chemotactic response. The various chemoreceptors all share a pathway which transfers information from the chemoreceptors to the flagella and which involves at least three components, because there are three complementation groups of "generally nonchemotactic" mutants, bacteria which are fully motile but nevertheless fail to respond to any attractant[3,20]. The mechanisms of chemoreception and response are all still unknown. As the chemoreceptor binds (or releases) an attractant, it could cause a change in conformation or electrical potential of the adjacent cell membrane. This change could be propagated along the membrane to reach the flagella, and thus finally affect the direction of swimming. The idea that the membrane participates in this action is attractive because the flagella are in contact with it[21,22] and the presumed location[23] of the binding proteins in the periplasmic space (the region between the cell membrane and the cell wall) makes membrane contact possible.

Very little is known at the molecular level about mechanisms of chemoreception—or for that matter sensory reception in general—in any organism. We hope that some of the information gained from the study of chemoreception in bacteria can be applied to understanding sensory reception in higher organisms.

We thank Dr Winfried Boos for galactose binding protein mutants. Our galactose taxis mutants were isolated by Margaret M. Dahl and Mark S. Lubinsky. This research was supported by a US Public Health Service grant from the US National Institute of Allergy and Infectious Diseases. G. L. H. held a Wisconsin Alumni Research Foundation predoctoral fellowship.

* Alternatively, the galactose binding protein may be modified in this kind of mutant in such a way that binding of substrates is normal but a subsequent function is prevented. Genetic complementation tests will settle this.

Received November 15, 1970; revised February 4, 1971.

[1] Adler, J., *Science*, **153**, 708 (1966).
[2] Adler, J., *Science*, **166**, 1588 (1969).
[3] Armstrong, J. B., Adler, J., and Dahl, M. M., *J. Bact.*, **93**, 390 (1967).
[4] Hazelbauer, G. L., Mesibov, R. E., and Adler, J., *Proc. US Nat. Acad. Sci.*, **64**, 1300 (1969).
[5] Anraku, Y., *J. Biol. Chem.*, **243**, 3116 (1968).
[6] Anraku, Y., *J. Biol. Chem.*, **243**, 3123 (1968).
[7] Anraku, Y., *J. Biol. Chem.*, **243**, 3128 (1968).
[8] Neu, H. C., and Heppel, L. A., *J. Biol. Chem.*, **240**, 3685 (1965).
[9] Nossal, N. G., and Heppel, L. A., *J. Biol. Chem.*, **241**, 3055 (1966).
[10] Boos, W., *Europ. J. Biochem.*, **10**, 66 (1969).
[11] Boos, W., and Sarvas, M. O., *Europ. J. Biochem.*, **13**, 526 (1970).
[12] Boos, W., and Gordon, A. S., *Europ. J. Biochem.* (in the press).
[13] Wu, H. C. P., Boos, W., and Kalckar, H. M., *J. Mol. Biol.*, **41**, 109 (1969).
[14] Rotman, B., Ganesan, A. K., and Guzman, R., *J. Mol. Biol.*, **36**, 247 (1968).
[15] Buttin, G., *J. Mol. Biol.*, **7**, 164 (1963).
[16] Penrose, W. R., Nichoalds, G. E., Piperno, J. R., and Oxender, D. L., *J. Biol. Chem.*, **243**, 5921 (1968).
[17] Furlong, C. E., *Fed. Proc.*, **29**, 341 (1970).
[18] Weiner, J. H., Berger, E. A., Hamilton, M. N., and Heppel, L. A., *Fed. Proc.*, **29**, 341 (1970).
[19] Klein, W. L., Dahms, A. S., and Boyer, P. D., *Fed. Proc.*, **29**, 341 (1970).
[20] Armstrong, J. B., and Adler, J., *Genetics*, **61**, 61 (1969).
[21] Van Iterson, W., Hoeniger, J. F. M., and Van Zanten, E. N., *J. Cell Biol.*, **31**, 585 (1966).
[22] DePamphilis, M. L., and Adler, J., *J. Bacteriol.*, **105**, 396 (1971).
[23] Heppel, L. A., *Science*, **156**, 1451 (1967).
[24] Paulus, H., *Anal. Biochem.*, **32**, 91 (1969).
[25] Lowry, O. H., Rosebrough, N. J., Farr, A. L., and Randall, R. J., *J. Biol. Chem.*, **193**, 265 (1951).
[26] Adler, J., *J. Bact.*, **92**, 121 (1966).
[27] Adelberg, E. A., Mandel, M., and Chen, G. C. C., *Biochem Biophys. Res. Commun.*, **18**, 788 (1965).

Errata

In the article "Role of the Galactose Binding Protein in Chemotaxis of *E. coli* toward Galactose" by Gerald L. Hazelbauer and Julius Adler (*Nature New Biology*, **230**, 101; 1971), Fig. 3, which is printed below, was omitted.

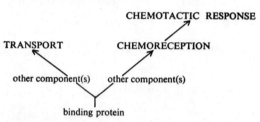

Fig. 3 Role of binding protein in chemoreception, and the relationship between chemoreception and transport. See text for explanation.

The authors wish to add that full restoration of galactose taxis (as in Fig. 2) has now been obtained with pure galactose binding protein—either from a wild type strain or from the galactose taxis mutant AW551.

The Co-Transport Theory

VI

Editor's Comments on Papers 29 Through 35

29 **Csaky and Thale:** *Effect of Ionic Environment on Intestinal Sugar Transport*

30 **Csaky, Hartzog, and Fernald:** *Effect of Digitalis on Active Intestinal Sugar Transport*

31 **Bihler and Crane:** *Studies on the Mechanism of Intestinal Adsorption of Sugars. V. The Influence of Several Cations and Anions on the Active Transport of Sugar* in vitro *by Various Preparations of Hamster Small Intestine*

32 **Bihler, Hawkins, and Crane:** *Studies on the Mechanism of Intestinal Adsorption of Sugars. VI. The Specificity and Other Properties of* Na+*-dependent Entrance of Sugars into Intestinal Tissue Under Anaerobic Conditions* in vitro

33 **Crane:** Na+*-Dependent Transport in the Intestine and Other Animal Tissues*

34 **Vidaver:** *Glycine Transport by Hemolyzed and Restored Pigeon Red Cells*

35 **Curran, Schultz, Chez, and Fuisz:** *Kinetic Relations of the* Na+*–Amino Acid Interaction at the Mucosal Border of the Intestine*

The viability of nearly all cells depends on their maintaining a constant internal ionic environment. As is well known, this function is fulfilled by active transport systems which generate and maintain ionic gradients across the cytoplasmic membrane. Many organisms have evolved methods of harnessing the energy stored in these gradients for specialized and diverse purposes, namely, as triggering devices for muscular contractions, as a means of intercellular communication, and, in certain instances at least, as a driving force for the active transport of sugars and amino acids.

Throughout the 1950s, Christensen and co-workers carried out extensive investigations of the effects of different ions on the uptake of amino acids by duck erythrocytes and Ehrlich ascites tumor cells (see Christensen, 1960, 1962). These studies led to the suggestion, in 1958, that the uptake of amino acids might be driven by the movement of cations across the membrane (Riggs, Walker, and Christensen, 1958). The first detailed formulation of the "co-transport theory," which postulated that sugar uptake in the intestine is coupled to the flux of Na+ ions across the brush border of mucosal cells, was presented by Crane, Miller, and Bihler (1961) at the Prague symposium on active transport in 1960. Subsequent work in many different laboratories has resulted in a great quantity of experimental evidence supporting the co-transport theory as it applies to sugar and amino acid uptake in the intestine and in various other cells and tissues (see review by Schultz and Curran, 1970). Although Na+-driven transport is associated primarily with animal cells, a recent report suggests that melibiose uptake in *S. typhimurium* (Stock and Roseman, 1971) may be coupled to the movement of Na+ ions into the cell.

The present-day version of the co-transport theory was formulated by Crane (Paper 33) for sugar transport in the intestine and by Vidaver (1964a) for amino acid uptake in erythrocytes. In essence, the theory states that both Na+ and the transported sugar (or amino acid) bind to a common carrier in such a way that the binding of Na+ facilitates the binding of the transport substrate. The ternary complex can diffuse through the membrane in either direction. The net dissociation of Na$^+$ on the side where its electrochemical potential is lowest brings about a decrease in affinity of the carrier for the transport substrate and release of the sugar (or amino acid). Thus the accumulation of sugars and amino acids against a concentration gradient is driven by an oppositely directed gradient of Na$^+$ ions. In vivo the sodium gradient is maintained by an independent active transport system involving, presumably, the Na$^+$-K$^+$-activated ATPase (Skou, 1957). Some of the evidence supporting this theory is summarized below.

Active transport of sugars in the intestine requires the presence of Na$^+$ at the mucosal surface (Riklis and Quastel, 1958; Csaky and Thale, Paper 29; Bihler and Crane, Paper 31). Furthermore, Na$^+$ is required even for transport processes not requiring energy, such as exchange and facilitated diffusion under anaerobic conditions (Bihler and Crane, Paper 31; Bihler, Hawkins, and Crane, Paper 32). This observation was of great importance in the development of the theory because it strongly suggested that Na$^+$ exerted its effect on the transport process itself rather than on some indirectly related cellular function. Inhibitors of active Na$^+$ transport block the uptake of sugars in the intestine (Csaky, Hartzog, and Fernald, Paper 30). In contrast, those sugars that are not normally transported against a concentration gradient do not require Na$^+$ to cross the mucosal membrane (Bihler and Crane, Paper 31). The presence of Na$^+$ is also required for the uptake of amino acids by a wide variety of animal cells (see Schultz and Curran, 1970).

In an ingenious experiment, Vidaver (Paper 34) showed that the direction of glycine transport in erythrocyte ghosts depends on the direction of the Na$^+$ gradient. Glycine is accumulated within the ghost when the external Na$^+$ concentration is higher than that in the interior; when the Na$^+$ gradient is reversed, glycine is actively transported *out* of the ghost. No net transport of glycine is observed when the Na$^+$ concentrations are equal on both sides of the membrane. Similarly, reversal of the Na$^+$ gradient across the wall of the intestine leads to active transport of sugars in the "wrong" direction, i.e., from the serosal to the mucosal surface (Crane (1964).

In pigeon erythrocytes, the K_m for glycine uptake is inversely proportional to the square of the external Na$^+$ concentration (Vidaver, 1964a). In the intestine, the K_m for alanine uptake decreases with increasing Na$^+$ concentration (Curran, Schultz, Chez, and Fuisz, Paper 35), as does the K_m for absorption of 3-deoxyglucose (Crane, Forstner, and Eichholz, 1965; Crane, Paper 33). Similar data have been reported for many other Na$^+$-dependent transport systems (see Schultz and Curran, 1970).

According to the co-transport theory, Na$^+$ should enter the cell along with the sugar or amino acid being transported. Schultz and Zalusky (1964) found that glucose causes an increase in the potential difference and the short-circuit current across the wall of rabbit intestine. These changes are thought to reflect an increase in the activity of the active transport system for Na$^+$ resulting from the influx of Na$^+$ into the mucosal cells. The increase in short-circuit current is dependent on the Na$^+$ concentration and is a saturable function of the sugar concentration, yielding a K_m comparable to the K_m for transport. The changes in potential and short-circuit current are eliminated by phlorizin, an inhibitor of sugar transport, and by ouabain, an inhibitor of active Na$^+$ transport. Similar results are found for alanine transport across the rabbit intestine (Schultz and Zalusky, 1965).

Curran, Schultz, Chez, and Fuisz (Paper 35) measured directly the influx of $^{22}Na^+$ into mucosal cells of the rabbit intestine and found that alanine transport stimulates Na^+ influx. The stoichiometry of the Na^+ and alanine influxes approaches a limiting value of $1:1$ at high Na^+ concentrations. Vidaver (1964b) found a stoichiometry of *two* Na^+ ions per glycine molecule transported in erythrocytes, results that are consistent with the observed dependence of K_m for transport on the inverse *square* of the Na^+ concentration in this system (Vidaver, 1964a).

In contrast to the agreement between theory and experiment presented above, several reports indicate that, in certain cases, active transport of amino acids and sugars occurs against a Na^+ gradient (Wheeler and Christensen, 1967; Margolis and Lajtha, 1968; Jacquez and Schafer, 1969; Kimmich, 1970). Jacquez and Schafer (1969) found that the magnitude of the Na^+ gradient in cold-shocked, ouabain-treated Ehrlich ascites tumor cells was insufficient to account for the observed levels of accumulation of α-aminoisobutyric acid. Better agreement was obtained if both Na^+ and K^+ gradients were considered. Eddy (1968) has suggested that the movements of both Na^+ and K^+ are coupled to amino acid uptake in Ehrlich ascites tumor cells. Potashner and Johnstone (1971), however, have reported that the levels of amino acid accumulation in this system are determined more by cellular ATP levels than by cation gradients.

Kimmich (1970) has published a detailed critique of the co-transport theory. Using isolated cells of chick intestinal epithelia, he found significant accumulation of galactose against a Na^+ concentration gradient, and has suggested that sugar transport in this system is driven by an energized form of the Na^+-K^+ ATPase. These results do not necessarily invalidate the co-transport theory, however, since compartmentalization or binding of Na^+ within the cell, or an electrical potential across the membrane, could lead to an electrochemical gradient of Na^+ markedly different from that calculated on the basis of the cellular Na^+ content. In this area, as in most other areas of membrane-transport research, the pursuit of clear-cut answers to seemingly straightforward questions is likely to be foiled by the inherent complexity of the experimental system.

References

Christensen, H. N. (1960). Reactive sites and biological transport. *Advan. Protein Chem.*, **15**, 234–314.

Christensen, H. N. (1962). *Biological Transport*. W. A. Benjamin, Menlo Park, Calif.

Crane, R. K. (1964). Uphill outflow of sugar from intestinal epithelial cells induced by reversal of the Na^+ gradient: its significance for the mechanism of Na^+-dependent active transport. *Biochem. Biophys. Res. Commun.*, **17**, 481–485.

Crane, R. K., G. Forstner, and A. Eichholz (1965). Studies on the mechanisms of intestinal absorption of sugars. X. An effect of Na^+ concentration on the apparent Michaelis constants for intestinal sugar transport, in vitro. *Biochim. Biophys. Acta*, **109**, 467–477.

Crane, R. K., D. Miller, and I. Bihler (1961). Restrictions on the possible mechanisms of intestinal active transport of sugars. In A. Kleinzeller and A. Kotyk (eds.), *Membrane Transport and Metabolism*. Academic Press, New York, pp. 439–449.

Eddy, A. A. (1968). The effects of varying the cellular and extracellular concentrations of sodium and potassium ions on the uptake of glycine by mouse ascites-tumour cells in the presence and absence of sodium cyanide. *Biochem. J.*, **108**, 489–498.

Jacquez, J. A., and J. A. Schafer (1969). Na$^+$ and K$^+$ electrochemical potential gradients and the transport of α-aminoisobutyric acid in Ehrlich ascites tumor cells. *Biochim. Biophys. Acta,* **193,** 368–383.

Kimmich, G. A. (1970). Active sugar accumulation by isolated intestinal epithelial cells. A new model for sodium-dependent metabolite transport. *Biochemistry,* **9,** 3669–3677.

Margolis, R. K., and A. Lajtha (1968). Ion dependence of amino acid uptake in brain slices. *Biochim. Biophys. Acta,* **163,** 374–385.

Potashner, S. J., and R. M. Johnston (1971). Cation gradients, ATP and amino acid accumulation in Ehrlich ascites cells. *Biochim. Biophys. Acta,* **233,** 91–103.

Riggs, T. R., L. M. Walker, and H. N. Christensen (1958). Potassium migration and amino acid transport. *J. Biol. Chem.,* **233,** 1479–1484.

Riklis, E., and J. H. Quastel (1958). Effects of cations on sugar absorption by isolated surviving guinea pig intestine. *Can. J. Biochem. Physiol.,* **36,** 347–362.

Schultz, S. G., and P. F. Curran (1970). Coupled transport of sodium and organic solutes. *Physiol. Revs.,* **50,** 637–718.

Schultz, S. G., and R. Zalusky (1964). Ion transport in isolated rabbit ileum. II. The interaction between active sodium and active sugar transport. *J. Gen. Physiol.,* **47,** 1043–1059.

Schultz, S. G., and R. Zalusky (1965). Interactions between sodium transport and active amino-acid transport in isolated rabbit ileum. *Nature,* **205,** 292–294.

Skou, J. C. (1957). The influence of some cations on an adenosine triphosphatase from peripheral nerves. *Biochim. Biophys. Acta,* **23,** 394–401.

Stock, J., and S. Roseman (1971). A sodium-dependent sugar co-transport system in bacteria. *Biochem. Biophys. Res. Commun.,* **44,** 132–138.

Vidaver, G. A. (1964a). Transport of glycine by pigeon red cells. *Biochemistry,* **3,** 662–667.

Vidaver, G. A. (1964b). Some tests of the hypothesis that the sodium-ion gradient furnishes the energy for glycine-active transport by pigeon red cells. *Biochemistry,* **3,** 803–808.

Wheeler, K. P., and H. N. Christensen (1967). Role of Na$^+$ in the transport of amino acids in rabbit red cells. *J. Biol. Chem.,* **242,** 1450–1457.

Copyright © 1960 by the Physiological Society, London
Reprinted from *J. Physiol.*, **151**, 59–65 (1960)

29

EFFECT OF IONIC ENVIRONMENT ON INTESTINAL SUGAR TRANSPORT

By T. Z. CSÁKY* AND MARGRETHE THALE

From the Institute of Biological Chemistry, University of Copenhagen, Copenhagen K, Denmark

(*Received* 18 *September* 1959)

Glucose is absorbed from the intestine by two different mechanisms, diffusion and active transport. The former is effected by a concentration difference between the lumen and the blood; the latter proceeds against a concentration gradient. Active transport hastens the absorption up to a certain limit of concentration and causes the glucose to be completely absorbed even if the concentration in the lumen falls below that in the blood. Whereas the experimental approach to diffusion is relatively simple and clear-cut, active transport must be studied in 'living' preparations such as surviving isolated loops of intestine (Fisher & Parsons, 1949; Darlington & Quastel, 1953; Wilson & Wiseman, 1954). In such a preparation the sugar can be present in identical initial concentrations on both serosal and mucosal sides: an increase in concentration of sugar on either side, not accounted for through changes in volume, indicates that active transport is involved. This kind of experimental approach in mammalian gut has several drawbacks. One of them is the quantitative recovery of the sugar. This is complicated by the rapid metabolism of part of the glucose by the surviving gut tissue (Fisher & Parsons, 1950). Instead of glucose, however, 3-0-methylglucose (3-methylglucose) can be used. This sugar behaves like glucose in regard to intestinal absorption (Csáky, 1942), but since it is not metabolized (Csáky & Wilson, 1956; Csáky & Glenn, 1957), it can be quantitatively recovered. Another tiresome factor in connexion with the study of sugar transport in surviving mammalian gut is that the preparation has to be kept at body temperature and well supplied with oxygen. This difficulty can be eliminated by the use of the intestine of a poikilothermic animal, which transports both glucose and 3-methylglucose at room temperature and requires less vigorous oxygenation (Csáky & Fernald, 1960).

* John Simon Guggenheim Memorial Fellow, 1958–59. Permanent address: Department of Pharmacology, University of North Carolina School of Medicine, Chapel Hill, N.C.. U.S.A.

This work was undertaken in order to study the effect of ionic environment on the active transport of 3-methylglucose from the small intestine. For this purpose a simple technique was developed for the study of active transport in the isolated intestine of the toad.

METHODS

Modified Ringer's solutions isosmotic with the plasma of frogs were used in this study (Boyle & Conway, 1941); the composition of the various solutions is shown in Table 1. The sugar was dissolved in these solutions to make concentrations as indicated in each type of experiment, but the starting concentration was always identical on both sides of the gut wall.

TABLE 1. Composition of media used in the experiments

		Final solutions made by mixing salt solutions on a volume basis					
Salt solutions	(m-mole/l.)	A	B	C	D	E	F
NaCl	116	85·0	—	—	—	—	—
KCl	117	2·1	2·1	2·1	2·1	2·1	2·1
CaCl$_2$	83	1·1	1·1	1·1	1·1	1·1	1·1
MgSO$_4$	185	1·2	1·2	1·2	1·2	86·2	1·2
NaHCO$_3$	112	1·5	1·5	1·5	1·5	1·5	1·5
Na$_2$HPO$_4$	124	1·0	1·0	1·0	1·0	1·0	1·0
NaH$_2$PO$_4$	124	2·2	2·2	2·2	2·2	2·2	2·2
LiCl	116	—	—	85·0	—	—	—
Na$_2$SO$_4$	97	—	85·0	—	—	—	—
Li$_2$SO$_4$	97	—	—	—	85·0	—	—
Choline Cl	116	—	—	—	—	—	85·0

Large toads (*Bufo bufo*) were decapitated and their spinal cords pithed. The small intestine was then removed and placed in the corresponding physiological solution after first rinsing through the lumen with a syringe. The gut was then mounted on the apparatus shown in Fig. 1. The volume of the mucosal compartment was 14 ml. and that of the serosal compartment 10 ml. At the beginning of the experiment and at certain intervals thereafter samples were taken from either mucosal or serosal compartments or both and the sugar content was analysed with the anthrone reagent (Mokrasch, 1954). The error of this analysis was < 2 %.

Phenol red in a final concentration of 10 mg/l. was routinely added to the mucosal compartment. This served as a convenient way to check 'leaks' and to calculate the movement of fluid, if any, from or to the mucosal compartment. Fluid movement was also determined at the end of each experiment by emptying the contents of both compartments and weighing the material. All experiments were carried out at room temperature (19 ± 2° C).

RESULTS

Fluid movement. In all experiments the movement of fluid was measured both with phenol red and by direct weighing at the end of the experiment. There was no measurable movement in any of the experiments described below.

Transport and recovery of glucose and 3-methylglucose. Table 2 shows the results of the experiments in which solution *A* was used on both mucosal and serosal sides. The initial concentrations of the two sugars were approximately the same on both sides, but after 4 hr the concentration had decreased on the mucosal and in some cases increased on the serosal side. Only 72 % of the glucose could be recovered at the end of the experiment. In the case of 3-methylglucose the recovery was better than 90 %. Consequently, the increase in the serosal concentration was more consistent with 3-methylglucose than with glucose, although more glucose than 3-methylglucose disappeared from the mucosal compartment.

In all subsequent experiments the concentration of 3-methylglucose in the serosal compartment was measured. The rate of increase in the concentration was expressed as 'transport' in μmole/hr.

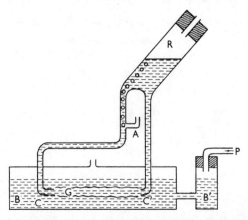

Fig. 1. Diagram of apparatus used in study of transport in surviving toad intestine. The gut (G) is mounted on cannulae C and C' through which the mucosal surface is perfused from reservoir R. Air bubbles through opening A keep the contents in steady circulation. The serosal surface is in contact with the liquid in bath B, the content of which is gently agitated with a small respiratory pump P connected through container B'.

TABLE 2. Transport and recovery of glucose and 3-methylglucose in the surviving isolated small intestine of the toad

| | Absolute amounts (mg) | | | | | | Total recovery after 4 hr (%) |
| | 0 hr | | | 4 hr | | | |
	Mucosal*	Serosal†	Total	Mucosal‡	Serosal‡	Total	
Glucose	5·80	4·18	9·98	3·34	3·64	6·98	69·9
Glucose	6·82	5·39	12·21	3·32	4·60	7·92	64·9
Glucose	7·58	5·23	12·81	4·98	5·64	10·62	82·9
						Mean	72·6
3-Methylglucose	7·52	5·28	12·80	6·33	6·28	12·61	98·5
3-Methylglucose	7·48	5·25	12·73	6·17	5·71	11·88	93·3
3-Methylglucose	6·89	5·23	12·12	5·29	5·84	11·13	91·8
						Mean	94·5

* In 14 ml. † In 10 ml. ‡ There was no measurable volume change in either compartment during 4 hr.

3-Methylglucose transport in the absence of sodium or chloride. Both mucosal and serosal surfaces were exposed to the same solutions (*A, B, C, E, F*). The results are summarized in Table 3. The replacement of chloride by sulphate (solution *B*) somewhat diminished the rate of active transport but did not completely abolish it. The replacement of Na by Li (solution *C*), Mg (solution *E*), or choline (solution *F*) completely abolished the active transport. There was no measurable active transport of xylose in solution *A*.

TABLE 3. Effect of the ionic composition of the medium on the transport of 3-methylglucose and xylose (4 and 6 hr experiments)

Solution	Principal cation	Principal anion	No. of experiments	Sugar transported (μmole/hr; mean $\pm$ S.D.)
		3-Methylglucose, 2·5 μmole/ml.		
A	Na	Cl	6	1·03 $\pm$ 0·38
B	Na	SO$_4$	4	0·41 $\pm$ 0·19
C	Li	Cl	5	0·0
E	Mg	SO$_4$	4	0·0
F	Choline	Cl	4	0·0
		Xylose, 2·5 μmole/ml.		
A	Na	Cl	4	0·0

TABLE 4. 3-Methylglucose transport in solution *C* followed by solution *A* in the same loop of intestine. Initial 3-methylglucose concentration, 2·5 μmole/ml.

	Solution (see Table 1)	Principal cation	Principal anion	Sugar transported (μmole/hr)
Expt. 1	C	Li	Cl	0·0
	A	Na	Cl	0·79*
Expt. 2	C	Li	Cl	0·0
	A	Na	Cl	0·92*

* There was no measurable transport during the first hour following the change of solution.

Reversibility of inhibition due to lack of sodium. The gut was placed first in solution *C* for 2 hr, during which time no sugar transport was observed. It was then transferred into another bath with solution *A* on both serosal and mucosal sides. There was a complete reversal of the inhibition, showing that the absence of Na did not permanently damage the sugar-transporting mechanism of the mucosa (Table 4).

Effects of lack of sodium on mucosal and serosal sides. The following experimental procedure was adopted. In one series of experiments the intestine was placed in solution *D*, which was in contact with the serosal surface, and solution *A* circulated through the lumen of the intestine. In another series of experiments the situation was reversed. The results shown in Table 5 indicate that when Na-containing solution was in contact with the mucosal surface, there was an active sugar transport regardless of the composition of solution on the serosal side; if the serosal surface was in contact with the Na-containing solution and the mucosal side bathed with the Li-containing solution, there was no measurable transport of 3-methylglucose.

TABLE 5. Transport of 3-methylglucose with solutions of different ionic composition on serosal and mucosal sides of the intestine (4 hr experiments). Initial 3-methylglucose concentration: 2·5 μmole/ml.

Solution		Principal cation	Principal anion	No. of experiments	Sugar transported (μmole/hr; mean $\pm$ S.D.)
D	Mucosal	Li	SO_4	3	0
B	Serosal	Na	SO_4		
B	Mucosal	Na	SO_4	6	0.63 ± 0.34
D	Serosal	Li	SO_4		

DISCUSSION

There was no appreciable net water transport in the preparations studied in this work. Since the original sugar concentration was approximately the same in the mucosal and serosal compartments, the progressive concentration changes, by definition, indicate an active transport of 3-methylglucose from the mucosal into the serosal compartment. Xylose, which is not actively transported from the mammalian gut, was not transported in unmodified Ringer's solution. The recovery of 3-methylglucose was better than 90 %.

The value of this method in studying active sugar transport is somewhat limited, inasmuch as experiments *in vitro* in general produce a situation different from that in the live animal. *In vitro*, the sugar molecule passes through the epithelial layer and is transferred into a compartment composed of connective tissue, muscle and serosal mesothelium. In the live animal the capillaries reach to the submucosal layer; thus the second compartment is eliminated. The significance of the compartment occupying the anatomical space between the mucosal epithelium and the 'free' serosal compartment is established by the finding that about 10 % of the 3-methylglucose cannot be recovered in the combined 'free' compartments on mucosal and serosal sides. However, this problem was not further pursued in this study.

It has been repeatedly shown that changes in ionic environment profoundly influence the intestinal glucose absorption both *in vitro* and *in vivo*. Sodium, potassium, calcium, and magnesium ions were found to be necessary for 'normal' absorption. The pertinent literature was summarized recently by Riklis & Quastel (1958). The same authors demonstrated that increasing the potassium concentration to 15·6 m-equiv/l. on both mucosal and serosal sides stimulates the transfer of glucose across the surviving guinea-pig intestine. When all the sodium on both sides was replaced by potassium there was practically no transfer of glucose, indicating that sodium ions are essential for glucose transfer.

The results of the present investigation indicate that sodium ions are essential for active sugar transport across the intestine. This is apparent from the observation that replacement of sodium by lithium, magnesium or choline completely inhibited active transport of 3-methylglucose. This inhibition is reversed by sodium, showing that the replacement of sodium did not permanently damage the epithelium. Moreover, it is apparent from the data presented here that sodium is essential only on the mucosal surface.

No attempt was made to investigate the concentration of sodium which is essential for active sugar transport. In unmodified Ringer's solution the bulk of the cations are in the form of sodium. Although the bulk of sodium was replaced in these experiments, this ion was still present in a concentration of 7·3 m-mole/l. in the media containing lithium, magnesium or choline as the principal cations (see Table 1). The lack of active sugar transport in these solutions indicates that more than 7·3 m-mole/l. of sodium is necessary for the sugar-transporting function.

Good evidence was recently presented by Curran & Solomon (1957) that sodium is actively transported in the gut from the mucosal to the serosal side. The nature of a sodium 'pump' was recently described in the frog skin by Koefoed-Johnsen & Ussing (1958). According to their concept the outward-facing membrane of the frog's skin epithelium is permeable to sodium, whereas the inward-facing membrane is practically impermeable to free sodium ions but sodium is 'pumped' across this membrane by an active process. The data presented here indicate that the intestinal sugar 'pump' does not function in the absence of sodium on the mucosal side.

SUMMARY

1. A method is described for the study of active sugar transport in the surviving isolated intestine of the toad by using the non-metabolized 3-methylglucose.

2. In unmodified frog Ringer's solution 3-methylglucose is actively transported from the mucosal to the serosal side.

3. Replacement of sodium by lithium, magnesium or choline abolishes active sugar transport. This effect is reversible if the original sodium concentration is restored.

4. Sodium is essential for active sugar transport on the mucosal but not on the serosal side.

The authors wish to thank Professor H. H. Ussing for his hospitality and for critical discussions. One of us (T.Z.C.) thanks the John Simon Guggenheim Memorial Foundation, New York, for a personal fellowship during the tenure of which this work was undertaken.

REFERENCES

BOYLE, P. & CONWAY, E. J. (1941). Potassium accumulation in muscle and associated changes. *J. Physiol.* **100**, 1–63.

CSÁKY, T. (1942). Über die Rolle der Struktur des Glucosemoleküls bei der Resorption aus dem Dünndarm. *Hoppe-Seyl. Z.* **277**, 47–57.

CSÁKY, T. Z. & FERNALD, G. W. (1960). The absorption of 3-methylglucose from the intestine of the frog (*Rana pipiens*). *Amer. J. Physiol.* (in the Press).

CSÁKY, T. Z. & GLENN, J. E. (1957). Urinary recovery of 3-methylglucose administered to rats. *Amer. J. Physiol.* **188**, 159–162.

CSÁKY, T. Z. & WILSON, J. E. (1956). The fate of 3-O-^{14}CH$_3$-glucose in the rat. *Biochim. biophys. acta*, **22**, 185–186.

CURRAN, P. F. & SOLOMON, A. K. (1957). Ion and water fluxes in the ileum of rats. *J. gen. Physiol.* **41**, 143–168.

DARLINGTON, W. A. & QUASTEL, J. H. (1953). Absorption of sugars from isolated surviving intestine. *Arch. Biochem.* **43**, 194–207.

FISHER, R. B. & PARSONS, D. S. (1949). A preparation of surviving rat small intestine for the study of absorption. *J. Physiol.* **110**, 36–46.

FISHER, R. B. & PARSONS, D. S. (1950). Glucose absorption from surviving rat small intestine. *J. Physiol.* **110**, 281–293.

KOEFOED-JOHNSEN, V. & USSING, H. H. (1958). The nature of the frog skin potential. *Acta physiol. scand.* **42**, 298–308.

MOKRASCH, L. C. (1954). Analysis of hexose phosphate and sugar mixtures with the anthrone reagent. *J. biol. Chem.* **208**, 55–59.

RIKLIS, E. & QUASTEL, J. H. (1958). Effects of cations on sugar absorption by isolated surviving guinea pig intestine. *Canad. J. Biochem. Physiol.* **36**, 347–362.

WILSON, T. H. & WISEMAN, G. (1954). The use of sacs of everted small intestine for the study of the transference of substances from the mucosal to the serosal surface. *J. Physiol.* **123**, 116–125.

Reprinted from *Am. J. Physiol.*, **200**, 459–460 (1961)

Effect of digitalis on active intestinal sugar transport[1]

30

T. Z. CSÁKY, H. G. HARTZOG III AND G. W. FERNALD

Department of Pharmacology, The University of North Carolina School of Medicine, Chapel Hill, North Carolina

Csáky, T. Z., H. G. Hartzog III and G. W. Fernald. *Effect of digitalis on active intestinal sugar transport.* Am. J. Physiol. 200(3): 459–460. 1961.—Ouabain (10^{-6} moles) and thevetin (10^{-7} moles) completely inhibit the active transport of 3-methylglucose across the intestine of the frog *in vitro*. Acetylstrophanthidin has a weak inhibitory action. The results are discussed in view of a possible link between active ionic and sugar transport mechanisms.

T HE MECHANISM RESPONSIBLE for the active transport of carbohydrates ('sugar pump') in the small intestine does not function in the absence of Na ions. Such an inhibition was recently demonstrated both *in vitro* in the intestine of the toad (1) and *in vivo* in the rat (2). Apparently a temporary lack of Na ions does not permanently damage the mucosa since the inhibition to the sugar pump is reversible. In view of these findings, one could assume that the sodium pump and sugar pump are closely related or may be identical.

Digitalis and similar drugs are known to inhibit active sodium transport in the red cells (3), kidney (4, 5), and in the surviving isolated frog skin (6). This paper describes the result of experiments indicating that digitalis inhibits the intestinal sugar pump in concentrations comparable to those which inhibit the sodium pump.

The experiments were carried out in the isolated surviving small intestine of the frog. The sugar used was 3-methylglucose. It was shown earlier that *in vitro* the small intestine of the frog transports 3-methylglucose against a gradient at room temperature without vigorous oxygenation (7). Moreover, this sugar is not metabolized by animal tissues (8, 9).

MATERIALS AND METHODS

The preparation of 3-methylglucose was the same as used in previous studies (7). Large frogs (*Rana pipiens*)

Received for publication November 21, 1960.

[1] This investigation was supported in part by a research grant from the National Institute of Neurological Diseases and Blindness, and by an institution grant from the American Cancer Society.

were used. The experimental procedure used was as previously described (7). Briefly, it consisted of injecting into an isolated loop of small intestine, ligated at both ends, the 3-methylglucose (about 10 mmole/l.) dissolved in frog Ringer's solution in which the chloride ion was replaced with sulfate. The replacement of chloride by sulfate in the Ringer's solution minimizes the transport of water in the surviving intestine. The loop was then placed in identical sugar-sulfate-Ringer's solution and shaken at $30°C$ for 6 hours. Samples were withdrawn from both the mucosal and serosal compartment, and the concentration of 3-methylglucose was determined after deproteinization (10) by the colorimetric procedure of Nelson (11).

The following digitalis preparations were used in these experiments: ouabain (G-strophanthin, U.S.P.-grade crystalline commercial preparation), crystalline thevetin, and acetylstrophanthidin. The concentration of the digitalis preparation was always identical in both mucosal and serosal compartments.

RESULTS

Since the presence of sulfate as principal anion results in a negligible net movement of fluid, the change of the sugar concentration in the serosal compartment was taken as a quantitative measure of active transport. It was found previously (7) that this change is always positive, viz. the sugar concentration increases on the serosal side of the intestine at $30°C$. At $0°C$ the change is negative; this is probably due to penetration of sugar into the intestinal tissue along with inhibition of the sugar pump. The effects of cardiac glucosides on the active transport of 3-methylglucose are presented in figures 1–3. The ordinate indicates the change in concentration at the end of 6 hours of 3-methylglucose in the serosal fluid expressed in percentage of the concentration at 0 hour. The abscissa shows the concentrations of the drugs employed. Each circle represents the result of an individual experiment, and the horizontal bars depict the mean value in each group.

Despite the individual variations it is clear that the cardiac glucosides inhibit active transport of 3-methylglucose. The rate of inhibition increases with the in-

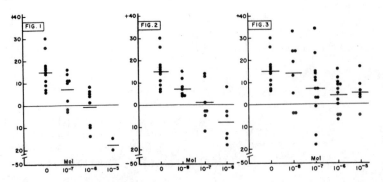

FIG. 1. Effect of ouabain on active intestinal transport of 3-methyl-glucose. *Abscissa:* concentration of glycoside; *ordinate:* end concentration of 3-methylglucose in serosal compartment expressed as % of initial concentration. 6-hr. incubation at 30°C.

FIG. 2. Effect of thevetin on active intestinal transport of 3-methylglucose. (*Abscissa, ordinate,* incubation period as in fig. 1.)

FIG. 3. Effect of acetylstrophanthidin on active intestinal transport of 3-methylglucose. (*Abscissa, ordinate,* incubation period as in fig. 1.)

creased concentration of the drug. Thevetin appears to be the most active drug, since a concentration of 10^{-7} moles decreased the average transport to almost nothing. Ouabain produces a similar effect in concentration of 10^{-6} moles. Acetylstrophanthidin, on the other hand, was considerably less effective. Complete inhibition of the average transport could not be achieved with concentrations as high as 10^{-5} moles, although the rate of transport compared with the controls gradually decreased with the increased concentration of the drug.

DISCUSSION

The experimental data presented in this paper indicate that digitalis inhibits active sugar transport in the isolated surviving small intestine of the frog. With thevetin and ouabain the inhibitory concentration is about the same as that found to inhibit active transport of sodium. Quantitatively there were marked differences among the individual drugs studied, thevetin and ouabain being more potent than acetylstrophanthidin.

The observed differences are not surprising as similar behavior is known among the digitalis-like drugs in their cardiac actions. Also very marked species differences have been noted, which may explain the weak inhibitory effect of acetylstrophanthidin on the frog intestine.[2]

The inhibition of active sugar transport by digitalis is a further evidence suggesting a probable link between the sodium pump and sugar pump. This connection may exist in the form of a functioning sodium pump which is essential for the function of the sugar pump. If this is the case, the primary effect of the digitalis would be on the ion transport, and the apparent effect on the sugar pump would be a secondary one.

An alternate theory would assume that all active transport mechanisms are driven by ATP and that a 'pump ATP-ase' is involved in every instance. This pump ATP-ase in the red cell membrane is known to be inhibited by digitalis (12). Finally, one should not neglect the possibility that one of the actions of digitalis is the uncoupling of oxidative phosphorylation, which, in turn, inhibits active transport. This possibility appears to be remote because in cardiac sarcosomes *in vitro* significantly higher concentrations (10^{-4} moles) are needed for uncoupling oxidative phosphorylation than those causing inhibition of active transport in the intact tissue (13). However, this latter possibility cannot be fully rejected before the concentrations of digitalis in various parts of intestinal tissue have been determined.

Thevetin and acetylstrophanthidin were generously supplied by Dr. K. K. Chen of Eli Lilly and Company.

REFERENCES

1. CSÁKY, T. Z. AND MARGRETHE THALE. *J. Physiol.* (*London*) 151: 59, 1960.
2. CSÁKY, T. Z. AND L. ZOLLICOFFER. *Am. J. Physiol.* 198: 1056, 1960.
3. SCHATZMANN, H. J. *Helvet. physiol. acta* 11: 346, 1953.
4. HYMAN, A. L., E. J. WILLIAM AND E. S. HYMAN. *Am. Heart J.* 52: 592, 1956.

[2] After submitting this paper the LD_{50} of acetylstrophanthidin and of ouabain were determined in frogs by injecting the drugs into the ventral lymph sac. The following values were obtained: acetylstrophanthidin 7.5 mg/kg; ouabain 0.20 mg/kg. In the cat, according to Dr. K. K. Chen (personal communication), the LD_{50} of acetylstrophanthidin is 0.1838 mg/kg and of ouabain 0.116 mg/kg.

5. SCHATZMANN, H. J., E. E. WINDHAGER AND A. K. SOLOMON. *Am. J. Physiol.* 195: 570, 1958.
6. KOEFOED-JOHNSEN, V. *Acta physiol. scandinav.* 42: Suppl. 145, 1957.
7. CSÁKY, T. Z. AND G. W. FERNALD. *Am. J. Physiol.* 198: 445, 1960.
8. CSÁKY, T. Z. AND J. E. WILSON. *Biochim. et biophys. acta* 22: 185, 1956.
9. CSÁKY, T. Z. AND J. E. GLENN. *Am. J. Physiol.* 188: 159, 1957.
10. SOMOGYI, J. *J. Biol. Chem.* 160: 61, 1945.
11. NELSON, N. *J. Biol. Chem.* 153: 375, 1944.
12. POST, R. L. *Fed. Proc.* 18: 121, 1959.
13. GOLDSCHMIDT, S. AND G. LAMPRECHT. *Hoppe-Seyl. Ztschr.* 307: 132, 1957.

Copyright © 1962 by Elsevier Publishing Company
Reprinted from *Biochim. Biophys. Acta*, **59,** 78–93 (1962)

31

STUDIES ON THE MECHANISM
OF INTESTINAL ABSORPTION OF SUGARS

V. THE INFLUENCE OF SEVERAL CATIONS AND ANIONS ON THE
ACTIVE TRANSPORT OF SUGARS, *IN VITRO,*
BY VARIOUS PREPARATIONS OF HAMSTER SMALL INTESTINE[*]

IVAN BIHLER AND ROBERT K. CRANE[**]

Department of Biological Chemistry, Washington University School of Medicine,
St. Louis, Mo. (U.S.A.)

(Received October 24th, 1961)

SUMMARY

Active sugar transport by hamster small intestine, *in vitro*, was studied with regard
to the influence of the ionic composition of the medium on the overall process. Active
transport was absolutely dependent upon Na^+. No other cation tested could replace
Na^+ and no other cation was specifically required. There was no specific requirement
for an anion. Also, active transport occurred in the presence of several anions which
do not penetrate tissues, indicating that trans-membrane movements of anions are
not involved in active sugar transport. The rate of active transport, the steady-state
tissue levels of accumulated sugar and the exchange rate of tritium-labelled sugar
in the steady-state condition were dependent on Na^+ and were apparently directly
proportional to the external Na^+ concentration. The conclusion is drawn that Na^+
directly affects the transport rate and that the steady-state levels of actively
transported sugars reflect a balance between inward active transport and outward
leakage and diffusion of the sugars. Measurements of actively transported sugar in
different parts of the intestinal wall show that the locus of Na^+ action is the brush-
border pole of the epithelial cell; the same pole at which it was previously shown that
the concentration difference of actively transported sugar is established.

INTRODUCTION

The active transport[***] of sugars by *in vitro* preparations of small intestine is absolutely dependent upon the presence of Na^+ in the bathing medium. RIKLIS AND QUASTEL[1],
using guinea-pig intestine, were apparently the first to make this observation. It has

---

[*] Preliminary reports have been made before a Symposium on Membrane Transport and
Metabolism, Prague, Czechoslovakia, August 22–27, 1960, and before the 45th Annual Meeting
of the American Society of Biological Chemists, Atlantic City, N. J., April 10–15, 1961.
[**] Present address: Department of Biochemistry, The Chicago Medical School, Chicago, Ill.
(U.S.A.).
[***] The term active transport is used here to mean the net movement or accumulation of a
compound against a concentration difference.

lately been confirmed by CSÁKY AND THALE[2] and CSÁKY AND ZOLLICOFFER[3] with the rat and the toad, respectively, and in our laboratory with the hamster. CSÁKY et al. have added the further information that Na^+ is required only at the mucosal surface of the intestinal wall and that inhibition of sugar transport due to the absence of Na^+ is readily reversed by its later addition.

The experiments in the present communication touch on several aspects of the subject; namely, the effect on transport of a variety of cations and anions, the tissue locus of Na^+ action and the effect of Na^+ concentration on the steady-state tissue levels of actively transported sugars, and on the rate of exchange of sugar between the external medium and the tissue.

METHODS AND MATERIALS

Incubation media

Throughout these studies the KREBS AND HENSELEIT[4] bicarbonate buffer at pH 7.4 and equilibrated with O_2-CO_2 (95:5) has been used to provide standard ionic conditions with which to compare the effect on active transport of numerous modifications. The ionic compositions of the KREBS-HENSELEIT medium and of a variety of other media which have been used are shown in Table I. Media which differed from the KREBS-HENSELEIT medium only with respect to their Na^+ concentration are not shown in Table I. These media were prepared by mixing KREBS-HENSELEIT medium with appropriate amounts of a Na^+-free medium. The K^+ medium (medium No. 2, Table I) was used for this purpose except when it is stated otherwise.

Tissue preparations

The experiments were performed by in vitro incubation of various preparations of hamster small intestine; namely, intestinal strips[6], suspensions of villi[6], everted sacs[7,8] and segments slit lengthwise to form a flat sheet. During their preparation, the tissues, except for villi, were immersed in the same medium as was later to be used for incubation. Villi lost their transporting ability when prepared in the K^+-medium and were for this reason always prepared in KREBS-HENSELEIT medium. Incubations were carried out at 37° in an atmosphere of O_2-CO_2 (95:5) when the medium contained bicarbonate or in 100% O_2 when it did not. Incubations were terminated and the tissues and media were processed for assay as described by CRANE AND MANDELSTAM[6].

In many of the experiments a value for the apparent extracellular space was obtained by measuring the distribution of [1,6-$^{14}C_2$]mannitol between the tissue and the medium. Mannitol has been used for the same purpose with erythrocytes[9] and the isolated, perfused heart[10] and has been shown not to be absorbed from the rat small intestine, in vivo[11].

Whenever it was necessary to measure the concentrations of sugar in different portions of the intestinal wall, the cutting technique of DICKENS AND WEIL-MALHERBE[12] as previously described[6] was used. With this technique, the segments are separated at about the level of the muscularis mucosae into a mucosal half which contains the villous portion of the wall held together by the muscularis mucosae and a serosal half which contains the submucosa, muscularis and serosa. The separated tissues were processed for assay in the usual way.

Biochim. Biophys. Acta, 59 (1962) 78–93

TABLE I

IONIC COMPOSITION OF THE VARIOUS MEDIA USED

All concentrations are given in mmoles/l.

---

1. KREBS–HENSELEIT bicarbonate

Na^+, 145; K^+, 6.6; Ca^{2+}, 2.5; Mg^{2+}, 1.9; Cl^-, 126; SO_4^{2-}, 1.9; HPO_4^{2-}, 1.9; HCO_3^-, 25.

Media based on KREBS–HENSELEIT *bicarbonate medium*

Designation	Difference from KREBS–HENSELEIT medium	
2. K^+ medium	$+ 145\ K^+$	$-145\ Na^+$
3. $Tris^+$ medium	$+ 145\ Tris^+$	$-145\ Na^+$
4. NH_4^+ medium	$+ 120\ NH_4^+$ $+ 25\ Tris^+$	$-145\ Na^+$
5. Li^+ medium	$+ 145\ Li^+$	$-145\ Na^+$
6. Ca^{2+} medium	$+ 79\ Ca^{2+}$ $+ 25\ Tris^+$ $+ 38\ Cl^-$	$-145\ Na^+$
7. Mg^{2+} medium	$+ 96\ Mg^{2+}$ $+ 47\ Cl^-$	$-145\ Na^+$
8. K^+-free medium	$+ 6.6\ Na^+$	$- 6.6\ K^+$
9. Iodomethane sulfonate medium No. 1	$+ 120\ IMS^-$	$-120\ Cl^-$
10. Ferrocyanide medium, No. 1 [*]	$+ 74\ Fe(CN)_6^{4-}$ $+ 66\ K^+$	$-126\ Cl^-$
11. Ferrocyanide medium, No. 2 [*]	$+ 45\ Fe(CN)_6^{4-}$ $+ 94\ Mannitol$	$-126\ Cl^-$
12. Na^+-free, ferrocyanide medium [*]	$+ 74\ Fe(CN)_6^{4-}$ $+ 217\ K^+$	$-126\ Cl^-$ $-145\ Na^+$

Other media

Designation	Total ionic composition
13. Iodomethane sulfonate medium, No. 2	Na^+, 150; IMS^-, 150; pH, 6.0
14. Mannitol medium	Mannitol, 308; pH, 6.7
15. Combined sulfate medium	Na^+, 192; K^+, 8; Mg^{2+}, 1; $Tris^+$, 20; SO_4^{2-}, 111; pH, 7.4
16. Na^+-sulfate medium	Na^+, 200; $Tris^+$, 20; SO_4^{2-}, 110; pH, 7.4
17. K^+-sulfate medium	K^+, 200; $Tris^+$, 20; SO_4^{2-}, 110; pH, 7.4
18. Mg^{2+}-sulfate medium	Mg^{2+}, 150; $Tris^+$, 20; SO_4^{2-}, 160; pH, 7.4

---

[*] The dissociation of ferrocyanide salts was assumed to be 65 % (see ref. 5). The concentrations of Na^+ and K^+ refer to the calculated concentration of the dissociated form only.

Methods of assay

Glucose was determined spectrophotometrically by means of the glucose oxidase reaction[13, 14], using the Glucostat reagent supplied by the Worthington Biochemical Corporation. 6-Deoxy-L-galactose was assayed by the methyl pentose method of DISCHE AND SHETTLES[15], ribose by the method of ROE AND RICE[16] and L-sorbose by the modified ketose method of ROE *et al.*[17]. When tritium-labelled compounds were

Biochim. Biophys. Acta, 59 (1962) 78–93

used as substrates, their radioactivity was assayed using the Packard Tri-Carb liquid scintillation spectrometer as previously described[18, 19]. When [1,6-$^{14}C_2$]mannitol was present along with a tritium-labelled compound, the radioactivity of the samples was measured in two different energy ranges and the activity of each isotope was calculated by a method of simultaneous equations which has a standard error of 4.7% for tritium and 8.4% for ^{14}C (see ref. 20).

Calculation of data

The data are reported in two ways: either as millimolar concentrations of sugar in the tissue and in the medium or as the ratio of these concentrations × 100; that is, as "per cent filling" of the tissue water space. The water content of the tissue was assumed to be 0.8 times the wet weight in all cases. This value was found by experiment to be the average water content of the separated mucosal and serosal halves, as well as of strips measured previously[6].

The mannitol space, as might have been expected, was not as great in experiments of short duration as it was with prolonged incubation. However, if it may be assumed that mannitol diffuses throughout the extracellular space of the tissue at about the same rate as sugars, the mannitol space may be taken to equal the extracellular space for sugars under the conditions of the particular experiment. Tissue concentrations were corrected on this assumption. No correction was made in experiments in which mannitol was omitted.

The methods used to calculate tissue sugar concentrations were like those previously used[6] differing only in the corrections for extracellular space.

RESULTS

Transport in cation-substituted media

The division of sugars and sugar analogs into two groups of compounds — those that are and those that are not actively transported — and the experimental basis for the division has been previously discussed[21]. In the experiments to be described, representative members of each of these two groups were used. Of the compounds used in this study, 1,5-anhydro-D-glucitol, 7-deoxy-D-glucoheptose and 6-deoxy-D-glucose, are actively transported; 1,5-anhydro-D-mannitol, 6-deoxy-L-galactose, L-sorbose, D-ribose and D-mannitol are not actively transported.

In the first experiments to be done, transport was measured in strips (Table II) and in villi (Table III) with the actively transported compounds, 1,5-anhydro-D-glucitol, 7-deoxy-D-glucoheptose and 6-deoxy-D-glucose, and was found to be absolutely dependent upon the presence of Na$^+$ in the medium. There was no active transport of these compounds when Na$^+$ was replaced by K$^+$, Li$^+$, NH$_4$$^+$, Tris$^+$, Mg^{2+}, Ca^{2+}, choline$^+$, guanidine$^+$ or mannitol. In fact, in the absence of Na$^+$, these ordinarily actively transported compounds did not come to equilibrium with the total water space of the tissue, but appeared to be restricted in distribution to a value lying between that of the mannitol space and complete equilibration; a distribution approximately the same as that observed with the compounds that are not actively transported. The entrance of the latter group of compounds into the tissue was not measurably influenced by the absence of Na$^+$ from the medium (Table III).

Tissue damage or deterioration is not a significant factor in the results shown in

Biochim. Biophys. Acta, 59 (1962) 78–93

TABLE II

THE INFLUENCE OF THE CATIONIC COMPOSITION OF THE INCUBATION MEDIUM ON THE
ACTIVE TRANSPORT OF SUGAR ANALOGS BY STRIPS OF HAMSTER INTESTINE

The compositions of the media used are given in Table I and are referred to in column 3, below.
The initial concentrations of the sugar analogs were, 6-deoxy-D-glucose, 5 mM, 1,5-anhydro-D-
glucitol, 5 mM and 7-deoxy-D-glucoheptose, 0.64 mM. The data in Expt. 4 were corrected for the
mannitol space, the others were not.

Expt. No.	Substrate	Medium used		Per cent filling	Duration of incubation (min)
		Principal cation	Reference to Table I		
1[*]	1,5-Anhydro-D-glucitol	Na$^+$	1	248	10
		K$^+$	2	31	
		Mg^{2+}	7	34	
		Ca^{2+}	6	28	
2	7-Deoxy-D-glucoheptose	Na$^+$	1	2960	20
		K$^+$	2	43	
		Mg^{2+}	7	26	
3[**]	7-Deoxy-D-glucoheptose	Na$^+$	1	947	15
		K$^+$	2	22	
		Guanidine$^+$	—	16	
		Choline$^+$	—	57	
4[***]	6-Deoxy-D-glucose	Na$^+$	1	330	10
		Tris$^+$	3	28	
		NH$_4^+$	4	24	
		Mannitol	14	21	
5	6-Deoxy-D-glucose	Na$^+$	1	217	10
		Li$^+$	5	22	
		K$^+$	2	6	
		Na$^+$	1	856	40
		Li$^+$	5	58	
		K$^+$	2	20	

[*] Analysis of the K$^+$ and Mg^{2+} media at the end of the experiment showed a Na$^+$ concentration
of 2 mM.
[**] The choline$^+$ and guanidine$^+$ media were prepared in concentrations the same as the Li$^+$
medium, Table I, No. 5.
[***] Analysis of the mannitol medium at the end of the experiment showed a Na$^+$ concentration
of 3 mM.

TABLE III

THE INFLUENCE OF THE CATIONIC COMPOSITION OF THE INCUBATION MEDIUM ON THE
ACTIVE TRANSPORT OF SUGAR ANALOGS BY
PREPARATIONS OF VILLI FROM HAMSTER SMALL INTESTINE

The compositions of the media used are given in Table I. All experiments were of 15 min duration.

Substrate	Substrate concentration (mM)	Per cent filling	
		KREBS-HENSELEIT	K$^+$ medium
6-Deoxy-D-glucose	4	593	27
7-Deoxy-D-glucoheptose	0.3	2450	25
1,5-Anhydro-D-glucitol	5	286	
1,5-Anhydro-D-mannitol	5	20	
6-Deoxy-L-galactose	5	46	48

Biochim. Biophys. Acta, 59 (1962) 78–93

Tables II and III. As is illustrated in Fig. 1, strips which were unable to accumulate 6-deoxy-D-glucose when incubated in the Tris+ medium completely recovered their normal transporting capacity when returned to KREBS–HENSELEIT buffer. The same

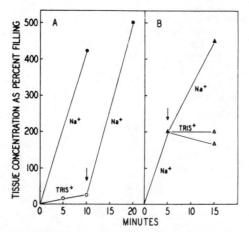

Fig. 1. Reversibility of ion effects on active transport of 6-deoxy-D-glucose by strips of everted hamster small intestine. The initial concentration of 6-deoxy-D-glucose was 5 mM. The data were corrected for mannitol space. The incubations were begun in a medium with the cation indicated and the tissue was transferred to the new medium at the time indicated by the arrow.

result was obtained when the K+ medium was used instead of the Tris+ medium. When Mg²⁺ medium was used, only partial recovery was observed during the time period of the experiment. However, hamster tissue survives incubation for periods not much in excess of one hour and it is likely that if it were able to survive longer

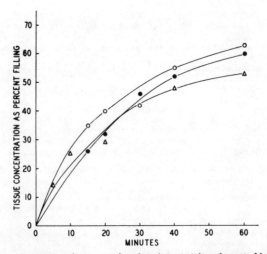

Fig. 2. Time course of entrance of sugar analogs into intact strips of everted hamster small intestine. The data are not corrected for the mannitol space. △—△, 5 mM 6-deoxy-L-galactose in KREBS–HENSELEIT medium; ●—●, 5 mM 6-deoxy-L-galactose in the K+-medium; ○—○, 0.3 mM 7-deoxy-D-glucoheptose in the K+-medium.

Biochim. Biophys. Acta, 59 (1962) 78–93

complete recovery would have been observed. CSÁKY AND THALE[2], using a preparation of toad intestine which does survive long periods of incubation, observed that the reduced transporting ability of this tissue following incubation in a Mg^{2+} medium was gradually but completely restored after being returned to a Na^+ medium.

With regard to the rate of entrance of sugar analogs into intestinal tissue in the absence of Na^+, there appears to be little difference between compounds which are actively transported and those that are not. As illustrated in Fig. 2, equilibration in the absence of Na^+ is attained slowly and at approximately the same rate for 7-deoxy-D-glucoheptose and 6-deoxy-L-galactose. Whether, with prolonged incubation, these compounds would have equilibrated with the total tissue water was not determined because the tissue did not remain viable long enough.

TABLE IV

THE INFLUENCE OF THE K^+ CONCENTRATION ON THE
ACTIVE TRANSPORT OF 7-DEOXY-D-GLUCOHEPTOSE IN STRIPS OF HAMSTER SMALL INTESTINE

The ionic compositions of the media are given in Table I. In the experiment with added ion-exchange resin, each flask contained 100 mg of sodium-loaded Dowex-50 resin having a total exchange capacity of about 250 μmoles. The initial concentration of 7-deoxy-D-glucoheptose was 0.6 mM. The duration of incubation was 15 min.

Expt. No.	Incubation medium	Per cent filling
1	KREBS–HENSELEIT	910
	KREBS–HENSELEIT + 8.4 mM KCl	670
	K^+-free medium (Table I, No. 8)	497
2	KREBS–HENSELEIT	850
	K^+-free medium	530
	K^+-free medium + ion-exchange resin	637

Na^+ is the only cation of the KREBS–HENSELEIT medium which appears to be essential for active transport. However, the rate of the process is influenced to some extent by the concentrations of some other cations. A few data showing the effects of K^+ are given in Table IV. In the absence of added K^+, that is, in a K^+-free medium, active transport was depressed. However, because liberation of intracellular K^+ from incubated tissue may be expected, it is likely that there was a small amount of K^+ present in this experiment, especially during the later period of incubation. An effort was made to capture any liberated K^+ by including a sodium-loaded ion-exchange resin in the contents of the incubation vessel. The rate of transport was not further reduced by the addition of resin, but seemed, on the contrary, to be somewhat improved.

Since RIKLIS AND QUASTEL[1] have reported an increased rate of glucose transport by guinea-pig intestine in the presence of 15 mM K^+, an experiment was performed in which KCl, in the solid form, was added to the KREBS–HENSELEIT buffer to achieve this concentration. In the presence of this added K^+, active transport by strips of hamster intestine was reduced.

The experiments with choline$^+$ and Li^+ (Table II, Expts. 3 and 5, respectively) appear to deserve some comment. Neither of these ions replaced Na^+ in supporting active transport, i.e., accumulation against a concentration difference. However,

Biochim. Biophys. Acta, 59 (1962) 78–93

both of them permitted a greater degree of penetration of sugar into the tissue than did K⁺, suggesting the possibility that these ions, choline⁺ and Li⁺, partially substitute for Na⁺ in some as yet unknown way.

Transport in anion-substituted media

As the data in Table V show, none of the anions of the KREBS–HENSELEIT buffer was specifically required for active transport of sugars to occur. Moreover, those data in Table V which show that active transport of sugar occurred when Cl⁻ and HCO_3^- were replaced by sulfate, ferrocyanide or iodomethane sulfonate indicate that trans-membrane or transcellular movements of anions are also not required. These ions which are much larger and, in two instances, carry more negative charges than either

TABLE V

THE INFLUENCE OF THE ANION COMPOSITION OF THE INCUBATION MEDIUM
ON ACTIVE TRANSPORT OF SUGAR ANALOGS BY STRIPS OF HAMSTER INTESTINE

The detailed compositions of the media are given in Table I. Normal composition refers to the composition of the KREBS–HENSELEIT buffer.

Substrate	Substrate concentration (mM)	Medium composition		Per cent filling	Duration of incubation (min)
		Principal cation	Principal anion		
7-Deoxy-D-glucoheptose	0.3	Normal	Normal	2130	15
		Normal −Ca²⁺	SO_4^{2-}	1935	
		Na⁺	SO_4^{2-}	1190	
		K⁺	SO_4^{2-}	31	
		Mg²⁺	SO_4^{2-}	56	
6-Deoxy-D-glucose	5	Normal	Normal	605	10
		Normal	Iodomethane sulfonate⁻	487	
		Normal	Normal	562	
		Na⁺	Iodomethane sulfonate⁻	383	
		Normal	Normal	387	
		Normal	$Fe(CN)_6^{4-}$	344	
		Normal + K⁺	$Fe(CN)_6^{4-}$	245	

Cl⁻ or HCO_3^- are believed to be restricted to the extracellular spaces of tissues[22]. Control experiments indicate that strips of hamster intestine are not different in this respect. For example, distribution of ferricyanide in incubated strips was measured and was found to be less than the distribution of mannitol. Ferricyanide, in the presence of which active transport also occurs, was chosen for these control experiments because ferrocyanide undergoes some oxidation during incubation and preparation for assay of tissue samples and cannot be measured with precision.

The influence of 4,6-dinitro-o-cresol on transport

It has been shown previously that intestinal active transport of sugar depends upon normally functioning aerobic metabolism and that reduction of energy supplies by the use of 4,6-dinitro-o-cresol or anaerobic conditions causes inhibition[6]. The data

Biochim. Biophys. Acta, 59 (1962) 78–93

TABLE VI

THE INFLUENCE OF 4,6-DINITRO-0-CRESOL ON THE ENTRANCE OF 6-DEOXY-L-GALACTOSE
AND THE ACTIVE TRANSPORT OF 7-DEOXY-D-GLUCOHEPTOSE
INTO STRIPS AND VILLI PREPARED FROM HAMSTER INTESTINE

The following concentrations of added substances were used: 7-deoxy-D-glucoheptose, 0.3 mM,
6-deoxy-L-galactose, 5 mM, and 4,6-dinitro-o-cresol, 0.075 mM. The duration of incubation was
30 min for strips and 15 min for villi.

Tissue preparation	Compounds added	Per cent filling	
		KREBS–HENSELEIT	K+ medium
Strips	7-Deoxy-D-glucoheptose	2950	47
	7-Deoxy-D-glucoheptose + 4,6-dinitro-o-cresol	522	63
	6-Deoxy-L-galactose	55	49
	6-Deoxy-L-galactose + 4,6-dinitro-o-cresol	52	47
Villi	7-Deoxy-D-glucoheptose	2120	55
	7-Deoxy-D-glucoheptose + 4,6-dinitro-o-cresol	800	33
	6-Deoxy-L-galactose	79	36
	6-Deoxy-L-galactose + 4,6-dinitro-o-cresol	87	63

in Table VI show that Na^+-dependent active transport is inhibited by 4,6-dinitro-
o-cresol. The slow entrance of the various compounds which occurs in the absence
of Na^+ appears, on the contrary, not to be appreciably influenced.

The influence of Na^+ concentrations on the rate of active transport

Insofar as can be determined, the rate of active transport appears to be a direct
function of the concentration of Na^+ added to the medium. An experiment illustrating

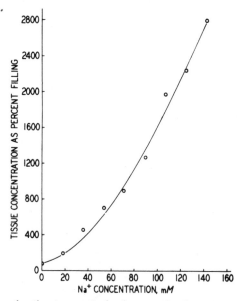

Fig. 3. Dependence of active transport of 7-deoxy-D-glucoheptose on the concentration of Na^+
in the medium. The initial medium concentration of 7-deoxy-D-glucoheptose was 0.3 mM. The
duration of incubation was 20 min. The results are corrected for the mannitol space.

Biochim. Biophys. Acta, 59 (1962) 78–93

TABLE VII

THE INFLUENCE OF Na$^+$ CONCENTRATIONS ON THE ACTIVE TRANSPORT OF SUGAR ANALOGS BY STRIPS OF HAMSTER SMALL INTESTINE

The detailed compositions of the media listed in column 2 are given in Table I to which the numbers in parentheses refer. Graded concentrations of Na$^+$ between 0 and 145 mM were obtained by mixing appropriate amounts of KREBS–HENSELEIT buffer with these media. Higher Na$^+$ concentrations were achieved in the ferrocyanide medium by replacement of K$^+$ with Na$^+$ and in the KREBS–HENSELEIT medium by addition of solid NaCl. The media marked with an asterisk were osmotically equivalent to one another.

Substrate	Medium	Duration of incubation (min)	Tissue accumulation as "per cent filling" at the following concentrations of Na$^+$ in mM									
			0	18	36	72	108	126	145	163	179	217
6-Deoxy-D-glucose	KREBS–HENSELEIT	10							366			
6-Deoxy-D-glucose	Tris$^+$ (No. 3)	10	24	123	187	250						
6-Deoxy-D-glucose	Mannitol (No. 14)	10	25		231	315	351					
6-Deoxy-D-glucose	Ferrocyanide (No. 10)	10				123	163		244	282		335
1,5-Anhydro-D-glucitol	K$^+$ (No. 2)	30			43	126	287	368				
1,5-Anhydro-D-glucitol	KREBS–HENSELEIT	20							313		353	
	Same + NaCl*											
	Same + mannitol*								244			

this point is shown in Fig. 3. The tissue concentration of 7-deoxy-D-glucoheptose increased in a nearly linear fashion with the medium Na+ concentration. Other experiments (Table VII) show that this relationship between the active transport of sugar and the Na+ concentration is independent of the actively transported sugar analog used, of the anionic composition of the medium and of the alternate cation or other substance used to replace Na+. Most striking is the fact that there is no indication from the data in Table VII, particularly in the experiment with a ferrocyanide medium, that the process of active transport can be saturated with respect to Na+. Each increment of Na+ up to the highest value tested, 217 mM, produced a corresponding increment in the rate of active transport.

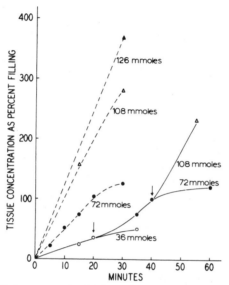

Fig. 4. Influence of the Na+ concentration in the medium on the active transport of 1,5-anhydro-D-glucitol by intact strips of hamster small intestine. At the times indicated by the arrows, the Na+ concentrations were adjusted to the final concentrations shown by the addition of solid NaCl. The initial medium concentration of 1,5-anhydro-D-glucitol was 5 mM. The data were corrected for the mannitol space.

It has been previously shown by CRANE AND MANDELSTAM[6] that the accumulation of actively transported sugar continues with time until a plateau value is reached beyond which there is no further increase with prolonged incubation. The experiment illustrated in Fig. 4 indicates that this plateau value is also a direct function of the medium Na+ concentration. Tissue incubated initially at a low Na+ concentration and capable of achieving only a proportionately low plateau level of accumulation responded rapidly to added Na+ with increased transport rates and higher plateau levels of accumulation.

Steady-state tissue levels and the exchange rate of transported sugar

Other experiments indicate that the plateau value described above is a steady-state condition resulting from a balance between inward active transport and outward leakage and diffusion of sugar. The relevant data are given in Table VIII. The experiment which yielded these data was performed as follows: Strips of hamster small intestine were incubated in media containing Na$^+$ and 1,5-anhydro-D-glucitol at the concentrations stated in the table for 40 min, a period of time long enough for the plateau value to be reached. At this point of time, a tracer amount of tritium-labelled

TABLE VIII

THE INFLUENCE OF Na$^+$ CONCENTRATIONS ON THE STEADY-STATE TISSUE CONCENTRATIONS
OF 1,5-ANHYDRO-D-GLUCITOL AND ON THE
EXCHANGE RATE OF 1,5-ANHYDRO-D-GLUCITOL IN THE STEADY-STATE

The technical aspects of the experiment are described in the text. The data are corrected for mannitol space. The medium concentration of 1,5-anhydro-D-glucitol was 5 mM.

Medium Na$^+$ (mM)	Tissue 1,5-AG (mM)	Exchange rate (mmoles/l tissue water/min)	Fraction exchanged (per cent/min)
145	18.0	0.69	3.8
108	13.8	0.51	3.7
72	9.6	0.37	3.8
36	4.0	0.14	3.6
0	2.1	0.07	3.2

1,5-anhydro-D-glucitol of high specific activity was added and incubation was continued. The distribution of the tritium-labelled compound was then measured in tissue samples recovered at 2 and 5 min after the addition of the tracer. The total quantity of 1,5-anhydro-D-glucitol in the tissue was measured in different samples that had been incubated from the beginning with the tritium-labelled compound and was found to remain unchanged during the 5-min period of the exchange measurement. The distribution of radioactivity found at the 2- and 5-min intervals was used to calculate the exchange rates given in Table VIII.

The data in Table VIII show that the exchange rate and the steady-state tissue concentrations of 1,5-anhydro-D-glucitol increase with the Na$^+$ concentration in the same way as does the initial transport rate (*cf.* Fig. 3). Moreover, the fraction of the total tissue sugar exchanged per unit time was constant over wide variations in the Na$^+$ levels. The results obtained in these exchange experiments seem to agree well with what would be expected from directly measured initial rates of transport. For example, at an exchange rate of 3.7%/min (Table VIII), the total sugar content of the tissue would be replaced in about 27 min. The plateau level in accumulation experiments is reached in about 35 min. Theoretically, these two periods of time should be the same. However, the point may be made that they are not markedly different, especially when one considers the effects of leakage. The exchange rates were measured under conditions where the tissue level was relatively high and only a small proportion of the total sugar was exchanged. Thus leakage of labelled sugar from the tissue was undoubtedly only a small factor in the measured rate. When the tissue concentration starts at zero as in the accumulation experiments, however, leakage occurs throughout incubation and is probably an important factor in the length of time required for the plateau level to be reached.

It is reasonable to conclude that the same process which is responsible for filling the tissue with sugar continues at approximately the same rate after the plateau

Biochim. Biophys. Acta, 59 (1962) 78–93

level has been reached. In other words, the results suggest that the plateau level is, in fact, a steady-state resulting from a balance between inward active transport and outward leakage and diffusion.

Site of entry of sugars and the locus of the action of Na+

Throughout the studies described above, the assumption has been implicit that the locus of Na+ action and the site of Na+-dependent entry of sugars with *in vitro* preparations is the brush-border pole of the epithelial cells in or near which the process of active transport has been previously shown to be located by McDougal, Little and Crane[23] and Csáky and Fernald[24]. The validity of this assumption is shown by the experiments described below.

These experiments all depend upon measurements of sugar concentrations in different parts of the intestinal wall during incubation. In performing them, the technique with small strips could not be used because such small pieces cannot be separated into mucosal and serosal halves. Longer segments had to be used. Also, some of the experiments were carried out with everted sacs. In both cases, the advantage of the strip technique of Crane and Mandelstam[6] in providing a random sample of tissue was lost. Comparisons between two segments, even adjacent ones, cannot be made because of the markedly different transport activity in different parts of the small intestine[6, 25].

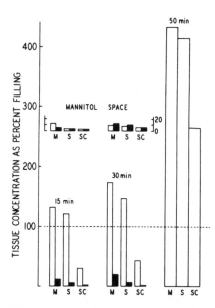

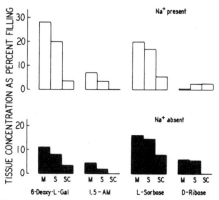

Fig. 6. Penetration of sugars which are not actively transported into everted sacs of hamster small intestine. The results are not corrected for mannitol space. 10 mM D-ribose, or 5 mM of the other sugars and tracer amounts of mannitol were present initially on the mucosal side only. Time of incubation was 30 min. Symbols as in Fig. 5. 1,5-AM, 1,5-anhydro-D-mannitol.

Fig. 5. Active transport of 1,5-anhydro-D-glucitol by everted sacs of hamster small intestine. The results are not corrected for mannitol space. 10 mM 1,5-anhydro-D-glucitol and tracer amounts of mannitol were present initially on the mucosal side only. Mannitol spaces, determined in the same sacs as the respective sugar analog, are drawn to the same scale. Mannitol space was not determined at the 50-min time period. The distance from the baseline to the top of the respective portion of the bar represents the value for percent filling. □—□, Krebs–Henseleit medium; ■—■, K+ medium; M, mucosal layer of the tissue; S, serosal layer of the tissue; SC, serosal compartment (contents of sac).

Biochim. Biophys. Acta, 59 (1962) 78–93

The time course of active sugar transport into the wall and into the serosal compartment of everted sacs which were initially filled with a sugar-free medium is shown in Fig. 5. As expected from the previous findings of McDougal, Little and Crane[23] the highest concentration of sugar, throughout the entire 50 min of incubation, was found in the mucosal layer of the preparation. These results confirm the suggestion of Crane and Mandelstam[6] that the underlying layers of the intestinal wall impose a quantitatively important, unphysiological barrier to the movement of sugars in everted sacs as well as the conclusion[23] that the overall kinetics of sugar movements in sacs reflect the characteristics of diffusion from high concentrations within the epithelial cells to the relatively lower concentrations in the underlying tissues and the serosal medium.

The penetration of four compounds, which are not actively transported, 6-deoxy-L-galactose, 1,5-anhydro-D-mannitol, L-sorbose and D-ribose into everted sacs is shown in Fig. 6. As was to be expected from the results of previous experiments with the strip technique, penetration was slow and, except for 6-deoxy-L-galactose relatively uninfluenced by Na^+. The tissue concentration was in all cases less than the concentration in the medium, even in the mucosal layer exposed directly to the sugar-containing medium. Also, with the exception of 6-deoxy-L-galactose, the mannitol space accounted for nearly the total quantity of the sugar analogs which entered the tissue. On the basis of these results, it would appear that 6-deoxy-L-galactose is not ideal as an example of a sugar which is not actively transported. Its use has been dictated by the ease and precision with which it may be measured by the Dische and Shettles method for methyl pentoses[15]. The reasons for its variance from the other compounds of this group are not clear.

It was also of interest to study further the penetration of mannitol, since the mannitol spaces of everted sacs exposed to this compound on the mucosal surface only are very small and are actually near the limit of accuracy of the methods of assay and calculation (cf. Figs. 5 and 6). Experiments were performed by fastening the mucosal half of a segment of rabbit small intestine between two lucite chambers of equal size. When mannitol was initially present only on the mucosal side, no [^{14}C]-mannitol could be detected in the serosal compartment after a period of 55 min at 37° and the tissue space filled was only 2.8%. However, when the mannitol was initially present in contact with the cut undersurface about 1% of [^{14}C]mannitol was found in the mucosal compartment after 70 min and the tissue space filled was 25%. Substantially the same results were obtained when the intact intestinal wall of rabbit small intestine was used. Thus, it appears that the mucosal surface is nearly impervious to mannitol and the conclusion seems justified that mannitol enters the extracellular space of the various in vitro preparations through their cut edges and the serosal membrane.

The fact that the penetration of sugars and sugar analogs into strips follows the patterns established by the above experiments with everted sacs is illustrated by experiments (Fig. 7) in which segments of intestine cut open to form flat sheets were used. These preparations expose both surfaces of the intestinal wall to the sugar-containing medium. In the presence of Na^+, 1,5-anhydro-D-glucitol was transported against a concentration difference and was accumulated to a higher concentration in the mucosal half of the intestinal wall. In the absence of added Na^+, the penetration of 1,5-anhydro-D-glucitol into the tissue was the same for both layers and not appreci-

Biochim. Biophys. Acta, 59 (1962) 78–93

ably different from the penetration by mannitol and 6-deoxy-L-galactose. The influence of the mucosal epithelium, *per se*, is shown by the result obtained when a segment of intestine was denuded of epithelial cells prior to incubation. This kind of preparation, which cannot carry out active transport of sugars was penetrated by 1,5-anhydro-D-glucitol to a concentration equal to that in the medium and penetration was not influenced by Na+.

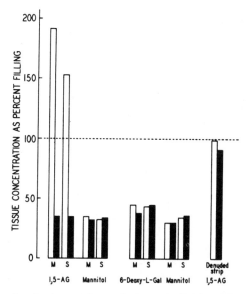

Fig. 7. Active transport and entrance of 1,5-anhydro-D-glucitol (1,5-AG) and 6-deoxy-L-galactose into intestinal segments slit open lengthwise and into segments denuded of epithelial cells. Period of incubation was 15 min. Symbols as in Fig. 5. The concentration of sugars was 5 mM.

DISCUSSION

Two points raised by the results of the present experiments seem worthy of further discussion. First, if quantitatively small effects are ignored, the conclusion seems inescapable that not only is Na+ required for active transport of sugars but it is the only such ion. Second, the question may be asked whether the entrance of Na+, alone or together with an anion, into the cells is a necessary part of its activity. On this point there is no direct evidence in the form of measurements of intracellular Na+ concentrations. From the data so far reported, it would appear possible for the action of Na+ to be limited to the external membrane of the cells, perhaps to a specific site on the cell surface through which actively transported sugars may cross the membrane at high rates. However, in the light of experiments reported in the following paper[26] and in a manuscript in preparation, strong consideration should be given to the possibility that a trans-membrane movement of Na+ is essential to active sugar transport. If this be so, then it would seem to follow that inward trans-membrane movement of ions associated with the active transport of sugars is limited to Na+. The experiments with impermeant anions appear to be conclusive on this point.

Biochim. Biophys. Acta, 59 (1962) 78–93

ACKNOWLEDGEMENTS

The authors are indebted to Miss M. L. ROBERTS for her competent technical assistance. This investigation was supported by grant No. G-11216 from the National Science Foundation.

REFERENCES

1 E. RIKLIS AND J. H. QUASTEL, *Can. J. Biochem. and Physiol.*, 36 (1958) 347.
2 T. Z. CSÁKY AND M. THALE, *J. Physiol. (London)*, 151 (1960) 59.
3 T. Z. CSÁKY AND L. ZOLLICOFFER, *Am. J. Physiol.*, 198 (1960) 1056.
4 H. A. KREBS AND K. HENSELEIT, *Z. physiol. Chem. Hoppe-Seyler's*, 210 (1932) 33.
5. G. FALK AND J. F. LANDA, *Am. J. Physiol.*, 198 (1960) 289.
6 R. K. CRANE AND P. MANDELSTAM, *Biochim. Biophys. Acta*, 45 (1960) 460.
7 T. H. WILSON AND G. WISEMAN, *J. Physiol. (London)*, 123 (1954) 116.
8 R. K. CRANE AND T. H. WILSON, *J. Appl. Physiol.*, 12 (1958) 145.
9 C. R. PARK, R. L. POST, C. F. KALMAN, J. H. WRIGHT, L. H. JOHNSON AND H. E. MORGAN, *Ciba Foundation Colloq. Endocrinol.*, 9 (1955) 249.
10 R. B. FISHER AND D. B. LINDSAY, *J. Physiol. (London)*, 131 (1956) 526.
11 R. HÖBER AND J. HÖBER, *J. Cellular Comp. Physiol.*, 10 (1937) 401.
12 F. DICKENS AND H. WEIL-MALHERBE, *Biochem. J.*, 35 (1941) 7.
13 A. S. KESTON, *Abstracts of Papers, 129th Meeting of the Am. Chem. Soc.*, 1956, p. 31C.
14 A. SAIFER AND S. GERSTENFELD, *J. Lab. Clin. Med.*, 51 (1958) 448.
15 Z. DISCHE AND L. B. SHETTLES, *J. Biol. Chem.*, 175 (1948) 595.
16 J. H. ROE AND E. W. RICE, *J. Biol. Chem.*, 173 (1948) 507.
17 J. H. ROE, J. H. EPSTEIN AND N. P. GOLDSTEIN, *J. Biol. Chem.*, 178 (1949) 939.
18 R. K. CRANE, G. R. DRYSDALE AND K. A. HAWKINS, *4th Annual Symposium on Tracer Methodology*, New England Nuclear Corp., Atomlight, No. 15, 1960, p. 4.
19 E. HELMREICH AND R. K. CRANE, in H. SCHWIEGK AND F. TURBA, *Künstliche radioaktive Isotope in Physiologie, Diagnostik und Therapie*, Band I, Springer-Verlag, Berlin, 1961, p. 644.
20 G. T. OKITA, J. J. KABARA, F. RICHARDSON AND G. V. LEROY, *Nucleonics*, 15 (1957) 111.
21 R. K. CRANE, *Physiol. Revs.*, 40 (1960) 789.
22 C. R. KLEEMAN, F. H. EPSTEIN, M. E. RUBIN AND A. F. LAMDIN, *Am. J. Physiol.*, 182 (1955) 548.
23 D. B. McDOUGAL, Jr., K. S. LITTLE AND R. K. CRANE, *Biochim. Biophys. Acta*, 45 (1960) 483.
24 T. Z. CSÁKY AND G. W. FERNALD, *Nature*, 191 (1961) 709.
25 R. B. FISHER AND D. S. PARSONS, *J. Physiol. (London)*, 119 (1953) 224.
26 I. BIHLER, K. A. HAWKINS AND R. K. CRANE, *Biochim. Biophys. Acta*, 59 (1962) 94.

Reprinted from *Biochim. Biophys. Acta*, **59**, 94–102 (1962)

94

32

STUDIES ON THE MECHANISM OF
INTESTINAL ABSORPTION OF SUGARS

VI. THE SPECIFICITY AND OTHER PROPERTIES OF
Na+-DEPENDENT ENTRANCE OF SUGARS INTO INTESTINAL TISSUE
UNDER ANAEROBIC CONDITIONS, *IN VITRO**

IVAN BIHLER, KENNETH A. HAWKINS** AND ROBERT K. CRANE**

Department of Biological Chemistry, Washington University School of Medicine,
St. Louis, Mo. (U.S.A.)

(Received October 24th, 1961)

SUMMARY

Measurements *in vitro* of anaerobic sugar entrance into the tissue of hamster small intestine have established the existence of a specific, Na+-dependent and energy-independent process mediating the rapid equilibration of certain sugars between the tissue and the medium. This process appears to be a component of the aerobic active-transport mechanism. It shares with active transport a dependence on the Na+ concentration in the medium, a location at the brush-border pole of the epithelial cells and an apparently identical substrate specificity. The entrance into the tissue of a new analog, 6-deoxy-1,5-anhydro-D-glucitol, was found to be Na+-dependent, thus confirming the previously derived specificity pattern of active transport.

INTRODUCTION

The absolute dependence of intestinal active transport of sugars, *in vitro*, on the presence of Na+ in the medium bathing the mucosal surface of the epithelial cells was discussed in the preceding paper[1] and various aspects of this phenomenon, as observed with hamster intestine, were examined. In the present paper, the subject is pursued further, with attention to the influence of Na+ on the entrance of various sugars and sugar analogs into intestinal tissue under conditions which prevent their active transport and consequently their accumulation against a concentration difference.

The significance of the present studies with regard to the mechanism of intestinal active transport of sugars resides in the fact that the results collectively suggest that the overall process of active transport has two components; namely, (a) a Na+-dependent and energy-independent entrance of sugars through the mucosal surface

---

* A preliminary report was made to the 45th Annual Meeting of the American Society of Biological Chemists, Atlantic City, N. J., April 10–15, 1961.
** Present address: Department of Biochemistry, The Chicago Medical School, Chicago, Ill. (U.S.A.).

of the epithelial cells and (b) a Na^+- and energy-dependent accumulation against a concentration difference. As the data will show, the substrate specificity of the overall process is exhibited by the first-named component.

A new compound, 6-deoxy-1,5-anhydro-D-glucitol (1,6-dideoxyglucose) has been synthesized. The effect of Na^+ on the entrance of this compound into hamster intestine adds significantly to the specificity studies previously reported from this laboratory[2,3] and elsewhere[4] which were interpreted to mean that all actively transported sugars share a single, common process[5-7].

MATERIALS AND METHODS

Incubation techniques

The present experiments were performed by *in vitro* incubation of small strips or rings of everted hamster intestine[8] and segments slit open lengthwise to form a flat sheet. The media used for incubation were KREBS–HENSELEIT bicarbonate buffer[9], pH 7.4 or its Na^+-free, K^+ modification described in the preceding paper[1]. Anaerobic conditions, when desired, were produced and maintained by a continuous flow of water-saturated N_2-CO_2 (95:5) directly into the media contained in the incubation flasks. In aerobic experiments, the gas space of the flasks contained O_2-CO_2 (95:5). Incubations were terminated and the tissues and media were processed for assay as previously described by CRANE AND MANDELSTAM[8].

Assay methods

Tritium-labelled sugars and sugar analogs were prepared by the WILZBACH direct exposure technique[10], purified and assayed with the Tri-Carb liquid scintillation spectrometer as previously described[11,12]. All the sugars used except D-xylose and 6-deoxy-L-galactose were available in the tritium-labelled form. D-[1-^{14}C]Xylose was purchased from the National Bureau of Standards and was assayed on the liquid scintillation spectrometer. 6-Deoxy-L-galactose was assayed by the DISCHE AND SHETTLES procedure for methyl pentoses[13].

Preparation of 6-deoxy-1,5-anhydro-D-glucitol

This previously unreported sugar analog was synthesized in the following way: 6-O-Tosyl-2,3,4-tri-O-acetyl-1,5-anhydro-D-glucitol was prepared by the procedure of BARKER[14]. Iodination with NaI in acetic anhydride yielded 6-iodo-6-deoxy-2,3,4-tri-O-acetyl-1,5-anhydro-D-glucitol which was reduced and deacetylated with lithium aluminum hydride (Metal Hydrides, Inc., Beverly, Mass. (U.S.A.)) to give 6-deoxy-1,5-anhydro-D-glucitol. The product was crystallized from ethanol, m.p. 147–148. Commercial elemental analysis gave the following results: Found: C, 48.52 %; H, 8.17 %. Calculated: C, 48.65 %; H, 8.17 %.

Calculations

A detailed description and discussion of the various calculations used is given in the preceding paper[1]. The data of the present paper are reported as "per cent filling" which is equal to the concentration of the sugar or sugar analog in the tissue × 100 divided by the concentration in the medium. The data were corrected for the proportion of tissue solids and, where noted, for the mannitol space measured in the same tissue sample. The techniques and calculations involved in exchange experiments are described by BIHLER AND CRANE[1].

RESULTS AND DISCUSSION

The experimental separation of entrance from accumulation

The influence of Na^+ and of energy-yielding processes on the entrance and accumulation of three representative compounds was measured in the experiments summarized by Table I. 6-Deoxy-D-glucose was used to represent the group of fourteen actively transported compounds and 6-deoxy-L-galactose was used to represent the larger group of compounds which are not actively transported[6]. The new compound, 6-deoxy-1,5-anhydro-D-glucitol was compared with these.

TABLE I

THE INFLUENCE OF Na^+ AND 4,6-DINITRO-*o*-CRESOL ON ACTIVE TRANSPORT AND ANAEROBIC ENTRANCE OF SUGARS INTO STRIPS OF HAMSTER INTESTINE

The concentration of the sugars was 5 mM and of 4,6-dinitro-*o*-cresol, 0.75 mM. The duration of incubation was 30 min in the aerobic and 10 min in the anaerobic experiment. The data are corrected for the mannitol space.

Gas phase	Compound	Medium	Per cent filling	
			Control	+ 4,6-dinitro-o-cresol
O_2–CO_2	6-Deoxy-D-glucose	KREBS–HENSELEIT	688	164
		K^+ medium	12	12
	6-Deoxy-1,5-anhydro-D-glucitol	KREBS–HENSELEIT	53	35
		K^+ medium	19	25
	6-Deoxy-L-galactose	KREBS–HENSELEIT	22	20
		K^+ medium	18	18
N_2–CO_2	6-Deoxy-D-glucose	KREBS–HENSELEIT	62	65
		K^+ medium	9	17
	6-Deoxy-1,5-anhydro-D-glucitol	KREBS–HENSELEIT	42	40
		K^+ medium	15	15
	6-Deoxy-L-galactose	KREBS--HENSELEIT	15	15
		K^+ medium	8	15

Under aerobic conditions, 6-deoxy-D-glucose and 6-deoxy-L-galactose behaved in the manner previously observed[1]; that is, 6-deoxy-D-glucose underwent Na^+-dependent active transport and was accumulated within the tissue against a concentration difference whereas 6-deoxy-L-galactose entered to only a small extent which was not appreciably influenced by the presence of Na^+. The new compound, 6-deoxy-1,5-anhydro-D-glucitol, behaved in a manner that was, in a sense, intermediate between the others. Like 6-deoxy-L-galactose, it was not accumulated against a concentration difference, but, like 6-deoxy-D-glucose, its entrance into the tissue was greatly increased in the presence of Na^+.

When the available tissue energy supplies were reduced by the addition of 4,6-dinitro-*o*-cresol, which inhibits aerobic phosphorylation[15], the accumulation of 6-deoxy-D-glucose and the entrance of 6-deoxy-1,5-anhydro-D-glucitol were inhibited, but the entrance of 6-deoxy-L-galactose was scarcely affected. Under anaerobic con-

Biochim. Biophys. Acta, 59 (1962) 94–102

ditions, neither 6-deoxy-D-glucose nor 6-deoxy-1,5-anhydro-D-glucitol was accumulated against a concentration difference, but the entrance of each was increased by Na+. As may have been expected, 4,6-dinitro-o-cresol had no effect under anaerobic conditions. These results suggest that uncoupling of active sugar transport from its energy supplies, though it eliminates accumulation against a concentration difference, does not measurably alter other properties of the system. To provide a further test of this proposition, the experiment illustrated in Fig. 1 was performed. In this experi-

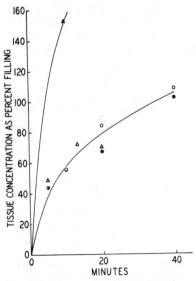

Fig. 1. The influence of high concentrations of 4,6-dinitro-o-cresol on sugar transport. The concentration of 1,5-anhydro-D-glucitol was 5 mM. The data were corrected for the mannitol space. KREBS–HENSELEIT medium was used. △—△, Anaerobic; ▲—▲, aerobic; O—O, aerobic, 4,6-dinitro-o-cresol 10^{-3} M; ●—●, aerobic, 4,6-dinitro-o-cresol $2 \cdot 10^{-3}$ M.

ment, the entrance of the ordinarily actively transported compound, 1,5-anhydro-D-glucitol was compared in the presence, under aerobic conditions, of concentrations of 4,6-dinitro-o-cresol which are ten to twenty times higher than those required to inhibit aerobic phosphorylation completely[15] and under anaerobic conditions. The entrance rates under these two conditions were indistinguishable.

The tissues used in these studies came from fasted animals and they contained minimal stores of glycogen (around 0.03% dry wt.). However, in spite of this small quantity the possibility remained that glycolysis could be a factor in maintaining the characteristics of Na+-dependent entrance under anaerobic conditions or in the presence of 4,6-dinitro-o-cresol. In other experiments carried out under anaerobic conditions, the addition of 25 mM sodium fluoride or 1 mM iodoacetate did not decrease the extent of entrance of either 6-deoxy-D-glucose or 1,5-anhydro-D-glucitol. These concentrations of sodium fluoride and iodoacetate are ordinarily inhibitory to glycolysis in other animal tissues[16], as well as the intestine[8,17], and their failure to influence Na+-dependent entrance is further assurance that the process is independent of tissue energy supplies.

Biochim. Biophys. Acta, 59 (1962) 94–102

The site of Na+-dependent anaerobic entrance

In the preceding paper[1], experiments were described in which analyses of the separated mucosal and serosal halves of incubated intestinal segments showed that the site of entry of sugars being actively transported and the locus of Na+ action were at the mucosal surface of the epithelial cells. Illustrated in Fig. 2 are similar

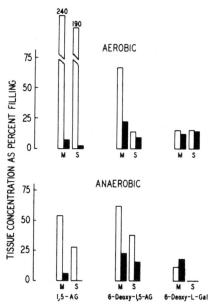

Fig. 2. The distribution of sugars in the mucosal and serosal halves of incubated intestinal segments. The segments were approx. 5 cm in length and were slit open lengthwise. The sugar concentrations were 5 mM. Incubation was for 15 min. The data were corrected for the mannitol space. The open bar is the value for KREBS-HENSELEIT medium, the shaded bar is for K+ medium. M, mucosal half; S, serosal half; 1,5-AG, 1,5-anhydro-D-glucitol.

experiments which were designed to test the proposition that Na+-dependent anaerobic entrance is, in these respects, identical to Na+-dependent active transport. Using the compounds, 1,5-anhydro-D-glucitol, 6-deoxy-1,5-anhydro-D-glucitol, and 6-deoxy-L-galactose, the relative concentrations observed in the mucosal and serosal halves of the tissue following incubation confirm the proposition.

The exchange rate under anaerobic conditions

The rate of exchange of 1,5-anhydro-D-glucitol was measured as described by BIHLER AND CRANE[1] and in the legend to Table II. As shown by the data in Table II, the rate of exchange was a direct function of the medium Na+ concentration. It was found in a separate experiment that 15 min of preincubation were enough for complete equilibration of 1,5-anhydro-D-glucitol between the medium and the tissue water at the highest concentration of Na+ used, 145 mM. The uniformity of the data would suggest that equilibration had occurred at the other Na+ concentrations as well, but the actual tissue concentrations of 1,5-anhydro-D-glucitol were not measured. Comparison of the data in Table II with those of Table VIII in the preceding paper[1]

Biochim. Biophys. Acta, 59 (1962) 94–102

where exchange was measured under aerobic conditions reveal a possibly important relationship between aerobic and anaerobic exchange rates. As shown in column 4, Table II, the ratio of these rates is nearly constant and it has an average value, for the limited data available, of about 2.2. What appears to be clear from this relationship is that Na^+ has proportionately the same effect anaerobically as aerobically. Whether aerobiosis always increases the rate of exchange to a similar degree or whether the degree of increase is a function of the sugar used must be decided by further experiment.

TABLE II

THE INFLUENCE OF THE CONCENTRATION OF Na^+ ON THE RATE OF EXCHANGE OF TRITIUM-LABELLED I,5-ANHYDRO-D-GLUCITOL WITH STRIPS OF HAMSTER INTESTINE AT EQUILIBRIUM, UNDER ANAEROBIC CONDITIONS

The strips were incubated for 15 min in media having the Na^+ concentration indicated and containing 5 mM unlabelled I,5-anhydro-D-glucitol. A tracer amount of the tritium-labelled analog was then added and incubation was continued for 10 min. The various media were prepared as described by BIHLER AND CRANE[1], and were equilibrated with N_2–CO_2 (95:5). The data are corrected for the mannitol space. The values for aerobic exchange rate are taken from BIHLER AND CRANE[1].

Na^+ concentration (mM)	Exchange rate (mmoles/l tissue water/min)	Exchange rate × 100 / Na^+ concentration	Aerobic exchange rate / Anaerobic exchange rate
145	0.29	0.20	2.38
108	0.22	0.20	2.32
72	0.18	0.25	2.06
36	0.07	0.20	2.0
0	0.06	—	1.16

TABLE III

THE SPECIFICITY OF Na^+-DEPENDENT ANAEROBIC ENTRANCE OF SUGARS AND SUGAR ANALOGS INTO STRIPS OF HAMSTER INTESTINE

The concentration of each sugar was 5 mM. The duration of incubation was 15 min. The data are not corrected for the mannitol space. The terms "actively transported" and "not actively transported" have been previously defined[6] and refer to whether or not a concentration difference was established during incubation of these compounds with everted sacs of hamster intestine.

Compound	Per cent filling		
	KREBS-HENSELEIT	K^+ medium	Difference
Actively transported			
I,5-Anhydro-D-glucitol	68	16	52
3-O-Methyl-D-glucose	40	16	24
D-Allose	48	15	33
6-Deoxy-D-glucose	75	28	47
Intermediate			
6-Deoxy-I,5-anhydro-D-glucitol	62	23	39
Not actively transported			
2-Deoxy-D-glucose	31	28	3
I,5-Anhydro-D-mannitol	25	15	10
D-Xylose	29	25	4
6-Deoxy-L-galactose	26	22	4

Biochim. Biophys. Acta, 59 (1962) 94–102

The specificity of Na⁺-dependent anaerobic entrance

As shown by the data in Table III, the pattern of specificity of Na⁺-dependent anaerobic entrance, insofar as it was tested, appears to be identical with the specificity of the overall process of Na⁺-dependent active transport which has been previously described[6]. Those sugars which are actively transported by the criterion of accumulation against a concentration difference under aerobic conditions respond to the addition of Na⁺ with a greatly increased entrance under anaerobic conditions. Those sugars which are not actively transported are not appreciably influenced by Na⁺. 6-Deoxy-1,5-anhydro-D-glucitol, as noted above, acts, in this regard, like an actively transported compound.

TABLE IV

THE INHIBITION OF THE ENTRANCE OF SUGAR ANALOGS INTO STRIPS OF HAMSTER INTESTINE
BY THE ADDITION OF A SECOND SUGAR OR SUGAR ANALOG

In all experiments the concentration of the test sugar analog was 5 mM. The duration of incubation was 10 min. The data are not corrected for the mannitol space.

Expt. No.	Gas phase	Test compound	Added compound	Per cent filling	
				KREBS-HENSELEIT	K⁺ medium
1	N₂–CO₂	6-Deoxy-D-glucose		95	22
			Glucose, 5 mM	55	
2	N₂–CO₂	6-Deoxy-1,5-anhydro-D-glucitol		54	31
			Glucose, 5 mM	39	32
3	N₂–CO₂	6-Deoxy-L-galactose		33	33
			Glucose, 5 mM	33	32
4	O₂–CO₂	6-Deoxy-1,5-anhydro-D-glucitol		60	33
			Glucose, 25 mM	36	26
			Galactose, 25 mM	44	26
5	O₂–CO₂	6-Deoxy-1,5-anhydro-D-glucitol		64	31
			6-Deoxy-L-galactose, 10 mM	58	39
			D-Ribose, 10 mM	67	35

Na⁺-dependent anaerobic entrance of one ordinarily actively transported sugar is also inhibited by the simultaneous addition of a second such sugar (Table IV). It has been previously shown that mutual inhibition occurs between actively transported sugars[5,7]. The data in Table IV suggest that Na⁺-dependent anaerobic entrance shares this characteristic.

Of particular interest is the inhibition of the entrance of 6-deoxy-1,5-anhydro-D-glucitol by glucose and galactose. Accumulation of 6-deoxy-1,5-anhydro-D-glucitol against a concentration difference has not been observed. Yet, by the criteria of Na⁺-dependent entrance and inhibition of entrance by other sugars, it appears to belong to the group of actively transported sugars. Moreover, as shown by the data in Table I, a portion of its rate of aerobic entrance appears to be sensitive to 4,6-dinitro-o-cresol. Further experiments along this line, such as the one illustrated in Fig. 3 in which the aerobic entrance rate is significantly faster than the anaerobic entrance rate, have given strong indications that there is an energy-dependent component in the entrance of 6-deoxy-1,5-anhydro-D-glucitol under aerobic conditions.

Biochim. Biophys. Acta, 59 (1962) 94–102

Because of the technical difficulties encountered in attempts to measure entrance quantitatively when there is no accumulation of sugar, it has not yet been possible to determine whether our failure to detect accumulation of 6-deoxy-1,5-anhydro-D-glucitol against a concentration difference means that it does not occur or merely

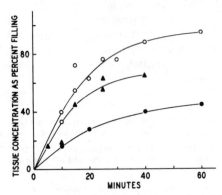

Fig. 3. The time course of entrance of 6-deoxy-1,5-anhydro-D-glucitol into strips under different conditions. The concentration of 6-deoxy-1,5-anhydro-D-glucitol was 5 mM. The data were corrected for the mannitol space. O—O, Aerobic, KREBS–HENSELEIT medium; ●—●, aerobic, K⁺ medium; ▲—▲, anaerobic, KREBS–HENSELEIT medium.

that its degree is too small for easy analytical detection. This problem of interpretation with which we are confronted is a general one and its solution requires information which we are not yet in a technical position to obtain. For example, in our calculations it is an implicit assumption that all parts of the epithelial cell are equally permeated by sugar that has entered through the mucosal surface. It may be emphasized that if the sugar were restricted in its distribution within the cell, it would be possible for a concentration difference between some portion of the cell and the medium to exist although analysis of the whole cells would fail to show it. In this regard the experiments of KIPNIS AND CORI[18] which may be interpreted as indicating a non-uniform distribution of sugar in diaphragm muscle appear to be relevant.

CONCLUSIONS

There are three conclusions relevant to the mechanism of intestinal absorption of sugars which appear to be possible on the basis of the data presented; namely: (a) intestinal active transport of sugars has two components; namely, Na⁺-dependent, energy-independent entrance and Na⁺-dependent, energy-dependent accumulation against a concentration difference. Whether or not these components are sequentially related, i.e., whether the latter is a distinct and separate second step in the overall process, is a question of considerable importance which cannot be answered unequivocally at the present time. (b) The specificity of the overall process of intestinal active transport of sugars resides in the entrance component. (c) The compound, 6-deoxy-1,5-anhydro-D-glucitol, is one of the group of actively transported compounds.

The special interest in finding that 6-deoxy-1,5-anhydro-D-glucitol is an actively transported compound is the strong support which is thus given to the previous

Biochim. Biophys. Acta, 59 (1962) 94–102

conclusion[5-7] that all actively transported compounds share a single, common process. All of the compounds which are presently known to be accumulated against a concentration difference have structure A, in the diagram below. However, in the tests that were previously made, there was no one compound which differed from glucose by

          A   B

more than the absence of one single, potentially reactive, hydroxyl group; *i.e.*, all of the compounds contained a hydroxyl group either at carbon 6 or at carbon 1. Thus, it could have been imagined that two chemically similar reactions, one occurring at carbon 6 and the other at carbon 1, could have accounted for the overall specificity of active transport. 6-Deoxy-1,5-anhydro-D-glucitol has the structure B, above. The fact that this compound is unable to react at either carbon 6 or carbon 1 serves to weld the previous observations into a uniform whole.

ACKNOWLEDGEMENTS

The authors are indebted to Miss M. L. ROBERTS for her competent technical assistance. This investigation was supported by grant No. G-11216 from the National Science Foundation.

REFERENCES

[1] I. BIHLER AND R. K. CRANE, *Biochim. Biophys. Acta*, 59 (1962) 78.
[2] R. K. CRANE AND S. M. KRANE, *Biochim. Biophys. Acta*, 20 (1956) 568.
[3] T. H. WILSON AND R. K. CRANE, *Biochim. Biophys. Acta*, 29 (1958) 30.
[4] T. H. WILSON AND B. R. LANDAU, *Am. J. Physiol.*, 198 (1960) 99.
[5] R. K. CRANE, *Biochim. Biophys. Acta*, 45 (1960) 477.
[6] R. K. CRANE, *Physiol. Revs.*, 40 (1960) 789.
[7] C. R. JORGENSEN, B. R. LANDAU AND T. H. WILSON, *Am. J. Physiol.*, 200 (1961) 111.
[8] R. K. CRANE AND P. MANDELSTAM, *Biochim. Biophys. Acta*, 45 (1960) 460.
[9] H. A. KREBS AND K. HENSELEIT, *Z. physiol. Chem. Hoppe-Seyler's*, 210 (1932) 33.
[10] K. E. WILZBACH, *J. Am. Chem. Soc.*, 79 (1957) 1013.
[11] R. K. CRANE, G. R. DRYSDALE AND K. A. HAWKINS, *4th Annual Symposium on Tracer Methodology*, New England Nuclear Corporation, Atomlight, No. 15, 1960, p. 4.
[12] E. HELMREICH AND R. K. CRANE, in H. SCHWIEGK AND F. TURBA, *Künstliche radioaktive Isotope in Physiologie, Diagnostik und Therapie*, Band 1, Springer-Verlag, Berlin, 1961, p. 644.
[13] Z. DISCHE AND L. B. SHETTLES, *J. Biol. Chem.*, 175 (1948) 595.
[14] S. B. BARKER, *Can. J. Chem.*, 32 (1954) 634.
[15] G. H. A. CLOWES, A. K. KELTCH, C. F. STRITTMATTER AND C. P. WALTERS, *J. Gen. Physiol.*, 33 (1950) 555.
[16] P. P. COHEN, in C. A. ELVEHJEM AND P. W. WILSON, *Respiratory Enzymes*, Burgess, Minneapolis, 1939, p. 137.
[17] W. WILBRANDT AND L. LASZT, *Biochem. Z.*, 259 (1933) 398.
[18] D. KIPNIS AND C. F. CORI, *J. Biol. Chem.*, 234 (1959) 171.

Na⁺-dependent transport in the intestine and other animal tissues[1]

33

ROBERT K. CRANE

Department of Biochemistry, Chicago Medical School, Chicago, Illinois

A POSSIBLE INVOLVEMENT of Na⁺ in the transmembrane movement of organic substrates appears to have been first noted as the result of a controversy on the mechanism of intestinal absorption of sugars which began when the Coris (13) and Wilbrandt and Lengyel (49) observed that glucose absorption was decreased in adrenalectomized animals and ended when Deuel (26), Althausen (2), Clark (12), and their various associates found that this decrease was prevented by providing sodium chloride in the drinking water. Some years later, Riklis and Quastel (41), working with one of the newly introduced in vitro techniques, found effects of monovalent cations which led them to conclude that "Na⁺ must be present for glucose absorption to take place." Since that time the Na⁺ ion effect has become one of the more completely documented phenomena (6, 23, 24) and the findings of recent years appear to have brought our knowledge to a point where it seems that Na⁺-dependent transport can be adequately explained, at least at the phenomenological level.

In previous symposia (15, 20), my colleagues and I have presented a concept of the brush border membrane of the epithelial cell as a digestive-absorptive surface in which the elements responsible for digestion and absorption are arranged in ordered proximity to one another, as in a mosaic. As a part of the over-all concept, we also proposed a mechanism for the specific energy-requiring process of the mucosa which enables the small intestine to carry out sugar active transport. This mechanism had the following experimental basis. Experiments under conditions of limited tissue energy supplies as compared to others under ordinary aerobic conditions identified the entry of sugars across the brush border membrane as the primary site of Na⁺ involvement in the over-all process (7). These results were interpreted as having "established the existence of a substrate-specific, Na⁺-dependent and energy-independent process mediating the rapid equilibration of sugars between the cells and the medium" (7). Along the lines of current mobile carrier concepts (50), interaction of actively transported sugars with a specific binding site on a mobile carrier

was postulated to account for the substrate specificity (14) of the over-all process and for the characteristics of phlorizin inhibition (4). It was also postulated that the carrier possessed a second binding site specific for Na⁺ and that binding of Na⁺ to this site was essential to the ability of the carrier to equilibrate sugar across the brush border membrane. Simultaneous movement of Na⁺ and sugar was assumed to occur as a direct consequence of these postulated interactions (20).

As we visualized it, the carrier system was of itself capable only of equilibration. The asymmetry required to achieve uphill accumulation of sugar was attributed to a gradient of Na⁺ concentration "downhill" into the cell maintained by the operation of an outwardly directed, energy-dependent Na⁺ pump present at a different locus in the same membrane. This location of a Na⁺ pump was assumed largely because of the probability that the brush border region, which can be isolated as a discrete entity (35), actually forms a subcompartment of the cell and the assumption is to some extent justified by the fact that a Na⁺, K⁺-dependent ATPase is present in the brush border region (A. Eichholz, unpublished observations and 47). However, the position of the pump is not of prime importance; it is essential for the hypothesis only that the local internal Na⁺ concentration in the region of the equilibrating carrier be maintained low relative to the medium.

Thus, the elements of the hypothesis which are specifically essential to Na⁺-dependent transport are only those shown in Fig. 1 where the mobile carrier is represented as possessing two specific binding sites—one for the substrate and one for Na⁺. This diagram is, of course, only a kinetic model and is not intended to illustrate any particular molecular properties. It has been drawn to illustrate the usual assumption that the carrier migrates freely between the interfaces of the cell membrane in either the free or the combined form (50) and our special assumption that substrate and Na⁺ interact with the specific binding sites on the carrier in approximately the same way in response to their respective local concentrations on both sides of the diffusion barrier.

There are now numerous experiments which have detected the interaction and movement of Na⁺ proposed by this hypothesis. For example, Bosackova and Crane

[1] Research on Na⁺-dependent transport in the author's laboratory has been supported by grants from the National Science Foundation.

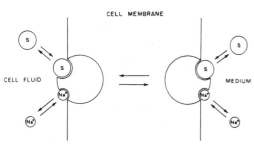

FIG. 1. Postulated mobile carrier and its interactions with substrate and Na⁺.

(8) have studied the effect of Na⁺ concentration on sugar transport in the presence of a variety of substances in replacement of Na⁺. Consistent with our previous results (6), the rate of sugar transport was found to be a function of the Na⁺ concentration of the medium. However, it was also found to be a function of the particular substance used to replace Na⁺; K⁺, Li⁺, Cs⁺, Rb⁺, and NH₄⁺ were all strongly inhibitory as compared to Tris⁺, choline⁺, or urea. On the basis of the hypothesis, we surmised that K⁺ and related ions inhibit sugar transport because they interfere with Na⁺-carrier interaction and concluded that one should expect to see inhibition of Na⁺ movement into the cell by K⁺. As a test, the movement of ²²Na into the cell in the presence of various concentrations of K⁺ and Tris⁺ was studied and was found to be inhibited. Moreover, there was a marked coincidence between the degree of inhibition of ²²Na influx and of 6-deoxyglucose influx. These results were interpreted as suggesting that K⁺ competes with Na⁺ for the specific cation binding site on the sugar carrier and that a K⁺-loaded carrier is relatively inefficient in equilibrating sugar.

Recently, a more direct demonstration of the interaction of Na⁺ with the carrier has been achieved. Among the various criteria which may be applied to test a presumed mobile carrier mechanism, the phenomenon of countertransport is the most critical and most widely accepted (50). Countertransport of sugars with the in vitro intestine was shown several years ago by Salomon and his colleagues who studied xylose and glucose (43) and more recently by Alvarado who has completed an extensive study of countertransport of a variety of actively transported sugars (3). These studies have shown that addition of a second substrate will induce a previously equilibrated substrate to efflux into the medium against its own concentration gradient.

Na⁺-carrier interaction has been demonstrated in much the same way. Since it is the downhill gradient of Na⁺ into the cell which is presumed to result in the "uphill" movement of sugar in the same direction, the essence of a proper test is to reverse the Na⁺ gradient and to observe whether sugar is then moved uphill out of the cell into the medium. For these experiments, tissue was incubated under nitrogen and in the presence of dinitrophenol to limit Na⁺ pumping and sugar

accumulation. After 5 min in the presence of 6-deoxyglucose and a Na⁺ concentration of 120 mEq/liter, 6-deoxyglucose had entered to a tissue/medium ratio of slightly over one. The preparation was then removed to a medium containing the same external concentration of 6-deoxyglucose but with Tris⁺ in replacement of Na⁺; 6-deoxyglucose moved out of the cells into the medium against its own concentration gradient (17).

As briefly described by Dr. Curran in the preceding paper (25) there is strong additional support for the hypothesis of Na⁺-carrier interaction from studies of the in vitro transmural potential. Addition of any one of the glucose group of actively transported sugars to the mucosal medium causes an immediate and rapid rise in the potential. This is the result to be expected were the movement of these sugars and of Na⁺ across the brush border membrane dependent upon one another as we have postulated. Many such studies (5, 33, 44) show the same phenomenon.

Although these experiments virtually demand the conclusion that Na⁺ interacts with the sugar carrier, some have stated (21, 22) that inhibition of sugar transport by K⁺ and related ions is caused not by an effect on the sugar carrier but by their entrance into the cell and consequent dilution of intracellular Na⁺. Bosackova and Crane (9) have measured the variations of intracellular Na⁺ under several different conditions of Na⁺ replacement and they have found sugar transport to vary independently.

On this basis, it may be concluded that accumulation of substrate by the equilibrating system is dependent upon the gradient of Na⁺ downhill into the cell as one, and doubtlessly the major, asymmetry and upon interference by the high intracellular K⁺ concentration with Na⁺-carrier interaction for outward movement as a second asymmetry. Although most results can be so explained those of two experiments cannot. In one the effect of Na⁺ concentration on the tissue/medium substrate ratio was tested by Bihler and Crane (6). At low Na⁺ concentrations, they found tissue/medium ratios less than one. The other was an in vivo perfusion study by Csaky (21) in which he detected no difference in the absorption of 3-methylglucose whether Na⁺ was or was not added to the perfusion fluid: To explain these observations, the system must possess some additional asymmetric property.

What we believe to be this additional asymmetric property has been reported by Crane et al. (18) and is depicted in Fig. 2 where it can be seen that Na⁺ has a profound influence on the affinity of sugar for the transport carrier; the lower the Na⁺ concentration, the lower the affinity of sugar. In fact, the apparent K_m is increased about 200-fold in going from Na⁺ of 145 mEq/liter to zero. The effect of K⁺ expected from the work of Bosackova and Crane is also found. As shown in Fig. 3, at a given concentration of Na⁺, replacement of Tris⁺, which does not interfere with Na⁺-carrier interaction, by K⁺ which does, causes a large additional decrease in affinity over that caused by lowering the Na⁺ concentration.

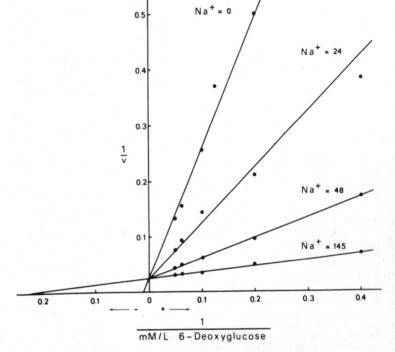

FIG. 2. Dependence of the rate of 6-deoxyglucose transport on the concentration of 6-deoxyglucose and Na⁺. (From Crane, Forstner, and Eichholz (18).)

These experiments on K_m were carried out by measurements of sugar accumulation. What appears to be precisely the same effect has also been demonstrated by Irving Lyon (33) who has measured substrate-dependent alterations in the transmural potential of rat jejunum. In Fig. 4 reciprocals of sugar-induced increases in transmural potential, which measure increased mucosal → serosal Na⁺ flux (33, 45), are plotted against the reciprocals of the sugar concentrations; the apparent K_m is high when Na⁺ concentration is low.

Our interpretation of these various phenomena in terms of the proposed carrier is shown in Fig. 5. Two forms of the carrier are assumed. One, the Na⁺-loaded carrier, as in our original postulate (Fig. 1), in which the conformation of the substrate binding site is such that affinity is high, and the other, a K⁺-loaded carrier in which the conformation of the carrier is assumed to be altered in such a way as to reduce affinity for the substrate. As a first approximation, one needs only two such species of the carrier to explain the results. At saturating Na⁺ concentrations, where affinity is maximal, only the Na⁺-loaded carrier is assumed to be present. Complete replacement of Na⁺ by K⁺ would be expected to convert all of the carrier to the K⁺-loaded variety and affinity would be minimal. Intermediate values of affinity can be considered to result from the presence of various proportions of these two species.

The consequences of these interactions are readily deduced from basic principles but can be made more clearly apparent by reference to a generalized kinetic plot as in Fig. 6. The two solid lines represent, respectively, the expected effect of substrate concentration on the inward movement of carrier from the Na⁺-rich medium and the outward movement of carrier from the Na⁺-poor cell contents. Assuming incubation to be started at a medium concentration indicated by the vertical dotted line, $1/S$ (medium), the rate of substrate movement into the cell is represented by the horizontal dotted line. Initially, with no substrate in the cell, the rate of outward sugar movement is zero. As substrate enters and its concentration rises, the rate of outward movement will increase along the upper solid line until equilibrium is established at the point of intersection with the horizontal dotted line. The concentration of substrate within the cell at equilibrium will be that found by dropping a vertical line from this point of intersection.

Reference to this plot makes several aspects of Na⁺-dependent transport clear. First, as has been found (19), the lower the concentration of substrate in the medium the higher will be the ratio of tissue/medium concentrations at equilibrium. Second, since the maximal rate is the same, or nearly the same, no matter what the Na⁺ concentration, Na⁺ will have no apparent effect if the

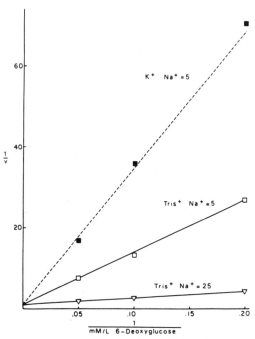

FIG. 3. Comparison of kinetic constants of 6-deoxyglucose transport with Tris+ or K+ as replacement for Na+. The open symbols are Tris+, the solid symbol is K+. The Na+ concentrations were as follows: ▽, 25 mEq/liter and ■, □, 5 mEq/liter. (Modified from Crane, Forstner, and Eichholz (18).)

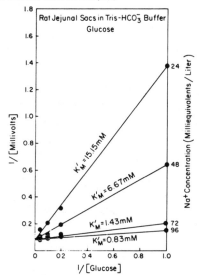

FIG. 4. Effect of Na+ concentration on the apparent K_m of D-glucose determined from transmural PD data. (From Lyon and Crane (33).)

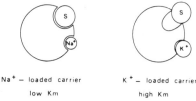

FIG. 5. Postulated species of the mobile carrier.

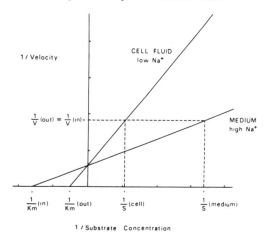

FIG. 6. Generalized kinetic plot to illustrate the effect of asymmetry of K_m on an equilibrating system.

substrate concentration is very high. This would appear to be the explanation for Csaky's experiments which led him to dispute (21, 22) direct interaction of Na+ with the substrate carrier. The concentration of 3-methylglucose he used was 145 mM and no effect of Na+ was to be expected. Third, when the Na+ concentration in the medium is made deliberately lower than that in the cell, equilibrium will be established at a tissue/medium ratio less than one (6).

The theoretical basis for viewing translocation of molecules through cell membranes against a concentration or electrochemical gradient as the consequence of metabolism-dependent asymmetries has been well described by Wilbrandt and Rosenberg (50) and by Mitchell (36). As we now view it, Na+-dependent transport is probably influenced by three asymmetries; namely, *1*) the inward, downhill Na+ gradient which is effective because the rate of carrier movement is dependent upon the interaction with Na+, *2*) the outward, downhill K+ gradient because intracellular K+ would be expected to interfere with Na+ interaction for outward movement of the carrier, and *3*) the gradient of substrate-carrier affinity which requires that equilibrium be established at a tissue/medium substrate ratio greater than one. All three of these asymmetries depend ultimately upon the energy-dependent translocation of Na+

out of the cell which has been reviewed by Curran (25) and which is probably a vectorial biochemical activity, but it is important to recognize that Na$^+$-dependent transport of itself is an equilibrating system and not a vectorial biochemical activity. In this regard, it is of special interest to note that, with this kind of system in the membrane, the rate of inward movement of substrate is not directly controlled by the asymmetry of the system and the cell is able to obtain essential nutrient at an adequate rate irrespective of the state of its available energy supplies.

With regard to the intestinal absorption of sugar, independent work by Alvarado (3) has confirmed the proposed interactions of Na$^+$ and K$^+$ using the countertransport phenomenon as a test. However, as we have suggested previously (16, 20), it may be anticipated that any carrier system—for any substrate, in any animal cell—with which Na$^+$ interacts as described, will unavoidably result in the accumulation of substrate against the concentration gradient; all animal cells are relatively depleted in Na$^+$. With this in mind, a review of recent literature on intestinal transport of various compounds and of Na$^+$-dependent transport in other kinds of cells has been undertaken with the result that this mechanism appears to be ubiquitous, although the quantitative proportions of the asymmetries may vary from cell to cell and with the substrate.

For example, recent experiments by the Rosenbergs (42), which were graciously made available prior to publication, have shown effects of Na$^+$ concentration on amino acid uptake by intestinal tissue which are in entire agreement with the results with sugars; i.e., the results demonstrate that the entrance of the substrate into the tissue is a step controlled by Na$^+$. In this connection, it can be recalled that Nathans, Tapley, and Ross (37) studied the accumulation of monoiodotyrosine by rat intestine in vitro and found it to require Na$^+$ and

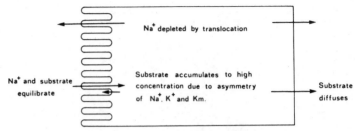

FIG. 7. Representation of the in vitro test system.

to be inhibited by K$^+$. They also found a graded inhibition by K$^+$ which they interpreted as a competitive inhibition between K$^+$ and monoiodotyrosine for the uptake process. On the basis of what we now know, these same data can just as easily be reinterpreted as showing a decrease in the affinity of monoiodotyrosine for the mobile carrier system as the Na$^+$ concentration is decreased.

The transport of bile salts studied by Lack and Weiner (32), Playoust and Isselbacher (39), and Holt (29) responds to Na$^+$ and to a variety of replacement ions in much the same way as does sugar uptake. With kidney cortex slices, Kotyk and Kleinzeller (30) have observed effects on sugar accumulation and Segal (28) and his co-workers have observed effects on amino acid accumulation that are highly suggestive of a unity of mechanism.

The accumulation of some amino acids by nuclei (1), lymphocytes (51), granulocytes (51), rat diaphragm (38), ascites tumor cells (31), and red cells (48) is Na$^+$-dependent. Much work in this area has been carried out by Christensen and his colleagues who described, in a series of publications (11), a complex association between amino acid transport and ion distribution. Although their experiments never clearly differentiated between the effects of K$^+$ and Na$^+$, they concluded that the effective ion was K$^+$ and that "K$^+$ efflux from the cell moved a carrier for neutral amino acids in a direction favorable to amino acid transport" (40). Understanding seems to have come to this field after Heinz and his co-workers (31) made an observation upon which later experiments on Na$^+$-dependent amino acid transport appear to have been based. Just as there had been found evidence for a Na$^+$-dependent step of sugar entry into intestine (7), Heinz and his co-workers found glycine entry into ascites tumor cells to be dependent upon Na$^+$. With regard to mechanism, this is the key observation for it suggests direct interaction of Na$^+$ with the substrate carrier. Vidaver (48), with this observation as a guide, has recently completed extensive studies of glycine

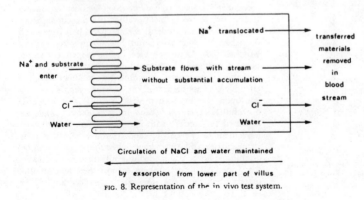

FIG. 8. Representation of the in vivo test system.

uptake by pigeon red blood cells which show interactions of Na^+ in most respects similar to those which have been described for sugar and the intestine.

Although the evidence for this mechanism is strong, some disagreement still exists. It would appear useful, then, with these in mind, to examine intestinal absorption in terms of the methods used to study it—with particular regard to the differences one might expect to see between in vitro and in vivo test systems. Figure 7 represents the view of the in vitro test system suggested by the experiments reviewed above. Na^+ and substrate are shown to equilibrate across the brush border membrane, intracellular Na^+ is depleted by translocation and substrate accumulates to high concentrations due to the asymmetries of Na^+, K^+, and K_m. Substrate moves into the underlying tissues and into the serosal compartment of everted sac preparations by diffusion from this high concentration. In studies of glucose transport, for example, it can be expected from this model that ouabain will inhibit glucose accumulation (20, 21) because it inhibits the Na^+ pump and prevents Na^+ depletion. The rate of glucose movement into the serosal compartment would then fall because the gradient of glucose would be less steep. However, glucose utilization would not necessarily be reduced—the equilibrating system can still allow glucose in at a rate exceeding its utilization. Thus, what has appeared to some investigators (10, 46) as evidence for a basal location of a sugar pump can be interpreted as an expected consequence of the use of an in vitro system.

In fact, it is not difficult to imagine that large cellular accumulations of sugar in the epithelial cell (34) may be seen only in the in vitro system. Such accumulation may occur mostly as a result of the normal biochemical events of the cell finding their expression in a substantially static system rather than in their usual dynamic environment. As an illustration of this suggestion, one possible representation of the in vivo system is shown in Fig. 8. Because transferred materials are continuously removed in the blood stream, one can imagine substrate entering but not accumulating because it may flow with the stream of water which in turn moves as a result of Na^+ pumping from the base of the cell (25). And this stream may itself be part of an internal circulation as suggested by Florey 25 years ago (27). Exsorption and insorption of salt and water which are known to occur need not take place in the same part of the cell nor, indeed, the same part of the villus. If there were a flow of salt and substrate-free water from the base of the villus into the lumen and a removal of salt, water, and substrate inward at the tip of the villus, measurements of in vivo transport might give an appearance suggestive of a different mechanism than has been described in vitro. Analysis of experimental results, especially in vivo, must take such possibilities into account.

REFERENCES

1. ALLFREY, V. G., R. MENDT, J. W. HOPKINS, AND A. E. MIRSKY. *Proc. Natl. Acad. Sci. U. S.* 47: 907, 1961.
2. ALTHAUSEN, T. L., E. M. ANDERSON, AND M. STOCKHOLM. *Proc. Soc. Exptl. Biol. Med.* 40: 342, 1939.
3. ALVARADO, F. *Biochim. Biophys. Acta* In press.
4. ALVARADO, F., AND R. K. CRANE. *Biochim. Biophys. Acta* 56: 170, 1962.
5. ASANO, T. *Am. J. Physiol.* 207: 415, 1964.
6. BIHLER, I., AND R. K. CRANE. *Biochim. Biophys. Acta* 59: 78, 1962.
7. BIHLER, I., K. A. HAWKINS, AND R. K. CRANE. *Biochim. Biophys. Acta* 59: 94, 1962.
8. BOSACKOVA, J., AND R. K. CRANE. *Biochim. Biophys. Acta* 102: 423, 1965.
9. BOSACKOVA, J., AND R. K. CRANE. *Biochim. Biophys. Acta* 102: 436, 1965.
10. CAPRARO, V., A. BIANCHI, AND C. LIPPE. *Arch. Sci. Biol., Bologna* 47: 238, 1963.
11. CHRISTENSEN, H. N. *Biological Transport.* New York: W. A. Benjamin, 1962.
12. CLARK, W. G., AND E. M. MACKAY. *Am. J. Physiol.* 137: 104, 1942.
13. CORI, C. F., AND G. T. CORI. *J. Biol. Chem.* 73: 555, 1927.
14. CRANE, R. K. *Physiol. Rev.* 40: 789, 1960.
15. CRANE, R. K. *Federation Proc.* 21: 891, 1962.
16. CRANE, R. K. 3rd Annual Meeting, Am. Soc. Cell. Biol. New York, N. Y., 1963, unpublished.
17. CRANE, R. K. *Biochem. Biophys. Res. Commun.* 17: 481, 1964.
18. CRANE, R. K., G. FORSTNER, AND A. EICHHOLZ. *Biochim. Biophys. Acta* In press.
19. CRANE, R. K., AND P. MANDELSTAM. *Biochim. Biophys. Acta* 45: 460, 1960.
20. CRANE, R. K., D. MILLER, AND I. BIHLER. In: *Symposium on Membrane Transport and Metabolism,* edited by A. Kleinzeller and A. Kotyk. London: Academic, 1961, p. 439.

21. CSAKY, T. Z. *Federation Proc.* 22: 3, 1963.
22. CSAKY, T. Z. *Gastroenterology* 47: 201, 1964.
23. CSAKY, T. Z., AND M. THALE. *J. Physiol., London* 151: 59, 1960.
24. CSAKY, T. Z., AND L. ZOLLICOFFER. *Am. J. Physiol.* 198: 1056, 1960.
25. CURRAN, P. *Federation Proc.* 24: 993, 1965.
26. DEUEL, H. J., JR., L. F. HALLMANN, S. MURRAY, AND L. T. SAMUELS. *J. Biol. Chem.* 119: 607, 1937.
27. FLOREY, H. W., R. D. WRIGHT, AND M. A. JENNINGS. *Physiol. Rev.* 21: 36, 1941.
28. FOX, M., S. THIER, L. ROSENBERG, AND S. SEGAL. *Biochim. Biophys. Acta* 79: 167, 1964.
29. HOLT, P. R. *Am. J. Physiol.* 207: 1, 1964.
30. KOTYK, A., AND A. KLEINZELLER. *Biochim. Biophys. Acta* 54: 367, 1961.
31. KROMPHARDT, H., H. GROBECKER, K. RING, AND E. HEINZ. *Biochim. Biophys. Acta* 74: 551, 1963.
32. LACK, L., AND I. M. WEINER. *Am. J. Physiol.* 200: 313, 1961.
33. LYON, I., AND R. K. CRANE. *J. Cell. Biol.* 23: 120A, 1964.
34. MCDOUGAL, D. B., K. D. LITTLE, AND R. K. CRANE. *Biochim. Biophys. Acta* 45: 483, 1960.
35. MILLER, D., AND R. K. CRANE. *Biochim. Biophys. Acta* 52: 293, 1961.
36. MITCHELL, P. *Biochem. Soc. Symp.* 22: 142, 1962.
37. NATHANS, D., D. F. TAPLEY, AND J. E. ROSS. *Biochim. Biophys. Acta* 41: 271, 1960.
38. PARRISH, J., AND D. M. KIPNIS. *J. Clin. Invest.* 43: 1994, 1964.
39. PLAYOUST, M. R., AND K. J. ISSELBACHER. *J. Clin. Invest.* 43: 467, 1964.
40. RIGGS, T. R., L. M. WALKER, AND H. N. CHRISTENSEN. *J. Biol. Chem.* 233: 1479, 1958.
41. RIKLIS, E., AND J. H. QUASTEL. *Can. J. Biochem. Physiol.* 36: 347, 1958.
42. ROSENBERG, I. H., A. L. COLEMAN, AND L. E. ROSENBERG. *Biochim. Biophys. Acta* 102: 161, 1965.

43. SALOMON, L., J. A. ALLUMS, AND D. E. SMITH. *Biochem. Biophys. Res. Commun.* 4: 123, 1961.

44. SCHULTZ, S. G., AND R. ZALUSKY. *Biochim. Biophys. Acta* 71: 503, 1963.

45. SCHULTZ, S. G., AND R. ZALUSKY. *J. Gen. Physiol.* 47: 1043, 1964.

46. SMITH, M. W. *J. Physiol., London* 175: 38, 1964.

47. TAYLOR, C. B. *Biochim. Biophys. Acta* 60: 437, 1962.

48. VIDAVER, G. A. *Biochemistry* 3: 795, 799 and 803, 1964.

49. WILBRANDT, W., AND L. LENGYEL. *Biochem. Z.* 204: 267, 1933.

50. WILBRANDT, W., AND T. ROSENBERG. *Pharmacol. Rev.* 13: 109, 1961.

51. YUNIS, A. A., G. K. ARIMURA, AND D. M. KIPNIS. *J. Lab. Clin. Med.* 62: 465, 1963.

Reprinted from *Biochemistry*, **3**, 759–799 (1964)

$$34$$

Glycine Transport by Hemolyzed and Restored Pigeon Red Cells*

GEORGE A. VIDAVER†

From the Department of Chemistry, Indiana University, Bloomington
Received December 18, 1963; revised March 19, 1964

Pigeon red cells can be lysed and restored by a modification of the method used for mammalian red cells. During lysis, glycine is equally distributed, and K^+ nearly so, between cell and lysing-fluid phases. Such preparations retain glycine and K^+ at 0–7° but lose about 32% of the K^+ in 40 minutes at 39°. The lysed and restored cells can build up glycine concentration gradients only if a Na^+ gradient exists across the cell membrane. If the normal gradient is reversed (Na^+ inside, K^+ outside) lysed and restored cells pump glycine out to an extent comparable to the accumulation occurring with a normal gradient. The Na^+ dependence is discussed in relation to Christensen's hypothesis that the energy for active transport of glycine arises from a cation gradient. This hypothesis is supported by the data, although others are not excluded.

Hypotonic hemolysis of mammalian red cells is a "reversible" process in that much of the impermeability

* The work described in this paper was supported by research grants to Professor F. Haurowitz from the National Science Foundation (NSF G16345) and the U. S. Public Health Service (NIH RG 1852), and by contracts of Indiana University with the Office of Naval Research (Nonr-3104[00]) and the Atomic Energy Commission (AEC AT[11-1]-209).

† Part of this work was done while the author held a N.S.F. postdoctoral fellowship.

of the cell membrane, lost during hemólysis, can be restored by restoring the ionic strength and then holding at 37° (Hoffman *et al.*, 1960; Whittam, 1962). Mammalian erythrocytes lysed and restored in this way can transport Na^+ actively (Hoffman, 1962). By use of this phenomenon the composition of the red-cell interior can be varied within wide limits.

Since mammalian red cells concentrate glycine very poorly in contrast to avian cells (Christensen *et al.*, 1952a) the technique of lysis and restoration was adapted to pigeon red cells. The procedure, some

Table I

Composition of Lysed and Restored Cells and Distribution of Glycine and K^+ between Cell and Lysate Fluid Phases[a]

Glycine/ml Pellet H_2O / Glycine/ml Lysate Supernatant H_2O	K^+/ml Pellet H_2O / K^+/ml Lysate Supernatant H_2O	Retention of [^{14}C]Glycine by Cells at 0–7° (%)	Extracellular Space (as % of pellet H_2O)	Loss of Dry Wt. on Lysis and Restoration (%)	Gain of Pellet H_2O on Lysis and Restoration (%)
0.98[b]	1.12[b]	70.0	23.0	52.3	38.9
0.95	1.18	71.6	18.4	55.0	29.0
1.01		75.1	17.5	58.9	55.8
1.02			24.8	51.9	29.7
				56.8	25.6
Mean 0.99	1.15	72.2	20.9	55.0	35.8

[a] Cells were lysed and restored as described under Methods. Glycine-distribution figures were obtained by determination of glycine or ^{14}C in the pellets and supernatants from annealed cell suspensions. For the first experiment glycine was determined chemically; for the others, [^{14}C]glycine was added to the lysing solution and ^{14}C in pellet and supernatant was determined by counting. Thick-sample plates were used (Vidaver, 1964). The K^+ distribution figures were obtained by flame-photometric analyses of picric acid extracts of pellets and supernatants from cells lysed and restored with solutions containing only Na^+. Cells lose 55% of their dry weight on lysis; therefore 0.55×0.325 g solid is dissolved in the *ca.* 3.3-ml lysate fluid. This is 5.4% (w/v); assuming a volume of 0.75 ml/g, 1 ml lysate fluid contains $1 - (0.054 \times 0.75)$ or 0.96 ml H_2O. Glycine and K^+ distribution figures were calculated on the basis of measured H_2O contents of the pellets and a value of 96% for H_2O/ml lysate fluid. "Retention of [^{14}C]glycine" is the amount of ^{14}C in the cells (cpm/ml pellet H_2O) after annealing but before washing, divided by the ^{14}C in the cells after washing and resuspension in the incubation media but before incubation, times 100. Most of the ^{14}C loss of 28% can be accounted for as ^{14}C washed out of the extracellular space (*ca.* 21%). Extracellular space was determined by diluting duplicate cell suspensions with high and low Na^+ diluents in the cold, immediately centrifuging, and determining Na^+ in pellets and supernatants, or by the equivalent procedure with ^{14}C-containing and ^{14}C-free diluents. Penetration of glycine or Na^+ is very slow in the cold. Loss and gain of dry weight and H_2O were determined by weighing. Values for different quantities on the same line are not necessarily data from the same cell samples. [b] An average value for two different samples obtained in the same experiment.

properties of lysed and restored cells, and some features of glycine transport by them are described. A glycine concentration gradient can be built up by such cells if there is a Na^+ gradient across the cell membrane. The direction of the Na^+ gradient determines the direction of net glycine transport.

The observations support the hypothesis of Christensen (Christensen *et al.*, 1952b; Riggs *et al.*, 1958) that the energy source for glycine-active transport is the cation gradient.

MATERIALS AND METHODS

Fresh cells were prepared, and glycine, Na^+, K^+, radioactivity, and extracellular space were all determined as in earlier work (Vidaver, 1964).

Cells were lysed and restored in the following way. Each aliquot of cell suspension containing 1–3 g of cells was centrifuged in the cold at 11,000 rpm for 15 minutes in a 15-ml cellulose nitrate Servall tube in a Servall SS1 head (Ivan Sorvall Inc., Norwalk, Conn.). The supernatant was removed and the pellet was distributed over the bottom of the tube with a stirring rod. For each gram of cells, 3 ml of lysing solution was quickly added and the tube contents was vigorously beaten with a stirring rod for 1 minute. For each gram of cells, 0.34 ml of restoring solution was then quickly added and the tube contents was vigorously mixed for 1 minute. The solutions and tube contents were kept at 5–6° during these operations. The tube was then immediately placed in a bath at 37–39° for 17 minutes ("annealed"). (This time is arbitrary; in a single trial, 8 and 34 minutes were equally effective.) The annealing step completes the restoration process. After the cells were chilled, centrifuged, and washed, they were ready for use. When several aliquots had been lysed and restored, they were held at 0° after the annealing step until all were ready for the remaining operations. The lysing solution contained 5 mM Na_2HPO_4 and/or K_2HPO_4, 5 mM NaH_2PO_4 and/or KH_2PO_4, 2 mM $MgSO_4$, 1 mM $CaCl_2$, and 1 mg/ml glucose. The restoring solution contained 1.38 M NaCl and/or KCl, 0.020 M $MgSO_4$, and 0.010 M $CaCl_2$.

When glycine-concentration ratios were determined chemically, incubations were done in shaken (*ca.* 80/minute) flasks for 40 or 100 minutes. For glycine entry and exit measurements with [^{14}C]glycine, incubation was for 10 minutes in centrifuge tubes. Incubation media (modified Krebs Ringer Phosphate, pH 7.2) and subsequent handling were as described for intact cells (Vidaver, 1964). Solutions and cells were not gassed. No color change (deoxygenation of hemoglobin) was observed in incubated cells.

RESULTS

Pellets of intact cells are the color of black cherries. Those of lysed and restored cells are a nearly uniform red-cherry color. If intact cells are mixed with lysed and restored ones and the mixture is centrifuged, the intact cells are deposited as a dark-red button at the bottom of the bright-red pellet. Three per cent of intact cells are easily seen. Incomplete lysis can therefore be determined visually.

As shown in Table I, glycine in the lysing solution distributes itself equally between intra- and extracellular H_2O; cell K^+ is nearly equally distributed. The cells lose about 55% of their dry weight and gain somewhat more than an equal weight of H_2O. The extracellular space is somewhat greater and more variable than that (10%) of intact cells.

The observed [^{14}C]glycine retention (Table I) by lysed and restored cells and the agreement between the calculated and observed values for K_i^+ (Table II) show that the bulk of the cells become relatively impermeable to glycine and K^+ during restoration.[1]

Lysed and restored cells have a limited capacity to retain K^+ at 39°. During 40 minutes of incubation (39 ± 0.5°), $K_i^+ - K_o^+$ decreased by 32% (Table II). The per cent loss was independent of K_i^+, K_o^+, and $(K_i^+ - K_o^+)$.

[1] The subscripts "*o*" or "*i*" after a symbol for or name of a substance mean that the substance represented is present in the medium or the cell, respectively. K_m is the glycine concentration giving half-maximal entry rate, as estimated from a Lineweaver-Burk plot.

TABLE II

POTASSIUM RETENTION AT 0–7° AND LOSS AT 39°[a]

Experiment Number	K_i^+, Calcd, before Incubation (μmoles/ml pellet H_2O)	K_i^+, Determined, before Incubation (μmoles/ml pellet H_2O)	K_i^+, Determined, after 40-min Incubation at 39° (μmoles/ml pellet H_2O)	K_o^+ (mM)	Loss of K_i^+ in 40 min at 39° (%)
1	76.6	68.7	51.0	6	28
	76.6	69.8	50.2	6	31
2	22	20	15	6	36
3	30.5	29.2	22.2	6	30
	56.2	51.8	36.5	11.7	38
	105.9	90.8	72.3	23	27
					Avg. 32

[a] The "determined" K_i^+ values were obtained by analysis of pellet extracts. The "K_o^+" values are K$^+$ added to media. "Per cent loss of K$^+$ at 39°" is $100 \times (K_i^+$, determined, of pellets before incubation $- K_i^+$, determined, of pellets after 40-min incubation at 39°) $\div (K_i^+$, determined, of pellets before incubation $- K_o^+$). The K_i^+ decrease is divided by $K_i^+ - K_o^+$ rather than K_i^+ because the net loss of K_i^+ should depend on the difference. Cell K$^+$ was also calculated from the equation, $K_i^+ = 147 \times$ fraction K$^+ + 5.4$. "Fraction K$^+$" is the sum of K$^+$ in cell H_2O (*ca.* 130 mM; Vidaver, 1964), lysing solution, and restoring solution, divided by the total Na$^+$ plus K$^+$ in these solutions. The figure 5.4 is the excess K$^+$ in cells lysed and restored with solutions containing only Na$^+$, over the theoretical amount for free distribution. The figure 147 is the total Na$_i^+ + K_i^+$ (152 μmoles/ml cell H_2O) in lysed and restored cells minus excess K$^+$. The numerical values used here (5.4 and 152) were obtained from other experiments. The determined values average 91% of the calculated values.

Some aspects of glycine transport are shown in Tables III and IV and Figure 1. The cases where Na$_i^+$ was not equal to Na$_o^+$ are illustrated by Table III and Figure 1. With Na$_i^+$ low and Na$_o^+$ high (the Na$^+$ gradient in the "normal" direction), glycine was accumulated (Table III). With Na$_i^+$ low and Na$_o^+$ low, but with an appreciable gradient in the normal direction, glycine was still accumulated (Fig. 1). With Na$_i^+$ high and Na$_o^+$ high, but still with a normal Na$^+$ gradient, glycine was also accumulated (experiment 5, Table III; Fig. 1). Note that this case shows that high K$_i^+$ is not necessary for glycine accumulation. With Na$_i^+$ high and Na$_o^+$ low, i.e., the reverse of the normal gradient, glycine is not accumulated, but is instead pumped out about as efficiently as it would be pumped in with a similar Na$^+$ gradient in the normal direction (experiments 3 and 4, Table III).

The cases where Na$_i^+$ equals Na$_o^+$ are illustrated by Table IV. Table IV shows that if the glycine concentrations are equal inside and out, and the Na$^+$ concentrations are equal inside and out, then glycine exit and entry rates are equal. This is independent of the numerical values of the Na$^+$ and glycine concentrations. That is, no pumping occurs in either direction in the absence of an Na$^+$ gradient.

DISCUSSION

The free distribution of solutes during lysis and the capacity of lysed and restored pigeon cells to pump glycine are analogous to the free solute distribution during lysis (Hoffman, 1958) and the capacity to pump Na$^+$ (Hoffman, 1962) previously reported for mammalian red cells. The avian and mammalian systems probably differ in their Ca^{2+} and/or Mg^{2+} requirement for restoration. In experiments prior to the introduction of the annealing step, it was observed that hemoglobin retention was very poor in the absence of Ca^{2+} and Mg^{2+} in the lysing and restoring solutions. Even in their presence it was poorer than with the annealing step included. Since the capacity for glycine-active transport in pigeon cells is absolutely dependent on the annealing step, the pigeon-cell system seems analogous to Hoffman's (1962) group II cells rather than group I cells.

It is assumed that the glycine-pump mechanism acting in lysed and restored cells is, or is part of, that

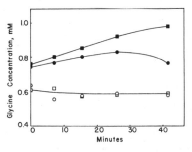

FIG. 1.—Glycine uptake by lysed and restored cells is plotted against time. One portion of the cells was lysed and restored with solutions containing only Na$^+$, the other with solutions containing only K$^+$. The calculated values for Na$_i^+$ prior to incubation were 126 and 3 μmoles/ml cell H_2O, respectively. Calculations were done as for the data of Table II. Aliquots of the Na$^+$-rich sample were incubated at 39° for varying periods of time in a slightly modified Krebs Ringer phosphate glucose medium (Vidaver, 1964) with Na$_o^+ = 140$ mM (the initial Na$_o^+ - $ Na$_i^+$ was therefore 14 mM) and the aliquots of the Na$^+$-poor sample were incubated in a medium with Na$_o^+ = 40$ mM (initial Na$_o^+ - $ Na$_i^+ = 37$ mM). The glycine values used were obtained by (duplicate) chemical analysis of media and pellets. Open symbols refer to glycine in the medium, solid ones to glycine in the cells. ○, ● Na$_o^+ - $ Na$_i^+ = 14$ mM; □, ■ Na$_o^+ - $ Na$_i^+ = 37$ mM.

acting in the intact-cell system, and that observations made on the lysed and restored system are at least qualitatively applicable to the intact-cell system. It is also assumed that Na$^+$, rather than K$^+$, is the active ion for lysed and restored cells as it is for intact ones (Kromphardt et al., 1963; Vidaver, 1964).

Figure 1 and Tables III and IV show that neither internal Na$^+$ (or K$^+$) alone nor external Na$^+$ (or K$^+$) alone controls glycine pumping, but that the Na$^+$ difference between inside and outside is the critical factor. Figure 1 shows that glycine accumulation can occur from a fairly low Na$^+$ medium if Na$_i^+$ is low enough to give a Na$^+$ gradient. Experiment 5 of Table III shows that considerable accumulation occurs even with fairly high Na$_i^+$ (and low K$_i^+$), again, provided a Na$^+$ gradient exists. Table IV shows that no

TABLE III
GLYCINE ACCUMULATION AND EXPULSION BY LYSED AND RESTORED CELLS[a]

Experiment Number	Sample	Cell Type	Initial Cell Na$^+$, Value Calculated (mM)	Na$^+$ in Medium (mM)	Incubation Time (min at 39°)	Initial Glycine in Medium (mM)	Initial Ratio, Glycine$_i$/ Glycine$_o$	Final Ratio, Glycine$_i$/ Glycine$_o$
1	a	Lysed and restored	24	140	40	0.4$_6$	1.4$_5$	2.7$_6$
	b	Lysed and restored	24	140	100	0.4$_6$	1.4$_5$	3.0$_2$
2	c	Lysed and restored	24	140	103	0.6$_3$	1.17	2.3$_1$
	d	Intact	17.5	140	103	0.6$_3$	6.0$_6$	8.2$_2$
3	e	Lysed and restored	24	140	40	0.6$_0$	1.59	2.3$_9$
	f	Lysed and restored	24	0	40	0.6$_0$	1.57	1/1.0$_8$
	g	Lysed and restored	115	0	40	0.6$_0$	1.43	1/2.0$_6$
	h	Lysed and restored	126	0	40	0.6$_0$	1.24	1/2.3$_4$
4	i	Lysed and restored	24	140	40	0.57	1.12	1.9$_5$
	j	Lysed and restored	126	0	40	0.57	1.07	1/1.9$_4$
5	k	Lysed and restored	115	140	40	0.6$_2$	1.6$_3$	1.7$_3$
	l	Lysed and restored	84	134	40	0.6$_2$	1.67	1.9$_1$
	m	Lysed and restored	24	125	40	0.6$_2$	1.6$_8$	2.57

[a] Cells were lysed and restored as described under Methods. For samples a–c, e, f, i, and m, the lysing solution contained only K$^+$ and the restoring solution K$^+$/Na$^+$ ratio was 4.31. For samples h and j, both solutions contained only Na$^+$. For samples g, k, and l, intermediate proportions of Na$^+$ were used. The initial cell Na$^+$ values for lysed and restored cells, in μmoles Na$^+$/ml cell H$_2$O before incubation, are calculated from the equation Na$_i^+$ = 152 − K$_i^+$. The value for K$_i^+$ was calculated as for Table II. The total Na$^+$ plus K$^+$ in lysed and restored cells is 152 μmoles/ml cell H$_2$O. The numerical value used here (152) was obtained from other experiments. Comparison of calculated and analytical values in other experiments gave reasonable agreement, e.g., the observed value for Na$_i^+$, third line, Table IV, is 37 mM. The calculated value is 35.5 mM. The K$_i^+$ data in experiment 3, Table II, are from experiment 4 in this table. Since lysed and restored cells are leaky at 39° with respect to K$^+$, the calculated values are taken to be, here, as adequate as analytical ones would be. The Na$_i^+$ value listed for the intact-cell sample (sample d) is simply the average Na$_i^+$ determined in other experiments (Vidaver, 1964). The glycine values used for the glycine$_i$/glycine$_o$ ratios were obtained by (duplicate) chemical analysis of media and pellets. (Second-decimal-place digits are given because, for some values, rounding off overstates the error.)

TABLE IV
EQUALITY OF GLYCINE EXIT AND ENTRY FROM EQUAL GLYCINE CONCENTRATIONS WHEN NO Na$^+$ GRADIENT EXISTS[a]

Glycine Inside (μmoles/ml cell H$_2$O)	Na$^+$ Inside (μmoles/ml cell H$_2$O)	Glycine Exit (μmoles/ml cell H$_2$O in 10 min, 39°)	Glycine Entry (μmoles/ml cell H$_2$O in 10 min, 39°)	Na$^+$ Outside (μmoles/ml medium)	Glycine Outside (μmoles/ml medium)
0.8$_2$	125	0.24	0.27	125	0.90
2.7$_5$	125	0.55	0.61	125	3.00
0.8$_2$	37	0.16	0.15	35	0.90
0.8$_2$	3.4	0.060	0.066	3.3	0.90

[a] Cells were lysed and restored as described under Methods. The desired cell-Na$^+$ concentration was obtained by use of the appropriate proportion of Na$^+$ in the lysing and restoring solutions. Either [^{14}C]- or [^{12}C]glycine was added to the lysing solution to produce the desired glycine and ^{14}C contents. Glycine entry and exit was measured to and from pairs of samples. One member of each pair had ^{14}C inside and ^{12}C outside; the other had ^{14}C outside and ^{12}C inside; otherwise the composition and incubation media of the two members of the pair were identical. Incubation was done in centrifuge tubes (see Methods). Radioactivity was determined from thick-sample plates. Exit and entry rates approximating initial rates were calculated as previously described (Vidaver, 1964), with the modification that the entry rates given here include a "re-exit" correction. A fraction of the internal glycine leaves the cells during incubation. The re-exit correction used was 0.5 times this fraction, as estimated from ^{14}C loss from the internally labeled member of the pair, times the measured ^{14}C entry into the other member. Values for concentrations are given in μmoles/average ml cell H$_2$O, using the average of the H$_2$O/pellet values for incubated and unincubated samples (corrected for extracellular space). Entry and exit rates are given in μmoles/unincubated ml cell H$_2$O. The glycine values used were obtained by (duplicate) chemical analysis of media and pellets.

pumping occurs in either direction in the absence of a Na$^+$ gradient regardless of the Na$^+$ and K$^+$ concentrations inside and outside. The data of Table III also show that the Na$^+$ gradient so far controls glycine transport that, on merely inverting the normal gradient, glycine is pumped out to an extent comparable to the accumulation occurring with a normal gradient.

It may be noted that K_m for glycine entry into intact cells depends on (Na$^+$)^2 rather than Na$^+$ (Vidaver, 1964). At Na$_o^+$ = 40 mM, K_m for intact cells is 0.7 mM, thus the considerable glycine accumulation with Na$_o^+$ = 40 mM (and Na$_i^+$ = 3 mM) (Fig. 1) is not unexpected.

The equality of glycine entry and exit rates from equal Na$^+$ and glycine concentrations implies that compartmentalization of Na$_i^+$ is not significant in lysed and restored cells.

The limited ability of lysed and restored cells to maintain internal K$^+$ (and presumably to exclude Na$^+$) probably explains in part the poor glycine accumulation by lysed and restored cells compared with intact ones. This leakiness also limits the significance of the numerical values for glycine concentration ratios.

One aim of the present work was to test Christensen's hypothesis that the energy for amino acid accumulation comes from the alkali metal-ion gradient (Christensen *et al.*, 1952b). This hypothesis has two forms. One is that internal K$^+$ exchanges for external amino acid. Hempling and Hare (1961), using the Ehrlich ascites system, concluded that the energy expenditure by K^+ efflux down the K^+ gradient was insufficient to account for the glycine influx at the glycine-concentration ratio maintained. The other form of the hypothesis is that there is a linked entry of Na^+ and amino acid, with the energy from Na^+ influx down its chemical-activity gradient furnishing the energy for the transport of glycine against its gradient (Riggs *et al.*, 1958). The dependence of glycine entry in Na_o^+ (Kromphardt *et al.*, 1963; Vidaver, 1964) and the apparent involvement of a complex containing two Na ions and one glycine in the glycine-entry process (Vidaver, 1964) made this form of the hypothesis attractive.

No pumping can occur in the absence of an energy source. If a Na^+ gradient is the energy source for the glycine pump, in its absence glycine exit and entry rates from equal glycine concentrations must be equal, regardless of what the glycine concentrations are, and regardless of what the Na^+ concentrations are. This appears to be the case (Table IV).

Some other type of energy source, such as ATP, should be unequally distributed between the inside and outside of the cell. Also, a pump mechanism adapted to operate between the different phases, cell interior and plasma, might be expected to have a polarity. In either case unequal entry and exit rates might be expected under the conditions of the experiment shown in Table IV. However, it is possible to devise a pump model using, e.g., internal ATP as an energy source which would operate equally effectively in the two directions, thus these data do not prove the Na^+ operated pump hypothesis.[2] Since the relationships found between glycine pumping and a Na^+ gradient, and the occurrence of a complex containing both glycine and Na^+ are required by any pump model with a Na^+ gradient as energy source, but only correspond to a special case of the (e.g.) ATP-powered-pump hypothesis, they are taken to support the former.

ACKNOWLEDGMENTS

The author wishes to thank Prof. Felix Haurowitz for his advice and support throughout the course of this work. He also wishes to thank Mr. Lee Van Tornhout for technical assistance.

REFERENCES

Christensen, H. N., Riggs, T. R., Fischer, H., and Palatine, I. M. (1952b), *J. Biol. Chem. 198*, 1.
Christensen, H. N., Riggs, T. R., and Ray, N. E. (1952a), *J. Biol. Chem. 194*, 41.
Hempling, H. G., and Hare, D. (1961), *J. Biol. Chem. 236*, 2498.
Hoffman, J. F. (1958), *J. Gen. Physiol. 42*, 9.
Hoffman, J. F. (1962), *J. Gen. Physiol. 45*, 837.
Hoffman, J. F., Tosteson, D. C., and Whittam, R. (1960), *Nature 185*, 186.
Kromphardt, H., Grobeker, H., Ring, K., and Heinz, E. (1963), *Biochim. Biophys. Acta 74*, 549.
Riggs, T. R., Walker, L. M., and Christensen, H. N. (1958), *J. Biol. Chem. 233*, 1479.
Vidaver, G. A. (1964), *Biochemistry 3*, 662.
Whittam, R. (1962), *Biochem. J. 184*, 110.

[2] Such a model is represented by the sequence: E_i + ATP $\xrightarrow{\text{fast}}$ E_i^*; $E_i^* \underset{\text{fast}}{\rightleftarrows} E_o^*$; E_o^* (or E_i^*) + G_o (or G_i) + 2 Na_o^+ (or 2 Na_i^+) $\underset{\text{fast}}{\rightleftarrows}$ $E^*Na_2G_o$ (or $E^*Na_2G_i$); $E^*Na_2G_o$ (or $E^*Na_2G_i$) $\xrightarrow{\text{slow}}$ $E^{**}G_o$ (or $E^{**}G_i$) + 2 Na_o^+ (or 2 Na_i^+); $E^{**}G_o$ (or $E^{**}G_i$) $\rightleftarrows E^{**}G_i$ (or $E^{**}G_o$) (this is the translocation step); $E^{**}G_i$ (or $E^{**}G_o$) $\xrightarrow{\text{slow}} E_i$ (or E_o) + G_i (or G_o); $E_i \underset{\text{fast}}{\rightleftarrows} E_o$. "E" is taken to be a mobile carrier in this model. The superscript asterisks represent "states" of E. "G" is glycine. All reactions except E_i + ATP $\rightarrow E_i^*$ might be catalyzed by the carrier ("E") is. and so be independent of location.

Copyright © 1967 by the Rockefeller University Press

Reprinted from *J. Gen. Physiol.*, **50**, 1261–1286 (1967)

Kinetic Relations of the

35   Na-Amino Acid Interaction at

the Mucosal Border of Intestine

PETER F. CURRAN, STANLEY G. SCHULTZ, RONALD A. CHEZ, and ROBERT E. FUISZ

From the Biophysical Laboratory, Harvard Medical School, Boston, Massachusetts. Dr. Chez's present address is the University of Pittsburgh School of Medicine, Pittsburgh, Pennsylvania. Dr. Fuisz's present address is the Peter Bent Brigham Hospital, Boston, Massachusetts

ABSTRACT   The relation between unidirectional influxes of Na and amino acids across the mucosal border of rabbit ileum was studied under a variety of conditions. At constant Na concentration in the mucosal bathing solution, amino acid influx followed Michaelis-Menten kinetics permitting determination of maximal influx and the apparent Michaelis constant, K_t. Reduction in Na concentration, using choline as substitute cation, caused an increase in K_t for alanine but had no effect on maximal alanine influx. The reciprocal of K_t was a linear function of Na concentration. Similar results were obtained for valine and leucine and these amino acids competitively inhibited alanine influx both in the presence and in the absence of Na. These results lead to a model for the transport system which involves combination of Na and amino acid with a single carrier or site leading to penetration of both solutes. The model predicts that alanine should cause an increase in Na influx and the ratio of this extra Na flux to alanine flux should vary with Na concentration. The observed relation agreed closely with predicted values for Na concentrations from 5 to 140 mM. These results support the hypothesis that interactions between Na and amino acid transport depend in part on a common entry mechanism at the mucosal border of the intestine.

In the preceding paper (1), a method for measuring unidirectional solute influx across the mucosal border of intestinal epithelial cells was described and some characteristics of L-alanine and Na influxes were presented. These studies provided the first unequivocal evidence that the flux of an actively transported amino acid from the mucosal solution into the cell is influenced by the Na concentration in the mucosal solution and is independent of the bulk cellular Na concentration. Further study of the influx processes is necessary in order to define more completely the transport mechanisms for non-

1261

electrolytes and the nature of the interaction with Na. Data on cell accumulation or transmural transport of amino acids or sugars do not provide this information because these phenomena depend in a rather complex way on events at both the mucosal and serosal borders of the cell. The present experiments are concerned primarily with the kinetics of L-alanine influx and the interaction between L-alanine and Na. In addition, measurements have been made of valine and leucine fluxes, and the interaction between these amino acids and alanine influx has been examined.

METHODS

The apparatus and methods used to measure amino acid and Na influxes have been described in detail in the preceding paper (1). The basic Ringer solution used contained 140 mM NaCl, 10 mM KHCO$_3$, 1.2 mM CaCl$_2$, 1.2 mM MgCl$_2$, 1.2 mM K$_2$HPO$_4$, and 0.2 mM KH$_2$PO$_4$. In most experiments, the tissue segments were bathed with solutions having the same Na concentration, but, from one experiment to another, Na concentration of the Ringer solution was altered by replacing NaCl with choline chloride. The tissue was preincubated for 30 min in Ringer's solution having the same Na concentration as that used in the influx measurements. The preincubation solution did not contain amino acid. In many of the experiments, fluxes were determined at four different amino acid concentrations on tissue from the same animal. Since the experimental arrangement permitted eight separate influx measurements, duplicate determinations at each concentration were possible. The solution for flux measurement (test solution) was prepared by adding ^{14}C-labeled L-alanine, L-valine, or L-leucine to the appropriate Ringer solution to give concentrations varying from 1.7 to 20 mM; this solution also contained inulin-^{3}H and ^{22}Na. In experiments testing the effect of valine or leucine on alanine influx, the alanine concentration was 3.3 mM and the concentration of the other amino acid was 20 mM. Methods of tissue extraction, of assaying the three isotopes, and of flux calculations have been described (1).

All influx values reported were obtained by measuring tracer uptake over a 60 sec interval. Since the present experiments involved higher alanine concentrations and higher fluxes than previously observed, a single experiment was performed to determine the time course of alanine-^{14}C uptake at 20 mM alanine and 140 mM Na. Tracer uptake was linear for more than 60 sec in agreement with observations at 5 mM alanine (1).

RESULTS

Kinetics of Alanine Influx

Alanine influx was determined as a function of alanine concentration at four Na concentrations ranging from zero (nominally Na-free medium) to 140 mM. The results of two individual experiments at the extreme Na concentrations are shown in Fig. 1. As expected from previous data (1), alanine influx is considerably lower in the absence of Na than in its presence. The influx increases with alanine concentration but shows a tendency toward saturation. The apparent hyperbolic relation between flux and concentration was confirmed by plotting the reciprocal of flux against the reciprocal of concentration[1] (Fig. 2). The resulting straight lines indicate that the alanine influx, J_A^i, can be described by a relation of the form

$$J_A^i = \frac{J_A^{im} [A]_m}{K_t + [A]_m} \qquad (1)$$

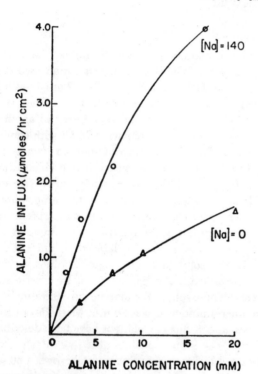

FIGURE 1. Alanine influx as a function of alanine concentration. Data at each Na concentration were obtained using tissue from a single animal and each point is the average of duplicate flux determinations.

in which $[A]_m$ is alanine concentration in mucosal solution and J_A^{im} is the maximal influx. K_t is the "apparent Michaelis constant" (the value of $[A]_m$ for which $J_A^i = J_A^{im}/2$). The fact that the intercepts of the lines shown in Fig. 2 are identical indicates that J_A^{im} is the same at the two Na concentrations while the difference in slopes shows that K_t is different.[2] This behavior was found at all Na concentrations tested and the average values of J_A^{im} and K_t

[1] Though it would have been desirable to examine alanine influx at higher alanine concentrations in order to obtain values closer to saturation, concentrations of amino acid greater than 20 mM were not employed because of the inhibitory effects reported by other investigators (21) and because it was desirable to maintain the test solution at approximately constant osmolarity and ionic strength.

[2] Alternative methods of plotting the data which weigh the points differently were also tried (36). Least squares analysis of the various lines showed no significant differences for J_A^{im} and K_t among the various methods.

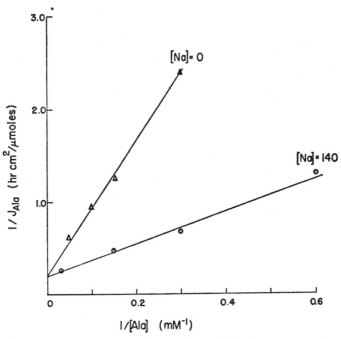

FIGURE 2.   Double reciprocal plot of data shown in Fig. 1.

for alanine are summarized in Table I. Three experiments, yielding 24 determinations of alanine influx, were carried out at each Na concentration and J_A^{im} and K_t were determined by least squares analyses of lines such as those shown in Fig. 2. The relatively high value of J_A^{im} at 70 mM Na appears to be due to animal variation rather than to Na concentration. The data in Table I show that there is no consistent variation of J_A^{im} with Na concentration, and fluxes consistent with a value for J_A^{im} as high as 13 μmoles/hr cm² were ob-

TABLE I

PROPERTIES OF THE AMINO ACID TRANSPORT SYSTEM

Amino acid	[Na]	K_t	J_A^{im}
	m M	*m M*	*μmoles/hr cm²*
Alanine	140	9.1 (7.2–11.0)*	6.1±0.9
	70	16.3 (11.5–21.2)	13.7±1.9
	22	31.2 (25.0–34.5)	7.8±1.5
	0	70.0 (60.0–79.6)	6.8±0.7
Valine	140	5.0 (3.3–6.4)	4.3±3.1
	0	31.5 (23.1–36.0)	6.5±0.8
Leucine	140	4.2 (3.0–5.7)	6.3±1.2
	0	29.0 (24.0–34.0)	4.5±1.3

* Range.

393

served in different experiments at 140 mM Na. In addition, one experiment was carried out in which J_A^{im} was estimated at 140 and 70 mM Na on tissue from a single animal; no significant effect of Na concentration on J_A^{im} was observed. We conclude, therefore, that variation in mucosal Na concentration does not significantly affect the maximum alanine influx. On the other hand, K_t increases markedly with reduction in Na concentration. Examination of these data indicates that the reciprocal of K_t is a linear function of Na con-

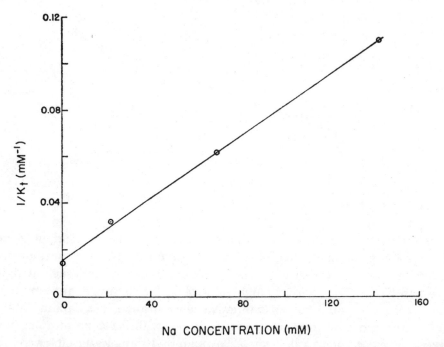

FIGURE 3. Reciprocal of the apparent Michaelis constant (K_t) for alanine as a function of Na concentration in the mucosal bathing solution.

centration, as shown in Fig. 3. Thus the relation between K_t and Na concentration can be described by an expression of the form

$$K_t = \frac{\alpha}{\beta + [Na]_m} \quad \text{or} \quad \frac{1}{K_t} = \frac{1}{\alpha}[Na]_m + \frac{\beta}{\alpha} \tag{2}$$

in which α and β are constants, and $[Na]_m$ is Na concentration in the mucosal solution. The values of α and β can be obtained from the slope and intercept of the line in Fig. 3.

Valine and Leucine Fluxes

The kinetics of valine and leucine influxes were investigated at 140 mM Na and in the absence of Na and were found to resemble the kinetics of alanine influx. Values of maximal flux and K_t were determined as described above; the results are summarized in Table I. Within experimental error, the maximum fluxes were independent of Na concentration and did not differ significantly from those obtained for alanine. The values of K_t for both valine and leucine were smaller than those observed for alanine but also showed a marked increase when Na was absent from the bathing medium.

Neutral amino acids are known to compete with each other in transport across the intestine (2, 3), and the competition appears to occur, at least in part, at the mucosal border (4). An examination of these competitive properties can provide evidence regarding modes of alanine entry. Experiments were carried out to test the effects of 20 mM valine or leucine on alanine influx at 140 mM Na and zero Na; the results are summarized in Table II. Valine and

TABLE II

INHIBITION OF ALANINE INFLUX BY LEUCINE AND
VALINE

		Alanine influx*			
Competitor	[Na]	Control	+ Competitor	K_I	K_t
	mM	*μmoles/hr cm²*		*mM*	*mM*
Leucine	140	1.20	0.27	3.9	4.2
(20 mM)	0	0.44	0.25	30.7	29.0
Valine	140	1.72	0.38	3.9	5.0
(20 mM)	0	0.26	0.16	31.0	31.5

* Alanine concentration was 3.3 mM in all experiments. Influx values are the average of six or more determinations under each condition.

leucine decreased alanine influx below control levels under both conditions, but the effect was greater at 140 mM Na. If the effect of these amino acids is assumed to be due entirely to competitive inhibition, the "K_I" for valine and leucine can be calculated (6) from the relation

$$\frac{J_A^{i\,'}}{J_A^i} = \frac{K_t + [A]_m}{K_t + [A]_m + \dfrac{K_t\,[I]_m}{K_I}}$$

in which $J_A^{i\,'}$ and J_A^i are alanine influxes in the presence and absence of competitor respectively and $[I]_m$ is the concentration of competitor in the mucosal solution. The values of K_t for alanine were taken as those in Table I. The resulting K_I's of valine and leucine, given in Table II, are nearly identical with the K_i's of these amino acids. Thus, the effects of valine and leucine are consistent with "classical" competition with alanine for the influx mechanism at both 140 mM and zero Na. These observations have two important implications. First, alanine influx does not take place by a mechanism not shared, at least in part, by valine and leucine; if it did, agreement between K_I and K_t would not be obtained. Second, amino acid entry in the absence of Na cannot be attributed to simple diffusion since, in addition to displaying saturation kinetics, it is subject to competitive inhibition. Indeed, the good agreement between K_I and K_t suggests that simple diffusion does not contribute significantly to alanine entry at 140 mM Na or at zero Na.

Oxender and Christensen (5) have suggested that neutral amino acid transport in ascites tumor cells is mediated by multiple entry routes having overlapping affinities. Their data indicate that satisfactory agreement between K_t and K_I may not be a sufficiently critical criterion for the presence of a single transport site. However, there is at present no compelling evidence for multiple systems in intestine so for simplicity the agreement between K_I and K_t will be interpreted as evidence for a single site. If subsequent information indicates that a single site is insufficient, the model presented below will have to be modified.

A Transport Model

The data reported above seem sufficient to warrant consideration of possible models which could describe the Na-alanine interaction at the mucosal border of the cell. A number of observations (7, 8) suggest that such a model should include a provision for the specific entry of Na into the cell together with alanine, and a variety of models involving combination of alanine and Na with a carrier or site located in the mucosal membrane were, therefore, examined. In addition, the model should provide for the following characteristics of alanine influx: (i) alanine influx is independent of cellular Na concentration (1); (ii) at constant Na concentration, alanine influx is given by equation 1; (iii) J_A^{im} is independent of $[Na]_m$; (iv) alanine influx can occur in the absence of Na (Fig. 1) by a mechanism similar to that involved in the presence of Na; (v) K_t is dependent on $[Na]_m$ as shown in equation 2.

Of the possibilities considered, a model of the type shown in Fig. 4 seemed particularly promising. A detailed analysis of the kinetic behavior of this system is given in Appendix A. As a first step, a complete solution for the initial alanine influx was derived assuming that the concentration of alanine in the cell was zero. This solution is extremely unwieldy and is not presented. It predicts that alanine influx should depend on cell Na concentration and, in view of point (i) above, cannot describe the present data. If P_1, P_2, and P_3 are assumed to be small relative to the rates of reaction with the carrier so that translocation is the rate-limiting step, the expression is simplified and J_A^i no longer depends on cell Na. However, the resulting solution (equation A4) predicts that J_A^{im} should be dependent on external Na concentration, and does not agree with point (iii). If we further assume that the permeabilities of all forms of the carrier are equal ($P_1 = P_2 = P_3 = P$), we obtain the following expression for J_A^i

$$J_A^i = \frac{C_t\,P\,[A]_m}{\dfrac{K_1\,K_2}{[Na]_m + K_2} + [A]_m} = \frac{J_A^{im}\,[A]_n}{K_t + [A]_m} \tag{3}$$

in which C_t is total carrier concentration; $K_1 = k_{-1}/k_1$ and $K_2 = k_{-2}/k_2$ where k_1 and k_{-1} are the forward and reverse rate coefficients of the reaction $A + X \rightleftharpoons XA$ and k_2 and k_{-2} are similar coefficients for $XA + Na \rightleftharpoons XANa$. The coefficient K_t is given by

$$K_t = \frac{K_1\,K_2}{[Na]_m + K_2} \tag{4}$$

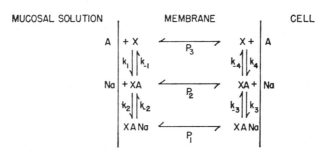

FIGURE 4.   Model system for alanine and Na transport across the mucosal border of intestine. A represents amino acid and X is a carrier molecule. The $k_{\pm i}$ are rate coefficients for the chemical reactions and the P_i are permeability coefficients.

which is the form required by equation 2. The values of the dissociation constants for alanine and Na, determined from the slope and intercept of the line in Fig. 3, are $K_1 = 70$ mM and $K_2 = 17$ mM.

Equation 3 fulfills the five requirements outlined above so that this model satisfies all the data so far considered on alanine influx. This agreement is not, however, particularly compelling evidence in favor of the model because the requirements are rather simple and could, no doubt, be satisfied by a number of other approaches. The proposed model would gain further support if it could predict other characteristics of the system which are independent of the observations employed in its construction. One such prediction, the linear relation between $1/K_t$ and $[Na]_m$, has already been noted (equation 4). Two additional predictions can be made. First, the model predicts that the unidirectional influx of amino acid will be unaffected by intracellular amino acid concentration. That is, there should be no significant transconcentration effect on influx. Data presented in the preceding paper (1) have shown the absence of a transconcentration effect on alanine influx. Second, a portion of the Na influx should be directly related to alanine influx and the coupling between these two movements can be predicted (equation A6). If we designate J'_{Na} as that portion of Na influx which is independent of alanine influx, the total Na influx, J^i_{Na}, should be given by

$$J^i_{Na} = \left[\frac{[Na]_m}{K_2 + [Na]_m} \right] J^i_A + J'_{Na} \qquad (5)$$

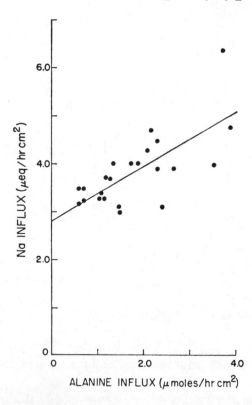

FIGURE 5. Relation between Na and alanine influxes for experiments at a Na concentration of 22 mM in the mucosal solution. Each point represents a single tissue sample in which both fluxes were determined simultaneously. The line was determined by least squares analysis.

Thus, the model predicts that J^i_{Na} will vary linearly with J^i_A at constant Na concentration and that the slope of this relation $(\Delta J^i_{Na}/\Delta J^i_A)$ will vary with Na concentration. Further, since the value of K_2 has been determined from the relation between K_t and $[Na]_m$, the slope can be predicted quantitatively.

Relation between Alanine and Na Fluxes

Since both alanine and Na influxes were measured simultaneously in many experiments, an evaluation of equation 5 is possible. Experiments such as those shown in Fig. 1 provide the necessary data since variation in alanine flux at constant Na concentration was obtained by changing alanine concentration. In each experiment there was a direct relation between Na influx and alanine influx which suggested a coupling between these processes. In order to evaluate this relation Na influx was plotted against the simultaneously measured alanine influx for all experiments at a given Na concentration. An example of the results is shown in Fig. 5 for experiments at 22 mM Na. At all four Na concentrations tested, (140, 70, 22, and 5 mM), there appeared to be a linear relation between the two fluxes and correlation coefficients for linear regression were statistically significant ($p < 0.01$) in all cases. The slopes and intercepts of the lines were determined by least squares analyses. Fig. 6 shows a comparison between experimentally determined slopes and those predicted

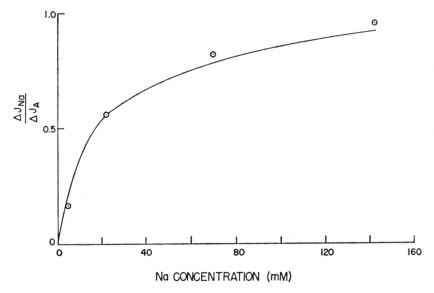

FIGURE 6. Ratio of alanine-dependent Na influx to alanine influx as a function of Na concentration in the mucosal solution. Points are experimental values given by the slopes of lines such as that shown in Fig. 5. The solid curve was calculated from equation 5 as described in the text.

by equation 5; the points are experimental values and the solid curve was calculated from $[Na]_m/(K_2 + [Na]_m)$ using $K_2 = 17$ mM as determined from the line in Fig. 3. The agreement is satisfactory and lends strong support to the proposed model.

The intercept of the line relating alanine and Na influxes provides information on the Na influx in the absence of alanine, J'_{Na} in equation 5. Fig. 7 shows J'_{Na} as a function of Na concentration; this portion of Na influx increases linearly with concentration at least up to 140 mM. The rate of active Na transport across isolated rabbit ileum is also a linear function of Na concentration over this range (9). In the presence of 5 mM alanine, the alanine-dependent Na influx makes up less than 15% of the total Na influx at all Na concentrations. Thus, unless alanine concentration is quite high, the major

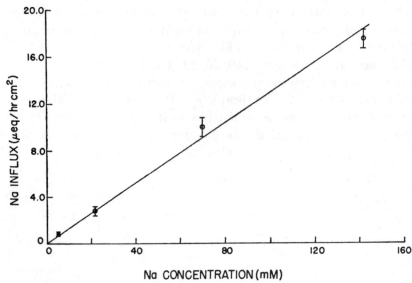

FIGURE 7. Alanine-independent Na influx, J'_{Na}, as a function of Na concentration in the mucosal solution.

portion of Na influx across the mucosal border of the cells takes place via a mechanism other than that proposed in Fig. 4.

Effects of Temperature

The influence of temperature on alanine and Na influxes was examined in several experiments in which one piece of tissue was kept at 37°C and the other at room temperature (23°C). The results are summarized in Table III.

TABLE III

EFFECT OF TEMPERATURE ON Na AND ALANINE INFLUXES

	Influx			
	Alanine*		Sodium‡	
	[Na] = 140	[Na] = 0	Ala-dependent	Ala-independent
	μmoles/hr cm²		*μmoles/hr cm²*	
37°	2.6	0.5	2.5	18.0
23°	0.3	0.1	0.3	11.4
Q_{10}	4.5	3.2	4.5	1.4
n§	8	6	6	6

* Alanine concentration was 5 mM in all experiments.
‡ Alanine-dependent and alanine-independent Na fluxes are defined by equation 5.
§ No. of observations.

The effect of temperature on alanine flux in the presence of Na was quite marked; a Q_{10} of 4.5 was calculated from the data. Since in the presence of 140 mM Na, approximately 1 Na enters with 1 alanine (Fig. 5), the Q_{10} of the alanine-dependent Na influx is also approximately 4.5. The effect of temperature on total Na flux was, however, much smaller, and if the total Na fluxes were corrected for alanine-dependent flux, a Q_{10} of 1.4 was calculated for the alanine-independent Na influx, J'_{Na}. At zero Na, alanine influx had a Q_{10} of 3.2. This value is not known accurately, however, because at room temperature alanine influx in the absence of Na was quite small so that experimental errors could have a rather marked effect on the calculated Q_{10}.

These observations on the effects of temperature on influx at the mucosal border may be compared with the findings of Schultz and Zalusky (9) on active Na transport across rabbit ileum. They found a Q_{10} of 4.8 for the sugar-stimulated increase in net Na transport (measured by short-circuit current) and a Q_{10} of only 1.6 for net Na flux in the absence of actively transported nonelectrolyte. These values are in good agreement with those observed here for the alanine and Na influx processes. The marked differences in response of Na fluxes to temperature in the presence and absence of alanine lend further support to the concept of two independent paths for Na influx as expressed in equation 5. The explanation of these rather high temperature coefficients is unclear at present but the phenomenon certainly merits further study since it may offer clues to the molecular mechanism of the transfer process.

DISCUSSION

Model for Amino Acid Transport

The model shown in Fig. 4 provides a satisfactory explanation for all the data presented in this and the preceding paper on amino acid and Na influxes across the mucosal border of the intestine. The available data rule out several other possible models involving a ternary complex between amino acid, Na, and a carrier (or site). The presence of two independent and parallel pathways for alanine influx, one Na-dependent and the other Na-independent, is rendered unlikely by the observation that J_A^{im} is the same in the presence and absence of Na; if there were two independent pathways, the maximum influxes should be additive. For this reason the first reaction in the proposed model cannot involve association with Na rather than amino acid since, under these conditions, there would be no flux of amino acid in the absence of Na. The same point would apply to a system involving an obligatory simultaneous reaction among Na, amino acid, and carrier. A model in which there is no preferred order for association of Na and amino acid with the carrier is extremely complex and has not been fully evaluated. Preliminary calculations suggest, however, that amino acid influx would depend on intracellular Na concentration in such a system. Thus the model presented seems the simplest one which can offer a reasonable explanation of the experimental observations.

The quantitative considerations of this model involve a number of assumptions which merit comment. The assumption that the steps involving transfer of carrier across the membrane are rate-limiting is based primarily on the fact that the general expression for alanine influx obtained without this assumption predicts dependence on cell Na concentration, a prediction which is contrary to our observations (1). However, the influence of cell Na on alanine influx would depend strongly on relative values of the rate coefficients for the chemical reactions. In particular, if k_3 were small, dependence of amino acid influx on cell Na concentration would be minimized. The general solution also has the form of Michaelis-Menten kinetics although the meanings of the quantities J_A^{im} and K_t are much more complex than for the simplified case. In this general case, J_A^{im} would depend on Na concentration in the mucosal solution, $[Na]_m$, and the relation between K_t and $[Na]_m$ would be different from that observed. Thus these points, together with failure to observe an effect of cell Na on amino acid influx, seem to offer reasonable support for the assumption that the carrier transfer processes are rate-limiting.

The assumption that the three permeability coefficients are equal is perhaps more difficult to justify on experimental grounds although it is reasonable in terms of the concept of translocation of a relatively large carrier molecule. The data are, however, adequate to demonstrate that if the model is correct, the coefficients cannot differ markedly. First, the prediction that there will be no transconcentration effect on influx arises directly from the assumption of equal permeability coefficients (see for example the discussion of Heinz and Durbin, 10). Appreciable differences in permeabilities would lead to either a decrease or an increase of influx in tissues preloaded with alanine, depending on the relative values of the permeability coefficients. Since no significant effect of preloading was observed (1), the coefficients must be similar. Second, the assumption that $P_1 = P_2$ leads to the prediction that J_A^{im} will be independent of $[Na]_m$ and is justified by the observation that there is no consistent variation of J_A^{im} with $[Na]_m$. However, if P_1 and P_2 do not differ markedly, the dependence of J_A^{im} on $[Na]_m$ might be sufficiently small to escape detection. Thus, the available evidence lends support to these assumptions but does not completely prove their validity.

In developing expressions for behavior of the model system, we have tacitly assumed that all forms of the carrier are uncharged or, alternatively, that there is no electrical potential difference across the mucosal border of the cell. The latter is clearly incorrect (11, 12, footnote 3) and since an ion (Na+) is assumed to associate with the carrier, at least one of the forms might be expected to carry

[3] Field, M., and P. F. Curran, unpublished observations.

a net charge. As discussed by Britton (13), inclusion of a potential difference in consideration of carrier models is not entirely straightforward. In the present case, the simplest approach would be to assume that the complex XANa is positively charged and that the effect of the electrical potential difference is confined to its influence on the translocation of this complex. If the flux of XANa is assumed to be described by the Goldman equation (14), the quantity $P_1([XA\text{Na}]_m - [XA\text{Na}]_c)$ in equations A2 must be replaced by $P_1\xi([XA\text{Na}]_m - [XA\text{Na}]_c e^{-\xi})/(1 - e^{-\xi})$ where $\xi = z_i \mathfrak{F}\Delta\psi/RT$, and $\Delta\psi$ is the electrical potential difference across the mucosal membrane. The resulting expression for amino acid influx contains terms dependent on the electrical potential difference. Even with the assumption that $P_1 = P_2 = P_3$, J_A^{im} should depend on both Na concentration and potential difference. The failure to observe a dependence of J_A^{im} on Na thus suggests that the potential difference may not influence the transport system unless the effects of changes in Na concentration and in potential difference cancel in a rather fortuitous manner. On the other hand, if the association of Na with the carrier involved an exchange with another cation such as hydrogen ion, none of the carrier forms would involve a net charge and a potential difference would have no effect. Further studies are clearly necessary to examine these factors in detail and there is no point at present in speculating on their possible implications in the transport mechanism.

Finally, the proposed model has been discussed mainly in terms of the mobile carrier hypothesis, but this concept need not be invoked to explain the experimental observations. The steps involved in transfer of materials across the membrane phase could involve chemical reactions or molecular rearrangements such as the "gate" mechanism suggested by Patlak (15). In this case, the P's would represent rate coefficients for the translocation process. If the steps involving transfer across the membrane were slow relative to the association-dissociation reactions at the surfaces, the kinetic description of the system would be unaltered and all considerations presented above would remain valid. Thus, the ability of the model to describe the kinetics of the Na-alanine interaction should not be taken as evidence for the existence of a mobile carrier.

Alanine-Dependent Na Influx

The addition of actively transported sugars or amino acids to the solution bathing the mucosal surface of rabbit ileum results in an immediate increase in the short-circuit current due to an increased rate of active Na transport from mucosa to serosa (7, 16). The increase in the short-circuit current is a saturable function of the mucosal alanine (or sugar) concentration (7). These observations supported the concept that a ternary complex involving Na is responsible for the influx of the nonelectrolyte (17). The present experiments provide the first clear demonstration of a coupled entry of amino acid and Na from the mucosal solution into the cell and the alanine-dependent Na influx is of sufficient magnitude to account for the increased rate of transmural Na transport in the steady state. Thus at 5 mM alanine, the steady-state influx of alanine across the mucosal border averages 2.2 μmoles/hr cm². Since the coupling ratio between alanine and Na is 0.9 at 140 mM Na (Fig. 6), the increased Na influx will be 2.0 μmoles/hr cm². The rate of net transmural alanine transport under similar conditions averages 1.3 μmoles/hr cm² (footnote 3), so that alanine efflux out of the cell across the mucosal border must be 0.9 μmole/hr cm². Assuming that the transport mechanism is symmetrical,[4] as shown in Fig. 4, the coupling ratio between alanine efflux and Na efflux in the presence of a cell Na concentration of 50 mM (18) is 0.75. The efflux of Na coupled to alanine would be (0.75) (0.9) = 0.7 μmole/hr cm², and the net alanine-dependent flux of Na across the mucosal border becomes 1.3 μmoles/hr cm². This prediction for the increased rate of Na transport in the presence of 5 mM alanine is in good agreement with the value of 1.5 μmoles/hr cm² observed by Schultz and Zalusky (Fig. 2, reference 7).

Alanine-Independent Na Influx

The conclusions drawn from the model of Na and amino acid interaction permit a more detailed examination of the data presented in the preceding paper on the effects of K and Li on Na and alanine influxes. Since Na influx can be described adequately by equation 5 and since the coefficient multiplying J_A^i in the expression has been evaluated, the effects of K and Li on the alanine-independent Na influx, J_{Na}', can be estimated. Table IV gives values of J_{Na}' at 22 mM Na under control conditions (118 mM choline) and in the presence of various other replacement solutes. As expected from the data given previously, replacement of choline by Tris or of choline chloride by mannitol does not alter J_{Na}' significantly. K causes a depression of the alanine-independent Na influx under all conditions tested. When K is used in the test solution only, so that the tissue is exposed to high K for approximately 60 sec, the effect on Na influx is small. Preincubation of the tissue in high K for 30 min causes a significant further inhibition of Na influx and, as shown in the last row of the table, the preincubation in high K has an effect even if the K concentration in the test solution is normal (12 mM). The effects of K on the alanine-independent Na influx are approximately the same as those observed on alanine influx (Fig. 5, reference 1). Since these two processes appear to be independent, this observation suggests that the decrease in fluxes caused by K is not specific but is the result of a general effect on the tissue. The effect of Li is of particular interest. The substitution of Li for

---

[4] The model shown in Fig. 4 is considered symmetrical if $K_1 = K_4$ and $K_2 = K_3$.

choline caused only a 18% inhibition of alanine influx (Fig. 4, reference 1) but resulted in a 60% inhibition of the alanine-independent Na influx. As discussed previously, the observation suggests that this portion of the Na influx may not be the result of a simple diffusion process; it may involve a carrier-mediated process or some other specific interaction with the membrane which is subject to inhibition by Li.

TABLE IV

Treatment	Na influx		n^*	p
	Control‡	Test		
	$\mu moles/hr\ cm^2$			
118 mM Tris	4.6	4.7	6	>0.1
236 mM mannitol	4.4	5.0	4	>0.1
118 mM Li	4.8	2.0	8	<0.01
130 mM K	4.1	3.4	10	>0.05
130 mM K (preincubation) 130 mM K (test)	4.7	2.4	6	<0.01
130 mM K (preincubation) 12 mM K (test)	4.1	2.8	4	<0.01

* No. of observations.
‡ Control tissue preincubated in solution containing 118 mM choline, 22 mM Na and uptake measured from same solution. Test tissue was preincubated in this same solution unless otherwise noted and uptake measured from solution shown in column 1. All solutions contained 22 mM Na. Control and test data for each condition were obtained on tissues from the same animals.

Net Transport and Tissue Accumulation of Amino Acids

The model presented in Fig. 4 makes no provision for the direct utilization of metabolic energy in the processes of amino acid entry. Since amino acids are accumulated by the mucosal cells to concentrations considerably above those in the bathing medium (5, 18, 19) and are transported across the intestine against substantial concentration differences (19–21), a consideration of energetics is necessary. Both cell accumulation and net transmural transport of amino acids are inhibited by metabolic poisons (22) but this observation does not necessarily demonstrate a *direct* coupling of amino acid transport to metabolism. The same results would be obtained if amino acid influx were coupled to the movement of another solute whose transport is, in turn, directly coupled to metabolic energy (see references 23 and 24 for a phenomenological description of such coupled processes). That amino acid transport "could be wholly driven by the energy inherent in the asymmetry of cellular alkali metal distribution" was suggested by Christensen and his coworkers (25), and the possibility of a ternary complex between Na, glycine, and carrier was entertained by these investigators. More recently Crane (26) has proposed that accumulation and, hence, net transport of sugars can be explained by the concentration differences of Na and K between the cell and its environment and does not require a direct coupling to metabolic energy. Vidaver (27) has suggested that glycine accumulation by pigeon erythrocytes is due to a Na concentration difference. The models involved are rather similar to that proposed here. However, the data available do not provide conclusive evidence that alanine transport by rabbit ileum can be accounted for by the Na concentration difference alone. This point may be illustrated by examining some simple thermodynamic implications of the system shown in Fig. 4. The most straightforward approach is the evaluation of the dissipation function Φ (or the entropy production)[5] of the system which must be positive if the process under consideration is to occur spontaneously (24). As shown in Appendix B, the dissipation function for the model depicted in Fig. 4 can be written simply

$$\Phi = J_A\Delta\mu_A + J_{Na}\Delta\mu_{Na} \qquad (6)$$

in which the J's are the *coupled* net fluxes of alanine and Na across the mucosal membrane and are positive in the direction mucosal solution to cell. The $\Delta\mu$'s are differences in chemical potentials across the membrane. As indicated above, the coupled net fluxes of Na and alanine across the membrane are approximately equal at 140 mM Na. If we assume that cellular concentrations determined on mucosal strips (18) apply to the condition of steady-state net transfer, equation 6 can be evaluated. With mucosal solution concentrations of 140 mM for Na and 5 mM for alanine, cellular concentrations were found to be approximately 50 mM Na and 40 mM alanine. Assuming that cellular and extracellular activity coefficients are equal, the dissipation function then becomes

$$\Phi = RTJ_A\{\ln 0.125 + \ln 2.80\} = -1.05RTJ_A$$

Since the dissipation function is negative, the proposed process cannot occur spontaneously. As indicated in Appendix B, if one of the carrier complexes is charged, $\Delta\mu_{Na}$ should be replaced by $\Delta\bar{\mu}_{Na}$. Assuming a potential difference of 15 mv, cell-negative (footnote 3), Φ becomes -0.49 RT J_A. Since Φ is still negative, a spontaneous process of net alanine transport under the conditions assumed seems impossible in terms of this simple model. There is, however, a net alanine flux across the mucosal border so that the correct dissipation function must involve terms in addition to the two indicated, or the evaluation of Φ must be incorrect. For example, the association-dissociation

---

[5] The dissipation function is given by the absolute temperature times the rate of entropy production due to irreversible processes occurring within a system.

reactions were considered to be at equilibrium and, thus, to make no contribution to the dissipation function. Since net flows of the reactions must occur, this assumption cannot be entirely correct but the contribution to Φ cannot be evaluated. Further, cell Na may be compartmentalized in such a way that the concentration at the membrane itself is considerably lower than the average value determined for total cell water. In order to give a positive dissipation function, the Na concentration at the cytoplasmic side of the mucosal border would have to be less than 17 mM (or less than 30 mM if the transport is influenced by the potential difference[6]). Finally, the intracellular Na and alanine concentrations pertaining under conditions of steady-state net transport may not be the same as those observed in mucosal strips even though the two conditions seem comparable.

The above considerations are all based on the assumption that the transport system is symmetrical, having the same properties at both sides of the membrane. Alternatively, the system may not be symmetrical. The proposal of Crane (26) (see also Kipnis and Parrish (28)) that the high cellular K influences the affinity of substrate for the carrier is a possible source of asymmetry. However, we have been unable to obtain any evidence for a major effect of extracellular K on alanine influx and there are indications that even the observed 20% inhibition may be a relatively nonspecific effect. Thus, a marked influence of the K concentration difference seems unlikely in the present case. Nonetheless, the possibility remains that the influence of K on the cellular side of the membrane is different from that on the outer side. In addition, other properties of the cytoplasm could contribute to an asymmetry. Finally, there is the clear possibility that a direct participation of metabolic energy is involved in providing the asymmetry necessary to account for tissue accumulation and net transport. For example, a metabolic alteration of the carrier at the inner surface which reduced its affinity for substrate would increase the accumulation ratio. In this case a new, detailed formulation of the model system would be required and the dissipation function of the system would take a different form. A more detailed understanding of the cellular compartment is clearly necessary for an explicit thermodynamic description. In addition, the influence of metabolic inhibitors on the influx process must be examined and information on outflux from the cell must be obtained before the process of accumulation can be fully understood.

Although the present model may not account quantitatively for the observed levels of amino acid accumulation by mucosal cells, the inhibitory effects of ouabain or a Na-free medium are readily explicable in terms of a symmetrical transport system and the Na concentration difference. When

---

[6] The cell Na concentration of 50 mM was calculated after correction for the Na content of the inulin space of mucosal strips. Since, as noted previously (18), inulin may underestimate the extracellular space, it is likely that the true intracellular Na concentration is less than 50 mM.

mucosal strips are incubated in the presence of ouabain or in a Na-free medium, the difference between intracellular and extracellular Na concentrations is essentially abolished within 30 min (18). Examination of Fig. 4 or equation A7 shows that if $K_1 = K_4$ and $K_2 = K_3$, and if $[Na]_m = [Na]_c$ the system will reach a steady state when $[A]_c = [A]_m$ (that is, net flux across the mucosal border will be zero when the concentration ratio for alanine reaches unity). Thus the inhibition of amino acid accumulation and transmural transport against a concentration difference by Na-free medium, ouabain, or metabolic inhibitors may be attributed to the fact that under all these conditions $[Na]_c \cong [Na]_m$.

This explanation for the inhibitory effect of ouabain is consistent with a number of experimental observations. Ouabain inhibits transmural transport of sugars (29) and amino acids (30) only when it is present in the serosal bathing solution even though the mechanisms responsible for transmural transport of sugars and amino acids against concentration differences are located in the brush border (4). While it is possible that some ouabain crosses the serosal membrane and gains access to the nonelectrolyte transport mechanisms, two observations mitigate against this possibility. First, when ouabain is placed in the serosal solution, very little enters the mucosal cells and virtually none appears in the mucosal solution (29). Second, in squid axon (31) and erythrocytes (32) there is evidence suggesting that the site of ouabain action on cation transport is at the extracellular surface of the membrane. While these observations do not rule out an intracellular action of ouabain on nonelectrolyte transport in intestine, a much simpler explanation is suggested by the finding that ouabain inhibits active Na transport across rabbit ileum only when it is present in the serosal solution (9). Thus, inhibition of the mechanism by which Na is extruded from the cell could abolish the difference between $[Na]_c$ and $[Na]_m$ and thereby inhibit sugar and amino acid transport. The effect of ouabain may be restricted only to its well known role as an inhibitor of cation transport and no direct action on nonelectrolyte transport need be postulated.

Other Studies

It is somewhat difficult to compare these studies quantitatively with other investigations of interrelations between Na and nonelectrolyte transport by intestine. The present experiments are unique in that they provide data on unidirectional influx across the mucosal border. Other studies involving measurement of tissue accumulation or transmural transport do not provide unequivocal information on this influx process. In fact, extrapolation of data on accumulation or transport to conclusions regarding the transport mechanism may be misleading. Thus, the careful studies of Mathews and Laster (21) on transmural transport indicated that the maximal transport rates for alanine, valine, and leucine were in the ratio 1.0:0.56:0.29 while the data shown in Table I indicate that, for the mucosal entry step, the maximal rate is the same, within experimental error, for all three compounds. While this difference could be due in part to species variation, it probably also reflects the fundamental difference between mucosal influx and transmural flux. Even measurements of tissue uptake over relatively short time intervals (5–10 min) do not provide data which are directly comparable to the present measurements for two reasons. First, the tissue is exposed to fluid on both sides and there is no clear way of separating uptake across the serosal side from that at the mucosal side. Second, in the presence of Na, sufficient amino acid enters the cells in 5–10 min to make estimation of a unidirectional influx impossible. In 10 min, alanine has reached approximately 75% of its steady-state concentration in mucosal strips (18). Thus, such experiments actually measure a complex relation between influx and outflux at the mucosal and serosal borders. Coefficients such as K_t or maximal rates calculated from these data are extremely difficult to interpret as Mathews and Laster (21) have pointed out. The studies most similar to the present ones are those of Rosenberg et al. (19) examining the effect of Na on the steady-state influx and efflux of the model amino acid 1-aminocyclopentane-5-carboxylic acid (ACPC) in segments of rabbit jejunum. A decrease in Na concentration of the bathing medium caused a marked depression of ACPC influx into the tissue. On the assumption that the major route of ACPC entry into the tissue was via the mucosal border, their results are similar to those reported here. However, quantitative comparisons cannot be made until information on the relative contributions of the mucosal and serosal borders to total influx into tissue segments has been evaluated.

Studies on several nonepithelial systems have yielded data on amino acid transport which are similar to the present observations but in no case do precisely the same factors seem to be involved. Vidaver (27) has suggested that his observations on the influence of Na on glycine transport by pigeon erythrocytes could be explained by a system involving a carrier or site which combines with two Na ions and one amino acid molecule. His data appear to be fitted best by a model in which the site must first combine with the Na ions. The studies of Kipnis and Parrish (28) on the influence of Na on amino acid transport by rat diaphragm and isolated lymph node cells also seem qualitatively similar to our observations with respect to effects of changes in Na concentration. They have suggested that amino acid accumulation is the result of Na and K concentration differences as proposed by Crane (26) for sugars in intestine. Finally, the transport of amino acids by Ehrlich ascites cells is strongly dependent on Na (33). In contrast to the present observations, Wheeler et al. (34) have recently reported that both the K_t and the maximal velocity for influx of aminoisobutyric acid are reduced by lowered Na concentration. Interpretation of observations on amino acid transport by ascites cells may be complicated by the possibility that there are multiple modes of transport (5) but this conclusion has been questioned by Jacquez and Sherman (35). In intestine, we have been unable to obtain any evidence that more than one mechanism is involved in the transport of alanine, valine, and leucine across the mucosal membrane.

It is not clear whether the differences between these other tissues and intestine indicate the presence of different transport mechanisms or whether they represent different expressions of the same basic mechanism. Certainly the Na dependence of amino acid influx is present in all cases and many observations seem compatible with variants of the model shown in Fig. 4. For example, as indicated by equation A4, the present model predicts, under certain conditions, that both K_t and J_A^{im} would vary with Na concentration. It would be of interest to have information on the influence of amino acids on Na influx in these tissues since such data have proved particularly useful in suggesting and supporting the present model for the intestinal transport system.

APPENDIX A

We wish to examine the kinetic behavior of the model shown in Fig. 4 under steady-state conditions. Under these conditions, the concentrations of all components in the membrane phase are constant at all points of the phase so that, for example,

$$\frac{d\,[X]}{dt} = 0$$

and at the mucosal boundary

$$\frac{d\,[X]_m}{dt} = -k_1\,[A]_m\,[X]_m + k_{-1}\,[XA]_m + P_3\,[X]_c - P_3\,[X]_m = 0 \qquad (\text{A1})$$

Similar expressions may be written for the other components of the membrane phase at both the "m" (mucosal solution) and "c" (cell) boundaries. The net steady-state velocity of the system, J_A, is given by

$$J_A = P_1([XA]_m - [XA]_c) + P_2([XANa]_m - [XANa]_c)$$

Following the approach discussed by Hearon (37), the set of equations of which equation A1 is an example can be used to form the set

$$\begin{aligned}
J_A &= w_1[X]_m - w_{-1}[XA]_m \\
&= w_2[XA]_m - w_{-2}[XANa]_m + P_2([XA]_m - [XA]_c) \\
&= P_1([XANa]_m - [XANa]_c) + P_2([XA]_m - [XA]_c) \\
&= w_3[XANa]_c - w_{-3}[XA]_c + P_2([XA]_m - [XA]_c) \\
&= w_4[XA]_c - w_{-4}[X]_c \\
&= P_3([X]_c - [X]_m)
\end{aligned} \qquad (\text{A2})$$

in which

$$w_1 = k_1[A]_m \qquad w_2 = k_2[Na]_m \qquad w_3 = k_{-3} \qquad w_4 = k_{-4}$$

$$w_{-1} = k_{-1} \qquad w_{-2} = k_{-2} \qquad w_{-3} = k_3[Na]_c \qquad w_{-4} = k_4[A]_c$$

The $k_{\pm i}$ are forward and backward rate coefficients for the various reactions and the P_i are permeability coefficients. In addition, the total amount of carrier is assumed to be constant so that

$$2C_t = [X]_m + [X]_c + [XA]_m + [XA]_c + [XANa]_m + [XANa]_c \qquad (A3)$$

in which C_t is total carrier concentration. The factor 2 enters because the mean concentration of each form has been defined by relations of the type $[X]_{mean} = ([X]_m + [X]_c)/2$. The set of equations A2 together with equation A3 may be solved simultaneously for J_A. Since we are, at present, interested in the initial alanine influx into cells which have not been preloaded, the condition $[A]_c = 0$ has been imposed. The resulting expression for J_A will then represent the unidirectional influx from mucosal solution to the cell, J_A^i. The general solution of this set is straightforward but tedious; the resulting expression for J_A^i contains 32 terms in the denominator involving products of the w's and P's. An expression of this complexity is of little use in quantitative considerations of alanine flux. In addition, the general expression for J_A^i contains terms in $[Na]_c$ in the denominator while our observations have indicated that J_A^i is independent of $[Na]_c$. While it is impossible to evaluate the relative importance of these terms in $[Na]_c$, their presence suggests that the general expression for J_A^i might not fit the experimental data and we have, therefore, examined the influence of some simplifying assumptions.

The diffusion of carrier may be assumed to be much slower than the postulated association-dissociation reactions; that is, P_1, P_2, $P_3 < w_i$, w_{-i} for all i. This assumption leads to considerable simplification and the resulting expression for J_A^i is

$$J_A^i = \frac{2C_t P_3 k_1 \{P_1 k_2 [Na]_m + P_2 k_{-2}\} [A]_m}{2P_3 k_{-1} k_{-2} + k_1 [A]_m \{(P_1 + P_3)k_2 [Na]_m + (P_2 + P_3)k_{-2}\}} \qquad (A4)$$

This expression could satisfy much of the present data but, depending on the relative values of P_1, P_2, and P_3, the maximal influx rate, J_A^i, could depend on $[Na]_m$. Since the data suggest that such a dependence does not occur, the additional assumption that the three permeability coefficients are equal has been made. This assumption is also necessary to insure that there be no transconcentration effect (see the analysis of Heinz and Durbin, 10). Equation A4 then reduces to

$$J_A^i = \frac{C_t P [A]_m}{\dfrac{K_1 K_2}{[Na]_m + K_2} + [A]_m} \qquad (A5)$$

in which $P = P_1 = P_2 = P_3$, $K_1 = k_{-1}/k_1$, and $K_2 = k_{-2}/k_2$. The use of equation A5 in analysis of the present data is discussed in the text.

411

An expression of the form of A5 can be obtained without assuming that all three permeability coefficients are equal. If $P_1 = P_2$ and $P_1 \neq P_3$, the term $C_t P$ becomes $2C_t P_1 P_3/(P_1 + P_3)$ and K_1 becomes $2K_1 P_3/(P_1 + P_3)$. For the present, we shall retain the simpler assumption of uniform permeabilities since none of the basic considerations presented here is affected by choice of these two possibilities.

The unidirectional Na influx arising from the system shown in Fig. 4 should be given by $P_2[XA\,\mathrm{Na}]_m$. In solving the set A2 for J_A^i using the simplifying assumptions, the following expression is obtained:

$$[XA\mathrm{Na}]_m = \frac{k_2\,[\mathrm{Na}]_m\,J_A^i}{P(k_2\,[\mathrm{Na}]_m + k_{-2})}$$

The total measured Na influx, J_{Na}^i, is given by the flux via the carrier system plus a flux J_{Na}' through other paths so that

$$J_{\mathrm{Na}}^i = \left[\frac{[\mathrm{Na}]_m}{[\mathrm{Na}]_m + K_2}\right] J_A^i + J_{\mathrm{Na}}' \qquad (\mathrm{A6})$$

As in equation A5, $K_2 = k_{-2}/k_2$. An experimental test of equation A6 is described in the text.

If the restriction that $[A]_c = 0$ is removed, the set of equations A2 may be solved for the rate of *net* transfer, J_A. Using the assumptions that the permeation steps are rate-limiting and that $P_1 = P_2 = P_3 = P$, the resulting expression is

$$J_A = \frac{C_t\,P\{K_3\,K_4([\mathrm{Na}]_m + K_2)\,[A]_m - K_1\,K_2([\mathrm{Na}]_c + K_3)[A]_c\}}{\begin{array}{c}K_1\,K_2\,K_3\,K_4 + K_3\,K_4([\mathrm{Na}]_m + K_2)[A]_m + K_1\,K_2([\mathrm{Na}]_c + K_3)[A]_c \\ + ([\mathrm{Na}]_m + K_2)([\mathrm{Na}]_c + K_3)[A]_m\,[A]_c\end{array}} \quad (\mathrm{A7})$$

in which $K_3 = k_3/k_{-3}$ and $K_4 = k_4/k_{-4}$.

APPENDIX B

The entropy production, d_iS/dt, or dissipation function, $\Phi = T\,d_iS/dt$, in the system illustrated in Fig. 4, may be evaluated easily under steady-state conditions (24). Since the steps involving transfer across the membrane are rate-limiting, the chemical reactions at the boundaries may be assumed to be approximately in equilibrium and their contribution to entropy production within the membrane may be neglected. Thus, to a first approximation

$$\Phi = J_{XA}\Delta\mu_{XA} + J_{XA\mathrm{Na}}\Delta\mu_{XA\mathrm{Na}} + J_X\Delta\mu_X \qquad (\mathrm{B1})$$

The J's are positive in the direction mucosal solution to cell and Δ indicates mucosal solution minus cell. At the mucosal solution side of the membrane, the condition of equilibrium for the chemical reactions requires that

$$\mu_A' + \mu_X' = \mu_{XA}'$$

$$\mu_{\mathrm{Na}}' + \mu_{XA}' = \mu_{XA\mathrm{Na}}'$$

and at the opposite side

$$\mu_A'' + \mu_X'' = \mu_{XA}''$$

$$\mu_{Na}'' + \mu_{XA}'' = \mu_{XANa}''$$

Thus,

$$\Delta\mu_{XA} - \Delta\mu_X = \Delta\mu_A \tag{B2}$$

$$\Delta\mu_{XANa} - \Delta\mu_X = \Delta\mu_{Na} + \Delta\mu_A \tag{B3}$$

Further, since the carrier is confined to the membrane phase,

$$J_{XA} + J_{XANa} + J_X = 0 \tag{B4}$$

Introducing equations B2, B3, and B4 into equation B1 and rearranging yields

$$\Phi = (J_{XA} + J_{XANa})\Delta\mu_A + J_{XANa}\Delta\mu_{Na} \tag{B5}$$

Assuming that there are boundary equilibria between the solutions and the membrane phases (24), $\Delta\mu_A$ and $\Delta\mu_{Na}$ may be taken as the differences between the two solution phases and equation B5 may be written

$$\Phi = J_A\Delta\mu_A + J_{Na}\Delta\mu_{Na} \tag{B6}$$

in which J_A and J_{Na} are the coupled net fluxes of amino acid and Na. If $X A$Na is assumed to be positively charged, similar arguments including electrical potentials show that

$$\Phi = J_A\Delta\mu_A + J_{Na}\Delta\bar{\mu}_{Na} \tag{B7}$$

in which $\bar{\mu}_{Na}$ is the electrochemical potential of Na. Thus under steady-state conditions the dissipation function for the proposed model can be expressed simply in terms of flows and forces external to the membrane.

We would like to acknowledge the valuable technical assistance of Mrs. Carol Blum.

This work was supported by a Public Health Service Research Grant (AM-06540) from the National Institute of Arthritis and Metabolic Diseases.

Dr. Schultz was an Established Investigator of the American Heart Association.

Dr. Curran was supported by a Public Health Service Research Career Program Award (AM-K3-5456) from the National Institute of Arthritis and Metabolic Diseases.

Dr. Chez was a Public Health Service trainee in reproductive physiology supported by a training grant from the National Institute of Child Health and Human Development (5-T1-HD-38-07).

Dr. Fuisz was a Special Research Fellow of the National Heart Institute.

Received for publication 6 July 1966.

REFERENCES

1. SCHULTZ, S. G., P. F. CURRAN, R. A. CHEZ, and R. E. FUISZ. 1967. Alanine and sodium fluxes across mucosal border of rabbit ileum. *J. Gen. Physiol.* **50**:1241.

2. WILSON, T. H. 1962. Intestinal Absorption. W. B. Saunders Company, Philadelphia.

3. FINCH, L. R., and F. J. R. HIRD. 1960. The uptake of amino acids by isolated segments of rat intestine. II. A survey of affinity for uptake from rates of uptake and competition for uptake. *Biochim. Biophys. Acta.* **43**:278.

4. KINTER, W. B., and T. H. WILSON. 1965. Autoradiographic study of sugar and amino acid absorption by everted sacs of hamster intestine. *J. Cell Biol.* **25** (2, Pt. 2):19.

5. OXENDER, D. L., and H. N. CHRISTENSEN. 1963. Distinct mediating systems for the transport of neutral amino acids by the Ehrlich cell. *J. Biol. Chem.* **238**:3686.

6. DIXON, M., and E. C. WEBB. 1964. Enzymes. Academic Press, Inc., New York. 2nd edition. 318.

7. SCHULTZ, S. G., and R. ZALUSKY. 1965. Interactions between active sodium transport and active amino acid transport in isolated rabbit ileum. *Nature.* **205**:292.

8. ESPOSITO, DI G., A. FAELLI, and V. CAPRARO. 1964. Influence of the transport of amino acids on glucose and sodium transport across the small intestine of the albino rat incubated *in vitro. Experientia.* **20**:122.

9. SCHULTZ, S. G., and R. ZALUSKY. 1964. Ion transport in isolated rabbit ileum. I. Short-circuit current and Na fluxes. *J. Gen. Physiol.* **47**:567.

10. HEINZ, E., and R. P. DURBIN. 1957. Studies of the chloride transport in the gastric mucosa of the frog. *J. Gen. Physiol.* **41**:101.

11. GILLES-BAILLIEN, M., and E. SCHOFFENIELS. 1965. Site of action of L-alanine and D-glucose on the potential difference across the intestine. *Arch. Intern. Physiol. Biochim.* **73**:355.

12. WRIGHT, E. M. 1966. The origin of the glucose dependent increase in the potential difference across tortoise small intestine. *J. Physiol., (London).* **185:** 486.

13. BRITTON, H. G. 1966. Fluxes in passive, monovalent and polyvalent carrier systems. *J. Theoret. Biol.* **10**:28.

14. GOLDMAN, D. E. 1943. Potential, impedance and rectification in membranes. *J. Gen. Physiol.* **27**:36.

15. PATLAK, C. S. 1956. Contributions to the theory of active transport. *Bull. Math. Biophys.* **18**:271.

16. SCHULTZ, S. G., and R. ZALUSKY. 1964. Ion transport in isolated rabbit ileum. II. The interaction between active sodium and active sugar transport. *J. Gen. Physiol.* **47**:1043.

17. CRANE, R. K. 1962. Hypothesis for mechanism of intestinal active transport of sugars. *Federation Proc.* **21**:891.

18. SCHULTZ, S. G., R. E. FUISZ, and P. F. CURRAN. 1966. Amino acid and sugar transport in rabbit ileum. *J. Gen. Physiol.* **49**:849.

19. ROSENBERG, I. H., A. L. COLEMAN, and L. E. ROSENBERG. 1965. The role of sodium ion in the transport of amino acids by intestine. *Biochim. Biophys. Acta.* **102**:161.

20. LIN, E. C. C., H. HAGIHIRA, and T. H. WILSON. 1962. Specificity of the transport system for neutral amino acids in the intestine. *Am. J. Physiol.* **202**:919.

21. MATTHEWS, D. M., and L. LASTER. 1965. Kinetics of intestinal active transport of five neutral amino acids. *Am. J. Physiol.* **208**:593.

22. FRIDHANDLER, L., and J. H. QUASTEL. 1955. Absorption of amino acids from isolated surviving intestine. *Arch. Biochem. Biophys.* **56**:424.

23. KEDEM, O. 1961. Criteria of active transport. *In* Membrane Transport and Metabolism. A. Kleinzeller and A. Kotyk, editors. Czechoslovak Academy of Sciences, Prague. 87.

24. KATCHALSKY, A., and P. F. CURRAN. 1965. Nonequilibrium Thermodynamics in Biophysics. Harvard University Press, Cambridge.

25. RIGGS, T. R., L. M. WALTER, and H. N. CHRISTENSEN. 1958. Potassium migration and amino acid transport. *J. Biol. Chem.* **233**:1479.

26. CRANE, R. K. 1965. Na-dependent transport in the intestine and other animal tissues. *Federation Proc.* **24**:1000.

27. VIDAVER, G. A. 1964. Glycine transport by hemolyzed and restored pigeon red cells. *Biochemistry.* **3**:795.

28. KIPNIS, D. M., and J. E. PARRISH. 1965. Role of Na and K on sugar (2-deoxyglucose) and amino acid (α-aminoisobutyric acid) transport in striated muscle. *Federation Proc.* **24**:1051.

29. CSÁKY, T. Z., and Y. HARA. 1965. Inhibition of active intestinal sugar transport by digitalis. *Am. J. Physiol.* **209**:467.

30. CSÁKY, T. Z. 1963. A possible link between active transport of electrolytes and non-electrolytes. *Federation Proc.* **22**:3.

31. CALDWELL, P. C., and R. D. KEYNES. 1959. Effect of ouabain on the efflux of sodium from a squid giant axon. *J. Physiol.*, *(London).* **148**:8P.

32. WHITTAM, R. 1962. The asymmetrical stimulation of a membrane adenosine triphosphatase in relation to active cation transport. *Biochem. J.* **84**:110.

33. KROMPHARDT, H., H. GROBECKER, K. RING, and E. HEINZ. 1963. Über den Einfluss von Alkali-Ionen auf den Glyzintransport in Ehrlich-Ascites-Tumorzellen. *Biochim. Biophys. Acta.* **74**:549.

34. WHEELER, K. P., Y. INUI, P. F. HOLLENBERG, E. EAVENSON, and H. N. CHRISTENSEN. 1965. Relation of amino acid transport to sodium ion concentration. *Biochim. Biophys. Acta.*, **109**:620.

35. JACQUEZ, J. A., and J. H. SHERMAN. 1965. The effect of metabolic inhibitors on transport and exchange of amino acids in Ehrlich ascites cells. *Biochim. Biophys. Acta*, **109**:128.

36. DOWD, J. E., and D. S. RIGGS. 1965. A comparison of estimates of Michaelis-Menten kinetic constants from various linear transformations. *J. Biol. Chem.* **240**:863.

37. HEARON, J. Z. 1952. Rate behavior of metabolic systems. *Physiol. Rev.* **32**:499.

415

Author Citation Index

Abelson, P. H., 119, 145
Abraham, D., 243
Abrams, A., 298
Adelberg, E. A., 261, 317, 337
Adler, J, 281, 282, 337
Aksamit, R. R., 282
Alexander, D. P., 80
Allfrey, V. G., 383
Allison, D. P., 253
Allums, J. A., 384
Alpers, D. H., 298
Althausen, T. L., 383
Alvarado, F., 383
Ames, B. N., 298, 317, 325, 333
Ames, G. F., 325, 333
Amos, H., 142
Anderson, B. E., 215, 246, 252, 261,
    275, 301, 309, 317
Anderson, E. M., 383
Anderson, R. L., 253
Anraku, A. B., 261
Anraku, Y., 282, 298, 308, 309, 317,
    325, 333, 337
Anson, M. L.,41
Anton, D. N., 261
Appel, S. H., 298
Arimura, G. K., 384
Armstrong, J. B., 337
Arrhenius, S.A., 228
Asano, T., 383
Asensino, C., 185, 197, 206, 253
Ashwell, G., 309
Audureau, A., 119
Avigad, G., 185, 197, 206, 253

Baldwin, A. N., 301
Bandurski, R. S., 290
Bang, O., 63, 80
Barany, E., 88

Bardawill, C. J., 253
Barker, H. A., 144
Barker, S. B., 377
Bärlund, H., 5, 6
Barnes, E. M., Jr., 94, 220, 228
Barrett, J. J., 118
Barrett, J. T., 142
Barron, E. S. G., 41, 63
Barry, G. W., 242
Bartsch, G., 298
Beckmith, J. R., 317
Berg, P., 298, 301
Berger, E. A., 337
Bergmann, E. D., 142
Bergmann, F. H., 298, 301
Berman, D. T., 261
Berman, M., 234
Bertsch, L., 298
Bhattacharyya, P., 94, 220
Bianchi, A., 383
Bihler, I., 197, 342, 368, 377, 383
Bojarska, A., 253
Bolton, B., 145
Bolton, E. T., 119, 142, 298
Bonner, D. M., 142, 290, 317
Bonner, J., 11
Boos, W., 282, 317, 337
Borrero, L. M., 41
Bosackova, J., 383
Boyer, P. D., 215, 243, 337
Boyle, P., 350
Bray, G. A., 325
Brenner, S., 308
Britten, R. J., 118, 119, 142, 145, 185,
    246, 290, 298
Britton, H. G., 414
Brockman, R. W., 309
Bronner, F., 228, 229
Brooks, S. C., 8

417

Subject Index

Acid phosphatase, 295, 297
Adenosine-5'-monophosphate, cyclic (cyclic AMP), 234
Adenosine-5'-triphosphate (ATP), 92–94, 146, 158–159, 207, 210, 213, 236, 237, 238, 252, 352, 389
Adsorption isotherm, 69, 86
Aerobacter aerogenes, 233, 247–253
Alanine, uptake by intestinal mucosa cells, 390–415
  inhibition by valine and leucine, 394–396
  sodium dependence, 390–394
  stimulation of sodium flux, 403–404
Amino acids (*see also names of individual amino acids*)
  binding proteins, 278, 280, 299–301, 318–333
  chemotaxis, 281, 337
  co-transport with sodium, 340–341, 382, 390–415
  intestinal absorption, 382, 405–410
  osmotic shock, 297
  transport
    bacteria, 93, 116, 134–138, 299–301, 318–333
    bacterial membrane vesicles, 93, 207
    intestinal mucosa cells, 390–415
    kidney, 382
    pigeon erythrocytes, 385–389
Ascites tumor cells, 342, 382
Azide, in *E. coli*
  ONPG hydrolysis, 173–177
  transport, 91, 98–100, 134, 173–177, 182, 186–197, 213–214, 217, 218

Binding, of transport substrates inside cell, 81, 86–88, 107–108, 117, 123–125, 142

Binding proteins, Part V
  dissociation constants, 279, 284, 321, 300, 329–330, 335
  function in transport, 280–281, 289, 301, 307, 315–317, 336–338
  galactose, 302–309, 310–317, 334–338
  genetics, 279–280, 288–289, 310–317, 318–325
  histidine, 318–333
  kinds, 278
  leucine–isoleucine–valine, 299–301
  sulfate, 278, 283–290

Carbohydrates (*see* Galactose transport, β-Galactosides, Glucose, Phosphotransferase system, *and* Sugars)
Carbon source for growth, and β-galactoside transport in *E. coli*, 156–157, 161, 169–170, 172–173, 179, 182
car⁻ mutants, 233, 256–257, 260
Carrier, Part II
  β-galactoside, isolation, 198–206
Carrier-mediated transport, Part II
  kinetics, 44, 53–55, 69–70, 74–76
Catalytic mechanism of transport, 107–109, 117, 123–125, 147
Chara ceratophylla, permeability to nonelectrolytes, 6–9
Chemical potential, 29
Chemotaxis, in *E. coli*, 281, 334
  galactose binding protein, 334–338
  maltose binding protein, 336–337
  osmotic shock, 335
  relation to transport, 334, 337
  restoration, in shocked cells, 335–336, 338
p-Chloromercuribenzoate, 58, 154, 213, 214, 219, 226

426

431